SOCIETÀ ITALIANA DI FISICA

RENDICONTI
DELLA
SCUOLA INTERNAZIONALE DI FISICA
«ENRICO FERMI»

XCII Corso

a cura di N. Cabibbo
Direttore del Corso
VARENNA SUL LAGO DI COMO
VILLA MONASTERO
26 Giugno - 6 Luglio 1984

Particelle elementari

1987

SOCIETÀ ITALIANA DI FISICA
BOLOGNA - ITALY

ITALIAN PHYSICAL SOCIETY

PROCEEDINGS
OF THE
INTERNATIONAL SCHOOL OF PHYSICS
«ENRICO FERMI»

Course XCII

edited by N. Cabibbo
Director of the Course
VARENNA ON LAKE COMO
VILLA MONASTERO
26 June - 6 July 1984

Elementary Particles

1987

NORTH-HOLLAND
AMSTERDAM · OXFORD · NEW YORK · TOKYO

Copyright © 1987, by Società Italiana di Fisica

All rights reserved. No part of this publication may be reproduced, stored in a retrieval system, or transmitted, in any form or by any means, electronic, mechanical, photocopying, recording or otherwise, without the prior permission of the copyright owner.

PUBLISHED BY:

North-Holland Physics Publishing
a division of
Elsevier Science Publishers B.V.
P.O. Box 103
1000 AC Amsterdam
The Netherlands

SOLE DISTRIBUTORS FOR THE USA AND CANADA:

Elsevier Science Publishing Company Inc.
52, Vanderbilt Avenue
New York, N.Y. 10017
U.S.A.

Technical Editor
P. PAPALI

Library of Congress Cataloging-in-Publication Data

Elementary particles.

(Proceedings of the International School of Physics "Enrico Fermi" ; course 92)
   At head of title: Italian Physical Society.
   Includes index.
   1. Particles (Nuclear physics)--Congresses.
I. Cabibbo, N. II. Societa italiana di fisica.
III. Series: International School of Physics "Enrico Fermi." Proceedings of the International School of Physics "Enrico Fermi" course 92.
QC793.E44 1987   539.7'21   87-18598
ISBN 0-444-87075-X

Proprietà Letteraria Riservata
Printed in Italy

INDICE

N. CABIBBO – Preface . . . . pag. XI

Gruppo fotografico dei partecipanti al Corso   fuori testo

P. HANSEN – Results from the UA1 and UA2 experiments

   1. Introduction . . . . »   1
   2. The UA1 and UA2 detectors . . . . »   1
   3. Jets and QCD . . . . »   2
      3`1. Two-jet dominance . . . . »   2
      3`2. Two-jet transverse momentum . . . . »   3
      3`3. Two-jet angular distribution . . . . »   3
      3`4. Structure functions and jet cross-section . . . . »   4
      3`4. Two-jet mass distribution . . . . »   6
      3`6. Jet fragmentation . . . . »   7
   4. W, Z and the standard model . . . . »   8
      4`1. W production cross-section . . . . »   8
      4`2. W momentum distribution . . . . »   9
      4`3. W decay asymmetry . . . . » 10
      4`4. Z mass and the standard model . . . . » 10
      4`5. Z width and the number of neutrinos . . . . » 12
      4`6. Search for $W \to t + \bar{b}$ . . . . » 13
   5. Exotic phenomena . . . . » 14
      5`1. Radiative Z decays . . . . » 14
      5`2. Monojets . . . . » 15
      5`3. High-p_T W's . . . . » 16

G. MARTINELLI – Experimental test and theoretical predictions for electroweak processes.

   1. Introduction . . . . » 19
   2. The standard model . . . . » 19

3. Low-energy experimental tests of the standard model . . . pag. 22
  3˙1. Leptonic processes . . . . » 22
  3˙2. Lepton-hadron reactions . . . . » 26
4. Radiative corrections to low-energy processes . . . . . . » 32
  4˙1. Renormalization group applied to the effective weak Hamiltonians . . . . » 33
  4˙2. Corrections to the evaluation of $R^{\nu,\bar{\nu}}$ . . . . » 34
5. Measurements of the W and Z^0 masses at the SPS collider and comparison with theoretical predictions . . . . » 39

R. RÜCKL – Weak decays of heavy quark states.

1. Introduction . . . . » 43
2. The effective weak Hamiltonian . . . . » 44
3. Inclusive decays: the spectator model . . . . » 50
  3˙1. B-decays . . . . » 53
  3˙2. D, F decays . . . . » 56
4. Nonasymptotic effects . . . . » 58
  4˙1. Interference effects in charmed-meson decays . . . . » 58
  4˙2. Annihilation processes in charmed-meson decays . . . » 59
  4˙3. Inclusive charmed-baryon decays . . . . » 61
  4˙4. Inclusive bottom decays . . . . » 63
5. Two-body decays . . . . » 64
6. Conclusions . . . . » 68

R. PETRONZIO – Compositeness.

1. Introduction . . . . » 76
2. Composite Higgs . . . . » 78
3. Composite W, Z . . . . » 84
4. Composite fermions . . . . » 92
5. Phenomenology . . . . » 98
6. Concluding remarks . . . . » 103

W. BUCHMÜLLER – Some aspects of supersymmetric composite models of quarks and leptons.

1. Are quarks and leptons composite? . . . . » 105
2. Quasi-Goldstone fermions . . . . » 106
3. The coset space $U_6/SU_2 \times U_4$ . . . . » 109
4. A U_6 model . . . . » 112
5. Towards a realistic preon model . . . . » 117

M. RONCADELLI – Neutrino masses 1984.

Introduction . . . . pag. 121
1. Generalities on fermion masses . . . . » 122
2. Varieties of neutrino masses . . . . » 124
3. A glimpse to experiment . . . . » 128
4. A concrete proposal . . . . » 130
Appendix . . . . » 132

J. R. PRIMACK – Dark matter, galaxies and large-scale structure in the Universe.

1. Matter . . . . » 140
   1`1. Sizes . . . . » 140
   1`2. Galaxies . . . . » 141
   1`3. Groups and clusters . . . . » 152
   1`4. Superclusters and voids . . . . » 157
2. Gravity . . . . » 161
   2`1. Cosmology . . . . » 161
   2`2. General relativity . . . . » 163
   2`3. Friedmann universes . . . . » 166
   2`4. Comparison with observations . . . . » 169
   2`5. Growth and collapse of fluctuations . . . . » 177
   2`6. Inflation and the origin of fluctuations . . . . » 186
   2`7. Is the gravitational force $\propto r^{-1}$ at large r? . . . . . . » 190
3. Dark matter . . . . » 192
   3`1. The hot big bang . . . . » 192
   3`2. The dark matter is probably not baryonic . . . . » 194
   3`3. Three types of DM particles: hot, warm and cold . . » 200
   3`4. Galaxy formation with hot DM . . . . » 200
   3`5. Galaxy formation with warm DM . . . . » 207
4. Cold dark matter . . . . » 210
   4`1. Cold-DM candidates . . . . » 211
   4`2. Galaxy and cluster formation with cold DM . . . . . » 214
   4`3. N-body simulations of large-scale structure . . . . . . » 223
   4`4. Summary and prospect . . . . » 228

S. FERRARA – Supersymmetry and supergravity.

1. Introduction . . . . » 242
2. The challenge of supersymmetry . . . . » 245
3. The $N = 1$ supergravity theory . . . . » 250
4. Tensor calculus for $N = 1$ supergravity . . . . » 254
5. Manifestly invariant actions and transformation laws . . . » 262
6. Final form of the Lagrangian, transformation laws and gauged R-symmetry . . . . » 269
7. The super-Higgs effect in $N = 1$ supergravity . . . . » 276

A. SAGNOTTI – Ultraviolet divergences and supersymmetric theories.

1. Introduction and survey . . . . pag. 280
2. Superspace, superfields and the superspace power counting . » 288
3. The breakdown of the power counting . . . . » 300
4. Discussion . . . . » 309
Appendix . . . . » 310

J. B. KOGUT – From asymptotic freedom to fermion computer simulations: lectures in lattice gauge theory.

Introductory remarks . . . . » 315
1. Field theory, statistical mechanics and the transfer matrix . » 315
2. Lattice gauge theory . . . . » 322
3. Weak-coupling perturbation theory, asymptotic freedom, the renormalization group and continuum limits . . . . . . » 329
4. Topological excitations in spin and gauge theories . . . . » 340
5. Hamiltonian lattice gauge theory, flux tube dynamics and continuum string models . . . . » 353
6. Lattice fermions . . . . » 365
7. Fermion computational methods . . . . » 379

F. KARSCH – The deconfinement transition in finite-temperature lattice gauge theory.

1. Introduction . . . . » 389
2. Thermodynamics of SU_N gauge theories . . . . » 390
3. The SU_N deconfinement transition . . . . » 393
   3`1. The order of the deconfinement transition . . . . . . » 393
   3`2. Scaling properties of thermodynamic quantities . . . » 397
4. Deconfinement in the presence of virtual quarks . . . . » 398
5. Conclusions . . . . » 401

G. PARISI – Spin glasses.

1. Introduction . . . . » 404
2. Spin glasses . . . . » 404
3. The Sherrington-Kirkpatrick model . . . . » 409
4. The solution . . . . » 412
5. The physical interpretation . . . . » 416

A. E. TERRANO – Computers and theoretical physics.

1. The architecture of conventional computers . . . . pag. 422
2. Special-purpose machine design . . . . » 428
3. Elements of modern programming . . . . » 435
4. Algebraic and interactive programs . . . . » 443
Appendix . . . . » 449

G. CHARPAK – Selected topics in detector physics.

1. High-accuracy detectors . . . . » 456
  1`1. Bubble chambers . . . . » 457
  1`2. Streamer chambers . . . . » 457
  1`3. MWPC, drift chambers and solid-state detectors . . . » 459
2. Calorimetry . . . . » 462
  2`1. The various calorimeters . . . . » 462
  2`2. Liquid ionization detectors (LIDs) at room temperature » 463
  2`3. Liquid photocathodes and the coupling of BaF_2 to wire chambers . . . . » 465
  2`4. The high-density projection chamber . . . . » 467
3. Progress in particle identification . . . . » 467
4. Conclusion . . . . » 468

K. JOHNSEN – Introduction to accelerator physics.

1. Introduction . . . . » 470
2. Short description of a few circular accelerators . . . . . . » 471
  2`1. The cyclotron . . . . » 471
  2`2. The betatron . . . . » 473
  2`3. The synchrotron . . . . » 473
  2`4. Collider beams . . . . » 474
3. Transverse motion in weak-focusing accelerators . . . . . » 476
  3`1. The radial motion . . . . » 476
  3`2. The vertical motion . . . . » 478
  3`3. Solutions . . . . » 478
  3`4. Stability . . . . » 478
  3`5. Acceptance/emittance . . . . » 479
  3`6. Momentum compaction . . . . » 481
  3`7. Summary of the important concepts introduced so far » 482
4. Introduction to alternating-gradient focusing . . . . » 482
  4`1. A segment of a magnet as a focusing element . . . . » 482
    4`1.1. Point lens . . . . » 484
    4`1.2. Drift length . . . . » 484
    4`1.3. The relation between vertical and horizontal focusing of an element with high field gradient » 484

	4'2.	Doublet	pag.	485
	4'3.	Simple description of an AG accelerator	»	486
	4'4.	Stability criterion	»	486
	4'5.	Simple example of an AG system	»	487
	4'6.	Comments on the elements of the transfer matrix over a period of an AG structure	»	490
5.	Longitudinal particle motion.	»	491
	5'1.	Travelling-wave representation of the accelerating field	»	492
	5'2.	The equations for the longitudinal motion	»	493
	5'3.	Stability. Transition energy	»	494
	5'4.	The adiabatic solution	»	496
6.	Representative examples of accelerator systems for hadrons	»	497
7.	Some special features of electron machines	»	498
	7'1.	Radiated power and associated losses	»	499
	7'2.	Radiation damping	»	500
		7'2.1. The vertical betatron oscillations	»	500
		7'2.2. The horizontal betatron oscillation	»	501
		7'2.3. Longitudinal oscillations	»	501
		7'2.4. The relation between the three partition numbers J	»	502
		7'2.5. Summary of the differences between electron and proton machines	»	502
8.	Prospects for the future	»	502

Preface.

This volume contains the lectures given at the XCII Course of the Enrico Fermi School of Physics, held in Varenna in Summer 1984.

The course was organized with the aim of giving a comprehensive panorama of the more recent developments in theoretical and experimental high-energy physics.

Many different subjects were covered during the period of the school. Among them the phenomenology of the standard model, with particular emphasis on the latest experimental results from the CERN $Sp\bar{p}S$ collider, neutrino physics, supersymmetry, supergravity, composite models and cosmology.

Some of the lectures were dedicated to fields other than high-energy physics such as spin glasses, a new and very rapidly developing subject, and to specialized computers in theoretical physics.

The basic working principles of e^+e^- storage rings and hadron colliders were also reviewed.

I thank the speakers for their unvaluable collaboration and the students for their warm and lively participation.

I want to express our gratitude to the Italian Physical Society, who gave us all its support for the success of the School.

I also acknowledge P. PAPALI and the whole staff of the Italian Physical Society for their assistance during the School and in the preparation of this volume.

N. CABIBBO

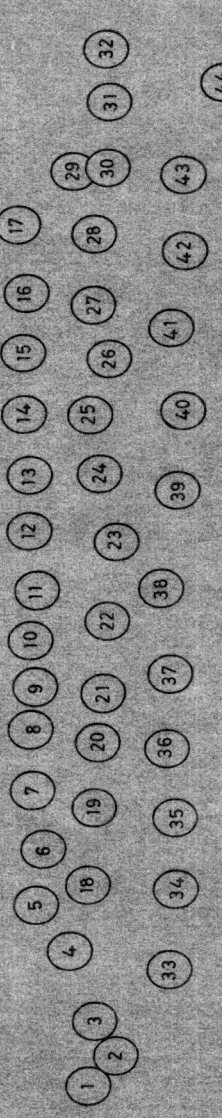

1) J. Pashale-Rad
2) E. Franco
3) M. Corni
4) G. P. Stroll
5) P. Debu
6) O. Nicrosini
7) N.-E. Behlil
8) G. Mussardo
9) G. Gamberini
10) A. Treves
11) F. Ravanini
12) S. Catani
13) Y. Kitazawa
14) G. melegari
15) M. Meschini
16) G. Ridolfi
17) R. Laflamme
18) A. M. Allega
19) L. Ramello
20) A. Vladikas
21) B. Mole
22) F. Gibrat
23) E. L. Olgil
24) F. Cardone
25) A. Inciochitti
26) J. Sloan
27) R. Petronzio
28) P. Hansen
29) K. Jensen
30) F. Karson
31) M. Coti Zelati
32) N. Primack
33) A. Zocca
34) G. Farrar
35) G. M. Prosperi
36) G. Martinelli
37) G. Charpak
38) T. A. Terrano
39) N. Cabibbo
40) E. Mazzi
41) J. Primack
42) S. Ferrara
43) P. Mauri
44) F. Pittino

SOCIETÀ ITALIANA DI FISICA

SCUOLA INTERNAZIONALE DI FISICA «E. FERMI»

XCII CORSO - VARENNA SUL LAGO DI COMO - VILLA MONASTERO - 26 Giugno - 6 Luglio 1984

Results from the UA1 and UA2 Experiments.

P. HANSEN

CERN-EP - 1211 Geneva 23, Switzerland

1. – Introduction.

By using its existing accelerator complex to collide protons with antiprotons at $\sqrt{s} = 540$ GeV, CERN has made it possible to get a first look into a new energy regime without building a major new facility. Since the first operation in December 1981 three runs have taken place at the SPS Collider with steadily increasing luminosity.

This review covers results from the two largest experiments at the Collider, UA1 and UA2, obtained during the two last runs in which an integrated luminosity of 130 nb^{-1} was collected. The physics results from these runs include progress in understanding of high-p_T jet production and the discovery of the intermediate vector bosons as predicted by the standard model.

The latest reported findings could hint at new phenomena beyond the standard model. Hopefully some of the vagueness connected with these findings will disappear after the coming run of September-December 1984 in which a higher luminosity (perhaps 200 nb^{-1}) is expected as well as higher beam energy ($\sqrt{s} = 630$ GeV).

2. – The UA1 and UA2 detectors.

The UA1 and UA2 experiments [1, 2] both make use of large detectors with the aim of measuring leptons and jets over most of the solid angle. Both detectors include a tracking device closest to the interaction region, followed by a total-absorption calorimeter for electromagnetic showers, which is followed in turn by a calorimeter absorbing hadronic showers.

UA1 has the advantage of a large tracking detector in a region with a 0.7 T horizontal magnetic field. This measures momenta of charged particles and enables detection of muons that penetrate the hadron calorimeter and leave a signal in the surrounding muon chambers. Furthermore UA1 has advantage in covering polar angles down to near 0° with respect to the beam.

UA2 has a much smaller tracking detector and no magnetic field in the central region. The resulting loss of electron identification power is partly compensated by a tungsten converter plate followed by a preshower detector that localizes early electromagnetic showers. UA2 has advantage in its well-segmented calorimeter that covers uniformly the central region. It has a disadvantage in its lack of coverage of polar angles smaller than 20° and its incomplete coverage in the region between 20° and 40°. This region is occupied by magnetic spectrometers and electromagnetic calorimeters.

3. – Jets and QCD.

3`1. *Two-jet dominance*. – One of the immediate results from the Collider experiments was the clear evidence of two-jet dominance at high $\sum E_T$ [3]. This was seen at the same time at the ISR [4].

In UA2 jets are defined [5] as groups of adjacent calorimeter cells with energy exceeding 0.5 GeV. In fig. 1 it is shown how, at high $\sum E_T$, almost all the transverse energy gets carried away by two such clusters. The clusters themselves appear well defined (see fig. 1 for a widely published example), to a degree that was surprising for most people. It is conceivable that the detectors would have been built somewhat differently, had this been seriously anticipated. However, the possibility of clearly defining jets has opened up a rich field of physics, even to the point, as we shall see, of doing spectroscopy with jets.

The dominating configuration of two high-p_T jets at large $\sum E_T$ is believed to result from a hard collision between two proton constituents. These can be either quarks or gluons. The following subsections will present results concerning the dynamics of quarks and gluons obtained from studying two-jet systems.

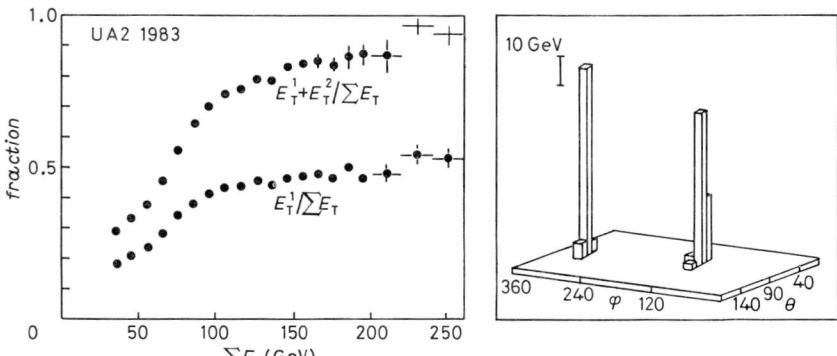

Fig. 1. – Two-jet dominance. Left: the fraction of $\sum E_T$ carried by the two most energetic clusters. Right: the E_T depositions in the highest $\sum E_T$ event recorded in 1982.

3'2. *Two-jet transverse momentum.* – The intrinsic transverse momentum of the partons in the nucleons and the initial-state gluon bremsstrahlung are expected to contribute to the transverse momentum of a two-jet system, p_T^{jj}.

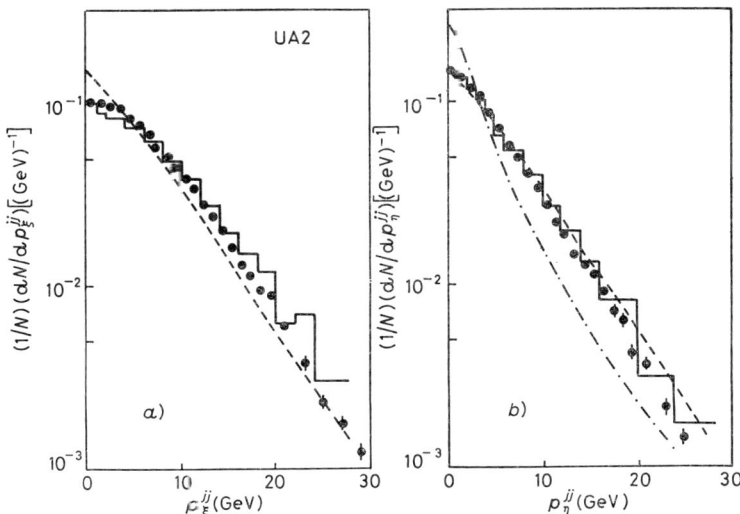

Fig. 2. – Two-jet p_T components. See text for details.

Figure 2 shows the distribution of $|p_\xi^{jj}|$ (the component of p_T^{jj} parallel to the jet axis) and $|p_\eta^{jj}|$ (the perpendicular component). The r.m.s. of the two distributions are measured by UA2 to 9.1 GeV and 7.5 GeV, respectively [5], since the perpendicular component is much less influenced by mismeasurements of jet energies. The dashed line in the figure shows a QCD prediction [6] and the full line represents the prediction after instrumental effects have been taken into account. The agreement with the data is good. The result is sensitive to the assumption that gluons contribute to the jet production and that they radiate more than quarks because of their larger colour charge. As an illustration of this, the dash-dotted line in fig. 2 shows the distribution in the case that gluons radiate like quarks.

3'3. *Two-jet angular distribution.* – It has been pointed out [7] that in lowest-order QCD the jet cross-section is expected to factorize into an overall structure function $F(x)$ (sum of all the quark and gluon densities) and an effective parton-parton cross-section $d\sigma/d\cos\theta^*$:

$$(1) \quad \frac{d^3\sigma}{dx_1\, dx_2\, d\cos\theta^*} = \frac{S(x_1, x_2)}{x_1 x_2} \frac{d\sigma}{d\cos\theta^*} = \frac{F(x_1)}{x_1} \frac{F(x_2)}{x_2} \frac{d\sigma}{d\cos\theta^*},$$

where x_1 and x_2 are the fractions of the beam energy carried by the colliding partons (given by m^{jj} and p_L^{jj} of the two-jet system) and θ^* is their c.m.s. scattering angle.

In spite of the average over a wide range of subprocesses and Q^2's, the factorization of eq. (1) has been shown to hold experimentally [8, 9]. The c.m.s. scattering angle can be approximately determined as long as p_T^{jj} is small compared to m^{jj} [10].

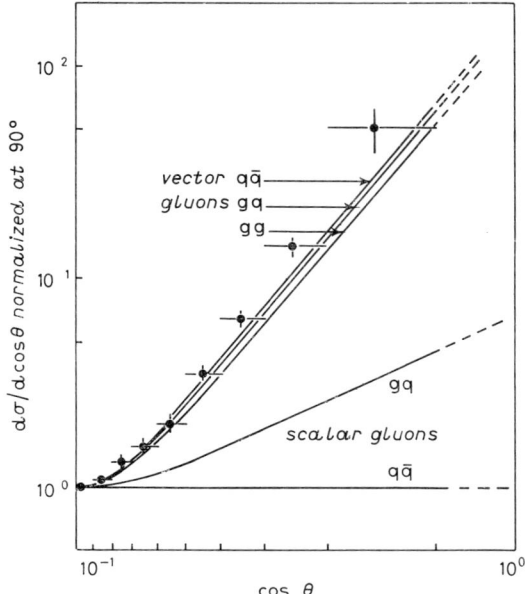

Fig. 3. – $\cos\theta^*$, UA1 two-jet scattering angle in the c.m. of the system.

In fig. 3 the measured angular distribution of UA1 [8] is shown together with predictions from various processes. The difference between QCD subprocesses is not large enough to allow distinction between them, but the distribution that would result from scalar gluon exchange is clearly out of line with the data. The fact that the angular distribution of the data exceeds the predictions for small angles is attributed to the increase in the coupling constant α_s for decreasing momentum transfer.

3'4. *Structure functions and jet cross-section.* – Having now determined the angular distribution, we can extract the effective structure function $F(x)$ from eq. (1) and the results are shown in fig. 4. The two experiments agree with each other for $x > 0.10$.

Because of its larger colour charge the gluons contribute more than the quarks to the average:

(2) $$F(x) = g(x) + (4/9)q(x) .$$

Figure 4 shows $F(x)$ measured at low energy [11] and extrapolated to $q^2 = 2000$ (GeV)^2. This extrapolation agrees reasonably well with the data (considering the absence of higher-order corrections) and again the presence of gluons is necessary in order to avoid disagreement with the data at low x.

The structure functions are partly responsible for the shape of the inclusive jet cross-sections, shown in fig. 5 together with the predicted rate from QCD

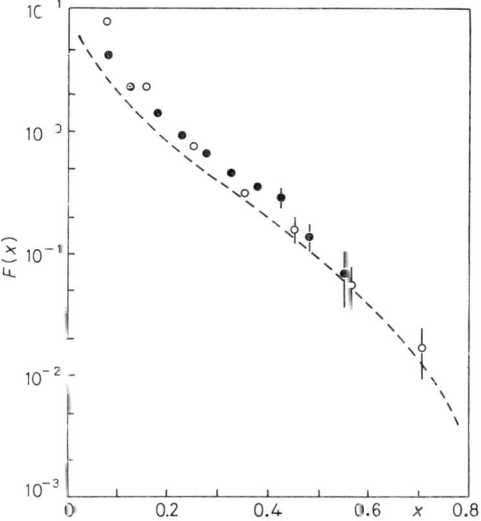

Fig. 4. – Effective structure function. The dashed line si computed from ν data [11] taking violations into account, • UA2, ○ UA1×$\sqrt{2}$.

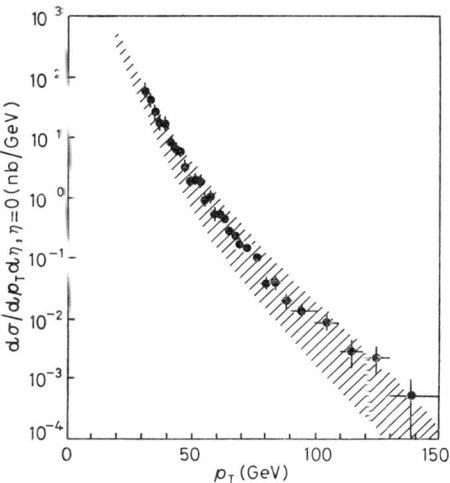

Fig. 5. – Inclusive jet cross-section, UA2, $\bar{p}p \to \text{jet}+X$, $\sqrt{s} = 540$ GeV. The shaded area covers a range of QCD predictions.

The predictions cover a broad band because of uncertainties in structure functions, scale, fragmentation, etc., but follow nevertheless the data over 5 orders of magnitude. It has been pointed out [12] that, in order to fit the jet cross-section, not only the presence of gluons is necessary, but also the existence of triple gluon vertices.

3`5. Two-jet mass distribution. – We finish the story on jet production by turning to the mass distribution of two-jet systems as shown on fig. 6. Here the eyes focus on possible bumps. In the region of the W and Z boson mass a fit to two Gaussians (with N_Z fixed at 0.41 N_W) on a smooth background yields $N_W = 108^{+80}_{-70}$ and $M_W = (75 \pm 10)$ GeV. This result carries no evidence for anything, but does at least not contradict the standard-model prediction of 150 W's decaying into jets, that get picked up by the UA2 apparatus.

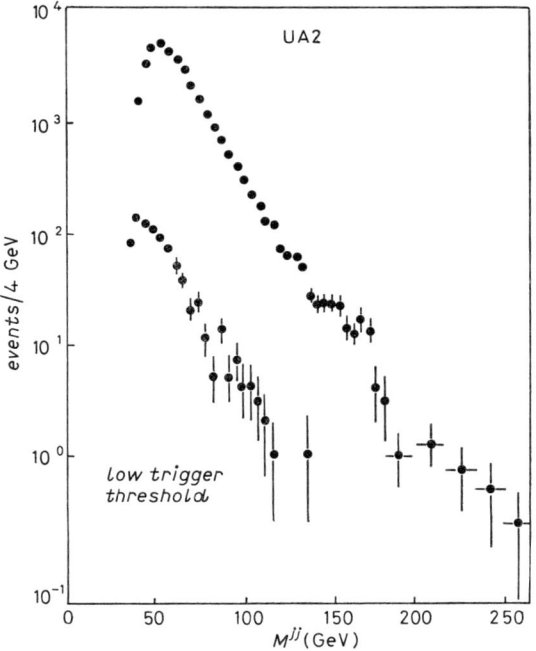

Fig. 6. – Two-jet mass distribution. Data were taken with two hardware $\sum E_T$ thresholds.

There seems to be some evidence for a structure at $m^{jj} \approx 150$ GeV. To see if this is really significant, a smooth-background fit was made giving $\chi^2/\text{n.d.f.} = 40/27$. Adding a free Gaussian on top of the background gives $\chi^2/\text{n.d.f.} = 28/25$ with 69 ± 17 events in the peak centred at 147 GeV. A maximum-likelihood analysis was then made for both the two fits. From the ratio of the likelihoods of the fits with and without the Gaussian, the probability that a statistical fluctuation has generated the enhancement was estimated to 0.5 %. Clearly more data are needed to confirm this result.

3`6. *Jet fragmentation*. – From the results on jet production we have seen the need of including gluons in the beam of partons that scatter into jets. It is, therefore, interesting to compare the fragmentation of these jets to the one of quark jets from e^+e^- data. According to QCD inspired models [13] gluon jets are expected to fragment in a softer way than quark jets because of the larger colour charge of the gluon.

The comparison with e^+e^- data has a problem in separating jet particles from particles related to spectator partons. An analysis [14] of the charged multiplicity in jets avoids this problem by subtracting at all azimuths the track density found at $\Delta\varphi = 90°$ from the jet axis and doing the same thing for the e^+e^- data. The resulting « jet core » multiplicity is shown in fig. 7 as a function of m^{jj}.

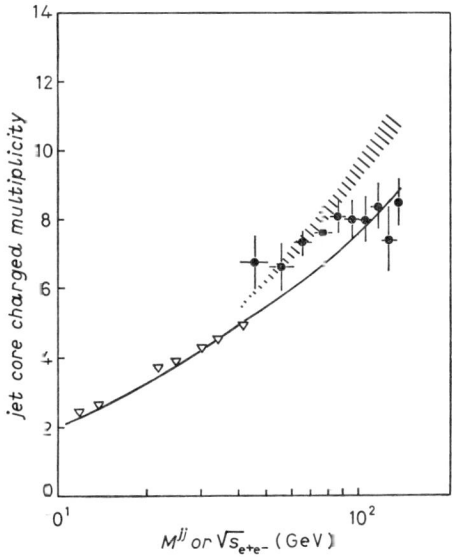

Fig. 7. – « Jet core » charged multiplicity. Black points are data from two-jet events, triangles are corresponding TASSO [15] data and curves are QCD inspired predictions (see text).

The extrapolation from TASSO data [15] suggests that the jets from $p\bar{p}$ collisions have a higher average multiplicity than the e^+e^- jets. A QCD parton shower model [13] has been tuned to agree with the e^+e^- data for quark jets and its prediction is shown in the figure for both quark and gluon jets (the prediction for gluon jets is shown with its theoretical uncertainty). At low two-jet mass the data seem to agree with the gluon prediction, while at higher mass they agree more with the quark prediction. This is consistent with the structure functions [11] in which the fraction of gluons varies from 75% to 30% as m^{jj} varies from 40 GeV to 140 GeV.

In the perturbative sector of the jet fragmentation, QCD predicts another jet to be radiated off one of the scattered jets in roughly a fraction α_s of the two-jet events. In order to estimate the level of third-jet activity, UA1 [8] has studied the parameter p_{out}, the momentum component out of the plane defined by the trigger jet and the beam axis. This is shown in fig. 8 for clusters found by the UA1 jet algorithm. It is clearly necessary to introduce a model that includes third jets, such as the QCD model shown in the figure.

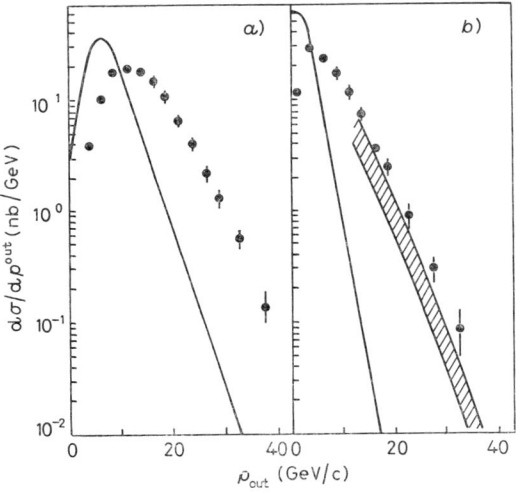

Fig. 8. – UA1 p_{out} distribution: a) all cells, ──── 2-jet MC; b) jet cells, ──── 2-jet MC, ▧▧▧ 3-jet MC (partons).

4. – W, Z and the standard model.

4.1. W *production cross-section.* – The motivation for building the Collider was not primarily to study jets, but to find the intermediate vector bosons (IVB) of the standard model. These would be produced by quark-antiquark annihilation and be observed through their leptonic decay modes. Such decays has indeed been observed [16, 17] as illustrated in fig. 9, although this figure shows a signature for the decay $W \to e\nu$ in a way that was rather surprising from an experimental point of view. It shows a clear Jacobian peak in the sum of all transverse momenta of seen particles (missing p_T) that indicates non-interacting neutrinos with energies around 40 GeV.

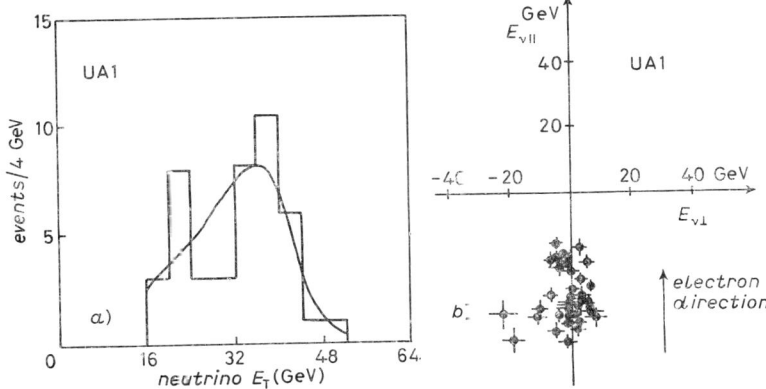

Fig. 9. – Missing p_T in events with a $W \to e\nu$ candidate.

A corresponding peak is also seen in the electron spectra and in the case of UA1 also in the muon spectra [18]. With the help of the clear neutrino signal it has been possible to separate the background from the signal under this peak, and thus determine the cross-section for W production and subsequent decay:

$$\sigma_{W \to e\nu} = (0.53 \pm 0.08 \text{ (stat)} \pm 0.09 \text{ (sys)}) \text{ nb} \quad \text{(UA1)},$$

$$\sigma_{W \to e\nu} = (0.53 \pm 0.10 \text{ (stat)} \pm 0.10 \text{ (sys)}) \text{ nb} \quad \text{(UA2)}.$$

These measurements are in good agreement with each other and also with the estimate from QCD calculations [19].

4'2. *W momentum distribution*. – The mass of the W has been determined by UA1 by fitting the distribution of « transverse mass » of electrons and neutrinos (where the neutrino momentum is given the value of the missing p_T vector) and by UA2 by fitting the electron momentum distribution with the results

$$M_W = (80.9 \pm 1.5 \text{ (stat)} \pm 2.4 \text{ (sys)}) \text{ GeV}/c^2 \quad \text{(UA1)},$$

$$M_W = (83.1 \pm 1.9 \text{ (stat)} \pm 1.3 \text{ (sys)}) \text{ GeV}/c^2 \quad \text{(UA2)}.$$

The latter method requires as input the longitudinal and transverse momentum distribution of the produced W as well as the expected $V-A$ structure of the decay. By using again the missing p_T to represent the neutrino it has been possible to measure the W momentum distribution as shown in fig. 10. Theoretical predictions for the transverse component are due to calculations [19, 20] of nonleading gluon emission from the initial state. The predicted probability of finding a W with $p_T > 25$ GeV is about 6% [19]. We shall in

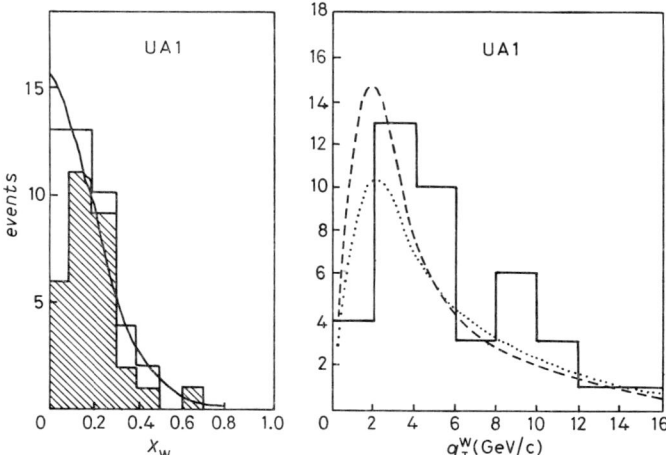

Fig. 10. – Longitudinal and transverse W momenta. Left: Feynman x of the W sample compared with the ISAJET model; ▨ unambiguous (unambiguous means a uniquely determined p_L^ν, where otherwise the lowest solution was chosen). Right: the p_T of the W's is compared with the QCD prediction [20] for two different choices of structure functions.

subsect. 6`3 discuss the observation of some events that possibly contain W's produced at much higher p_T.

4`3. *W decay asymmetry.* – The $V-A$ nature of the W decay cannot be uniquely determined without measuring the helicities of the outgoing leptons. But it can be verified that the W has spin one and that it is produced polarized in the antiproton direction, as a consequence of having both V and A components. From this it follows that positrons from the decay tend to be emitted in the antiproton direction and electrons in the proton direction. Figure 11 shows that the data indeed obey such an asymmetry.

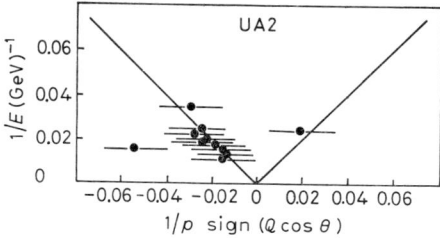

Fig. 11. – Forward-backward asymmetry in W → eν decays where the e is detected in one of the UA2 forward spectrometers.

4`4. *Z mass and the standard model.* – The evidence for a neutral partner to the W from neutrino elastic scattering was the first decisive argument in favour of the standard model. At the Collider this neutral boson has been directly observed throught its production q-q̄ annihilation and subsequent decay into an electron or muon pair. Figure 12 shows the mass distribution of the combined dilepton sample (as available at the time of writing) from the two experiments. The shape at low masses is dominated by different thresholds and efficiencies and these events will not be discussed here. But the remarkable feature of the plot is the long empty gap from 30 to 80 GeV followed by a structure around 90 GeV that contains 8 e⁺e⁻ pairs from UA2 [21], 4 e⁺e⁻ pairs from UA1 [22] and 5 μ⁺μ⁻ pairs from UA1 [23]. Two of the electron pairs and one muon pair are accompanied by a hard photon which has been included in the mass.

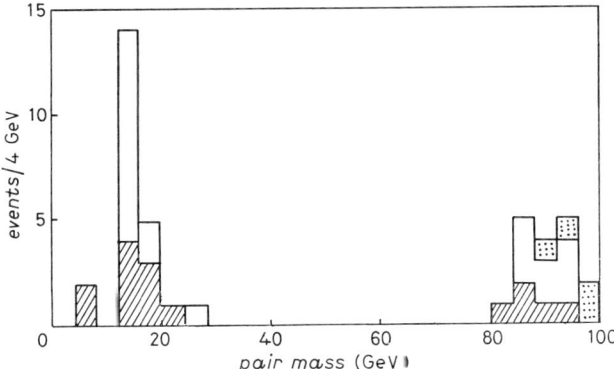

Fig. 12. – Mass spectrum of lepton pairs: ▨ μμ, UA1; ─── ee(γ), UA2; ▦ ee(γ), UA1.

The mass of the resonance is measured by the two experiments:

$$M_Z = (93.9 \pm 2.9) \text{ GeV}/c^2 \qquad \text{(UA1)},$$

$$M_Z = (92.7 \pm 2.2) \text{ GeV}/c^2 \qquad \text{(UA2)}.$$

We can now use the standard model and the value of $\sin^2\theta_W$ measured in neutrino scattering to predict the IVB masses [24]:

(3) $$M_W = A/\sin\theta_W = (82.4 \pm 1.1) \text{ GeV}/c^2,$$

(4) $$M_Z = M_W/\cos\theta_W = (93.3 \pm 0.09) \text{ GeV}/c^2,$$

where $A = 38.65$ GeV including higher-order corrections. This is in excellent agreement with the combined experimental result

$$M_W = (82.2 \pm 1.8) \text{ GeV}/c^2 \qquad \text{(UA1+UA2)},$$

$$M_Z = (93.1 \pm 1.8) \text{ GeV}/c^2 \qquad \text{(UA1+UA2)}.$$

Assuming now the validity of the standard model, we can use the relations (3), (4) to make an alternative measurement of $\sin^2\theta_W$:

$$\sin^2\theta_W = 0.221 \pm 0.010 \qquad \text{(UA1+UA2)}.$$

We can also check the prediction

(5) $$\varrho = M_W/M_Z \cos\theta_W = 1,$$

which results from a minimal Higgs structure of two complex scalar fields. The experimental result is

$$\varrho = 1.00 \pm 0.03 \qquad \text{(UA1+UA2)}.$$

We note that these measurements already have a precision comparable with the higher-order corrections of the field theory.

4˙5. *Z width and the number of neutrinos*. – The width of the Z^0 resonance provides a limit on the number of light neutrinos since each such species contributes $\Delta\Gamma_Z \approx 180$ MeV. The measured widths have the (90% confidence level) limits

$$\Gamma_Z < 9 \text{ GeV}/c^2 \qquad \text{(UA1)},$$

$$\Gamma_Z < 6.5 \text{ GeV}/c^2 \qquad \text{(UA2)},$$

from the combination of which we get

$$\Delta N_\nu < 15 \qquad \text{(UA1+UA2)},$$

where the Δ means in addition to the 3 more or less well-known species.

The ratio of the number of observed $Z \to ee(\gamma)$ to $W \to e\nu$ decays provides additional information on the total width if certain extra assumptions are made:

(6) $$\begin{cases} R = \dfrac{\sigma(Z)}{\sigma(W)} \dfrac{B(Z \to ee)}{B(W \to e\nu)} = \dfrac{\sigma(Z)}{\sigma(W)} \dfrac{\Gamma(Z \to ee)}{\Gamma(W \to e\nu)} \dfrac{\Gamma_W}{\Gamma_Z} = \dfrac{A}{\Gamma_Z}, \\ R_{\text{exp}} = 0.130 \pm 0.036, \end{cases}$$

where we have used the combined result for R_{exp}. The value of A has some theoretical uncertainty [19]:

$$A = 0.295 \pm 0.027,$$

which gives the 90% (95%) confidence level limits

$$\Gamma < 3.35 \ (3.56) \text{ GeV} \qquad (\text{UA1}+\text{UA2}),$$

$$\Delta N_\nu < 4 \ (5) \qquad (\text{UA1}+\text{UA2}).$$

These limits are the best given so far on the number of light neutrino species.

4'6. *Search for* $W \to t + \bar{b}$. – The expected partner to the b quark, the top t, has been intensively searched for at the Collider. If the mass of the t is about 40 GeV, then the best chance of observing it would be through the process

(7) $$W^* \to t + \bar{b}, \quad t \to b \ell^+ \nu_1.$$

Here the $\bar{b}$ quark would carry the largest momentum among the decay products. Due to the large top mass there is a chance that the lepton may be sufficiently separated from the second b jet to be clearly identified as an electron or muon.

After applying tight electron identification criteria, the UA1 sample of electrons accompagnied by jets (and not coming from W and Z leptonic decays) can be broken down into the following categories [25]:

electron + 1 jet: 14 events;

electron + 2 jets: 3 events;

electron + > 2 jets: 2 events;

where the electron has $p_T > 12$ GeV, the first jet $p_T > 10$ GeV and the other jets $p_T > 7$ GeV.

A similar study has been made for muons with the result

muon + 1 jet: 5 events;

muon + 2 jets: 3 events;

muon + > 2 jets: 1 event.

The lepton + two jets events are candidates for $W \to t + \bar{b}$ decays followed by a semi-leptonic decay of the t. The background is estimated on the basis of a larger sample where the electrons are replaced by π^0's. Two quantities are used to characterize the two samples: p_T^{out}, the « electron » momentum out of the plane of the first jet and the beam axis, and θ_{J2}, the polar angle of the second jet in the c.m. of the system. Since the background would result mainly from 2-jet events with some hard initial-state bremsstrahlung, it peaks at small p_T^{out} and small θ_{J2}. The signal events turn out to have large p_T^{out} and large θ_{J2}

which makes the background estimate small (≈ 0.1 events). For the muon events the background coming mainly from $K \to \mu\nu$ decays is estimated to 0.4 events.

In fig. 13 the three-body mass $M(\ell\nu j_2)$ is plotted against the four-body mass $M(\ell\nu j_1 j_2)$, where the missing p_T vector is set equal to the neutrino momentum. The events seem to cluster around the W mass, so the configurations of the 6 events are consistent with the decay (7). If that interpretation is correct, then the top mass would fall in the range 30 GeV $< m_t < 50$ GeV. A puzzling aspect of this story is the fact that only about 3 events are expected from W decays if the top mass is 40 GeV. Again the next run is eagerly awaited to provide better understanding of these events.

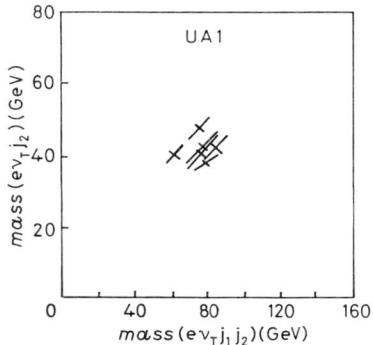

Fig. 13. – Events with an electron and two hets. $\bar{p}^\nu$ is set equal to the missing p_T vector.

5. – Exotic phenomena.

5'1. Radiative Z decays. – Among the events interpreted as Z^0 decays by UA2 one event contained 3 clearly separated electromagnetic clusters in the calorimeter, of which only 2 had an associated charged track pointing to it. This event they interpreted as a $Z \to e^+e^-\gamma$ decay, and later on also UA1 reported two events of this kind. The parameters of these three events are shown in table I.

The degree to which these events are « strange » depends on how they are viewed. The most conservative approach is to emphasize that the γ in the $\mu\mu\gamma$ event is not necessary to make the Z^0 mass (so it need not be a decay product) and then compute the probability that a photon from internal bremsstrahlung is clearly identified as an additional cluster in 2 events out of 12. This probability is about 6%, which seems acceptable. On the other hand, if only the configurations more extreme than the observed ones are taken into account, then the probability of seeing 2 of those is only 0.2%.

TABLE I. – $\ell^+\ell^-\gamma$ configurations.

	E^γ (GeV)	$\delta(\ell\gamma)$ (degrees)	$m(\ell^+\ell^-)$ (GeV/c²)	$m_{HI}(\ell\gamma)$	$m_{LO}(\ell\gamma)$	$m(\ell\ell\gamma)$
UA1 eeγ	38.8	14±4	42.7±2.4	88.8±2.5	4.6±1.0	98.7±5.0
UA2 eeγ	24.4	31±1	50.4±1.7	74.7±1.8	9.1±0.3	90.6±1.9
UA1 $\mu\mu\gamma$	23.3	8	71^{+37}_{-12}	52^{+28}_{-9}	5.0±0.4	88^{+16}_{-15}

If these events are not due to a statistical fluctuation, they would imply some degree of departure from the standard model. Using assumptions that preserve as much as possible of the standard model [26] the observation of one nontrivial $e^+e^-\gamma$ event would imply the presence of between 3 and 24 qqγ events at the same mass and between zero and 6 $\nu\nu\gamma$ events.

Using a sample of isolated single photon candidates UA2 can put a limit of 6 events containing $Z \to \nu\nu\gamma$ decays (90% confidence level) for photons emitted in the central region with $E^\gamma > 24$ GeV [27]. They also find less than 20 events containing $Z \to q\bar{q}\gamma$ decays where the photon is emitted in the central region with $p_T^\gamma > 15$ GeV/c. The candidate events for the latter process contain a photon candidate and two jets, and among those no structure is seen that distinguishes them from background events that have a jet in place of the photon candidate.

In UA1 one event has been found [28] where an electromagnetic cluster ($p_T = 55$ GeV) is accompanied by large missing p_T (≈ 40 GeV). This event is plotted as event H in fig. 14. Its interpretation in terms of $Z \to \nu\nu\gamma$ decay is, however, made difficult by the presence of a jet with $p_T = 12$ GeV at exactly opposite azimuth. Another unbalanced photonlike cluster (event G in fig. 14) points to a region where the track reconstruction is inefficient and could, therefore, well be a $W \to e\nu$ decay.

5.2. *Monojets*. – If all particle momenta in an event were measured without errors, then the sum of transverse components would be zero. In UA1 measurement errors introduce a r.m.s. deviation from zero of $\sigma \approx 0.7\sqrt{\sum E_T}$ in two-jet events, where $\sum E_T$ is the scalar sum of transverse energies. A sample of events containing a jet and missing p_T larger than 4σ has been studied [28].

After excluding events where the missing p_T vector points towards regions of incomplete detector coverage and events where $\sum \bar{p}_T$ of the other particles points opposite to the jet, the sample consists of 6 events ($A \to F$). The missing p_T spectrum of these events, which is shown in fig. 14, is interpreted as due to penetrating particles balancing the jets. One of the events (event F) can be interpreted in terms of $W \to \tau\nu$, whereas that explanation is not available for the other events. The event with the largest missing p_T (≈ 66 GeV) contains a high-p_T muon helping to make the missing p_T.

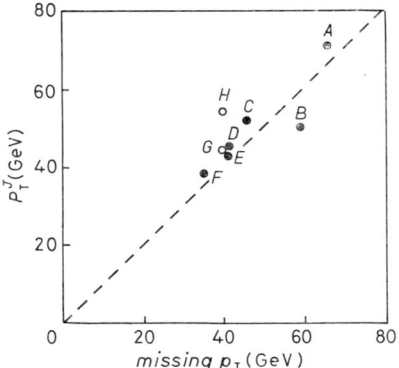

Fig. 14. – Jet p_T vs. p_T^{miss} in the UA1 sample of unbalanced jets: • single jet, ○ « photon ».

In UA2 a similar search [29] has left one event only with missing p_T larger than 50 GeV that lacks a natural explanation. It cannot be excluded, though, that this event is due to a beam-halo interaction in the calorimeter. This places a limit of 53 pb (90% confidence level) on unbalanced jets with $p_T > 50$ GeV in the UA2 central calorimeter.

5'3. *High-p_T W's*. – Another kind of unexplained events has been reported by UA2 [30] containing $W \to e\nu$ candidates associated with large jet activity. This jet activity causes some of the events to have extraordinarily large missing p_T, which excludes an interpretation in terms of background. The transverse views of the events are shown in fig. 15 of which the first three are well separated from background.

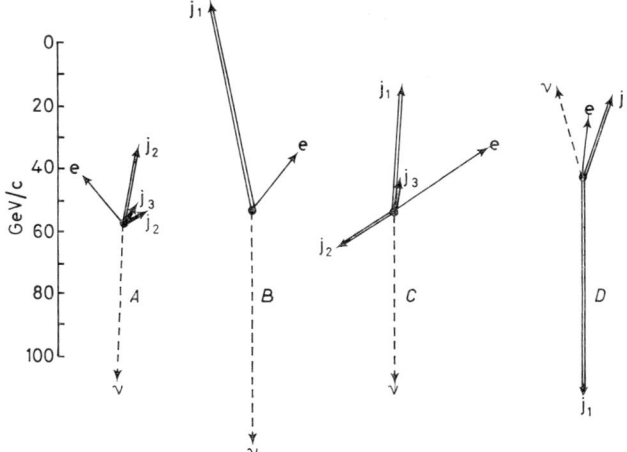

Fig. 15. – UA2 events containing electrons with high-p_T jets and large p_T^{miss}.

The jet activity in these events seems too large to allow an interpretation in terms of W's with jets from initial-state bremsstrahlung. If the W mass is assigned to the ev pairs in events A to C, then the mass of the evj systems would cluster around 160 GeV, but this mass region is also favoured by kinematical constraints. Finally it must be pointed out that the missing p_T in all the strange events mentioned could come from penetrating particles other than normal neutrinos. Suggestions have been made of photinos [31, 32] and heavy neutrinos [33].

* * *

I have used much material, while preparing this talk, from other talks given this Summer by J. D. DOWELL, P. JENNI, A. G. CLARK, A. R. WEIDBERG and L. DI LELLA. Their help is gratefully acknowledged.

REFERENCES

[1] J. TIMMER: in *Proceedings of the III Moriond Meeting on* $p\bar{p}$ *Physics* (Ed. Frontières, Paris, 1983), p. 593.
[2] B. MANSOULIE: in *Proceedings of the III Moriond Meeting on* $p\bar{p}$ *Physics* (Ed. Frontières, Paris, 1983), p. 609.
[3] UA2 COLLABORATION (M. BANNER et al.): *Phys. Lett. B*, **118**, 203 (1982).
[4] The AXIAL FIELD SPECTROMETER COLLABORATION (T. AKESSON et al.): *Phys. Lett. B*, **118**, 185, 193 (1982).
[5] UA2 COLLABORATION (P. BAGNAIA et al.): *Z. Phys. C*, **20**, 117 (1983).
[6] M. GRECO: *Z. Phys. C*, **26**, 567 (1985).
[7] F. HALZEN and P. HOYER: *Phys. Lett. B*, **130**, 326 (1983).
[8] UA1 COLLABORATION (G. ARNISON et al.): *Phys. Lett. B*, **136**, 294 (1984).
[9] UA2 COLLABORATION (P. BAGNAIA et al.): *Phys. Lett. B*, **144**, 283 (1984).
[10] J. C. COLLINS and D. E. SOPER: *Phys. Rev. D*, **16**, 2219 (1977).
[11] H. ABRAHAMOWICZ et al.: *Z. Phys. C*, **12**, 289 (1982); **13**, 199 (1982); **17**, 233, (1983).
[12] W. FURMANSKI and H. KOWALSKI: DESY 84/035.
[13] G. MARCHESINI and B. R. WEBBER: *Nucl. Phys. B*, **238**, 1 (1984).
[14] UA2 COLLABORATION (P. BAGNAIA et al.): *Phys Lett. B*, **144**, 291 (1984).
[15] M. ALTHOFF et al.: *Z. Phys. C*, **22**, 307 (1984).
[16] UA1 COLLABORATION (G. ARNISON et al.): *Phys. Lett. B*, **126**, 398 (1983).
[17] P. BAGNAIA et al.: *Z. Phys. C*, **24**, 1 (1984).
[18] UA1 COLLABORATION (G. ARNISON et al.): *Phys. Lett. B*, **134**, 469 (1984).
[19] G. ALTARELLI, R. K. ELLIS, M. GRECO and G. MARTINELLI: *Nucl. Phys. B*, **246**, 12 (1985).
[20] C. T. H. DAVIES, B. R. WEBBER and W. J. STIRLING: *Nucl. Phys. B*, **256**, 285 (1985).
[21] UA2 COLLABORATION (P. BAGNAIA et al.): *Phys. Lett. B*, **129**, 130 (1983).
[22] UA1 COLLABORATION (G. ARNISON et al.): *Phys. Lett. B*, **126**, 398 (1983).
[23] UA1 COLLABORATION (G. ARNISON et al.): *Phys. Lett. B*, **147**, 241 (1984).

[24] W. MARCIANO: talk given at the *Workshop on p$\bar{\text{p}}$ Collisions, Bern, March 1984*.
[25] M. DELLA NEGRA: talk given at the *Symposium on Multiparticle Dynamics, Lund, June 1984*.
[26] M. J. DUNCAN and M. VELTMAN: *Phys. Lett. B*, **139**, 310 (1984).
[27] A. G. CLARK: talk given at the *Symposium on Multiparticle Dynamics, Lund, June 1984*.
[28] UA1 COLLABORATION (G. ARNISON *et al.*): *Phys. Lett. B*, **139**, 115 (1984).
[29] J. P. REPELLIN: talk given at the *International Conference on High Energy Physics, Leipzig, July 1984*.
[30] UA2 COLLABORATION (P. BAGNAIA *et al.*:) *Phys. Lett. B*, **139**, 105 (1984).
[31] H. HABER and G. KANE: *Phys. Lett. B*, **142**, 212 (1984).
[32] J. ELLIS and H. KOWALSKI: *Phys. Lett. B*, **142**, 441 (1984).
[33] M. GRONAU and J. L. ROSNER: *Phys. Lett. B*, **147**, 217 (1984).

Experimental Tests and Theoretical Predictions for Electroweak Processes.

G. MARTINELLI

CERN - Geneva, Switzerland
Istituto Nazionale di Fisica Nucleare - Laboratori Nazionali di Frascati, Italia

1. – Introduction.

One of the most important advances in elementary-particle physics in the recent past has been the development and the experimental confirmation of the electroweak gauge theory based on the $SU_2 \times U_1$ model due to GLASHOW, WEINBERG and SALAM (GWS) [1-3]. The first important prediction of the model was that neutral currents should exist, as subsequently confirmed by experiments. More recently, the observation at the CERN collider of isolated, high-transverse-momentum electron and lepton pairs led to the discovery of W's and Z^0 bosons predicted by the GWS model with approximately the expected masses [4, 5].

At low energy the main experimental tests of the electroweak theory are based on the determination of neutral-current couplings to fundamental fermions. At collider energies, tests of the standard model are based on measurements of masses and decay rates of intermediate vector bosons (IVB). Radiative corrections to couplings and masses are quite sizable in the GWS model [6]. Therefore, precise measurements test the theory at the level of quantum corrections in much the same way that $g-2$ measurements in QED.

In sect. **2**, I will briefly recall the basic ingredients of the standard model and I will define the relevant parameters. Low-energy processes which enter into the determination of neutral-current couplings to fermions (in particular $\sin^2\theta_W$) are presented in sect. **3**. Radiative corrections to these processes are discussed in sect. **4**. In sect. **5** the measurements of the W and Z^0 masses at the SPS collider are described and compared with theoretical predictions including one-loop radiative corrections.

2. – The standard model.

The standard model is a gauge theory which unifies weak and electromagnetic interactions. The model is based on an SU_2 group of weak isospin and a U_1 group of weak hypercharge. This symmetry is spontaneously broken with the help of scalar mesons called Higgs scalars: in the resulting theory we find two charged and one neutral massive vector bosons, $W^\pm$ and Z^0, and one massless neutral vector boson, the photon. The Lagrangian which describes the interaction of these vector bosons with fermion fields is given by

(1)
$$\mathscr{L} = \overbrace{\frac{g}{\sqrt{2}}(J_\mu^- W_\mu^+ + J_\mu^+ W_\mu^-)}^{\text{weak charged currents}} + \overbrace{\frac{g}{\cos\theta_W}(J_\mu^3 - \sin^2\theta_W J_\mu^{\text{e.m.}})Z_\mu}^{\text{weak neutral currents}} + \underbrace{g\sin\theta_W J_\mu^{\text{e.m.}} A_\mu}_{\text{electromagnetic currents}},$$

where $J_\mu^{1,\dots,3}$ are the weak-isospin currents, $J_\mu^{\text{e.m.}}$ is the electromagnetic current, $\text{tg}\,\theta_W = g'/g$. g and g' are the couplings associated with the gauged SU_2 and U_1 group, respectively.

From eq. (1), it follows that

(2)
$$e = g\sin\theta_W,$$

where e is the electric change. The currents in eq. (1) are given once the isospin and hypercharge assignment of fermions is known. To describe present phenomenology, we put the left-handed fermions into isospin doublets:

(3)
$$\begin{cases} \begin{pmatrix} \nu_e \\ e^- \end{pmatrix}_L, \begin{pmatrix} \nu_\mu \\ \mu^- \end{pmatrix}_L, \begin{pmatrix} \nu_\tau \\ \tau^- \end{pmatrix}_L, & \text{leptons}, \\ \begin{pmatrix} u \\ d' \end{pmatrix}_L, \begin{pmatrix} c \\ s' \end{pmatrix}_L, \begin{pmatrix} t \\ b' \end{pmatrix}_L, & \text{quarks}, \end{cases}$$

where u, c, t are the charge $+\tfrac{2}{3}$ quarks and d', s', b' are related by a unitary transformation to the mass eigenstate charge $-\tfrac{1}{3}$ d, s and b quarks. Right-handed fermions are weak-isospin singlets. W exchanges give rise to charged-current processes like, for example, the muon decay:

(4)
$$\mu^- \to e^- + \bar{\nu}_e + \nu_\mu.$$

This process is illustrated in fig. 1. Since the momentum transfer $|q^2| \ll M_W^2$, the amplitude for this process can be obtained by an effective Hamiltonian of the form

(5)
$$H^{\text{eff}} = \frac{g^2}{8 M_W^2} \{[\bar{\nu}_\mu \gamma_\mu (1-\gamma_5)\mu][\bar{e}\gamma_\mu(1-\gamma_5)\nu_e] + \text{h.c.}\}.$$

We can then relate the coupling g and the W mass to the Fermi constant G_F:

(6)
$$\frac{G_F}{\sqrt{2}} = \frac{g^2}{8 M_W^2}, \qquad M_W^2 = \frac{\pi\alpha}{\sqrt{2}\, G_F \sin^2\theta_W},$$

where α is the fine-structure constant.

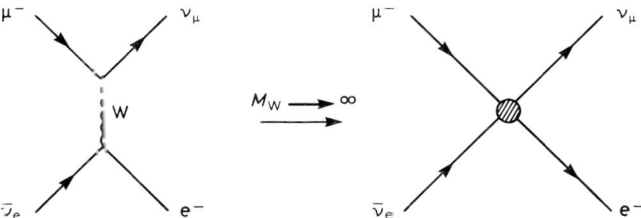

Fig. 1. – Relevant diagram for the muon decay. For $M_W \to \infty$, the diagram is described by an effective four-fermion Hamiltonian.

In analogy with this particular case, the low-energy effective Hamiltonian for weak interactions has the form

(7) $$H^{\text{eff}} = \frac{4 G_F}{\sqrt{2}} [J_\mu^+ J_\mu^- + \varrho (J_\mu^3 - \sin^2 \theta_W J_\mu^{\text{e.m.}})^2].$$

In eq. (7), we introduce the parameter ϱ [7]:

(8) $$\varrho = \frac{M_W^2}{M_Z^2 \cos^2 \theta_W} = \frac{\sum_i [I^{(i)}(I^{(i)}+1) - I_3^{(i)2}] \lambda_{(i)}^2}{2 \sum_i I_3^{(i)2} \lambda_{(i)}^2},$$

where the sums are over the representations of scalar fields with weak isospin $I^{(i)}$; $I_3^{(i)}$ and $\lambda_{(i)}$ are the third component of the weak isospin and the vacuum expectation value of the neutral member of the (i) Higgs multiplet. In the GWS model with only doublets of Higgs scalars, $\varrho = 1$.

For a complete determination of the weak-current couplings (excluding fermion and Higgs masses and mixing angles) we have to fix four (three if ϱ is chosen to be 1) constants. Two of them are usually taken to be

(9) $$\begin{cases} \alpha^{-1} \simeq 137.036 & \text{from Josephson effect or } g_e - 2 \text{ experiments,} \\ G_F \simeq 1.16634 \cdot 10^{-5} \, (\text{GeV})^{-2} & \text{from the } \mu \text{ lifetime.} \end{cases}$$

Including radiative corrections, the μ lifetime is given by

(10) $$\tau_\mu^{-1} = \frac{G_F^2}{192 \pi^3} m_\mu^5 \left(1 - \frac{8 m_e^2}{m_\mu^2}\right) \left[1 - \frac{\alpha}{2\pi}\left(\pi^2 - \frac{25}{4}\right) + O\left(\alpha^2 \ln \frac{m_\mu}{m_e}\right)\right].$$

The other parameters to be fixed could be chosen to be $\sin^2 \theta_W$ and ϱ taken from low-energy neutral and charged leptonic processes or, alternatively, the M_W and M_Z masses measured at the SPS Collider. In the next section we analyse some of the low-energy experiments used to measure $\sin^2 \theta_W$ and ϱ.

3. – Low-energy experimental tests of the standard model.

3'1. *Leptonic processes.* – The cleanest method to measure $\sin^2\theta_W$ and ϱ at low energy is certainly the study of pure leptonic processes, since for these interactions theoretical predictions are completely unambiguous. Pure leptonic reactions experimentally accessible are

$$(11) \quad \begin{cases} \text{I)} \quad e^+e^- \to \mu^+\mu^-, & \mu \simeq 35 \text{ GeV}, \\ \text{II)} \quad \left.\begin{array}{l}\nu_\mu e \to \nu_\mu e \\ \bar{\nu}_\mu e \to \bar{\nu}_\mu e\end{array}\right\}, & \mu \simeq 0.3 \text{ GeV}, \\ \quad\quad \left.\begin{array}{l}\nu_e e \to \nu_e e \\ \bar{\nu}_e e \to \bar{\nu}_e e\end{array}\right\}, & \mu \simeq 10^{-3} \text{ GeV}. \end{cases}$$

The figures reported above refer to the typical scales at which these processes are observed.

I) $e^+e^- \to \mu^+\mu^-$. Weak neutral-current effects in e^+e^- annihilation appear as deviations from pure QED predictions. At the present maximum available energy (PETRA $\sqrt{s} \simeq 35$ GeV) the interferences between the exchange of a photon and the Z^0 (see fig. 2) give rather small contributions to the cross-sections. In most of the cases, as for example for Bhabha scattering, the effect of this interference cannot yet be established. One of the best results is obtained by looking at the forward-backward asymmetry in $e^+e^- \to \mu^+\mu^-$. The asymmetry is defined as

$$(12) \quad A_{\mu\mu} = \frac{\sigma(\theta<\pi/2) - \sigma(\theta>\pi/2)}{\sigma(\theta<\pi/2) + \sigma(\theta>\pi/2)},$$

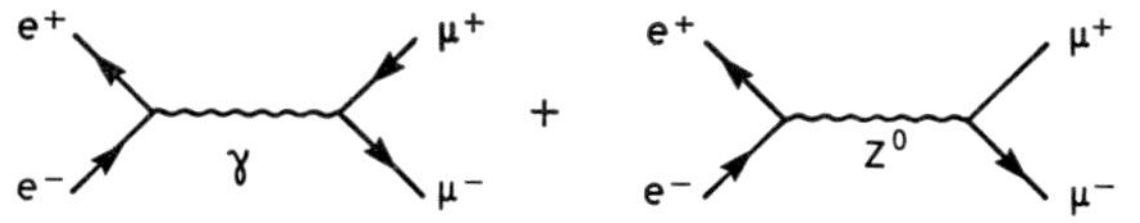

Fig. 2. – Lowest-order diagrams for $e^+e^- \to \mu^+\mu^-$.

where θ is the angle between the μ^- and the outgoing e^- beam. The Mark-J collaboration observed experimentally (see also fig. 3)

$$(13) \quad A_{\mu\mu} = (-8.1 \pm 1.6)\% \quad \sqrt{s} = 34.6 \text{ GeV}, \quad 0 \leqslant |\cos\theta| \leqslant 0.8.$$

Pure QED predicts $A_{\mu\mu}^{\text{QED}} \simeq +1.4\%$. Thus one gets

$$A_{\mu\mu}^W = (-9.5 \pm 1.6)\%,$$

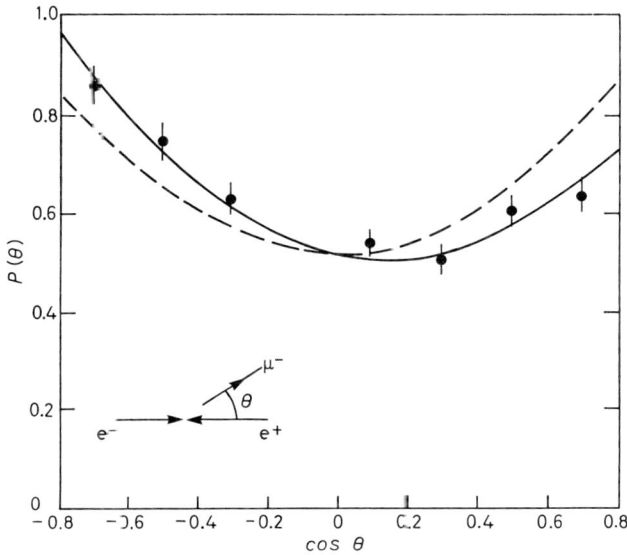

Fig. 3. – We report the number of events for the process in fig. 2 ($e^+e^- \to \mu^+\mu^-$) as a function of the angle θ defined in the figure (see also sect. 3). The data are from the Mark-J Collaboration, $\sqrt{s} = 34.6$ GeV.

TABLE I. – *The experimental results for the forward-backward asymmetry $A_{\mu\mu}^{\exp}$ in $e^+e^- \to \mu^+\mu^-$ are compared with theoretical predictions ($A_{\mu\mu}^{\th}$) obtained by assuming $\sin^2\theta_W \simeq 0.22$ and $\varrho = 1$.*

	$\sqrt{s}$ (GeV)	$A_{\mu\mu}^{\exp}(\%)$	$A_{\mu\mu}^{\th}(\%)$
MARK-J	34.6	$-11.7 \pm 1.7 \pm 1$	-8.8
CELLO	34.2	-6.4 ± 6.4	-8.6
JADE	33.5	$-12.8 \pm 3.8 \pm 1$	-8.2
MAC	29	$-5.8 \pm 1.0 \pm 0.3$	-5.9
MARK-II	29	$-7.1 \pm 1.7 \pm 0.5$	-5.9
PLUTO	34.7	$-13.4 \pm 3.1 \pm 1$	-8.8
TASSO	34.5	$-9.8 \pm 2.3 \pm 0.5$	-8.8

to be compared with standard-model theoretical prediction

(14) $$A_{\mu\mu}^{\text{W-th}} = (-7.7 \pm 0.4)\%.$$

In table I, results from different experiments are compared with theoretical predictions. From these measurements, one can extract a value for $\sin^2\theta_W$. For example, the Mark-J collaboration found

$$\sin^2\theta_W = 0.15^{+0.04}_{-0.02} \quad \text{assuming } M_{Z^0} = (93 \pm 2) \text{ GeV}$$

and

(15) $$0.11 \leqslant \sin^2 \theta_W \leqslant 0.23 \qquad (95\% \text{ c.l.}).$$

As appears from the above results, the determination of $\sin^2\theta_W$ from e^+e^- annihilation is at present rather poor.

II) $\nu(\bar\nu)e \to \nu(\bar\nu)e$. Signals for $\nu_\mu(\bar\nu_\mu)e$ scattering have been obtained in bubble chamber (low rate, clearer signals) and counter experiments using high-energy neutrino beams. $\bar\nu_e e$ processes have been studied using reactor-produced $\bar\nu_e$ beams.

The effective Hamiltonian (cf. the diagrams in fig. 4a), b)) which governs neutrino-initiated leptonic processes can be written as

(16) $$H = \frac{G_F}{\sqrt{2}} [\bar\nu \gamma_\mu (1-\gamma_5) \nu] J^e_\mu,$$

where

(17) $$\begin{cases} J^e_\mu = \varepsilon_L(e)[\bar e \gamma_\mu (1-\gamma_5) e] + \varepsilon_R(e)[\bar e \gamma_\mu (1+\gamma_5) e] = \bar e \gamma_\mu (g^e_V - g^e_A \gamma_5) e, \\ g^e_{V,A} = \varepsilon_L(e) \pm \varepsilon_R(e). \end{cases}$$

The standard-model expressions of $g^e_{V,A}$ (including the piece coming from the W exchange of fig. 4b)) for different processes are reported in table IIa). From the Hamiltonian of eq. (16), one easily derives the cross-section of the reaction $\nu(\bar\nu)e \to \nu(\bar\nu)e$:

(18) $$\begin{cases} \dfrac{d\sigma}{dE_e} = \dfrac{G_F^2 M_e}{2\pi} \left[A + B \left(1 - \dfrac{E_e}{E_\nu}\right)^2 - C \dfrac{M_e E_e}{E_\nu^2} \right], \\ \sigma \simeq \dfrac{G_F^2 M_e E_\nu}{2\pi} \left(A + \dfrac{B}{3}\right), \qquad E_\nu \gg M_e, \\ G_F M \simeq 10^{-41} \text{ cm}^2 \text{ (GeV)}^{-1}. \end{cases}$$

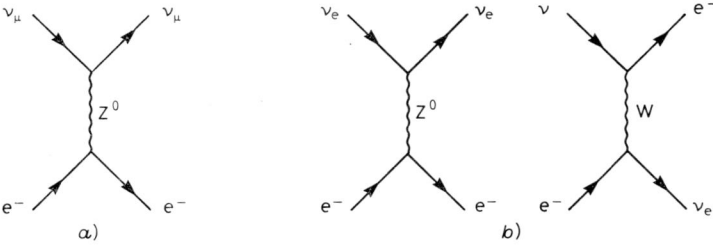

Fig. 4. – a) Relevant diagrams for $\nu_\mu e$ scattering and b) $\nu_e e$ scattering.

TABLE II. – *The standard-model expressions for $g^e_{V,A}$ defined in eq. (17) are given in a). In b), we report the coefficients A, B and C appearing in the νe cross-sections in terms of $g^e_{V,A}$.*

a)

	g^e_V	g^e_A
$\nu_\mu e$	$-\frac{1}{2} + 2\sin^2\theta_W$	$-\frac{1}{2}$
$\bar{\nu}_\mu e$	$-\frac{1}{2} + 2\sin^2\theta_W$	$+\frac{1}{2}$
$\nu_e e$	$+\frac{1}{2} + 2\sin^2\theta_W$	$+\frac{1}{2}$
$\bar{\nu}_e e$	$+\frac{1}{2} + 2\sin^2\theta_W$	$-\frac{1}{2}$

b)

	νe	$\bar{\nu}$e
A	$(g^e_V + g^e_A)^2$	$(g^e_V - g^e_A)^2$
B	$(g^e_V - g^e_A)^2$	$(g^e_V + g^e_A)^2$
C	$g^{e^2}_V - g^{e^2}_A$	$g^{e^2}_V - g^{e^2}_A$

M_e is the electron mass; $E_{\nu,e}$ are the initial-neutrino and final-electron energies, respectively, and the expressions for A, B and C are given in table IIb). In table III, I report some experimental results taken from ref. [9].

In fig. 5, the regions of the g^e_V-g^e_A plane for neutral currents allowed by a fit to the data (ref. [9]) are shown. Notice that a precise determination of the $\nu_e(\bar{\nu})_e$ cross-section would give the sign of the axial coupling of the electron to the Z^0, as a consequence of the interference between neutral and charged currents (fig. 4b)). We observe that, also for leptonic reactions, the data are at present not precise enough to allow a determination of $\sin^2\theta_W$ with an accuracy comparable to what can be obtained from semi-leptonic processes (discussed in the following subsection).

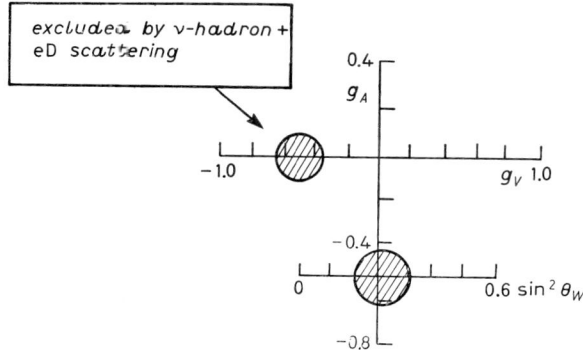

Fig. 5. – Regions of leptonic couplings allowed by the fit of ref. [9] to νe scattering data. The region indicated by the arrow is excluded by the ν-hadron and eD scattering experimental results.

TABLE III. – *Experimental results for the* νe *reaction cross-sections.*

Experiment	$\nu(\bar\nu)e$ candidates	Background	σ/E_ν (units 10^{-42} cm² (GeV)⁻¹)
GGM-CERN-PS, ν_μ (BLIETSCHAU et al.)	1	0.3 ± 0.1	< 3 (90% c.l.)
AP, ν_μ (FAISSNER et al.)	32	20.5 ± 2.0	1.1 ± 0.6
GGM'-CERN-SPS, ν_μ (ARMENISE et al.)	9	0.5 ± 0.2	$2.4^{+1.2}_{-0.9}$
CB-FNAL, ν_μ (CNOPS et al.)	11	0.5 ± 0.5	1.8 ± 0.8
CHARM, ν_μ (JONKER et al.)	11	4.5 ± 1.4	2.6 ± 1.6
VMWOF, ν_μ (HEISTERBERG et al.)	46	12	1.4 ± 0.3
GGM-CERN-PS, $\bar\nu_\mu$ (BLIETSCHAU et al.)	3	0.4 ± 0.1	$1.0^{+2.1}_{-0.9}$ (90% c.l.)
AP, $\bar\nu_\mu$ (FAISSNER et al.)	17	7.4 ± 1.0	2.2 ± 1.0
GGM'-CERN-SPS, $\bar\nu_\mu$ (BERTRAND et al.)	0	< 0.03	< 2.7
FMMS-FNAL, $\bar\nu_\mu$ (BERGE et al.)	0	0.2	—
BEBC-TST-CERN-SPS, $\bar\nu_\mu$ (ARMENISE et al.)	1	0.5 ± 0.2	< 3.4
			σ (10^{-46} cm²)
REINES et al., $\bar\nu_e$ (1.5 MeV $< E_e <$ 3.0 MeV)			7.6 ± 2.2
REINES et al., $\bar\nu_e$ (3.0 MeV $< E_e <$ 4.5 MeV)			1.86 ± 2.2

3'2. Lepton-hadron reactions. – In this subsection, we concentrate our attention on the information coming from neutrino-hadron and polarized electron-hadron deep inelastic scattering.

$$(19) \quad \begin{cases} \text{III)} & \begin{matrix} \nu_\mu + \mathcal{N} \to \mu^- + X \\ \bar\nu_\mu + \mathcal{N} \to \mu^+ + X \end{matrix} \Bigr\}, \quad \mu \simeq (1 \div 10) \text{ GeV}, \\ & \begin{matrix} \nu_\mu + \mathcal{N} \to \nu_\mu + X \\ \bar\nu_\mu + \mathcal{N} \to \bar\nu_\mu + X \end{matrix} \Bigr\}, \quad \mu \simeq 1.4 \text{ GeV}, \\ \text{IV)} & e^-_{L,R} + D \to e^- + X, \quad \mu \simeq 1.4 \text{ GeV}. \end{cases}$$

Let us discuss first neutrino charged-current and neutral-current induced processes.

III) The effective Hamiltonian for charged-current interactions ($\nu N \to \mu X$) was already given in eq. (7):

(20) $$H = \frac{4G_F}{\sqrt{2}} J_\mu^+ \bar{J}_\mu,$$

where

(21) $$J_\mu = [\bar{u}, \bar{c}, \bar{t}]_L \gamma_\mu U_{KM} \begin{bmatrix} d \\ s \\ b \end{bmatrix}_L + [\bar{\nu}_e, \bar{\nu}_\mu, \bar{\nu}_\tau] \gamma_\mu \begin{bmatrix} e^- \\ \mu^- \\ \tau^- \end{bmatrix}_L.$$

U_{KM} is the Kobayashi-Maskawa mixing matrix (assuming $m_\nu = 0$, $U_{KM}^{lept} = \hat{I}$). In the valence approximation (*i.e.* the nucleon made only by valence u and d quarks), the active part of $H^{c.c.}$ is given by

(22) $$H^{c.c.} = \frac{G_F}{\sqrt{2}} [\bar{\mu}\gamma_\mu(1-\gamma_5)\nu_\mu][\bar{u}\gamma_\mu(1-\gamma_5)d] + \text{h.c.} \qquad (\cos\theta_c = 1).$$

From eq. (22) in the naive parton model approximation, one finds for ν scattering on isoscalar targets (fig. 6)

(23) $$\begin{cases} \dfrac{d\sigma_{c.c.}^{\nu\mu\,I=0}}{dx\,dy} = \dfrac{2G_F^2 M_P E_\nu}{\pi}[xu(x) + xd(x)], \\ \dfrac{d\sigma_{c.c.}^{\bar\nu\mu\,I=0}}{dx\,dy} = \dfrac{2G_F^2 M_P E_{\bar\nu}}{\pi}[xu(x) + xd(x)](1-y)^2. \end{cases}$$

In eq. (23) $E_{\nu,\bar\nu}$ is the incident-neutrino (antineutrino) energy; M_P is the nucleon mass,

$$x = \frac{Q^2}{2M_P\nu}, \qquad y = E_h/E_{\nu(\bar\nu)}; \qquad E_h = \text{final hadronic energy}, \qquad 0 \leq x, y \leq 1,$$

where $Q^2 = |(P_\nu - P'_\nu)^2|$, $\nu = E_\nu - E'_\nu$.

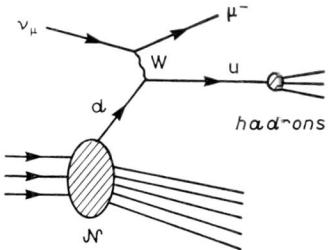

Fig. 6. – The neutrino scatters a d-quark off the nucleon. The final state is constituted by hadronic fragments of the nucleon and of the scattered quark + the muon.

On the other hand, for neutral-current processes

$$(24) \qquad H^{\text{n.c.}} = \frac{G_F}{\sqrt{2}} [\bar{\nu}_\mu \gamma_\mu (1-\gamma_5) \nu_\mu] J_\mu^H,$$

where

$$(25) \qquad J_\mu^H = \sum_f \bar{q}_f \gamma_\mu (g_V^f - g_A^f \gamma_5) q_f.$$

In eq. (25), we define (cf. eqs. (17)) $g_{V,A}(u,d) = \varepsilon_L(u,d) \pm \varepsilon_R(u,d)$.

The expressions for $\varepsilon_{L,R}(u,d)$ in the standard model are reported in table IV. From eqs. (24) and (25), one computes

$$(26) \qquad \begin{cases} \dfrac{d\sigma^{\nu_\mu I=0}_{\text{n.c.}}}{dx\,dy} = \dfrac{2G_F^2 M_P E_\nu}{\pi} \big[(|\varepsilon_L(u)|^2 + |\varepsilon_L(d)|^2) + \\ \qquad\qquad\qquad + (|\varepsilon_R(u)|^2 + |\varepsilon_R(d)|^2)(1-y)^2\big][xu(x) + xd(x)], \\[6pt] \dfrac{d\sigma^{\bar{\nu}_\mu I=0}_{\text{n.c.}}}{dx\,dy} = \dfrac{2G_F^2 M_P E_{\bar\nu}}{\pi} \big[(|\varepsilon_L(u)|^2 + |\varepsilon_L(d)|^2)(1-y)^2 + \\ \qquad\qquad\qquad + (|\varepsilon_R(u)|^2 + |\varepsilon_R(d)|^2)\big][xu(x) + xd(x)]. \end{cases}$$

TABLE IV. – *Neutral-current couplings of* u *and* d *quarks in the standard model.*

$\varepsilon_L(u)$	$\varepsilon_L(d)$	$\varepsilon_R(u)$	$\varepsilon_R(d)$
$\frac{1}{2} - \frac{2}{3}\sin^2\theta_W$	$-\frac{1}{2} + \frac{1}{3}\sin^2\theta_W$	$-\frac{2}{3}\sin^2\theta_W$	$\frac{1}{3}\sin^2\theta_W$

From eq. (26), we see that measurements of deep inelastic c.c. and n.c. reactions on isoscalar targets will only allow the determination of the left-right handed couplings $g_{L,R}$ defined as

$$(27) \qquad g_L = \big[|\varepsilon_L(u)|^2 + |\varepsilon_L(d)|^2\big]^{\frac{1}{2}}, \qquad g_R = \big[|\varepsilon_R(u)|^2 + |\varepsilon_R(d)|^2\big]^{\frac{1}{2}}.$$

The allowed regions of coupling constants may be represented as annuli in the $\varepsilon_L(u)$-$\varepsilon_L(d)$ and $\varepsilon_R(u)$-$\varepsilon_R(d)$ planes. Combining eqs. (23) and (26), one easily derives

$$(28) \qquad \begin{cases} R^\nu = \dfrac{\sigma^{\nu_\mu I=0}_{\text{n.c.}}}{\sigma^{\nu_\mu I=0}_{\text{c.c.}}} = |\varepsilon_L(u)|^2 + |\varepsilon_L(d)|^2 + \dfrac{1}{3}\big(|\varepsilon_R(u)|^2 + |\varepsilon_R(d)|^2\big), \\[8pt] R^{\bar\nu} = \dfrac{\sigma^{\bar\nu_\mu I=0}_{\text{n.c.}}}{\sigma^{\bar\nu_\mu I=0}_{\text{c.c.}}} = \dfrac{1}{3}\big(|\varepsilon_L(u)|^2 + |\varepsilon_L(d)|^2\big) + |\varepsilon_R(u)|^2 + |\varepsilon_R(d)|^2. \end{cases}$$

In the GWS model, eqs. (28) can be rewritten as (cf. table IV)

$$(29) \qquad R^\nu = \frac{1}{2} - \sin^2\theta_W + \frac{20}{27}\sin^4\theta_W, \qquad R^{\bar\nu} = \frac{1}{2} - \sin^2\theta_W + \frac{20}{9}\sin^4\theta_W.$$

We also note that for $\sin^2\theta_W \simeq 0.23$

(30) $$\frac{\delta R^\nu}{\delta \sin^2\theta_W} \simeq -0.67, \qquad \frac{\delta R^{\bar\nu}}{\delta \sin^2\theta_W} \simeq -0.02.$$

The very simple expressions given in eq. (29) are modified in a more accurate analysis which takes into account

    i) neutron excess in the target,

    ii) the sea ($q\bar{q}$ pairs and gluons) content of the nucleon,

    iii) effects of scaling violations on parton densities induced by strong interactions,

    iv) order α_s (α_s = strong-interaction coupling constant) corrections,

    v) the fact that experimentally the longitudinal structure function defined as $F_L = F_2 - 2 \times F_1$ is different from zero unlike the naive parton model.

All these effects slightly modify the annuli in the ε_L and ε_R planes. Another important effect, coming from the esperimental cuts in E_h, will be discussed in the next section. The results of the analysis of ref. [9] for $R^{\nu,\bar\nu}$ are reported in fig. 7a), b). A more complete analysis which included other processes like

    i) scattering on nonisoscalar targets (*e.g.*, $\nu p \to \mu X$),

    ii) inclusive π production

(31) $$R^{+/-}_{\nu,\bar\nu} = \frac{\sigma[\nu(\bar\nu)+\mathcal{N} \to \nu(\bar\nu)\pi^+ X]}{\sigma[\nu(\bar\nu)+\mathcal{N} \to \nu(\bar\nu)\pi^- X]},$$

    iii) elastic $\nu(\bar\nu)$-nucleon scattering

reduces the allowed regions of fig. 7b) to those shown in fig. 8. Moreover, of the two possible regions, the one corresponding to an almost isoscalar neutral current is excluded by this last set of data. Theoretically, predictions for processes i)-iii) are rather poor and will not be discussed here. From an overall fit of the data to $\sin^2\theta_W$ and ϱ, one obtains the results given in the first line of table V.

TABLE V. – *In the first line, we report the results for* $\sin^2\theta_W$ *and* ϱ *from neutrino-hadron deep-inelastic-scattering data. The first value of* $\sin^2\theta_W$ *is obtained by fixing* $\varrho = 1$. *The other two figures are obtained by a simultaneous fit of* ϱ *and* $\sin^2\theta_W$. *The second line gives the value of* $\sin^2\theta_W$ *from e-D polarized scattering.*

$\sin^2\theta_W$	ϱ	$\sin^2\theta_W$
0.234 ± 0.011	0.999 ± 0.025	0.232 ± 0.027
0.223 ± 0.015	—	—

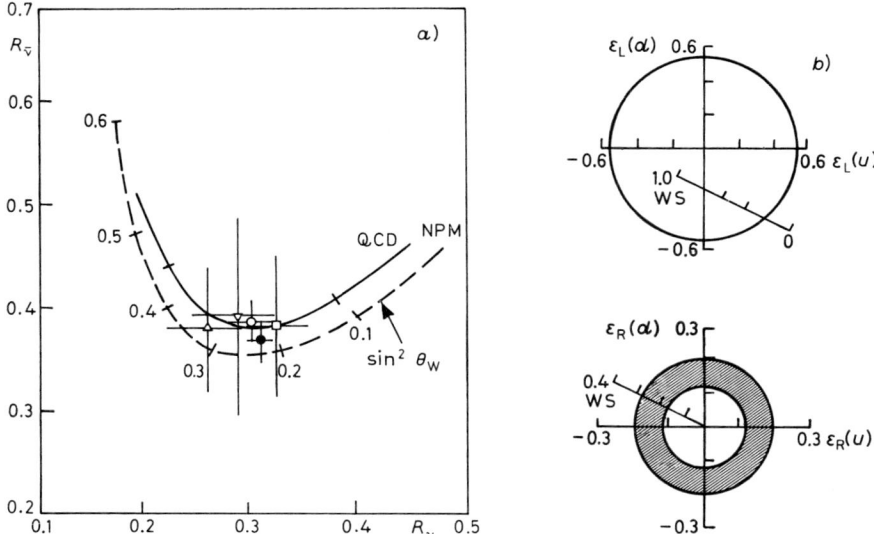

Fig. 7. – Results from the analysis of ref. [9]. In a), I report the theoretical predictions for $R^\nu - R^{\bar\nu}$ as a function of the value of $\sin^2\theta_W$. The full line refers to a complete computation which includes scaling violation of parton densities. The dashed line refers to the naive parton model result. The data cluster in the interval $0.2 \leqslant \sin^2\theta_W \leqslant 0.3$: • CDHS, ○ CHARM SPS, □ BEBC, △ GGM PS, ▽ HPWF FNAL. b) reports the allowed region in the ε_L and ε_R planes from the data shown in a).

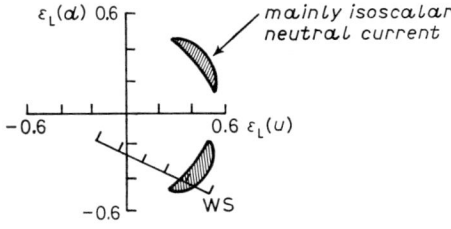

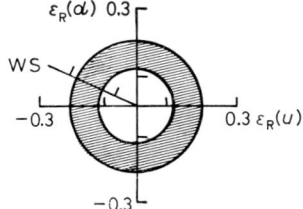

Fig. 8. – Same as fig. 7b) but including experimental results from $\nu(p, n) \to \nu X$. The mainly isoscalar neutral-current region is excluded by inclusive pion production.

iv) $e_{L,R} + \mathcal{N} \to e + X$: reactions which involve electron-hadron interactions have a unique position as tests of the parity-violating nature of neutral currents. The cleanest results came from the measurement of the asymmetry in the scattering of polarized electron from a deuteron target:

$$(32) \qquad e_{L,R} + D \to e + X .$$

The effective Hamiltonian responsible for weak interference effects can be written as

$$(33) \quad H = \frac{G_F \varrho}{2\sqrt{2}} [(\bar{e}\gamma_\mu\gamma_5 e)(V_\mu \bar{\mu}\gamma_\mu \mu + V_u \bar{u}\gamma_\mu u + V_d \bar{d}\gamma_\mu d) + $$
$$+ (\bar{e}\gamma_\mu e)(A_\mu \bar{\mu}\gamma_\mu\gamma_5 \mu + A_u \bar{u}\gamma_\mu\gamma_5 u + A_d \bar{d}\gamma_\mu\gamma_5 d) + $$
$$+ (\bar{e}\gamma_\mu\gamma_5 e)(C_\mu \bar{\mu}\gamma_\mu\gamma_5 \mu + C_u \bar{u}\gamma_\mu\gamma_5 u + C_d \bar{d}\gamma_\mu\gamma_5 d)] .$$

The expressions for V_i and A_i which enter into the process (32) are given in table VI for the standard model. The Hamiltonian in eq. (33) is also responsible for the forward-backward asymmetry in $e^+e^- \to \mu^+\mu^-$ previously discussed. In the present case, we define the asymmetry as

$$(34) \qquad A = \frac{d\sigma_R - d\sigma_L}{d\sigma_R + d\sigma_L} .$$

TABLE VI. – *The couplings V and A appearing in eq. (33) of the text.*

	V	A
u	$1 - \frac{8}{3}\sin^2\theta_W$	$1 - 4\sin^2\theta_W$
d	$-1 + \frac{4}{3}\sin^2\theta_W$	$-1 + 4\sin^2\theta_W$
μ	$-1 + 4\sin^2\theta_W$	$-1 + 4\sin^2\theta_W$

In general, one expects

$$(35) \qquad A = -\frac{Q^2 G_F}{2\sqrt{2}\alpha}\left[a + \frac{1-(1-y)^2}{1+(1-y)^2}b\right] .$$

G_F, Q^2, y and α have already been defined. The standard model (+ naive parton model in the valence approximation) predicts

$$(36) \qquad \begin{cases} a = \frac{3}{5}(2V_u - V_d) = 2\sin^2\theta_W - \frac{9}{10} , \\ b = \frac{3}{5}(2A_u - A_d) = -\frac{9}{10}(1 - 4\sin^2\theta_W) . \end{cases}$$

Notice that, since $\sin^2\theta_W \simeq \frac{1}{4}$, one should find a very small dependence of A on y. Figures 9a) and b) show the experimental results for these asymmetry

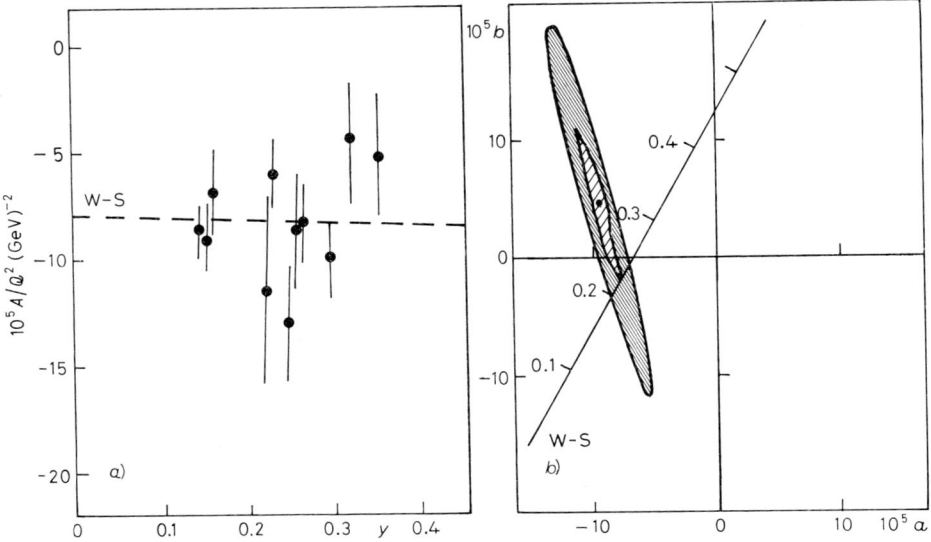

Fig. 9. – The total asymmetry in polarized eD scattering is reported in a) against y. The dashed line represents the GWS prediction with $\sin^2\theta_W \sim 0.224$. In b), the coefficients a and b are given separately. The shaded area corresponds to 95% c.l.

measurements. From the data, one obtains [9] for $\sin^2\theta_W(\varrho = 1)$ the value reported in table V (second line).

Modifications of eqs. (36) mainly come from the inclusion of effects due to the badly known sea densities in the nucleon. These corrections, which are not important for the y-dependent part of the asymmetry (since experimentally the value of y is always rather small), give an uncertainty of about 0.01 on $\sin^2\theta_W$. One can safely say

$$0.207 \leqslant \sin^2\theta_W \leqslant 0.227.$$

One could reduce these uncontrolled effects by going to larger values of x where the sea is suppressed.

4. – Radiative corrections to low-energy processes.

In the previous section, we discussed neutral-current processes used to determine the parameters of the standard model. The analysis was performed using relations derived in the Born approximation. In this limit, for each experiment, one should find a universal value for $\sin^2\theta_W$. Inclusion of higher-order corrections changes the Born relations found in sect. **3**:

(37) $$\sin^2\theta_W|_{\exp} = \sin^2\theta_W(\lambda) + \delta_{\exp}.$$

$\sin^2\theta_\text{W}(\lambda)$ is a theoretical universal quantity defined at a scale λ (in the following, we will assume for simplicity $\lambda \simeq M_\text{Z,W}$) in a given renormalization scheme. δ_exp is a process (experiment)-dependent correction. A priori, we expect for δ_exp the following hierarchy (relative to the size) of corrections:

$$(38) \quad \begin{cases} \alpha \ln(M^2/\mu^2) \simeq 10\alpha\,, \\ \dfrac{\alpha}{\sin^2\theta_\text{W}} \simeq 5\alpha\,, \\ [\alpha \ln(M^2/\mu^2)]^2 \simeq 0.8\alpha\,, \\ \alpha \ldots\,. \end{cases}$$

μ is the energy scale of the process under consideration ($\mu^2 \simeq 0.3\,(\text{GeV})^2$), $M \simeq M_\text{W,Z} \simeq 80\,\text{GeV}$. Further contributions of order $O(\alpha \ln(m_t^2/\mu^2))$, where m_t is a typical fermion mass, will also be considered. From eq. (38), we see that the most important corrections arise from leading-logarithm ($\sim \alpha \cdot \ln(M^2/\mu^2)$) contributions; a precise control of $O(\alpha^2 \ln^2(M^2/\mu^2))$ terms is also needed since they are comparable to $O(\alpha)$ terms.

A very simple and universal treatment of the leading-logarithm corrections (and subsequently of the $O(\alpha^2 \ln^2(M^2/\mu^2))$ corrections) can be obtained in the framework of the renormalization group. This approach will be discussed in subsect. 4'1. In 4'2, we will consider in detail radiative corrections, which enter into the determination of $R^{\nu,\bar\nu}$. The renormalization group methods exposed for this particular case can be easily extended to any other process. Details of the renormalization group approach can be found in ref. [6].

4'1. *Renormalization group applied to the effective weak Hamiltonians.* – In the limit $M_\text{W,Z} \to \infty$ ($\mu \ll M_\text{W,Z}$), the bare effective Hamiltonian for weak interactions can be written as

$$(39) \quad H^{(0)} = \sum_i c_i O_i\,,$$

where O_i are four fermion local operators. Examples of $H^{(0)}$ have been given in eqs. (5), (16), (20), (24) and (32). We choose for O_i a basis of operators which are multiplicately renormalized under e.m. corrections. As an explicit example, I report in fig. 10 the one-loop diagrams which renormalize the weak Hamiltonian given in eq. (22). The diagrams are computed with an ultraviolet momentum cut-off Λ. The result has the form

$$(40) \quad (O_i)^\text{renormalized} = (1 + \alpha\gamma_i \ln \Lambda^2)(O_i)^\text{bare}\,.$$

The scale of the ultraviolet cut-off corresponds to the mass of the intermediate vector bosons. The renormalization group states that the effective Hamiltonian

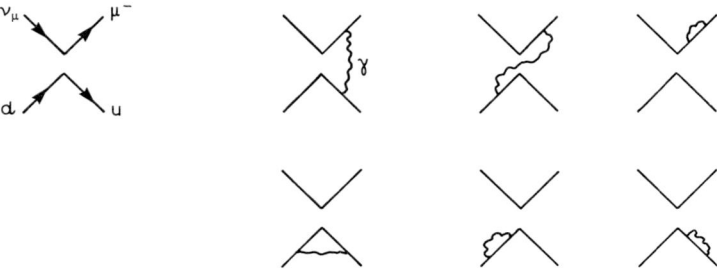

Fig. 10. – One-loop diagrams which enter into the renormalization of the Hamiltonian in eq. (22).

at a scale $\mu \ll M$ is expressed in the following way:

$$H(\mu) = \sum_i c_i(\mu)\, O_i(\mu)\,. \tag{41}$$

The coefficient $c_i(\mu)$ in eq. (41) satisfies the renormalization group equation

$$\left[-\frac{\partial}{\partial t} + \beta(\alpha)\frac{\partial}{\partial \alpha} + \Gamma_i(\alpha)\right] c_i(\mu) = 0\,, \qquad t = \ln(M^2/\mu^2)\,. \tag{42}$$

$O_i(\mu)$ are the operators renormalized at the scale μ. $\beta(\alpha)$ and $\Gamma_i(\alpha)$ are the coupling constant β-function and the anomalous dimension of the operators O_i:

$$\beta(\alpha) = \frac{\partial}{\partial \ln \mu^2}\alpha(\mu) = \beta_0 \alpha^2 + \ldots\,, \qquad \Gamma_i(\alpha) = \alpha \gamma_i + \ldots\,. \tag{43}$$

The solution of eq. (42) for the coefficients c_i has the form

$$c_i(t) = c_i(t=0,\alpha(t))\exp\left[\int_0^t \Gamma[\alpha(t')]\,\mathrm{d}t'\right] \simeq c_i^{(0)}\left[\frac{\alpha(M)}{\alpha(\mu)}\right]^{\gamma_i/\beta_0}(1+\ldots)\,. \tag{44}$$

We now apply these renormalization group equations to the computation of perturbative corrections to the determination of $R^{\nu,\bar{\nu}}$.

4.2. *Corrections to the evaluation of $R^{\nu,\bar{\nu}}$.* – We start by considering the corrections to the Hamiltonian responsible for neutrino charged-current reactions:

$$H^{(0)} = \frac{G_F^{(0)}}{\sqrt{2}}[\bar{\mu}\gamma_\mu(1-\gamma_5)\nu_\mu][\bar{u}\gamma_\mu(1-\gamma_5)d] + \text{h.c.} \tag{45}$$

After a straightforward evaluation of the diagrams in fig. 10, one finds for the operator in eq. (45)

$$\gamma = \frac{1}{2\pi}. \tag{46}$$

Then

$$H(\mu) = \frac{\mathcal{G}_F^{(0)}}{\sqrt{2}} \left[\frac{\alpha(M)}{\alpha(\mu)}\right]^{\gamma/b_Q} [\bar{\mu}\gamma_\mu(1-\gamma_5)\nu_\mu][\bar{u}\gamma_\mu(1-\gamma_5)d], \tag{47}$$

where $b_Q = \sum_f (Q_f^2/3\pi)$ and

$$\frac{\alpha(M)}{\alpha(\mu)} = \frac{1}{1 - \alpha(\mu) b_Q \ln(M^2/\mu^2)}. \tag{48}$$

Leading-logarithm corrections renormalize the strength of the charged-current amplitude.

For neutral currents, the relevant bare Hamiltonian is

$$H^{(0)} = \frac{G_F^{(0)}}{\sqrt{2}} \varrho \, [\bar{\nu}_\mu \gamma_\mu (1-\gamma_5) \nu_\mu] \left[J_\mu^Z + \frac{2}{\varrho} \bar{\mu}\gamma_\mu(1-\gamma_5)\mu \right], \tag{49}$$

where

$$J_\mu^Z = \sum_f \bar{f}\gamma_\mu[\tau_3^f - 4\sin^2\theta_W(M)Q^f - \tau_3^f \gamma_5]f,$$

$\tau^f = \pm 1$ and Q^f is the electric charge for the fermion of flavour f. Only the last three diagrams of fig. 10 are present in this case. However, they cannot give leading-logarithm contributions because both the vector and axial vector currents are conserved in the massless limit. The leading-logarithm corrections come in this case from a new type of diagram, the so-called « penguin » diagram, as illustrated in fig. 11. The fermions which loop in the diagram

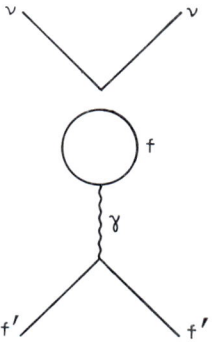

Fig. 11. – « Penguin » diagram which renormalizes the neutral-current Hamiltonian of eq. (49).

of fig. 11 give a factor

$$\sim \ln \Lambda^2 (q_\mu q_\nu - g_{\mu\nu} q^2) \, . \tag{50}$$

The $q_\mu q_\nu$ terms give zero when contracted with the conserved e.m. current; the term proportional to $g_{\mu\nu} q^2$ kills the photon propagator ($\sim 1/q^2$), giving a local four-fermion operator. Thus, for each looping flavour, neglecting isospin and charge factors, one obtains

$$\text{« penguin »} = -\frac{\alpha}{3\pi} \ln \Lambda^2 [\bar{\nu}_\mu \gamma_\mu (1 - \gamma_5) \nu_\mu] [\bar{f}' \gamma_\mu f'] \, . \tag{51}$$

It is clear from fig. 11 and eq. (51) that « penguin » diagrams cannot renormalize the fermion axial vector part of the Hamiltonian. Let us rewrite eq. (49) in the following way:

$$H^{(0)} = \frac{G_F^{(0)}}{\sqrt{2}} \varrho \left[\frac{4 b_z}{b_Q} - \frac{2}{\varrho} \frac{1}{3\pi b_Q} - 4 \sin^2 \theta_W(M) \right] \cdot \\
\cdot [\bar{\nu}_\mu \gamma_\mu (1 - \gamma_5) \nu_\mu] \left[\sum_f (\bar{f} \gamma_\mu Q' f) \right] + H_\perp \, , \tag{52}$$

where

$$b_z = \frac{1}{4} \sum_f \frac{\text{tr}\,(\tau_3^f Q')}{3\pi} \, , \qquad b_Q = \sum_f \frac{\text{tr}\,(Q')^2}{4\pi} \, ,$$

$H_\perp$ is the piece of the Hamiltonian not renormalized by « penguin » diagrams. In fact,

$$\tau_3^f - \frac{\sum_f \text{tr}\,(\tau_3^f Q')}{\sum_f \text{tr}\,(Q')^2} Q'$$

is orthogonal to Q_f. The operator $O_Q = [\bar{\nu}_\mu \gamma_\mu (1 - \gamma_5) \nu_\mu] \left[\sum_f (\bar{f} \gamma_\mu Q' f) \right]$ is multiplicatively renormalized. From eq. (51), one finds

$$(O_Q)^{\text{renormalized}} = [1 - \alpha b_Q \ln \Lambda^2] (O_Q)^{\text{bare}} \, , \qquad C_{O_Q} = \left[\frac{\alpha(M)}{\alpha(\mu)} \right]^{-1} , \tag{53}$$

then

$$H = \frac{G_F^{(0)}}{\sqrt{2}} \varrho \left[\frac{4 b_z}{b_Q} - \frac{2}{\varrho} \frac{1}{3\pi b_Q} - 4 \sin^2 \theta_W(M) \right] \frac{\alpha(\mu)}{\alpha(M)} O_Q + H_\perp \, . \tag{54}$$

It is possible to write the renormalized Hamiltonian of eq. (54) in the same form as the bare one (eq. (49)), provided we redefine the value of $\sin^2 \theta_W$:

$$\sin^2 \theta_W(\mu) = \sin^2 \theta_W(M) \frac{\alpha(\mu)}{\alpha(M)} - \frac{\alpha(\mu) - \alpha(M)}{\alpha(M)} \frac{1}{b_Q} \left(b_z - \frac{1}{6\pi \varrho} \right) . \tag{55}$$

The effect of the leading-logarithm corrections is to shift the value of $\sin^2\theta$ from the scale M to μ. From eq. (55), we derive the following renormalization group equation for the weak coupling $\alpha_W = \alpha/\sin^2\theta_W$:

$$\text{(56)} \qquad \frac{\partial}{\partial \ln \mu^2} \alpha_W^{-1}(\mu^2) = -\left(b_Z - \frac{1}{6\pi\varrho}\right).$$

The term $\sim b_Z$ is just the expected contribution to the renormalization of the $SU_{2,W}$ coupling constant coming from fermion loops. The extra $1/\varrho$ term is usually absent in the ordinary evolution equation for this coupling.

Besides the leading-logarithm correction just considered, we have to take into account other potentially large corrections (cf. eq. (38)) of order $\alpha_W = \alpha/\sin^2\theta_W$. These corrections are more easily estimated in the limit

$$\text{(57)} \qquad \alpha \to 0, \quad \sin^2\theta_W \to 0, \quad \alpha_W \text{ fixed}.$$

In this limit, the photon coincides with the U_1 gauge boson, Z^0 coincides with W^3 and the bare weak Hamiltonian has the form

$$\text{(58)} \qquad H^{(0)} = \frac{G_F^{(0)}}{\sqrt{2}} \boldsymbol{J}_\mu \cdot \boldsymbol{J}_\mu,$$

where

$$\boldsymbol{J}_\mu = \sum_f \left[\bar{f}\gamma_\mu \frac{1-\gamma_5}{2} \boldsymbol{\tau} f\right].$$

Fig. 12. – Diagrams contributing to the $O(\alpha_W)$ corrections.

The corrected Hamiltonian at order α_W is obtained by computing the contribution coming from the double-W-exchange diagrams of fig. 12. Because of the global SU_2 symmetry of the Hamiltonian of eq. (58), the corrected Hamiltonian must have the form

$$\text{(59)} \qquad H = \frac{G_F^{(0)}}{\sqrt{2}} [(1+\varepsilon_1) \boldsymbol{J}_\mu \cdot \boldsymbol{J}_\mu + \varepsilon_2 J_\mu^0 J_\mu^0],$$

where

$$J_\mu^0 = \sum_f \left[\bar{f}\gamma_\mu \frac{1-\gamma_5}{2} f\right]$$

is the $SU_{W,2}$ singlet current. ε_1 simply redefines the Fermi constant: $G_F =$

$= G_F^{(0)}(1 + \varepsilon_1)$. ε_2 has been evaluated in ref. [6]:

$$\varepsilon_2 = -\frac{9}{16}\frac{\alpha_W}{\pi}. \tag{60}$$

ε_2 will affect the value of neutral-current couplings to fermions.

In the previous section, we reported the expression for $R^{\nu,\bar{\nu}}$ in the Born approximation:

$$R^\nu = \frac{1}{2} - \sin^2\theta_W + \frac{20}{27}\sin^4\theta_W. \tag{61}$$

The radiative corrections previously considered modify eq. (61) in the following way:

$$R^\nu = \frac{1}{F^2}\left(\frac{1}{2} - \sin^2\theta_W(\mu) + \frac{20}{27}\sin^4\theta_W(\mu) - \frac{1}{3}\sin^2\theta_W(\mu)\varepsilon_2\right), \tag{62}$$

where the factor $F = [\alpha(M)/\alpha(\mu)]^{\gamma/b_Q}$ comes from charged-current corrections, $\sin^2\theta_W \to \sin^2\theta_W(\mu)$ from neutral currents and ε_2 comes from $O(\alpha_W)$ corrections. From eq. (62), for small perturbative corrections, we find

$$\delta s^2 = \frac{R^\nu \delta F^2 + (1/3)\sin^2\theta_W \varepsilon_2}{(40/27)\sin^2\theta_W - 1}. \tag{63}$$

For $R^\nu \sim 0.31$ and $s_0^2 \simeq 0.23$ at $\mu = 1.4$ GeV one obtains $\delta s^2 \simeq 0.004$ and

$$\sin^2\theta_W(M) \sim 0.217. \tag{64}$$

The result of a complete analysis, taking into account the $O(\alpha)$ corrections (ref. [6]), gives

$$\sin^2\theta_W(M)_{\overline{MS}} = 0.215 \pm 0.015. \tag{65}$$

Before leaving the argument of perturbative corrections, I want to mention another important correction, not considered above, which arises because of the experimental cuts in the y distribution and should be taken into account in the experimental analysis. This correction is due to the diagrams shown in fig. 13 which are singular in the muon mass ($\sim \alpha \ln(m_\mu^2/\mu^2)$) and in the quark mass [10].

The singularity in the muon mass cancels exactly only when we integrate over the final momentum of the μ and is obviously absent for $\nu_\mu + \mathcal{N} \to \nu_\mu + X$. It is present, however, in the data because of the experimental cuts in the y distribution:

$$E_h > E_h^{cut} \simeq 10 \text{ GeV}, \qquad \varepsilon \leqslant y \leqslant 1,$$

Fig. 13. – These diagrams give a contribution which is singular in the fermion mass ($\sim \alpha \ln(m_f^2/\mu^2)$). The analogous diagram in which the photon is emitted by the u-quark does not generate logarithms since we integrate over all the possible final states.

where

(66) $$\varepsilon = (E_h^{\text{cut}} - M_P)/E_\nu \,.$$

These corrections affect the determination of $F^{\nu,\bar\nu}$ and consequently the estimate of $\sin^2\theta_W$: the data should be corrected in the experimental analysis for these radiative effects. If the cut in y, ε is reasonably small, we can give *a posteriori* an estimate of the size of these corrections using the formula [10]

(67) $$\frac{d\sigma}{dx\,dy}\bigg|_{\text{exp}} = \frac{d\sigma^{(0)}}{dx\,dy} + \frac{\alpha}{2\pi} \ln\left(\frac{2E_\mu^*}{m_\mu}\right)^2 \cdot$$

$$\cdot \int_0^1 dz \frac{1+z^2}{1-z} \left[\frac{y\,\Theta[z-1+y(1-x)]}{z(y+z-1)} \frac{d\sigma^{(0)}}{dx\,dy}\bigg|_{\bar x,\bar y} - \frac{d\sigma^{(0)}}{dx\,dy} \right],$$

where

$$\bar x = \frac{xy}{y+z-1}, \quad 1-\bar y = \frac{1-y}{z}, \quad E_\mu^* = \left(\frac{1}{2} M_N E_\nu\right)^{\frac{1}{2}}\bigg/(1-y+xy)\,.$$

One expects a shift of $\sin^2\theta_W$ due to these contributions of $\sim -4\cdot 10^{-3}$. Finally, we observe that mass singularities connected with entering parton legs (the second diagram of fig. 13) can be reabsorbed in the definition of the nonscaling parton densities.

5. – Measurements of the W and Z⁰ masses at the SPS collider and comparison with theoretical predictions.

The discovery of the $W^\pm$ and Z^0 bosons and the measurement of their masses (and in the future of their widths) allows a new extraordinary test of the standard model.

From the IVB masses, it is possible to give a new definition of $\sin^2\theta_W$ [6] ($\varrho = 1$):

(68) $$\sin^2\theta_W(M) = \sqrt{1 - M_W^2/M_Z^2}\,.$$

The advantage of the definition of $\sin^2\theta_W$ in eq. (68) is that one does not have to extrapolate $\sin^2\theta_W$ from low-energy processes up to a large scale; this would require the evolution of the vacuum polarization diagrams of fig. 14 for the IVB which are affected by large theoretical incertitudes mainly due to light quark masses. On the contrary, the extrapolation of the e.m. coupling constant α (photon vacuum polarization) can be done on a more solid basis by using as input the experimentally measured cross-section for e^+e^- into hadrons. The disadvantage of the definition in eq. (68) is its sensitivity to small errors in $M_{Z,W}$. This definition differs from the running $\sin^2\theta_W$ in the $\overline{\mathrm{MS}}$ renormalization scheme (eq. (25)) by $\sim 0.6\%$ (in the leading-logarithm approximation, different definitions coincide).

Fig. 14. – Vacuum polarization diagram.

In the leading-logarithm approximation, we have

(69) $$\begin{cases} M_W^2 = \dfrac{\pi\alpha(M)}{\sqrt{2}\,G_F \sin^2\theta_W(M)}, \\[2mm] M_Z^2 = \dfrac{M_W^2}{\cos\theta_W(M)}, \\[2mm] \Gamma/M_W = \text{unrenormalized}. \end{cases}$$

Finite corrections modify eqs. (69) by terms of order of α_W and α. Using for $\sin^2\theta_W$ the definition of eq. (68), one finds at all orders in α

(70) $$M_W = \cos\theta_W(M)\, M_Z = \left[\frac{\pi\alpha(M)}{\sqrt{2}\,G_F}\right]^{\frac{1}{2}} \frac{1}{\sin\theta_W(M)} (1+\delta_M).$$

δ_M includes all finite corrections and it has been evaluated at order $O(\alpha)$ [6]. In fig. 15, we report the bare and corrected curves in the (M_Z, M_W)-plane for $m_{\text{Higgs}} \sim 100$ GeV and $m_t = 20$ GeV. The point $\sin^2\theta_W \sim 0.217$ is explicitly indicated on the curve. A detailed analysis shows that leading-logarithm corrections are positive and of order of 4%, while finite corrections are of opposite sign and $\sim 0.4\%$, confirming that the most important contribution comes from the leading piece. To compare the results obtained at the SPS collider with low-energy results, it is convenient to write eq. (7) as follows:

(71) $$M_W = \left[\frac{\pi\alpha(0)}{\sqrt{2}\,G_F}\right]^{\frac{1}{2}} \frac{1}{\sin\theta_W(M)} \left[\frac{\alpha(M)}{\alpha(0)}\right]^{\frac{1}{2}} (1+\delta_M) =$$
$$= \left[\frac{\pi\alpha(0)}{\sqrt{2}\,G_F}\right]^{\frac{1}{2}} \frac{1}{\sin\theta_W(M)} (1+\Delta).$$

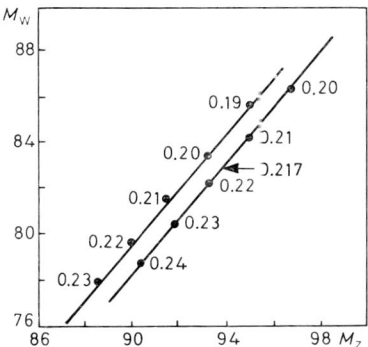

Fig. 15. – M_W vs. M_Z. The upper line refers to the Born relation. The lower line includes all $O(\alpha)$ corrections. The point $\sin^2\theta_W = 0.217$ is explicitly indicated.

Thus we can express the W mass in terms of low-energy parameters $\alpha(0)$ from the Josephson effect and G_F from μ decay. Using the computed value for Δ_r one predicts

$$(72) \quad \begin{cases} M_W = \dfrac{38.65 \pm 0.04}{\sin\theta_W(M)} \text{ GeV} = (83.0 \pm 2.8) \text{ GeV}, \\ M_Z = \dfrac{M_W}{\cos\theta_W(M)} \text{ GeV} = (93.0 \pm 2.2) \text{ GeV}, \end{cases}$$

where we used $\sin^2\theta_W(M) = 0.217 \pm 0.014$. These numbers should be compared with the experimental results of the UA1 and UA2 groups:

$$(73) \quad \begin{cases} M_W = (80.9 \pm 1.5 \pm 2.4) \text{ GeV} \\ M_Z = (95.6 \pm 1.5 \pm 2.9) \text{ GeV} \\ (\sin^2\theta_W(M) = 0.284) \end{cases} \text{UA1}, \\ \begin{cases} M_W = (83.1 \pm 1.9 \pm 1.3) \text{ GeV} \\ M_Z = (92.7 \pm 1.7 \pm 1.4) \text{ GeV} \\ (\sin^2\theta_W(M) = 0.196) \end{cases} \text{UA2}.$$

In brackets, I report the value of $\sin^2\theta_W(M)$ obtained by direct use of eq. (68). If we use eqs. (72) and the definition of $\varrho = M_W^2/M_Z^2 \cos^2\theta_W$, we find

$$(74) \quad \begin{cases} \sin^2\theta_W = 0.228 \\ \varrho = 0.928 \end{cases} \text{UA1}, \\ \begin{cases} \sin^2\theta_W = 0.216 \\ \varrho = 1.025 \end{cases} \text{UA2}.$$

All experimental results are compatible, within still rather large errors, with theoretical predictions based on low-energy data.

What about the Higgs?

* * *

I acknowledge the hospitality of the Theoretical Division at CERN where part of this manuscript has been prepared.

REFERENCES

[1] S. L. GLASHOW: *Nucl. Phys.*, **22**, 579 (1961).
[2] S. WEINBERG: *Phys. Rev. Lett.*, **19**, 1264 (1967); **27**, 1688 (1971).
[3] A. SALAM: *Elementary Particle Theory*, edited by N. SVARTHOLM (Almquist and Wiksells, Stockholm, 1969), p. 367.
[4] UA1 COLLABORATION (G. ARNISON et al.): *Phys. Lett. B*, **122**, 103 (1983); **126**, 398 (1983); **129**, 398 (1983).
[5] UA2 COLLABORATION (G. BANNER et al.): *Phys. Lett. B*, **122**, 476 (1983); **129**, 130 (1983).
[6] A. SIRLIN: *Phys. Rev. D*, **22**, 971 (1980); W. MARCIANO and A. SIRLIN: *Phys. Rev. D*, **22**, 2695 (1980); F. ANTONELLI, M. CONSOLI and G. CORBO: *Phys. Lett. B*, **91**, 90 (1980); M. VELTMAN: *Phys. Lett. B*, **91**, 95 (1980); A. SIRLIN and W. MARCIANO: *Nucl. Phys. B*, **189**, 442 (1981); C. H. LLEWELLYN SMITH and J. F. WHEATER: *Phys. Lett. B*, **105**, 486 (1981); *Nucl. Phys. B*, **208**, 185 (1982); F. ANTONELLI and L. MAIANI: *Nucl. Phys. B*, **186**, 269 (1981); D. YU. BARDIN, P. CH. CHRISTOVA and O. M. FEDORENKO: *Nucl. Phys. B*, **197**, 1 (1982); G. G. ROSS, C. H. LLEWELLYN SMITH and J. F. WHEATER: *Nucl. Phys. B*, **177**, 263 (1981); S. BELLUCCI, M. LUSIGNOLI and L. MAIANI: *Nucl. Phys. B*, **189**, 329 (1981). For a more complete list of references see also M. A. B. BÉG and A. SIRLIN: *Phys. Rep.*, **88**, 1 (1982); M. CONSOLI, S. LO PRESTI and L. MAIANI: *Nucl. Phys. B*, **223**, 474 (1983).
[7] B. W. LEE: in *Proceedings of the XVI International Conference on High Energy Physics*, edited by J. D. JACKSON and A. ROBERTS, Vol. 4 (Fermi National Acc. Lab., Batavia, Ill., 1972), p. 266.
[8] See, for example, H. H. CHEN and B. W. LEE: *Phys. Rev. D*, **5**, 1874 (1972).
[9] J. KIM, P. LANGACKER, M. LEVINE and H. H. WILLIAMS: *Rev. Mod. Phys.*, **53** 211 (1981).
[10] A. DE RÚJULA, R. PETRONZIO and A. SAVOY-NAVARRO: *Nucl. Phys. B*, **154**, 394 (1979).

Weak Decays of Heavy Quark States.

R. RÜCKL

CERN - Geneva, Switzerland

1. – Introduction.

In the framework of the standard $SU_{2,L} \times U_1$ electroweak theory [1] all unstable fermions decay in a universal manner, with their relative lifetimes solely determined by their masses and weak mixing. Accordingly, the lifetime of the τ-lepton or « free » charm quark can be inferred from the muon lifetime:

$$\tau_\mu \simeq \frac{192\pi^3}{G_F^2 m_\mu^5} \simeq 2.18 \cdot 10^{-6} \, \text{s}, \tag{1}$$

simply by rescaling the mass and counting the number of open decay channels:

$$\tau_\tau \simeq \frac{1}{5}\left(\frac{m_\mu}{m_\tau}\right)^5 \tau_\mu \simeq 3.3 \cdot 10^{-13} \, \text{s} \tag{2}$$

and ($m_c \simeq 1.5$ GeV)

$$\tau_c \simeq \frac{1}{5}\left(\frac{m_\mu}{m_c}\right)^5 \tau_\mu \simeq 7 \cdot 10^{-13} \, \text{s}. \tag{3}$$

These free-field predictions are in nice agreement with the measured τ and (average) charmed-particle lifetimes [2].

Considering the fact that, because of colour confinement, quarks exist only within strongly bound hadronic systems and, hence, any weak process involving quarks is necessarily accompanied by strong interactions, free-quark estimates may appear somewhat premature. Yet, the respectable results quoted above have a good reason, namely the asymptotic-freedom property of $SU_{3,c}$ colour interactions. On the other hand, the typical energy scales of bottom and, in particular, charm decays are relatively moderate: 5 GeV and 2 GeV, respectively. It is, therefore, not too surprising that the more detailed decay properties of charmed hadrons indeed indicate an appreciable impact of the environment of light quarks and soft gluons, in addition to short-distance gluon effects.

The interplay of strong and weak forces is not only an important issue in the theoretical understanding of weak decays of heavy quarks [3], but plays also a crucial role in the $\Delta I = \frac{1}{2}$ rule, in CP violation, in the determination of weak couplings and mixing angles and numerous other trials to which the standard model is put [4]. Unfortunately, due to the lack of quantitative methods to deal with confinement aspects of QCD from first principles, it is not yet possible to analyse the effects of strong interactions in a straightforward way. Instead, one usually assumes (with more or less good justification) that the dynamics of a given process can be separated in long- and short-distance aspects and treats these aspects individually. More specifically, in weak decays the short-distance dynamics is described by an effective weak Hamiltonian H_W^{eff}, which can be rigorously calculated using the well-established short-distance techniques of QCD. The long-range effects of strong interactions, on the other hand, are absorbed in the hadronic matrix elements of H_W^{eff}, which determine the decay amplitides, $T_{if} = \langle f|H_W^{\text{eff}}|i\rangle$. The latter can only be estimated in certain, sometimes rather crude, approximations. Clearly, the latter constitutes the weak point in the present understanding of heavy-quark decays and, therefore, deserves special attention.

My lecture is organized as follows. I first discuss in some detail the modifications of the bare weak Hamiltonian due to hard-gluon interactions. All further considerations are based on the QCD corrected Hamiltonian. The second part deals with the phenomenology of inclusive charm and bottom decays. Here, the focus is on the asymptotic approximation known as the spectator model, and on the nonspectator effects visible in the charm decay data. In the third part, I rather briefly tackle the more difficult problem of exclusive (two-body) decays, pointing out the main difficulties in the present understanding (see note added in proofs).

2. – The effective weak Hamiltonian.

In the minimal $SU_{2,L} \times U_1$ model, flavour-changing neutral currents [5, 6] do not exist at the tree level. Hence, heavy-flavour decays proceed exclusively via charged-current interactions mediated by the W$^\pm$ bosons. The fundamental couplings of the latter to fermions is given by the Lagrangian

(4)
$$L_{\text{CC}} = \frac{g}{2\sqrt{2}} (W_\mu^+ J_-^\mu + W_\mu^- J_+^\mu),$$

where

(5)
$$J_-^\mu = (J_+^\mu)^\dagger = (\overline{uct})\gamma^\mu(1-\gamma_5) V \begin{pmatrix} d \\ s \\ b \end{pmatrix} + (\overline{\nu_e \nu_\mu \nu_\tau})\gamma^\mu(1-\gamma_5)\begin{pmatrix} e^- \\ \mu^- \\ \tau^- \end{pmatrix}$$

is the standard $V-A$ charged current in the six-flavour version [6]. For simplicity and since J^μ_- is evidently a colour singlet, the colour indices are suppressed in eq. (5). The symbol V denotes the Kobayashi-Maskawa matrix [6]

$$(6) \qquad V = \begin{pmatrix} V_{ud} & V_{us} & V_{ub} \\ V_{cd} & V_{cs} & V_{cb} \\ V_{td} & V_{ts} & V_{tb} \end{pmatrix},$$

which relates the weak and mass eigenstates in the quark sector. The present knowledge about quark mixing is reviewed in detail in ref. [7]. A crude but, as far as the essential pattern is concerned, very transparent approximation for V reads

$$(7) \qquad |V| \sim \begin{pmatrix} 1 & s & s^3 \\ s & 1 & s^2 \\ s^3 & s^2 & 1 \end{pmatrix},$$

where $s \sim \sin\theta_c \simeq 0.23$.

From eq. (4) one readily derives the effective Hamiltonian to second order in the weak coupling g:

$$(8) \qquad H^{\text{eff}}_W = \frac{g^2}{8i} \int d^4x \, D_{\mu\nu}(x, m^2_W) \, T[J^\mu_+(x) J^\nu_-(0)] + \text{h.c.}].$$

Here, $D_{\mu\nu}(x, m^2_W)$ is the W propagator and T denotes time ordering. In the free-field approximation and taking the limit $m_W \to \infty$, which is justified

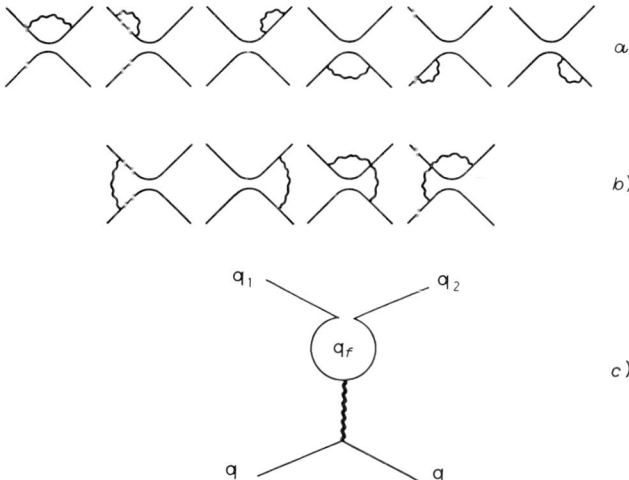

Fig. 1. – $O(\alpha_s)$ gluon corrections to the bare weak Hamiltonian.

for charm and bottom decays since m_c, $m_b \ll m_W$, eq. (8) yields the familiar effective Hamiltonian

$$H_W^{(0)} = \frac{G_F}{\sqrt{2}} \left(J_+^\mu(0) J_{-\mu}(0) + \text{h.c.} \right), \tag{9}$$

describing local current-current interactions.

The above bare Hamiltonian must now be corrected for strong interactions. The lowest-order gluon exchange contributions are illustrated in fig. 1. As can be seen from these diagrams, strong interactions not only renormalize the weak couplings and quark masses, etc. (fig. 1a)), but also induce new effective four-fermion interactions with different colour (fig. 1b)) and/or Lorentz structure (fig. 1c)). Obviously, semi-leptonic operators are affected only by corrections to the quark current of the kind shown in fig. 1a). I shall, therefore, concentrate on the nonleptonic part of the weak Hamiltonian given in eq. (8). The local free-field approximation of the latter reads

$$H_{NL}^{(0)} = \frac{G_F}{\sqrt{2}} (\bar{U} V D)_L (\bar{D} V^+ U)_L, \tag{10}$$

where

$$U = \begin{pmatrix} u \\ c \\ t \end{pmatrix}, \quad D = \begin{pmatrix} d \\ s \\ b \end{pmatrix} \tag{11}$$

and

$$(\bar{\psi}\psi)_L \equiv \bar{\psi}\gamma^\mu(1-\gamma_5)\psi. \tag{12}$$

The QCD corrected weak Hamiltonian H_{NL}^{eff} can be derived using short-distance expansion [8] and renormalization group techniques [9]. In short, the T-product of quark currents in eq. (8) is expanded in terms of local operators with, in general, divergent coefficient functions:

$$T[J_+^\mu(x) J_-^\nu(0)] = \sum_k c_k(x; g_s, m_W, m_q, ..., \mu) O_k^{\mu\nu}(\mu). \tag{13}$$

Here, g_s and m_q are the renormalized QCD coupling constant and quark masses, respectively, and μ is the normalization scale. Note that the operators on the right-hand side of eq. (13) have perfectly regular matrix elements $\langle f|O_k^{\mu\nu}(\mu)|i\rangle$. Because of the heavy mass of the W boson, only contributions to the integral in eq. (8) from short distances ($|x| < 1/m_W$) are important. Hence, the current-current product, eq. (13), is dominated by the local operators with the most singular coefficients, that is [3] dimension six, four-quark operators $O_k^{(6)}$. Substituting eq. (13) in eq. (8) and taking the limit $m_W \to \infty$, one gets, schematically,

$$H_{NL}^{eff} = \frac{G_F}{\sqrt{2}} \sum_k c_k(\alpha_s, m_\mu, m_q, ..., \mu) O_k^{(6)}(\mu). \tag{14}$$

Obviously, the weak amplitude $T_{\mathrm{if}} = \langle f|H_{\mathrm{NL}}^{\mathrm{eff}}|i\rangle$, being a measurable quantity, should not depend on the arbitrary scale μ chosen to normalize the theory. In other words, the μ-dependence of the coefficients c_k and the operators $O_k^{(6)}$ in eq. (14) must compensate each other. This requirement entails renormalization group equations [3] for the coefficients c_k,

$$(15) \qquad \left(\mu\frac{\partial}{\partial\mu} + \beta(g_s)\frac{\partial}{\partial g_s} + \sum_q \delta_q(g_s) m_q \frac{\partial}{\partial m_q} - \gamma_k(g_s) + \ldots\right) c_k = 0 ,$$

where $\beta(g_s)$ and $\delta_q(g_s)$ control the running of the coupling constant of QCD and the quark masses, respectively, and $\gamma_k(g_s)$ are the anomalous dimensions of the operators $O_k^{(6)}$. The coefficients c_k of $H_{\mathrm{NL}}^{\mathrm{eff}}$ can thus be obtained by solving eq. (15).

Let us first study the limit of zero quark masses and consider the case of realistic quark masses later. In the massless theory with the full SU_6-flavour symmetry, the expansion eq. (14) consists of only two operators [10]:

$$(16) \qquad H_{\mathrm{NL}}^{\mathrm{eff}} = \frac{G_{\mathrm{F}}}{\sqrt{2}}\left(c_+\left(\alpha_s, \frac{m_{\mathrm{W}}}{\mu}\right) O_+ + c_-\left(\alpha_s, \frac{m_{\mathrm{W}}}{\mu}\right) O_-\right)$$

with

$$(17) \qquad O_\pm = \tfrac{1}{2}[(\overline{U} V D)_{\mathrm{L}} (\overline{D} V^+ U)_{\mathrm{L}} \pm (\overline{U} V U)_{\mathrm{L}} (\overline{D} V^+ D)_{\mathrm{L}}] ,$$

where U and D are as defined in eq. (11). The operators $O_\pm$ belong to different SU_6-flavour representations, to wit $O_+ \sim 405$ and $O_- \sim 189$, and hence do not mix under renormalization. Solving then eq. (15) in the one-loop approximation for the β-function and anomalous dimensions $\gamma_\pm(g_s)$,

$$(18) \qquad \begin{cases} \beta(g_s) = -b_f \dfrac{g_s^3}{16\pi^2} , & b_f = 11 - \tfrac{2}{3} f , \\[6pt] \gamma_\pm(g_s) = -d_\pm \dfrac{g_s^2}{16\pi^2} , & d_- = -2 d_+ = 8 , \end{cases}$$

one obtains the well-known leading logarithmic (LL) result [10, 11]

$$(19) \qquad c_\pm\left(\alpha_s, \frac{m_{\mathrm{W}}}{\mu}\right) = \left(\frac{\alpha_s(\mu^2)}{\alpha_s(m_{\mathrm{W}}^2)}\right)^{d_\pm/2 b_f} ,$$

where

$$(20) \qquad \alpha_s(Q^2) = 4\pi \bigg/ \left(b_f \ln \frac{Q^2}{\Lambda_{\mathrm{QCD}}^2}\right) .$$

The next-to-leading logarithmic (NLL) approximation [12] yields

$$(21) \qquad c_\pm\left(\alpha_s, \frac{m_{\mathrm{W}}}{\mu}\right) = \left(\frac{\alpha_s(\mu^2)}{\alpha_s(m_{\mathrm{W}}^2)}\right)^{d_\pm/2 b_f} \left(1 + \frac{\alpha_s(\mu^2) - \alpha_s(m_{\mathrm{W}}^2)}{\pi} \varrho_{f\pm}\right) ,$$

where

(22) $$\alpha_s(Q^2) = \frac{4\pi}{b_f \ln(Q^2/\Lambda^2_{\text{QCD}})}\left(1 - \left(102 - \frac{38}{3}f\right)\frac{\ln\ln(Q^2/\Lambda^2_{\text{QCD}})}{b_f^2 \ln(Q^2/\Lambda^2_{\text{QCD}})}\right)$$

is the running coupling constant in the so-called $\overline{\text{MS}}$-scheme ($\Lambda_{\text{QCD}} = \Lambda_{\overline{\text{MS}}}$). The (renormalization scheme independent) coefficients $\varrho_{f\pm}$ are given by

(23) $$\begin{cases} \varrho_{f+} = \left(-\frac{221}{24} + \frac{5}{9}f\right)\frac{1}{b_f} + \left(51 - \frac{19}{3}f\right)\frac{1}{b_f^2}, \\ \varrho_{f-} = \left(\frac{263}{12} - \frac{10}{9}f\right)\frac{1}{b_f} + \left(-102 + \frac{38}{3}f\right)\frac{1}{b_f^2}. \end{cases}$$

The numerical significance of these corrections to c, b and t decays is illustrated below for f (= number of flavours) = 6, $\Lambda_{\overline{\text{MS}}} = 250$ MeV and $\alpha_s^{\text{LL}}(\mu^2) = \alpha_s^{\text{NLL}}(\mu^2)$:

(24) $$\mu(\text{GeV}) = \begin{cases} 2 \\ 5, \\ 40 \end{cases} \quad \begin{matrix} \text{LL/NLL} \\ 0.77/0.73 \\ c_+ = 0.85/0.82, \\ 0.97/0.96 \end{matrix} \quad \begin{matrix} \text{LL/NLL} \\ 1.69/1.90 \\ c_- = 1.39/1.49 \\ 1.07/1.08 \end{matrix}$$

A few comments may suffice:

i) The modifications of the bare coefficients, $c_\pm(0, m_W/\mu) = 1$ (see eq. (10)), are sizable at moderate mass scales and decrease with increasing scale μ as expected from asymptotic freedom.

ii) The next-to-leading corrections change the LL results only little and, moreover, reinforce the inequality $c_- > c_+$.

iii) Referring to the SU_3-flavour classification of the operators $O_\pm$, c_- being larger than c_+ is called **8**-enhancement in strange-particle decays [10] and **6**-enhancement in charm decays [11].

iv) The QCD corrections induce flavour-changing neutral-current interactions described by the operator $\frac{1}{2}(c_+ - c_-)(\overline{U}VU)_L(\overline{D}V^+D)_L$ of eq. (17).

It is further interesting to note that $H_{\text{NL}}^{\text{eff}}$, eqs. (16) and (17), can be rewritten in terms of charged-current operators only. Using the relation

(25) $$(\bar{\psi}_1 \lambda^a \psi_2)_L (\bar{\psi}_3 \lambda^a \psi_4)_L = -\frac{2}{3}(\bar{\psi}_1 \psi_2)_L (\bar{\psi}_3 \psi_4)_L + 2(\bar{\psi}_1 \psi_4)_L (\bar{\psi}_3 \psi_2)_L,$$

which is a consequence of Fierz identities and $SU_{3,c}$ colour algebra, one finds

(26) $$H_{\text{NL}}^{\text{eff}} = \frac{G_F}{\sqrt{2}}\left[\frac{2c_+ + c_-}{3}(\overline{U}VD)_L(\overline{D}V^+U)_L + \frac{c_+ - c_-}{4}(\overline{U}\lambda^a VD)_L(\overline{D}\lambda^a V^+U)_L\right].$$

This form very clearly exhibits the effects of hard-gluon exchanges anticipated from the lowest-order diagrams of fig. 1: renormalization of the bare four-quark interactions (see eq. (10)) and induction of local interactions of colour octet currents. Penguin type operators (fig. 1c)) are absent in the flavour symmetry limit since the contributions from the various quark flavours in the loop of fig. 1c) exactly cancel due to the generalized GIM mechanism [5, 6] or, equivalently, due to the unitarity of the Kobayashi-Maskawa matrix:

$$\sum_{q_f} V_{q_i q_f} V^*_{q_1 q_f} = 0. \tag{27}$$

This cancellation, however, is upset if the realistic quark masses are taken into account as explained next.

Generally, when one evolves $H^{\text{eff}}_{\text{NL}}$ from some asymptotic scale to the physical scale μ of a given process, one crosses several quark thresholds according to the quark masses m_q. In the region $\mu > m_q$, the flavour q and all lighter quarks may be considered massless. At $\mu < m_q$, however, the quark flavour q decouples from the effective theory. As a result, the number of excited flavours f diminishes and the GIM mechanism is partly put out of operation [13]. These effects can approximately be taken into account by renormalizing $H^{\text{eff}}_{\text{NL}}$ region by region, choosing f appropriately. For illustration, at $m_b > \mu > m_c$ one has [14]

$$c_\pm \simeq \left(\frac{\alpha_s(\mu^2)}{\alpha_s(m_b^2)}\right)^{d_\pm/2b_4} \left(\frac{\alpha_s(m_b^2)}{\alpha_s(m_t^2)}\right)^{d_\pm/2b_5} \left(\frac{\alpha_s(m_t^2)}{\alpha_s(m_W^2)}\right)^{d_\pm/2b_6}, \tag{28}$$

where $b_f = 11 - \frac{2}{3} f$ changes with the number of excited flavours in each region. Numerically, eq. (28) differs very little from the coefficients in the massless limit (see eqs. (19) and (24)). Furthermore, the penguin operators can roughly be estimated [3, 13] by evaluating the lowest-order diagram shown in fig. 1c). At $m_t < \mu < m_b$ the t-quark flavour is removed from the effective theory which gives rise to the interaction ($m_{c,s,d,u} \simeq 0$)

$$\frac{G_F}{\sqrt{2}} V_{tb} V^*_{tq} \left(-\frac{\alpha_s(\mu^2)}{12\pi} \ln \frac{m_t^2}{\mu^2}\right) (\bar{q} \lambda^a b)_L \sum_f (\bar{q}_f \lambda^a q_f)_V. \tag{29}$$

Here, q = s or d and the sum runs over the excited quark flavours. All other quark loop contributions cancel by GIM. Similarly, at $m_b < \mu < m_c$ the decoupling of the b-quark induces the operator ($n_{s,d,u} \simeq 0$)

$$\frac{G_F}{\sqrt{2}} V_{ub} V^*_{cb} \left(-\frac{\alpha_s(\mu^2)}{12\pi} \ln \frac{m_b^2}{\mu^2}\right) (\bar{u} \lambda^a c)_L \sum_f (\bar{q}_f \lambda^a q_f)_V. \tag{30}$$

From the above estimates one can see that in heavy-flavour decays penguins are not important for several reasons:

1) small coefficients: $(\alpha_s(\mu^2)/12\pi) \ln(m_{t,b}^2/\mu^2) \sim 0.03$ to be compared to $c_{\pm} \sim O(1)$;

2) small quark mixing: $V_{tb}V_{ts}^* \sim s^2$, $V_{tb}V_{td}^* \sim s^3$ and $V_{ub}V_{cb}^* \sim s^5$, while the main bottom and charm decays are proportional to s^2 and 1, respectively;

3) no appreciable enhancement [3, 15] of penguin matrix elements relative to $\langle f|O_{\pm}|i\rangle$.

Thus, only in rather special circumstances, penguin operators may have some influence. As one example, they alter the pattern of multiply Cabibbo-suppressed B-decays [15]. This contrasts with the essential role played by penguins in the explanation of the $\Delta I = \frac{1}{2}$ rule in strange-particle decays [13].

The effects of quark masses outlined above can be rigorously taken into account in the operator product expansion and renormalization group approach. However, flavour symmetry breaking leads to a proliferation of operators with independent coefficients and to mixing of some of these operators under renormalization. This makes the procedure somewhat clumsy. Also the renormalization group equation becomes more complicated (see eq. (15)) due to the scale dependence of quark masses and nonzero anomalous dimensions of currents. The rigorous results [13, 15] essentially confirm the conclusions drawn from eqs. (28)-(30).

Let us then summarize: the QCD corrected Hamiltonian for heavy-flavour decays is, to a good approximation, given by eq. (9) with the nonleptonic part, eq. (10), modified as detailed in eqs. (16)-(24). The leading logarithmic results for the coefficients $c_{\pm}(\alpha_s, m_W/\mu)$ are affirmed by the next-to-leading corrections. Numerically, one has

$$(31) \qquad c_+ \simeq 0.74, \quad c_- \simeq 1.8, \quad \Lambda_{\text{QCD}} \simeq 250 \text{ MeV},$$

for charm decays, and

$$(32) \qquad c_+ \simeq 0.85, \quad c_- \simeq 1.4, \quad \Lambda_{\text{QCD}} \simeq 250 \text{ MeV},$$

for bottom decays.

3. – Inclusive decays: the spectator model.

In order to link theory with experiment, one must eventually find ways to calculate hadronic matrix elements of the effective weak Hamiltonian. This proves not too difficult for inclusive decays of hadrons which contain a sufficiently heavy quark Q. Since the typical momentum transfer in the decay of such a system is of order of m_Q, one may neglect soft hadronic interactions and bound-state effects, once the heavy-quark mass is much larger than the ordinary hadronic scales represented by light constituent masses, confinement radius, bound-state wave functions, etc. Furthermore, the inclusive sum of the hadronic final states may be associated with free quark states carrying the large energy m_Q, similarly as $e^+e^- \to$ hadrons is dual to $e^+e^- \to q\bar{q}$. The resulting parton description is illustrated in fig. 2. Quite obviously, the decay

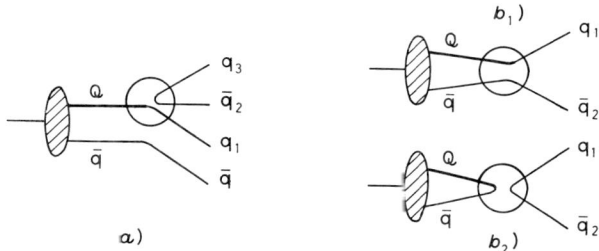

Fig. 2. – Parton description of inclusive heavy-meson decay via a) heavy-quark dissociation, b_1) W exchange, b_2) weak annihilation. Circles indicate H_W^{eff} including hard-gluon corrections.

mechanisms involving a light quark constituent (fig. 2b)) are suppressed by bound-state wave functions with respect to the dissociation of the heavy quark (fig. 2a)) and are, therefore, negligible in the asymptotic limit under consideration. More precisely, the relative contributions of the processes shown in fig. 2 to pseudoscalar-meson decay rates are given by [3]

$$(33) \qquad \frac{\Gamma_{b1}}{\Gamma_a} = \frac{1}{9}\frac{\Gamma_{b2}}{\Gamma_a} \simeq 24\pi^2 \left(\frac{f_P}{m_Q}\right)^2 \frac{m_{q_1}^2 + m_{q_2}^2}{m_Q^2} \frac{1}{9},$$

where only the leading terms in Λ/m_Q are kept, Λ being a typical hadronic scale. The parameter f_P is the meson decay constant and characterizes the overlap of the constituent quarks. For $P = \pi, K, D, F, B$ data or theoretical estimates [16] indicate that

$$(34) \qquad f_P \sim (100 \div 300) \text{ MeV}.$$

As can be seen from eq. (33), W exchange and weak annihilation are suppressed by the small constituent overlap (2nd factor), by helicity conservation (3rd factor) and, in the case of W exchange, by the colour degrees of freedom (4th factor). Thus, despite the appreciable phase space enhancement (1st factor), these mechanisms are already subdominant in charm decays and totally negligible in bottom decays. Also baryon decays will ultimately be dominated by heavy-quark dissociation. However, since W exchange in baryons is not forbidden by helicity conservation, this mechanism may play an important role for not too heavy baryons.

The above arguments suggest that the inclusive decay properties of sufficiently heavy quark states are essentially determined by weak decays of the «free» heavy quark corrected for hard-gluon interactions. The light quark constituents merely act as passive spectators. This is the essence of the spectator model [17]. In practice, one usually takes into account also gluon bremsstrahlung [12, 18], as well as some nonasymptotic effects such as phase space distortions [18, 19] due to finite quark masses and the Fermi motion [20] inside heavy bound states. The resulting semi- and nonleptonic decay widths are summarized below:

$$(35) \quad \Gamma_{\text{SL}} = \sum_{l,q} |V_{Qq}|^2 \, I\left(\frac{m_q}{m_Q}, \frac{m_l}{m_Q}, 0\right)\left(1 - \frac{2}{3}\frac{\alpha_s(m_Q^2)}{\pi} f\right)\frac{G_F^2 m_Q^5}{192\pi^3},$$

$$(36) \quad \Gamma_{\text{NL}} = \sum_{q_i} |V_{Qq_1}|^2 \, |V_{q_2 q_3}|^2 \, I\left(\frac{m_{q_1}}{m_Q}, \frac{m_{q_2}}{m_Q}, \frac{m_{q_3}}{m_Q}\right) \cdot$$
$$\cdot \frac{2c_+^2 + c_-^2}{3}\left(1 + \frac{2}{3}\frac{\alpha_s(m_Q^2)}{\pi} h\right)\frac{G_F^2 m_Q^5}{64\pi^3}.$$

Here, $I(x, y, z)$ is the three-body phase space factor [19] normalized such that $I(0, 0, 0) = 1$. The next factors arise from radiative gluon corrections to the decay widths and, in the nonleptonic case, from the short-distance modifications of $H_{\text{NL}}^{\text{eff}}$ specified in eqs. (16) and (17). In the massless limit [12],

$$(37) \quad f = \pi^2 - \frac{25}{4} = 3.6$$

and

$$(38) \quad h = \begin{cases} 2.2 & (m_c = 1.5 \text{ GeV}, f = 4), \\ 0.6 & (m_b = 5 \text{ GeV}, f = 5). \end{cases}$$

The last factors in eqs. (35) and (36) are the «free» quark decay widths directly inferable from the μ-decay formula (see eqs. (1)-(3)). As mentioned above, one can be a bit more sophisticated and parametrize the effective mass m_Q of the heavy quark in terms of the light constituent masses and the relative bound-state momenta [20]. In the case of a [$Q\bar{q}$]-meson with mass M_P, energy-momentum conservation requires

$$(39) \quad m_Q^2 = M_P^2 + m_q^2 - 2M_P \sqrt{|\boldsymbol{q}|^2 + m_q^2}.$$

Effective decay spectra and widths are then obtained from the results corresponding to a fixed value of m_Q (such as eqs. (35) and (36)) by averaging the latter over a Gaussian distribution of the Fermi momentum $|\boldsymbol{q}|$.

Although the spectator model itself is well defined and makes unambiguous predictions for asymptotically heavy quarks, in the mass range of charm and bottom one is still bothered by uncertainties. These are mainly connected with the question which effective quark masses one should use. The three-body phase space is particularly sensitive to masses. In accordance with the spirit of the spectator model, one would argue that current quark masses are the correct choice. On the other hand, if the average hadron multiplicity is as low as in charm decays, it may be more appropriate to use somewhat larger effective masses, in the extreme, constituent quark masses. For bottom decays, which are more asymptotic, this problem is less serious. Furthermore, a careful analysis of the semi-leptonic decay spectra reduces the uncertainty by a remarkable amount [19]. Semi-leptonic decays are conceptually simpler and considerably less subject to nonasymptotic effects than nonleptonic decays. Without going into details of such an analysis (these can been found in ref. [21]), I quote the resulting constraints on the quark masses. The usual procedure is to fit the electron spectra calculated [20] from $c \to se\nu_e$ and $b \to ce\nu_e$ to the ones measured in $D \to e\nu X$ [22] and $B \to e\nu X$ [23], respectively, letting the mass of the spectator quark m_{sp} and the width p_F of the Fermi momentum distribution, $\sim \exp[-|q|^2/p_F^2]$, vary between 0 and ~ 300 MeV with $m_{sp} + p_F \simeq 300$ MeV. Good agreement between theory and experiment is obtained for [21]

(40) $$\begin{cases} m_c \sim (1.5 \div 1.7) \text{ GeV}, \\ m_c - m_s \sim (1.1 \div 1.3) \text{ GeV}, \end{cases}$$

and

(41) $$\begin{cases} m_b \sim (4.9 \div 5.1) \text{ GeV}, \\ m_b - m_c \sim (3.3 \div 3.4) \text{ GeV}, \end{cases}$$

respectively. These constraints are taken into account in the following brief examination of the spectator model for inclusive B, D and F decays.

3`1. *B-decays*. – So far, the spectator model is consistent with all known features of inclusive B-decays [24]. However, as a word of warning, the present data refer to an average of B$^\pm$ (60%) and $\overset{(-)}{B^0}$ (40%) decays and may, therefore, hide nonspectator effects similarly as it was the case in the early data on average D-decays. Furthermore, B-decays involve two new weak mixing parameters V_{cb} and V_{ub}, which are constrained [7] by the observed mixing in the u-, d-, s- and c-sector and by the unitarity of the Kobayashi-Maskawa matrix, but only with large uncertainties. From a comparison of the experimental semi-leptonic electron spectrum with the two hypotheses, $b \to ce\nu_e$ and $b \to ue\nu_e$, one obtains the important additional bound [21]

(42) $$\left| \frac{V_{ub}}{V_{cb}} \right| < 0.116 \qquad (90\% \text{ c.l.}).$$

For definiteness, $|V_{ub}/V_{cb}| = 0.1$ is used throughout the following discussion. Finally, in order to explore the « flexibility » of the spectator model, two extreme sets of quark masses are considered:

$$
(43) \quad \begin{cases} m_{u,d} \simeq 0 & 0.35 \quad \text{GeV} \\ m_s \simeq 0.15 & 0.5 \quad \text{GeV} \\ m_c \simeq 1.4 & 1.8 \quad \text{GeV} \\ m_b \simeq 4.8 & 5.2 \quad \text{GeV}. \end{cases} \quad \begin{array}{c} \text{(I)} \quad \text{(II)} \end{array}
$$

For both sets, $m_b - m_c \simeq 3.4$, in agreement with eq. (41).

With the above specifications, eqs. (35) and (36) yield [3] the following semileptonic branching ratio and lifetime:

$$
(44) \quad \mathrm{BR}_{e,\mu} \simeq \begin{cases} 12\% \ \text{(I)}, \\ 15\% \ \text{(II)}, \end{cases}
$$

and

$$
(45) \quad \tau_b \cdot |V_{cb}|^2 \simeq \begin{cases} 2.9 \cdot 10^{-15} \ \text{s} \ \text{(I)}, \\ 3.1 \cdot 10^{-15} \ \text{s} \ \text{(II)}. \end{cases}
$$

Comparison of eq. (44) with the world averages [2]

$$
(46) \quad \mathrm{BR}_e = (13.0 \pm 1.3)\% \quad \mathrm{BR}_\mu = (12.4 \pm 3.5)\%
$$

shows that

i) current-type quark masses are favoured by the data as they are by theory, and

ii) QCD corrections are absolutely needed in order to reconcile the spectator model with the data: putting $\alpha_s = 0$ would increase $B_{e,\mu}$ to $(15 \div 18)\%$.

From now on the current quark masses of eq. (43) (I) are used. Substituting the average B lifetime [2],

$$
(47) \quad \tau_B = (1.4 \pm 0.4) \cdot 10^{-12} \ \text{s},
$$

measured by Mac [25] and Mark II [26], into eq. (45) (I), one derives

$$
(48) \quad |V_{cb}| \simeq 0.035 \div 0.065,
$$

a result which triggered a lot of theoretical speculations. It is interesting that the prediction of the spectator model on the lifetime is not very sensitive to a particular choice of quark masses as long as one consistently uses either current or constituent masses. To conclude, the presently quoted semi-leptonic branching ratios of $(12 \div 13)\%$ are perfectly consistent with the spectator model. However, a markedly smaller value could no longer be accommodated.

Further support for the spectator model comes from the semi-inclusive decays $B \to (K, D^0, J/\psi) + X$. To the extent that the average number of strange quarks per B-decay is a measure for the average number of kaons per B-decay the prediction [3]

(49) $$\langle n_s \rangle \simeq 1.4$$

is nicely confirmed by the experimental result [27]

(50) $$\langle n_K \rangle = 1.45 \pm 0.1 .$$

Secondly, the inclusive D^0 momentum distribution is found [24, 27] to be remarkably similar to the charm quark ($m_c = 1.86$ GeV) distribution expected from the semi-leptonic decay $b \to c e \nu_e$. This indicates that the D^0's originating in nonleptonic B-decays are also dominantly produced via b-quark dissociation, $b \to c \bar{u} d$, and that there is little communication between the $\bar{u}d$ and the $c\bar{q}_{\text{spectator}}$ system (see fig. 2a)). This picture is further corroborated by the branching ratio observed [24] for $B \to J/\psi + X$:

(51) $$\text{BR}(B \to J/\psi + X) = (1^{+0.5}_{-0.4})\% \quad \text{or} \quad < 1.6\% \quad (90\% \text{ c.l.}).$$

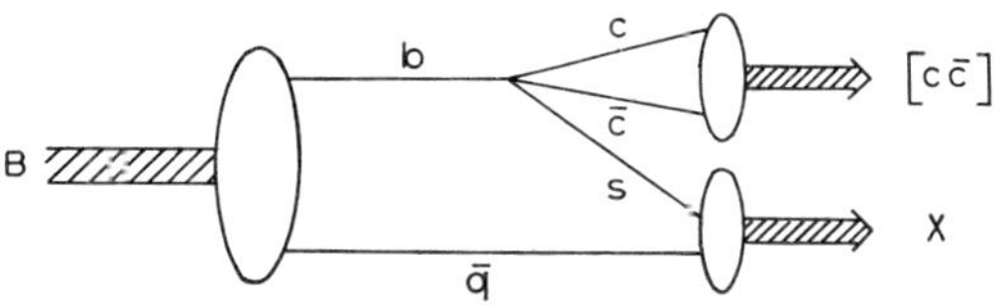

Fig. 3. – Quark diagram of the decay $B \to J/\psi + X$.

The relevant quark diagram is shown in fig. 3. Applying the transformation eq. (25) to eqs. (16) and (17), one readily derives the appropriate effective Hamiltonian:

(52) $$H_{\text{NL}}^{c\bar{c}} = \frac{G_F}{\sqrt{2}} V_{cb} V_{cs}^* \left[\frac{2c_+ - c_-}{3} (\bar{c}c)_L (\bar{s}b)_L + \frac{c_+ + c_-}{4} (\bar{c}\lambda^a c)_L (\bar{s}\lambda^a b)_L \right].$$

If there is indeed no interaction between the $c\bar{c}$ current and the rest of the diagram of fig. 3, the matrix element $\langle XJ/\psi | H_{\text{NL}}^{c\bar{c}} | B \rangle$ factorizes and one obtains [28, 29]

(53) $$\langle XJ/\psi | H_{\text{NL}}^{c\bar{c}} | B \rangle \simeq \frac{G_F}{\sqrt{2}} V_{cb} V_{cs}^* \frac{2c_+ - c_-}{3} \langle J/\psi | \bar{c}\gamma^\mu c | 0 \rangle \bar{u}_s \gamma_\mu u_b .$$

The matrix element $\langle J/\psi|\bar{c}\gamma^\mu c|0\rangle$ is directly measured [30] in e⁺e⁻ annihilation. Using this information and eqs. (53) and (45) (I), one predicts [28] the branching ratio

$$\text{BR}(B \to J/\psi\, X) \simeq \left(\frac{2c_+ - c_-}{3}\right)^2 \cdot 12\% \,. \tag{54}$$

This result also includes contributions from the cascade decays $B \to \chi$, $\psi' \to J/\psi$. The first factor in eq. (54) is due to the colour mismatch in fig. 3. Neglecting short-distance effects ($c_+ = c_- = 1$), the $\bar{c}s$ system is produced in a colour singlet configuration or, equivalently, the $s\bar{q}$ and $c\bar{c}$ systems form colour singlets only $\frac{1}{3}$ of the time. Hence, the decay rate $B \to J/\psi X$ is suppressed by a factor $\frac{1}{9}$. In that case, eq. (54) gives $\text{BR}(B \to J/\psi X) \simeq 1.3\%$ in agreement with the experimental result eq. (51). QCD corrections further decrease the branching ratio [28]. Unfortunately, the leading logarithmic correction,

$$1 \to (2c_+ - c_-)^2 \,, \tag{55}$$

is not reliable because of an accidental cancellation of the above factor at $c_- = c_+^{-2} = 2c_+ = 1.59$, a value which is not very different from $c_-(5\text{ GeV}) \simeq 1.4$ (see eq. (32)). This uncertainty, however, does not vitiate the strong experimental evidence that $B \to J/\psi + X$ is colour suppressed in accordance with the parton picture of fig. 3. I conclude by repeating the initial statement: the spectator model provides a consistent description of inclusive B-decays, at least at the level of the present experimental knowledge.

3'2. D, F *decays*. – Because of the early onset of asymptotic freedom, one originally believed the spectator model to represent a good approximation already of charm decays [11, 31]. This prejudice was particularly backed up by the progress made in understanding the $\Delta I = \frac{1}{2}$ rule in kaon and hyperon decays [10, 13] in terms of short-distance corrections to the weak Hamiltonian and a simple quark parton description of the amplitudes similar to fig. 2. Meanwhile, it has become clear [32] that charmed-particle decays are considerably more complex than anticipated.

According to the spectator model, all weakly decaying charmed hadrons should have almost equal lifetimes and semi-leptonic branching ratios. For the following two sets of masses,

$$(56) \quad \begin{cases} m_{u,d} \simeq 0.15 & 0.3 \quad \text{GeV} \\ m_s \simeq 0.3 & 0.4 \quad \text{GeV} \\ m_c \simeq 1.6 & 1.7 \quad \text{GeV}\,, \end{cases} \quad \text{(I)} \quad \text{(II)}$$

both consistent with the constraints eq. (40) from the semi-leptonic electron spectrum, eqs. (35) and (36) yield [3]

(57) $$\mathrm{BR}_{e,\mu} \simeq \begin{cases} 13\% \text{ (I)}, \\ 19\% \text{ (II)}, \end{cases}$$

and

(58) $$\tau_c \simeq \begin{cases} 6 \cdot 10^{-13} \text{ s (I)}, \\ 7.5 \cdot 10^{-13} \text{ s (II)}. \end{cases}$$

Here, $|V_{cs}|^2 \simeq |V_{ud}|^2 \simeq 0.95$ and $|V_{cd}|^2 \simeq |VV_{us}|^2 \simeq 0.05$ have been used. Because of the considerable sensitivity to a particular choice of quark masses, these results are only indicative for what could be considered consistent with the spectator model.

However, the uncertainty in the above predictions is not the main problem. The discovery of substantially differing lifetimes of D and F mesons [2],

(59) $$\begin{cases} \tau(\mathrm{D}^\pm) = (9.2^{+1.7}_{-1.2}) \cdot 10^{-13} \text{ s}, \\ \tau(\overset{(-)}{\mathrm{D}^0}) = (4.4^{+0.8}_{-0.6}) \cdot 10^{-13} \text{ s}, \\ \tau(\mathrm{F}^\pm) = (1.9^{+1.3}_{-0.7}) \cdot 10^{-13} \text{ s}, \end{cases}$$

shows that the spectator model is not the whole story for charm decays. Non-asymptotic effects involving the light constituent quarks are still important. Considering the relatively small charm quark mass, this is not too surprising. Similar differences as in the lifetimes are seen in the semi-leptonic branching ratios [2] of $\mathrm{D}^\pm$ and $\overset{(-)}{\mathrm{D}^0}$:

(60) $$\mathrm{BR}_e(\mathrm{D}^\pm) = (19^{+4}_{-3})\%, \quad \mathrm{BR}_e(\overset{(-)}{\mathrm{D}^0}) = (5.3^{+2.9}_{-1.3})\%.$$

Since the Cabibbo-allowed semi-leptonic decays have $\Delta I = 0$, the corresponding rates must be identical for $\mathrm{D}^\pm$ and $\overset{(-)}{\mathrm{D}^0}$, whence eq. (60) implies

(61) $$\frac{\mathrm{BR}_e(\mathrm{D}^+)}{\mathrm{BR}_e(\mathrm{D}^0)} \simeq \frac{\tau(\mathrm{D}^+)}{\tau(\mathrm{D}^0)} \simeq 3.6^{+2.1}_{-1.1}.$$

The last argument, together with the fact that the semi-leptonic electron spectrum is perfectly consistent with $c \to se^+\nu_e$ [21], also identifies the non-leptonic decays as responsible for the lifetime differences.

An interesting question is which, if any, of the charmed mesons is «normal» from the point of view of the spectator model. Because of the uncertainties in both model and data, one cannot give a totally clear answer. It seems, however, that the D^0 is the black sheep. Whereas the $\mathrm{D}^\pm$ lifetime and semi-leptonic branching ratio can still be accommodated in the spectator model (although they appear somewhat on the large side), it is virtually impossible to obtain a semi-leptonic branching ratio smaller than 10%, as it is observed for the D^0. In sum, one arrives at the conclusion that, in inclusive charm decays, the spectator model fails by roughly a factor 2 to 3.

4. – Nonasymptotic effects.

The presence of light quarks and gluons in heavy-quark bound states can in various ways give rise to lifetime differences and other modifications of the asymptotic decay pattern. One possibility are nonspectator interactions of the kind illustrated in fig. 2b) and commonly referred to as «annihilation» processes [33]. It may well be that the quark model estimate, eq. (33), which led to the neglect of these mechanisms, is oversimplified. Another pre-asymptotic effect is the interference of identical quarks [34] among the light constituents and the decay products of the heavy quark (see fig. 2a)) enforced by Pauli's exclusion principle. Both of the above effects depend strongly on QCD bound-state properties and are, therefore, difficult to quantify. However, one can certainly gain some qualitative insight in the essential physics points.

4˙1. *Interference effects in charmed-meson decays.* – On the Cabibbo-allowed level, the nonleptonic decay of the charm quark transmutes D and F mesons into four-quark systems with the flavour composition indicated below:

$$[c\bar{q}] \to [s\bar{d}u\bar{q}] \,. \tag{62}$$

In contrast to the D^0 and F^+ final states, which do not contain identical quarks, the D^+ final state contains two $\bar{d}$-quarks. Consequently, when computing the decay rate for the D^+, one must antisymmetrize the amplitude with respect to the $\bar{d}$-quarks in accordance with Fermi statistics.

It is instructive, although maybe not completely reliable in a quantitative sense, to perform this calculation using a nonrelativistic quark model. With $H_{\rm NL}^{\rm eff}$ as given in eqs. (16) and (17), and in the limit $p_{\rm F} \ll m_q \ll m_c$, where $p_{\rm F}$ characterizes the mean Fermi momentum, one derives [34]

$$\Gamma_{\rm NL}(D^+ \to s\bar{d}u\bar{d}) \simeq (2c_+^2 + c_-^2)\frac{G_{\rm F}^2 m_c^5}{192\pi^3} + (2c_+^2 - c_-^2)\frac{G_{\rm F}^2 m_c^2}{\pi}|\varphi(0)|^2 \,. \tag{63}$$

The first term is the usual spectator result in leading-logarithm approximation (see eq. (36)), whereas the second term arises from Pauli interference. Several comments are in order:

i) The interference is destructive as a consequence of the QCD short-distance corrections $(2c_+^2 - c_-^2 \simeq -2.1)$.

ii) The amount of interference is determined by the value of the wave function at the origin, $\varphi(0)$. Thus the ratio $\Gamma_{\text{interference}}/\Gamma_{\text{spectator}}$ scales like $(f_D/m_c)^2$, as can be seen if one substitutes the nonrelativistic relation

(64) $$f_D^2 \simeq 12|\varphi(0)|^2/m_c$$

in eq. (63).

iii) For $f_D \sim f_\pi$ and $m_c \sim m_D$, eq. (63) gives, numerically,

(65) $$\frac{\Gamma_{\text{int}}}{\Gamma_{\text{spect}}} \simeq \frac{c_-^2 - 2c_+^2}{c_-^2 + 2c_+^2} 16\pi^2 \left(\frac{f_D}{m_c}\right)^2 \simeq 0.5 \ .$$

Using instead Gaussian [35], Coulombic [36] or bag model [36, 37] wave functions one obtains ratios in the range

(66) $$\frac{\Gamma_{\text{int}}}{\Gamma_{\text{spect}}} \simeq 0.05 \div 0.2, \ 0.1 \div 0.2 \text{ and } 0.15 \div 0.4 \ ,$$

respectively.

The above results suggest that Pauli interference effects increase the lifetime and semi-leptonic branching ratio of the D^+ by about 20% with respect to the spectator model predictions. Although the latter and also the data are not yet precise enough to provide clear evidence for this effect, one can see a slight tendency in this direction if one compares eqs. (57)-(60). Interferences also occur in Cabibbo-suppressed D^+ and F^+ decays [3]. However, the D^0 and main F^+ decays are unaffected. Hence, the short D^0 and F^+ lifetimes indicate the existence of further nonspectator interactions.

4'2. *Annihilation processes in charmed-meson decays*. – Earlier in the discussion, it has been argued that the annihilation processes shown in fig. 2b) may be dismissed because of wave function and helicity suppression. The damping by the bound-state wave function is unavoidable. In contrast, helicity suppression is a combined effect of the $V - A$ structure of weak interactions and the quark model used to estimate the annihilation amplitudes. This approximation may be misleading [33]. Since hadronic bound states contain gluons and since gluons carry spin and colour, the $c\bar{q}$ system inside a pseudoscalar meson P could have a sizable probability to be in a colour singlet or octet, spin-1 state. Weak annihilation from these states, however, is not inhibited by helicity conservation and, hence, would occur with rates of the order of $(f_P/M_P)^2$ times a large phase space factor (see eq. (33)) relative to the rates of the spectator process (fig. 2a)). Clearly, if this is the case, annihilation is not negligible in charm meson decays.

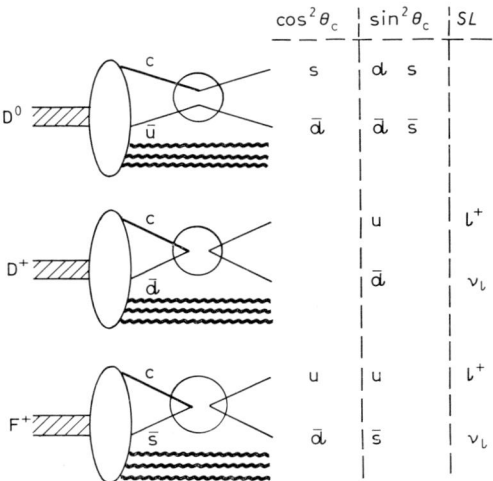

Fig. 4. – Gluon-enhanced annihilation processes in D and F meson decays. The wavy lines represent a gluonic component carrying spin 1.

From fig. 4, which indicates all possible annihilation processes, it is easy to deduce the qualitative effects annihilation would have on D and F meson decays:

1) enhancement of all nonleptonic $\overset{(-)}{D^0}$ decay modes,

2) enhancement of Cabibbo-suppressed $D^\pm$ decay modes,

3) enhancement of both nonleptonic and semi-leptonic $F^\pm$ modes.

We see that the D^0 and F^+ lifetimes are shortened and the semi-leptonic branching ratio of the D^0 is lowered with respect to the spectator model expectations. Exactly such a trend is observed experimentally as a glance at eq. (57)-(60) shows.

At this stage, it is mainly a quantitative question whether or not annihilation really explains the data. A firm answer requires a rather detailed understanding of QCD bound states. One can try a perturbative approach [38]: a gluon is radiated from the initial colour singlet, spin-0 $c\bar{q}$-state turning it into a colour octet, spin-1 system which then annihilates. This gives a rather small effect unless $f_{D,F} \gg f_\pi$:

$$(67) \quad \frac{\Gamma_{\text{ann}}}{\Gamma_{\text{spect}}} \simeq \frac{2\pi\alpha_s}{27(2c_+^2 + c_-^2)} \begin{cases} (c_+ + c_-)^2 \left(\dfrac{f_D}{m_u}\right)^2 \simeq 0.2 & \text{for } D^0, \\ (c_+ - c_-)^2 \left(\dfrac{f_F}{m_s}\right)^2 \simeq 0.03 & \text{for } F^+. \end{cases}$$

The above numbers correspond to $\alpha_s \simeq 0.5$ and $f_D/m_u \sim f_F/m_s \sim 1$, where m_u

and m_s are constituent masses. Moreover, the perturbative picture predicts $\tau(F^+) > \tau(D^0)$, at variance with experiment (see eq. (59)). Of course, perturbation theory may be inadequate. For this reason, one has also attempted various nonperturbative estimates [33, 39]. For example, performing a QCD multiple expansion of the gluon component [39] one can relate the annihilation probability to the gluon condensate $\langle 0|(\alpha_s/\pi)G^a_{\mu\nu}G^{a\mu\nu}|0\rangle \simeq 0.012$ (GeV)^4, which is known from the QCD sum rule analysis of the charmonium system [40]. The result,

$$(68) \qquad \frac{\Gamma_{\text{ann}}}{\Gamma_{\text{spect}}} \simeq \begin{cases} 1.5 & \text{for } D^0, \\ 0.7 & \text{for } F^+, \end{cases}$$

suggests that the three decay mechanisms illustrated in fig. 2 contribute roughly equally. A similar conclusion emerges from a straightforward phenomenological analysis [3] in which the annihilation amplitudes are directly determined from the data.

To summarize, the observed deviations of inclusive D^0 and F^+ decays from the decay pattern predicted by the spectator model may well be a pre-asymptotic effect originating in weak annihilation processes. However, one has so far no clear-cut theoretical proof. It is, therefore, very important to test further qualitative predictions of the annihilation hypothesis such as

a) $\text{BR}_{e,\mu}(F^+) \sim \text{BR}_{e,\mu}(D^+)$,

b) large $\text{BR}(F \to n\pi)$,

c) deviations of the electron spectrum in semi-leptonic F decays from the spectrum expected from the spectator process $c \to se\nu_e$,

d) unusually frequent Cabibbo-suppressed D^+ decays modes.

4'3. *Inclusive charmed-baryon decays*. – For sufficiently heavy quarks, the spectator model applies equally to baryons and mesons. Charmed-meson decays, on the other hand, show that the asymptotic regime is not yet reached at the charm scale. This leads one to expect pre-asymptotic effects also in charm baryon decays. Let us consider the $\Lambda_c^+ = [cdu]$ as an example. The decay of the c-quark produces a five-quark final state

$$(69) \qquad [cdu] \to s\bar{d}udu$$

which contains two u-quarks. These interfere according to the Pauli principle [41]. Furthermore, the Λ_c^+ can decay via W exchange between the c and d constituents [41, 42]

$$(70) \qquad [cdu] \to sud,$$

a process which is similar to the annihilation mechanisms illustrated in fig. 2b) for mesons. However, in contrast to the meson case, the decay indicated in eq. (70) is not helicity suppressed, even without invoking gluons. The reason is that in the Λ_c^+ the cd system is $\frac{1}{4}$ of the time in a spin-0 state. While the weak transition of a spin-0 fermion-antifermion system is helicity suppressed, it is perfectly allowed for a spin-0 fermion-fermion system.

A nonrelativistic quark model calculation, similar as the one which led to eq. (63), yields [41]

(71) $\quad \Gamma(\Lambda_c^+ \to s\bar{d}udu) \simeq (2c_+^2 + c_-^2) \frac{G_F^2 m_c^5}{192\pi^3} - c_+(2c_- - c_+) \frac{G_F^2 m_c^2}{4\pi} |\varphi_{cu}(0)|^2 ,$

where the first term is the spectator model result and the second term is due to Pauli interference. The W exchange contribution, on the other hand, is given by [42]

(72) $\quad \Gamma(\Lambda_c^+ \to suu) \simeq c_-^2 \frac{G_F^2 m_c^2}{2\pi} |\varphi_{cd}(0)|^2 .$

In the above, $|\varphi_{cq}(0)|^2$ characterizes the probability for the c-quark and a light constituent to be at the same point. This probability can be estimated from Σ_c^+-Λ_c^+ hyperfine splitting [43]:

(73) $\quad |\varphi_{cq}(0)|^2 \simeq \frac{9 m_u^2 m_c}{16\pi \alpha_s (m_c - m_u)} (M_{\Sigma_c^+} - M_{\Lambda_c^+}) \simeq 0.01 \; (\text{GeV})^3 .$

Using this number and taking into account phase space corrections, one finds [41]

(74) $\quad \Gamma_{\text{spect}} : \Gamma_{\text{int}} : \Gamma_{\text{W ex}} \sim 1 : 0.5 : 2 .$

Adding the spectator widths for the semi-leptonic decays, one finally gets [41]

(75) $\quad \text{BR}_{e,\mu}(\Lambda_c^+) \simeq (4 \div 7)\% , \quad \tau(\Lambda_c^+) \simeq (2.5 \div 3) \cdot 10^{-13} \text{ s}$

in agreement with experiment [2]:

(76) $\quad \text{BR}_e(\Lambda_c^+) = (4.5 \pm 1.7)\% , \quad \tau(\Lambda_c^+) = (2.3^{+1.0}_{-0.6}) \cdot 10^{-13} \text{ s} .$

A similar analysis is possible for Cabibbo-suppressed decays [41] and other weakly decaying baryons [3]. Qualitatively, one expects

(77) $\quad \tau(A^0 = [csd]) \leqslant \tau(\Lambda_c^+ = [cdu]) < \tau(A_s^0 = [css]) \leqslant \tau(A^+ = [csu]) .$

The recently reported [44] lifetime of the A^+,

(78) $\quad \tau(A^+) = (4.8^{+2.9}_{-1.8}) \cdot 10^{-13} \text{ s} ,$

fits nicely into the predicted pattern. As a final remark, if weak annihilation of charmed mesons is indeed enhanced by gluons, a mechanism similar to fig. 4 could also operate in charm baryon decays [45]. This possibility makes it even more important to test experimentally whether or not the straight valence quark model predictions described above are correct.

4'4. Inclusive bottom decays. – Before concluding the discussion of non-asymptotic effects, I should make a few comments on the relevance of Pauli interference and weak annihilation to bottom decays. In the valence quark approximation, the ratios of the interference terms (eqs. (63) and (71)) and the W exchange rate in baryon decays (eq. (72)) to the spectator model width scale like $|\varphi_Q(0)|^2/m_Q^3$, where $\varphi_Q(0)$ denotes the $Q\bar{q}$ and Qq wave functions at the origin. Thus, when going from charm to bottom decays, these effects are expected to decrease by roughly the factor

$$(79) \qquad \left|\frac{\varphi_b(0)}{\varphi_c(0)}\right|^2 \left(\frac{m_c}{m_b}\right)^3 \sim \left(\frac{m_c}{m_b}\right)^{2\div 3} \sim O(10^{-1}).$$

Note that $f_P \sim$ const implies $|\varphi_Q(0)| \sim m_Q$.

As far as the gluon-enhanced annihilation processes in meson decays are concerned, extrapolations from charm to bottom are more uncertain. The perturbative approach [38], eq. (67), suggests a rather slow decrease. On the other hand, it predicts small annihilation contributions anyway. In contrast, the QCD multipole estimate [39], eq. (68), as well as an analysis [45] based on evolution equations indicate

$$(80) \qquad \frac{\varGamma_{\text{ann}}(B)}{\varGamma_{\text{spect}}(B)} : \frac{\varGamma_{\text{ann}}(D)}{\varGamma_{\text{spect}}(D)} \sim 1:10.$$

Thus, even if one believes that $\varGamma_{\text{ann}}(D)/\varGamma_{\text{spect}}(D) \sim O(1)$ as required if one wants to explain the observed deviations from the spectator model in D-decays, one would not expect annihilation effects in B-decays larger than of the order of 10%. This extrapolation is also consistent with the scale dependence indicated in eq. (79), which should essentially apply to all processes damped by bound-state wave functions.

Although one has good reason to believe that pre-asymptotic effects play a minor role in bottom decays, only experiment can decide. Similarly, as in D^0 and D$^+$, weak annihilation and Pauli interference would be exposed in the inequalities [3]

$$(81) \qquad \begin{cases} \tau(B^\mp) > \tau(\overset{(-)}{B^0}), \\ BR_\ell(B^\mp) > BR_\ell(\overset{(-)}{B_0}), \\ BR(B^- \to J/\psi X) > BR(B^0 \to J/\psi X). \end{cases}$$

5. – Two-body decays.

It is quite obvious that bound-state properties and other aspects of strong interactions at typical hadronic scales play a more important role in exclusive decays than inclusively. As concerns two-body charm decays, one is at present still quite far from a clear qualitative picture, not to speak about a quantitative description. The original suggestion [17, 31] to calculate the relevant matrix elements of the effective weak Hamiltonian in a valence quark approximation appears to be inadequate in certain cases. In this approximation, the matrix element of a four-quark operator is factorized in a product of matrix elements of quark currents, similarly as indicated in eq. (53) for the semi-inclusive decay $B \to J/\psi X$. Let us derive the amplitude of $D^0 \to K^-\pi^+$ as an explicit example. The relevant weak Hamiltonian can be read off from eqs. (16) and (17):

$$(82) \quad H_{\rm NL}^{\rm eff} = \frac{G_{\rm F}}{\sqrt{2}} V_{\rm cs}^* V_{\rm ud} \left[\frac{c_+ + c_-}{2} (\bar{u}d)_{\rm L} (\bar{s}c)_{\rm L} + \frac{c_+ - c_-}{2} (\bar{u}c)_{\rm L} (\bar{s}d)_{\rm L} \right].$$

In the valence quark description, the above operators give rise to the processes depicted in fig. 5. Whereas in the first diagram the $u\bar{d}$ and $s\bar{u}$ quark pairs are

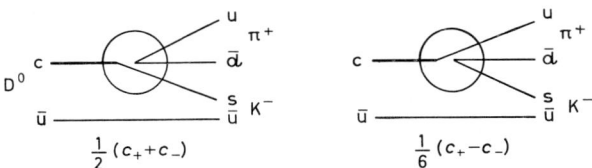

Fig. 5. – Valence quark description of the decay $D^0 \to K^-\pi^+$.

produced in colour singlet states and can directly form a π^+ and a K^- meson, respectively, the colour degrees of freedom are mismatched in the second diagram. Correspondingly, one must Fierz-transform the second operator in eq. (82) according to eq. (25), which gives $\frac{1}{3}(\bar{u}d)_{\rm L}(\bar{s}c)_{\rm L} + \frac{1}{2}(\bar{u}\lambda^a d)_{\rm L}(\bar{s}\lambda^a c)_{\rm L}$. Since colour-octet currents have zero matrix elements between colour-singlet states, one obtains

$$(83) \quad \langle K^-\pi^+ | H_{\rm NL}^{\rm eff} | D^0 \rangle =$$

$$= \frac{G_{\rm F}}{\sqrt{2}} V_{\rm cs}^* V_{\rm ud} \left(\frac{c_+ + c_-}{2} + \frac{c_+ - c_-}{6} \right) \langle \pi^+ |(\bar{u}d)_{\rm L}|0\rangle \langle K^- |(\bar{s}c)_{\rm L}|D_0\rangle =$$

$$= \frac{G_{\rm F}}{\sqrt{2}} V_{\rm cs}^* V_{\rm ud} \frac{2c_+ + c_-}{3} f_\pi [(M_{\rm D}^2 - M_{\rm K}^2) f_+(M_\pi^2) + M_\pi^2 f_-(M_\pi^2)],$$

where the usual definitions of meson decay constants (f_π) and vector form factors $(f_\pm(q^2))$ have been adopted.

A crucial consequence of the factorization of weak matrix elements and the insertion of the vacuum state is the suppression of certain channels due to colour mismatch (see discussion of B $\to$ J/ψX in 3'1). For the case at hand, this occurs in the decay $D^0 \to \bar{K}^0 \pi^0$. According to the quark diagrams of fig. 6

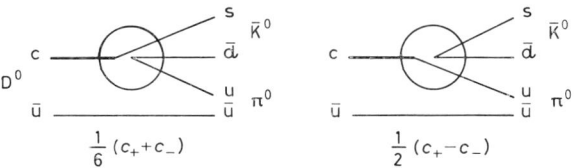

Fig. 6. – Valence quark description of the decay $D^0 \to \bar{K}^0 \pi^0$.

and the quark presentation of the π^0, $(1/\sqrt{2})(u\bar{u} - d\bar{d})$, the amplitude of $D^0 \to \bar{K}^0 \pi^0$ is given by

(84) $\langle \bar{K}^0 \pi^0 | H_{\rm NL}^{\rm eff} | D^0 \rangle =$

$= \frac{G_F}{\sqrt{2}} V_{cs}^* V_{ud} \left(\frac{c_+ + c_-}{6} + \frac{c_+ - c_-}{2} \right) \langle \bar{K}^0 | (\bar{s}d)_L | 0 \rangle \langle \pi^0 | (\bar{u}c)_L | D^0 \rangle =$

$= \frac{G_F}{\sqrt{2}} V_{cs}^* V_{ud} \frac{2c_+ - c_-}{3} \frac{f_K}{\sqrt{2}} [(M_D^2 - M_\pi^2) f_+(M_K^2) + M_K^2 f_-(M_K^2)] .$

One clearly sees that the small factor $(2c_+ - c_-)/3$ arises from colour mismatch in the contribution of the dominant operator of eq. (82) illustrated by the first diagram of fig. 6. Equations (83) and (84) imply a substantial suppression of $D^0 \to \bar{K}^0 \pi^0$ relative to $D^0 \to K^- \pi^+$. In the SU_3 limit,

(85) $\frac{\Gamma(D^0 \to \bar{K}^0 \pi^0)}{\Gamma(D^0 \to K^- \pi^+)} \simeq \frac{1}{2} \left(\frac{2c_+ - c_+}{2c_+ + c_-} \right)^2 = \frac{1}{18} \left(\frac{1}{200} \right) ,$

where the first number corresponds to the uncorrected ($c_+ = c_- = 1$) weak Hamiltonian, whereas the second number includes leading-logarithm QCD corrections ($c_+ = 0.74$, $c_- = 1.8$). The latter are extremely sensitive to the values of c_+ and c_- as pointed out in subsect. 3'1, eq. (55). SU_3 breaking effects [46] increase the above result by roughly a factor 2. At any rate, the approximation outlined above is in striking discrepancy with the measurement [47]

(86) $\frac{\Gamma(D^0 \to \bar{K}^0 \pi^0)}{\Gamma(D^0 \to K^- \pi^+)} = 0.73 \pm 0.35 .$

Another important consequence of factorization and vacuum saturation is the suppression of annihilation mechanisms such as

(87) $D^0 \to s\bar{d} \to \begin{cases} s(\bar{u}u) \bar{d} \to K^- \pi^+ , \\ s(\bar{d}d) \bar{d} \to \bar{K}^0 \pi^0 , \end{cases}$

where the ($\bar{u}u$) and ($\bar{d}d$) quark pairs are picked up from the vacuum. In analogy to eqs. (83) and (84), the corresponding amplitudes are given by

$$(88) \quad \langle K\pi|H_{\rm NL}^{\rm eff}|D^0\rangle \sim \frac{2c_+ - c_-}{3} \langle K\pi|(\bar{s}d)_{\rm L}|0\rangle\langle 0|(\bar{u}c)_{\rm L}|D_0\rangle \sim$$

$$\sim \frac{2c_+ - c_-}{3} f_{\rm D} \langle K\pi|p_{\rm D}^\mu \bar{s}\gamma_\mu d|0\rangle,$$

and vanish in the SU_3 limit because of conservation of vector currents:

$$(89) \quad p_{\rm D}^\mu \bar{s}\gamma_\mu d = (m_{\rm d} - m_{\rm s})\bar{s}d \xrightarrow{m_{\rm s}=m_{\rm d}} 0.$$

This shows that, in the valence quark approximation, annihilation processes contribute only at the level of SU_3-breaking effects. Moreover, the processes indicated in eq. (87) are colour-suppressed, so that the failure of the prediction eq. (85) on the ratio $\bar{K}^0\pi^0/K^-\pi^+$ can certainly not be blamed on the neglect of annihilation.

To save the day, a variety of ideas have been put forward. Without going into details, I mention a few of them. The simplest way out is to abandon the QCD values of $c_\pm$ and to assume [48] that, for some reason, $c_- \gg c_+$. One possibility are soft-gluon interactions. It has first been pointed out by FRITZSCH [49] that soft gluons may wash out the static colour structure of the diagrams drawn in fig. 5 and 6, and thus vitiate colour suppression. Clearly, *soft-gluon exchanges* between quark lines joining different mesons would spoil factorization and vacuum saturation of the amplitudes, and, therefore, undermine the basis of colour suppression in the valence quark approximation. One can further show [50] that the net effect amounts to a change of the ratio c_-/c_+ in eq. (85). Nonfactorization also invalidates the argument for neglecting *annihilation processes* [33]. The latter, on the other hand, give automatically $\bar{K}^0\pi^0/K^-\pi^+ = \frac{1}{2}$, since the final state in $D^0 \to \bar{s}d$ carries isospin $\frac{1}{2}$. Finally, the valence quark prediction, eq. (85), could be upset by *resonance effects* in the $K\pi$ system as pointed out by LIPKIN [51]. Decomposing the amplitudes according to the isospin of the final states and allowing for phases subject to final-state interactions, one has

$$(90) \quad \frac{\Gamma(D^0 \to \bar{K}^0\pi^0)}{\Gamma(D^0 \to K^-\pi^+)} = \left(\frac{A_{\frac{1}{2}}\exp[i\delta] + \sqrt{2}\,A_{\frac{3}{2}}\exp[i\delta']}{\sqrt{2}\,A_{\frac{1}{2}}\exp[i\delta] - A_{\frac{3}{2}}\exp[i\delta']}\right)^2.$$

Since the $I = \frac{1}{2}$ channel contains many K^* resonances, whereas the $I = \frac{3}{2}$ channel is exotic, the relative phase $\delta - \delta'$ may be large and could even reverse the cancellation, $A_{\frac{1}{2}} \sim -\sqrt{2}\,A_{\frac{3}{2}}$, should it exist at the quark level. We see that one can come up with a number of reasonable explanations for the apparent absence of colour suppression in $D^0 \to \bar{K}^0\pi^0$. Unfortunately, none of the above can be made quantitative at present.

With slight modifications due to SU_6 breaking [3], the valence quark prediction, eq. (85), also applies to the ratios $\bar{K}^0\rho^0/K^-\rho^+$ and $\bar{K}^{0*}\pi^0/K^{-*}\pi^+$. Here, the experimental situation [47] is unclarified, being consistent with $0 \div 0.1$ and $0 \div 1.1$ for the above ratios, respectively.

Turning to the Cabibbo-suppressed decay modes, $D^0 \to K^-K^+$ and $\pi^-\pi^+$, one encounters some minor problems. The valence quark approximation in the SU_3 limit predicts [2]

$$
(91) \quad \left. \begin{array}{l} \dfrac{\Gamma(D^0 \to K^-K^+)}{\Gamma(D^0 \to K^-\pi^+)} = \left|\dfrac{V_{us}}{V_{ud}}\right|^2 \\[2mm] \dfrac{\Gamma(D^0 \to \pi^-\pi^+)}{\Gamma(D^0 \to K^-\pi^+)} = \left|\dfrac{V_{cd}}{V_{cs}}\right|^2 \end{array} \right\} \simeq 0.046 \div 0.061 ,
$$

whereas, experimentally, one observes [47]

$$
(92) \quad \begin{cases} \dfrac{\Gamma(D^0 \to K^-K^+)}{\Gamma(D^0 \to K^-\pi^+)} = 0.113 \pm 0.03 , \\[2mm] \dfrac{\Gamma(D^0 \to \pi^-\pi^+)}{\Gamma(D^0 \to K^-\pi^+)} = 0.033 \pm 0.015 . \end{cases}
$$

SU_3-breaking effects [52] increase the ratios given in eq. (91) by roughly 70% and 30%, respectively. This leaves a discrepancy of a factor $1.2 \div 2$ between the valence quark approximation and the data. Also if annihilation processes are really important, as may be indicated by the absence of colour suppression, the predictions of eq. (91) would still be valid. In that case, the enhancement $K^-K^+/\pi^-\pi^+ \sim 3$ can fully be explained [53] by properly taking into account SU_3 breaking. I should, further, point out that the experimental result, eq. (92), is not inconsistent with the most general, model-independent SU_3 prediction [54], but requires a large difference in the magnitude of the relevant reduced matrix elements. Finally, the valence quark approximation yields the interesting relations [3]

$$
(93) \quad \frac{\Gamma(D^+ \to \bar{K}^0 K^+)}{\Gamma(D^+ \to \bar{K}^0 \pi^+)} \simeq \left(\frac{2c_+ + c_-}{4c_+}\right)^2 \frac{\Gamma(D^0 \to K^-K^+)}{\Gamma(D^0 \to K^-\pi^+)} \simeq (14 \div 18)\%
$$

and

$$
(94) \quad \frac{\Gamma(D^+ \to \pi^0\pi^+)}{\Gamma(D^+ \to \bar{K}^0\pi^+)} \simeq \frac{1}{2} \frac{\Gamma(D^0 \to K^-K^+)}{\Gamma(D^0 \to K^-\pi^+)} \simeq 1.5\% ,
$$

which are worth an experimental test. The above numerical predictions are obtained from the data given in eq. (92) and $c_- = c_+^{-2} = 1.8 \div 2.1$.

As exemplified in eqs. (83) and (84), the valence quark approximation does not only predict ratios of partial widths, but also the absolute magnitudes [46]. This obviously very important aspect of a critical appraisal of the present understanding can, unfortunately, not be considered further in this lecture (see, *e.g.*, ref. [55]). Also two-body baryon decays [56] must be left out.

To summarize, the valence quark approximation, or, more specifically, straight factorization and vacuum saturation of weak matrix elements, is at variance with the existing data on two-body D decays. The nonobservation of colour suppression constitutes the main problem and raises questions about the role of soft-hadron physics in exclusive decays. A convincing, even if only qualitative, answer has not yet emerged. In order to unravel the essential dynamical aspects, further and more accurate data are needed. In particular, tests of the valence quark predictions [46] on D^+- and F-meson decays would be very desirable. Also decays, such as $D^0 \to \bar{K}^0 \varphi$ and $F^+ \to \omega \pi^+$, which can proceed only via annihilation are of great interest. Finally, it is not clear whether or not two-body decays of B-mesons are appreciably simpler. On the one hand, one does certainly not expect D*-resonance effects in the 5 GeV range. Also, annihilation processes are rather unlikely, since the two final quarks move too fast in order to turn into two hadrons by sharing a slow quark pair created from the vacuum. However, one has no strong arguments against soft-gluon effects in typical decays like $B \to D\pi$. The case $B \to J/\psi X$ ($X = K$ or $K\pi$), shown in subsect. 3˙1 to exhibit colour suppression, is special in the sense that factorization and vacuum saturation should be more reliable due to the small size and the small gluon admixture of the J/ψ.

6. – Conclusions.

The present theoretical approach to weak decays of heavy quark states separates the underlying dynamics in short- and long-distance aspects. The former are incorporated in an effective weak Hamiltonian, whereas the latter are absorbed in the hadronic matrix elements of this Hamiltonian. Retaining this two-sided picture of the whole problem, one can summarize the status of understanding heavy-quark decays as follows.

The modifications of the bare weak Hamiltonian of the standard $SU_{2,L} \times U_1$ theory due to hard-gluon interactions have been rigorously calculated using operator product expansion and renormalization group techniques. It has been shown that the next-to-leading corrections have « normal » size and reinforce the nonleptonic enhancement found in leading-logarithm approximation. Also quark mass effects such as penguin operators are under control. Moreover, they turn out to play a rather unimportant role in charm and bottom decays. In short, the effective weak Hamiltonian part resides on a very solid basis.

As concerns the task of calculating weak matrix elements, I shall further distinguish between inclusive and exclusive (two-body) decays. Asymptotically, that is for sufficiently heavy quarks, inclusive decays are dominated by the short-distance properties of the «free» quark decays. Correspondingly, the spectator model should provide a good approximation. Indeed, the existing data on inclusive B-decays, both semi- and nonleptonic modes, are well described by the spectator model. It is further interesting and reassuring that the gluon corrections as predicted by QCD appear necessary to reconcile theory with experiment. For not-so-heavy quarks, on the other hand, one expects some influence of the internal structure of the heavy-quark bound states. Two prominent pre-asymptotic effects are identical particle interferences between the spectator quarks and the decay products of the heavy quark, and annihilation processes, possibly enhanced by the presence of soft gluons. Various estimates suggest that these effects may play a role in charm decays. Actually, the study of pre-asymptotic effects was initiated by the discovery of deviations from the spectator model in nonleptonic charmed meson and baryon decays. The observed differences in lifetimes and semi-leptonic branching ratios are in qualitative agreement with what one expects from Pauli interference and weak annihilation, indicating roughly equal strength of spectator and nonspectator contributions. This conclusion, however, still lacks a quantitative confirmation by a clear-cut theoretical calculation. In particular, reliable estimates of the annihilation contributions to D- and F-decays, which are negligible unless gluons annul the helicity suppression, require a deeper understanding of QCD bound states. Nevertheless, an offshot has been strong arguments that bottom decays should be short-distance dominated with pre-asymptotic effects not exceeding 10 % or so. To summarize, despite some open quantitative questions, the essential physics of inclusive decays appears to be understood.

In comparison, the situation in nonleptonic two-body decays is rather dim. One has not yet succeeded in developing, at least qualitatively, a consistent theoretical framework. The valence quark approximation, in which the matrix elements of the effective weak Hamiltonian are factorized in products of matrix elements of quark currents saturated by the vacuum state, seems to fail. The clearest counter-evidence comes from the absence of colour suppression in $D^0 \to \overline{K}^0 \pi^0$. This failure may be due to soft-gluon interactions and/or, unexpectedly, large annihilation contributions. In addition, final-state resonance effects may considerably confuse the matter. Also, flavour-symmetry breaking effects have properly to be taken into account as indicated by the Cabibbo-suppressed decays $D^0 \to K^-K^+$ and $\pi^-\pi^+$. It seems that, only when further and more accurate data become available, can one piece the puzzle together. Understanding two-body charm decays will not only have important implications on exclusive B-decays, but also illuminate the origin of the $\Delta I = \frac{1}{2}$ rule in kaon and hyperon decays. The point is that the current explanation of the $\Delta I = \frac{1}{2}$ rule as being mainly a penguin effect relies on either the soft-pion or the valence quark approximation. The latter is seriously questioned by D-decays.

As a final remark, the top quark, « predicted » by the standard model and awaiting its definite discovery, is so heavy that the inclusive decays of top flavoured hadrons [57] will clearly reflect the « free » quark flavour dynamics.

* * *

I would like to thank Prof. N. CABIBBO and Dr. G. MARTINELLI for inviting me to this stimulating and pleasant school. Furthermore, I am grateful for discussions with P. AVERY, L.-L. CHAU, P. FRANZINI, D. HITLIN, J. LEE-FRANZINI and B. STECH at the Europhysics Conference on Flavour Mixing in Erice.

Note added in proofs.

These lectures were presented in Summer 1984. Since then there has been substantial progress, in particular on the experimental side. The MARK III Collaboration [58] has reported results on more than a dozen D branching ratios and on inclusive D decay measurements. The pure annihilation decay $D^0 \to \overline{K}\varphi^0$ has been observed by various groups [59]. In case of B mesons, more data on inclusive decay properties have been collected and some exclusive branching ratios have been determined [60]. Not surprisingly, these new data led to a better theoretical understanding, in particular, of the physics of two-body decays [61]. However, not all questions posed in these lectures have so far found a totally satisfactory answer. The study of heavy-flavour decays is going on.

REFERENCES

[1] For reviews of the phenomenological aspects, see, for example, L. B. OKUN: *Leptons and Quarks* (North-Holland Publ. Co., Amsterdam, 1982); H. FRITZSCH and P. MINKOWSKI: *Phys. Rep.*, **73**, 67 (1981).
[2] PARTICLE DATA GROUP: *Rev. Mod. Phys.*, **56**, 1 (1984).
[3] A fairly detailed discussion can be found in R. RÜCKL: *Weak decays of heavy flavours*, Habilitationsschrift submitted to the University of Munich, CERN print (1983).
[4] See, for example, L.-L. CHAU: *Phys. Rep. C*, **95**, 1 (1983); A. J. BURAS: in *Proceedings of the Workshop on the Future of Intermediate Energy Physics in Europe*, edited by H. KOCH and F. SCHECK (University of Freiburg, Freiburg i.Br., FRG, 1984), p. 53; and various contributions in *Proceedings of the Europhysics Topical Conference on Flavor Mixing in Weak Interactions*, edited by L.-L. CHAU (Plenum Press, New York, N. Y., 1984).
[5] S. L. GLASHOW, J. ILIOPOULOS and L. MAIANI: *Phys. Rev. D*, **2**, 1285 (1970).
[6] M. KOBAYASHI and T. MASKAWA: *Prog. Theor. Phys.*, **49**, 652 (1973).

[7] K. KLEINKNECHT: in *Proceedings of the Europhysics Topical Conference on Flavor Mixing in Weak Interactions*, edited by L.-L. CHAU (Plenum Press, New York, N. Y., 1984), p. 459; J. LEE-FRANZINI: in *Proceedings of the Europhysics Topical Conference on Flavor Mixing in Weak Interactions*, edited by L.-L. CHAU (Plenum Press, New York, N. Y., 1984), p. 217; H. W. SIEBERT: in *Proceedings of the Europhysics Topical Conference on Flavor Mixing in Weak Interactions*, edited by L.-L. CHAU (Plenum Press, New York, N. Y., 1984), p. 37.

[8] K. G. WILSON: *Phys. Rev.*, **179**, 1499 (1969); *Phys. Rev. D*, **3**, 1818 (1971); W. ZIMMERMANN: *Ann. Phys. (N. Y.)*, **77**, 536, 570 (1973).

[9] E. C. G. STUECKELBERG and A. PETERMANN *Helv. Phys. Acta*, **26**, 499 (1953); M. GELL-MANN and F. E. LOW: *Phys. Rev.*, **95**, 1300 (1954); G. C. CALLAN: *Phys Rev. D*, **2**, 1541 (1970); K. SYMANZIK: *Commun. Math. Phys.*, **18**, 227 (1970)

[10] M. K. GAILLARD and B. W. LEE: *Phys. Rev. Lett.*, **33**, 108 (1974); G. ALTARELLI and L. MAIANI: *Phys. Lett. B*, **52**, 351 (1974).

[11] R. L. KINGSLEY, S. B. TREIMAN, F. WILCZEK and A. ZEE: *Phys. Rev. D*, **11**, 1919 (1975); J. ELLIS, M. K. GAILLARD and D. V. NANOPOULOS: *Nucl. Phys. B*, **100**, 313 (1975); G. ALTARELLI, N. CABIBBO and L. MAIANI: *Nucl. Phys. B*, **88**, 285 (1975).

[12] G. ALTARELLI, G. CURCI, G. MARTINELLI and R. PETRARCA: *Phys. Lett. B*, **99**, 141 (1981); *Nucl. Phys. B*, **187**, 461 (1981).

[13] A. I. VAINSHTEIN, V. I. ZAKHAROV and M. A. SHIFMAN: *Sov. Phys. JETP*, **45**, 670 (1977); M. A. SHIFMAN, A. I. VAINSHTEIN and V. I. ZAKHAROV: *Nucl. Phys. B*, **120**, 316 (1977); F. GILMAN and M. B. WISE: *Phys. Rev. D*, **20**, 2392 (1979); **27**, 1128 (1983); B. GUBERINA, D. TADIĆ and J. TRAMPETIĆ: *Nucl. Phys. B*, **152**, 429 (1979); F. BUCCELLA, M. LUSIGNOLI, L. MAIANI and A. PUGLIESE: *Nucl. Phys. B*, **152**, 461 (1979); B. GUBERINA and R. D. PECCEI: *Nucl. Phys. B*, **163**, 289 (1980).

[14] J. ELLIS, M. K. GAILLARD, D. V. NANOPOULOS and S. RUDAZ: *Nucl. Phys. B*, **131**, 285 (1977).

[15] B. GUBERINA, R. D. PECCEI and R. RÜCKL: *Phys. Lett. B*, **90**, 169 (1980); B. GUBERINA: *Fizika*, **16**, 49 (1984).

[16] L. MAIANI: in *Proceedings of the XXI International Conference on High Energy Physics*, edited by P. PETIAU and M. PORNEUF (Les Editions de Physique, Les Ulis, 1982), p. 631.

[17] M. K. GAILLARD, B. W. LEE and J. L. ROSNER: *Rev. Mod. Phys.*, **47**, 277 (1975); J. ELLIS, M. K. GAILLARD and D. V. NANOPOULOS: *Nucl. Phys. B*, **100**, 313 (1975).

[18] N. CABIBBO and L. MAIANI: *Phys. Lett. B*, **79**, 109 (1978); M. SUZUKI: *Nucl. Phys. B*, **145**, 420 (1978); A. ALI and E. PIETARINEN *Nucl. Phys. B*, **154**, 519 (1979); N. CABIBBO, G. CORBO and L. MAIANI: *Nucl. Phys. B*, **155**, 93 (1979); G. CORBO: *Phys. Lett. B*, **116**, 298 (1982); *Nucl. Phys. B*, **122**, 99 (1983); B. GUBERINA, R. D. PECCEI and R. RÜCKL: *Phys. Lett. B*, **91**, 116 (1980); *Nucl. Phys. B*, **171**, 333 (1980); Q. HOKIM and X. Y. PHAM: *Ann. Phys.*, **155**, 202 (1984).

[19] U. BAUR: Diploma Thesis, Universität München (1982); U. BAUR and H. FRITZSCH: *Phys. Lett. B*, **109**, 402 (1982); J. L. CORTES, X. Y. PHAM and A. TOUNSI: *Phys. Rev. D*, **25**, 188 (1982).

[20] A. ALI and E. PIETARINEN: *Nucl. Phys. B*, **154**, 519 (1979); G. ALTARELLI, N. CABIBBO, G. CORBO, L. MAIANI and G. MARTINELLI: *Nucl. Phys. B*, **208**, 365 (1982).

[21] J. LEE-FRANZINI: in *Proceedings of the Europhysics Topical Conference on Flavor Mixing in Weak Interactions*, edited by L.-L. CHAU (Plenum Press, New York, N. Y., 1984), p. 217. See also G. ALTARELLY: in *Proceedings of the 11 International*

Winter Meeting on Fundamental Physics, edited by A. FERRANDO (Instituto de Estudios Nucleares, Madrid, 1983), p. 295.

[22] W. BACINO, T. FERGUSON, L. NODULMAN, W. SLATER, H. TICHO, A. DIAMANT-BERGER, G. DONALDSON, M. DURO, A. HALL, G. IRWIN, J. KIRKBY, F. MERRITT, S. WOJCICKI, R. BURNS, P. CONDON, P. COWELL and J. KIRZ: *Phys. Rev. Lett.*, **43**, 1073 (1979).

[23] C. KLOPFENSTEIN, J. E. HORSTKOTTE, J. LEE-FRANZINI, R. D. SCHAMBERGER, M. SIVERTZ, L. J. SPENCER, P. M. TUTS, P. FRANZINI, K. HAN, E. RICE, D. SON, S. YOUSSEF, S. W. HERB, R. IMLAY, G. LEVMAN, W. METCALF, V. SREEDHAR, H. DIETL, G. EIGEN, E. LORENZ, G. MAGERAS, F. PAUSS and H. VOGEL: *Phys. Lett. B*, **130**, 444 (1983); A. CHEN, M. GODBERG, N. HORWITZ, A. JAWAHERY, P. LIPARI, G. C. MONETI, C. G. TRAHERN, H. VAN HECKE, M. S. ALAM, S. E. CSORNA, L. GARREN, M. D. MESTAYER, R. S. PANVINI, XIA YI, P. AVERY, C. BEBEK, K. BERKELMAN, D. G. CASSEL, J. W. DEWIRE, R. EHRLICH, T. FERGUSON, R. GALIK, M. G. D. GILCHRIESE, B. GITTELMAN, M. HALLING, D. L. HARTILL, S. HOLZNER, M. ITO, J. KANDASWAMY, D. L. KREINICK, Y. KUBOTA, N. B. MISTRY, F. MORROW, E. NORDBERG, M. OGG, A. SILVERMAN, P. C. STEIN, S. STONE, D. WEBER, R. WILCKE, A. J. SADOFF, R. GILES, J. HASSARD, M. HEMPSTEAD, K. KINOSHITA, W. W. MACKAY, F. M. PIPKIN, R. WILSON, P. HAAS, T. JENSEN, H. KAGAN, R. KASS, S. BEHRENDS, K. CHADWICK, J. CHAUVEAU, T. GENTILE, J. M. GUIDA, J. A. GUIDA, A. C. MELISSINOS, S. L. OLSEN, G. PARKHURST, D. PETERSON, R. POLING, C. ROSENFELD, E. H. THORNDIKE, P. TIPTON, D. BESSON, J. GREEN, R. G. HICKS, R. NAMJOSHI, F. SANNES, P. SKUBIC, A. SNYDER and R. STONE: *Phys. Rev. Lett.*, **52**, 1084 (1984).

[24] S. STONE: in *Proceedings of the 1983 International Symposium on Lepton and Photon Interactions at High Energies*, edited by D. G. CASSEL and D. L. KREINICK (Newman Laboratory of Nuclear Studies, Cornell University, Ithaca, N. Y., 1983), p. 203.

[25] E. FERNANDEZ, W. T. FORD, A. L. READ jr., J. G. SMITH, R. DE SANGRO, A. MARINI, I. PERUZZI, M. PICCOLO, F. RONGA, H. T. BLUME, H. B. WALD, R. WEINSTEIN, H. R. BAND, M. W. GETTNER, G. P. GODERRE, B. GOTTSCHALK, R. B. HURST, O. A. MEYER, J. H. MOROMISATO, W. D. SHAMBROOM, E. VON GOELER, W. W. ASH, G. B. CHADWICK, S. H. CLEARWATER, R. W. COOMBES, H. S. KAYE, K. H. LAU, R. E. LEEDY, H. L. LYNCH, R. L. MESSNER, S. J. MICHALOWSKI, K. RICH, D. M. RITSON, L. J. ROSENBERG, D. E. WISER, R. W. ZDARKO, D. E. GROOM, HOYUN LEE, E. C. LOH, M. C. DELFINO, B. K. HELTSLEY, J. R. JOHNSON, T. L. LAVINE, T. MARUYAMA and R. PREPOST: *Phys. Rev. Lett.*, **51**, 1022 (1983).

[26] N. S. LOCKYER, J. A. JAROS, M. E. NELSON, G. S. ABRAMS, D. AMIDEI, A. R. BADEN, C. A. BLOCKER, A. M. BOYARSKI, M. BREIDENBACH, P. BURCHAT, D. L. BURKE, J. M. DORFAN, G. J. FELDMAN, G. GIDAL, L. GLADNEY, M. S. GOLD, G. GOLDHABER, L. GOLDING, G. HANSON, D. HERRUP, R. J. HOLLEBEEK, W. R. INNES, M. JONKER, I. JURICIC, J. A. KADYK, A. J. LANKFORD, R. R. LARSEN, B. LECLAIRE, M. LEVI, V. LÜTH, C. MATTEUZZI, R. A. ONG, M. L. PERL, B. RICHTER, M. C. ROSS, P. C. ROWSON, T. SCHAAD, H. SCHELLMAN, D. SCHLATTER, P. D. SHELDON, J. STRAIT, G. H. TRILLING, C. DE LA VAISSIERE, J. M. YELTON and C. ZAISER: *Phys. Rev. Lett.*, **51**, 1316 (1983).

[27] P. AVERY: in *Proceedings of the Europhysics Topological Conference on Flavor Mixing in Weak Interactions*, edited by L.-L. CHAU (Plenum Press, New York, N. Y., 1984), p. 91.

[28] J. H. KÜHN, S. NUSSINOV and R. RÜCKL: *Z. Phys. C*, **5**, 117 (1980); J. H. KÜHN and R. RÜCKL: *Phys. Lett. B*, **135**, 477 (1984).

[29] M. B. WISE: *Phys. Lett. B*, **89**, 229 (1980); T. A. DEGRAND and D. TOUSSAINT: *Phys. Lett. B*, **89**, 256 (1980).

[30] B. H. WIIK and G. WOLF: *Electron-Positron Interactions, Springer Tracts in Modern Physics*, Vol. **86** (Springer, Berlin, 1979).

[31] D. FAKIROV and B. STECH: *Nucl. Phys. B*, **133**, 315 (1978); N. CABIBBO and L. MAIANI: *Phys. Lett. B*, **73**, 418 (1978).

[32] N. W. REAY: in *Proceedings of the 1983 International Symposium on Lepton and Photon Interactions at High Energies*, edited by D. G. CASSEL and D. L. KREINICK (Newman Laboratory of Nuclear Studies, Cornell University, Ithaca, N. Y., 1983), p. 244.

[33] W. BERNREUTHER, O. NACHTMANN and B. STECH: *Z. Phys. C*, **4**, 257 (1980); H. FRITZSCH and P. MINKOWSKI: *Phys. Lett. B*, **90**, 455 (1980); S. P. ROSEN: *Phys. Rev. Lett.*, **44**, 4 (1980); M. BANDER, D. SILVERMAN and A. SONI: *Phys. Rev. Lett.*, **44**, 7 (1980); (Erratum) *Phys. Rev. Lett.*, **44**, 962 (1980); V. BARGER, J. P. LEVEILLE and P. M. STEVENSON: *Phys. Rev. D*, **22**, 693 (1980).

[34] R. D. PECCEI and R. RÜCKL: in *Special Topics in Gauge Field Theories, Ahrenshoop Symposium* (Akademie der Wissenschaften der DDR, Berlin-Zeuthen, 1981), p. 8; T. KOBAYASHI and N. YAMAZAKI: *Prog. Theor. Phys.*, **65**, 775 (1981); M. A. SHIFMAN and M. B. VOLOSHIN: ITEP preprint-62 (1984); see also B. GUBERINA, S. NUSSINOV, R. D. PECCEI and R. RÜCKL: *Phys. Lett. B*, **89**, 111 (1979); Y. KOIDE: *Phys. Rev. D*, **20**, 1739 (1979); K. JAGANNATHAN and V. S. MATHUR: *Phys. Rev. D*, **21**, 3165 (1980).

[35] G. ALTARELLI and L. MAIANI: *Phys. Lett. B*, **118**, 414 (1982).

[36] H. SAWAYANAGI, K. FUJII, T. OKAZAKI and S. OKUBO: *Phys. Rev. D*, **27**, 2107 (1983).

[37] N. BILIĆ, B. GUBERINA and J. TRAMPETIĆ: *Nucl. Physi. B*, **248**, 261 (1984).

[38] M. BANDER, D. SILVERMAN and A. SONI: *Phys. Rev. Lett.*, **44**, 7, 962 (1980).

[39] K. SHIZUYA: *Phys. Lett. B*, **100**, 79 (1981); **105**, 406 (1981).

[40] M. SHIFMAN, A. VAINSHTEIN and V. I. ZAKHAROV: *Nucl. Phys. B*, **147**, 385, 448 (1979); L. J. REINDERS, H. R. RUBINSTEIN and S. YAZAKI: *Phys. Lett. B*, **94**, 203 (1980); **95**, 103 (1980); B. GUBERINA, R. MECKBACH, R. D. PECCEI and R. RÜCKL: *Nucl. Phys. B*, **184**, 476 (1981).

[41] R. RÜCKL: *Phys. Lett. B*, **120**, 449 (1983).

[42] V. BARGER, J. P. LEVEILLE and P. M. STEVENSON: *Phys. Rev. Lett.*, **44**, 226 (1980).

[43] A. DE RÚJULA, H. GEORGI and S. L. GLASHOW: *Phys. Rev. D*, **12**, 147 (1975).

[44] S. F. BIAGI, M. BOURQUIN, A. J. BRITTEN, R. M. BROWN, H. J. BURCKHART, A. A. CARTER, CH. DORÉ, P. EXTERMANN, M. GAILLOUD, C. N. P. GEE, W. M. GIBSON, J. C. GORDON, R. J. GRAY, P. IGO-KEMENES, P. JACOT-GUILLARMOD, W. C. LOUIS, T. MODIS, PH. ROSSELET, B. J. SAUNDERS, P. SCHIRATO, H. W. SIEBERT, V. J. SMITH, K.-P. STREIT, J. J. THRESHER, S. N. TOVEY and R. WEILL: *Phys. Lett. B*, **150** (1985).

[45] I. BIGI: *Z. Phys. C*, **9**, 197 (1981); see also I. BIGI: *Nucl. Phys. B*, **177**, 395 (1981); M. SUZUKI: *Nucl. Phys. B*, **177**, 413 (1981).

[46] D. FAKIROV and B. STECH: *Nucl. Phys. B*, **133**, 315 (1978).

[47] G. H. TRILLING: *Phys. Rep. C*, **75**, 57 (1981) R. SCHINDLER: Ph.D. Thesis, SLAC-report 219 (1979); D. HITLIN: in *Proceedings of the Europhysics Topical Conference on Flavor Mixing in Weak Interactions*, edited by L.-L. CHAU (Plenum Press, New York, N. Y., 1984), p. 361.

[48] M. KATUYA: *Phys. Rev. D*, **18**, 3510 (1978); M. KATUYA and Y. KOIDE: *Phys. Rev. D*, **19**, 2631 (1979); G. EILAM and M. GRONAU: *Phys. Lett. B*, **96**, 391 (1980); see also B. GUBERINA, S. NUSSINOV, R. D. PECCEI and R.RÜCKL: *Phys. Lett. B*,

89, 111 (1979); Y. KOIDE: *Phys. Rev. D*, **20**, 1739 (1979); K. JAGANNATHAN and V. S. MATHUR: *Phys. Rev. D*, **21**, 3165 (1980).

[49] H. FRITZSCH: *Phys. Lett. B*, **86**, 343 (1979).

[50] M. DESHPANDE, M. GRONAU and D. SUTHERLAND: *Phys. Lett. B*, **90**, 431 (1980); M. GRONAU and D. SUTHERLAND: *Nucl. Phys. B*, **183**, 367 (1981).

[51] H. J. LIPKIN: *Phys. Rev. Lett.*, **44**, 710 (1980).

[52] V. BARGER and S. PAKVASA: *Phys. Rev. Lett.*, **43**, 812 (1979).

[53] H. FRITZSCH and P. MINKOWSKI: *Nucl. Phys. B*, **171**, 413 (1980); I. BIGI: *Phys. Lett. B*, **90**, 177 (1980).

[54] R. L. KINGSLEY, S. B. TREIMAN, F. WILCZEK and A. ZEE: *Phys. Rev. D*, **11**, 1919 (1975); L.-L. CHAU WANG and F. WILCZEK: *Phys. Rev. Lett.*, **43**, 816 (1979).

[55] M. BONVIN and C. SCHMID: *Nucl. Phys. B*, **194**, 319 (1982).

[56] J. G. KÖRNER, G. KRAMER and J. WILLRODT: *Phys. Lett. B*, **78**, 492 (1978); (Erratum) **81**, 419 (1979); *Z. Phys. C*, **2**, 117 (1979); D. EBERT and W. KALLIES: *Phys. Lett. B*, **131**, 183 (1983).

[57] N. CABIBBO and L. MAIANI: *Phys. Lett. B*, **87**, 366 (1979).

[58] R. M. BALTRUSAITIS, J. J. BECKER, G. T. BLAYLOCK, J. S. BROWN, K. O. BUNNELL, T. H. BURNETT, R. E. CASSELL, D. COFFMAN, V. COOK, D. H. COWARD, S. DADO, C. DEL PAPA, D. E. DORFAN, G. P. DUBOIS, A. L. DUNCAN, K. F. EINSWEILER, B. I. EISENSTEIN, R. FABRIZIO, G. GLADDING, F. GRANCAGNOLO, R. P. HAMILTON, J. HAUSTER, C. A. HEUSCH, D. G. HITLIN, L. KOEPKE, W. S. LOCKMAN, P. M. MOCKETT, L. MOSS, R. F. MOZLEY, A. NAPPI, A. ODIAN, R. PARTRIDGE, J. PERRIER, S. A. PLAETZER, J. D. RICHMAN, J. R. ROEHRIG, J. J. RUSSELL, H. F. W. SADROZINSKI, M. SCARLATELLA, T. L. SHALK, R. H. SCHINDLER, A. SEIDEN, J. C. SLEEMAN, A. L. SPADAFORA, J. J. THALER, W. TOKI, B. TRIPSAS, F. VILLA, A. WATTENBERG, A. J. WEINSTEIN, N. WERMES, H. J. WILLUTZKI, D. E. WISINSKI and W. J. WISNIEWSKI: *Phys. Rev. Lett.*, **54**, 1976 (1985); R. M. BALTRUSAITIS, J. J. BECKER, G. T. BLAYLOCK, J. S. BROWN, K. O. BUNNELL, T. H. BURNETT, R. E. CASSELL, D. COFFMAN, V. COOK, D. H. COWARD, S. DADO, D. E. DORFAN, G. P. DUBOIS, A. L. DUNCAN, K. F. EINSWEILER, B.I. EISENSTEIN, R. FABRIZIO, G. GLADDING, R. P. HAMILTON, J. HAUSER, C. A. HEUSCH, D. G. HITLIN, L. KÖPKE, W. S. LOCKMAN, U. MALLIK, P. M. MOCKETT, R. F. MOZLEY, A. NAPPI, A. ODIAN, R. PATRIDGE, J. PERRIER, S. A. PLAETZER, J. D. RICHMAN, J. ROEHRIG, J. J. RUSSEL, H. F. W. SADROZINSKI, M. SCARLATELLA, T. L. SCHALK, R. H. SCHINDLER, A. SEIDEN, C. SIMOPOULOS, J. C. SLEEMAN, A. L. SPADAFORA, J. J. THALER, B. TRIPSAS, W. TOKI, F. VILLA, A. WATTENBERG, A. J. WEINSTEIN, N. WERMES, H. J. WILLUTZKI, D. WISINSKI and W. J. WISNIEWSKI: *Phys. Rev. Lett.*, **55**, 150 (1985); B. M. BALTRUSAITIS, J. J. BECKER, G. T. BLAYLOCK, J. S. BROWN, K. O. BUNNELL, T. H. BURNETT, R. E. CASSELL, D. COFFMAN, V. COOK, D. H. COWARD, S. DADO, D. E. DORFAN, G. P. DUBOIS, A. L. DUNCAN, K. F. EINSWEILER, B. I. EISENSTEIN, R. FABRIZIO, G. GLADDING, F. GRANCAGNOLO, R. P. HAMILTON, J. HAUSER, C. A. HEUSCH, D. G. HITLIN, L. KÖPKE, W. S. LOCKMAN, U. MALLIK, P. M. MOCKETT, R. F. MOSLEY, A. NAPPI, A. ODIAN, R. PARTRIDGE, J. PERRIER, S. A. PLAETZER, J. D. RICHMAN, J. ROEHRIG, J. J. RUSSEL, H. F. W. SADRISINSKI, M. SCARLATELLA, T. L. SCHALK, R. H. SCHINDLER, A. SEIDEN, C. SIMOPOULOS, J. C. SLEEMAN, A. L. SPADAFORA, I. E. STOCKDALE, J. J. THALER, B. TRIPSAS, W. TOKI, F. VILLA, A. WATTENBERG, A. J. WEINSTEIN, N. WERMES, R. J. WILLUTSKI, D. WISINSKI, W. J. WISNIEWSKI and G. WOLF: SLAC-PUB-3861 (1985); R. H. SCHINDLER (representing the MARK III Collaboration): invited talk at the 1985 SLAC Summer School, SLAC-PUB-3799, CALT-68-1307 (1985).

[59] H. ALBRECHT, U. BINDER, G. HARDER, A. PHILIPP, W. SCHMIDT-PARZEFALL, H. SCHRÖDER, H. D. SHULZ, R. WURTH, A. DRESCHER, B. GRÄWE, U. MATTHIESEN, H. SCHECK, J. SPENGLER, D. WEGENER, K. R. SCHUBERT, J. STIEWE, R. WALDI, S. WESELER, N. N. BROWN, K. W. EDWARDS, W. R. FRISKEN, CH. FUKUNAGA, D. J. GILKINSON, D. M. GINGRICH, M. GODDARD, P. C. H. KIM, R. KUTSCHKE, D. B. MACFARLANE, J. A. MCKENNA, K. W. MCLEAN, A. W. NILLSON, R. S. ORR, P. PADLEY, P. M. PATEL, J. D. PRENTICE, H. C. J. SEYWERD, B. J. STACEY, T.-S. YOON, J. C. YUN, R. AMMAR, D. COPPAGE, R. DAVIS, S. KANEKAL, N. KWAK, G. KERNEL, M. PLEŠKO, L. JÖNSSON, Y. OKU, A. BABAEV, M. DANILOV, A. GOLUTVIN, V. LUBIMOV, V. MATVEEV, V. NAGOVITSIN, V. RYLTSOV, A. SEMENOV, V. SHEVCHENKO, V. SOLOSHENKO, V. SOPOV, I. TICHOMIROV, YU. ZAITSEV, R. CHILDERS, C. W. DARDEN and H. GENNOW: *Phys. Lett. B*, **158**, 525 (1985); R. M. BALTRUSAITIS, J. J. BECKER, G. T. BLAYLOCK, J. S. BROWN, K. O. BUNNELL, T. H. BURNETT, R. E. CASSELL, D. COFFMAN, V. COOK, D. H. COWARD, S. DADO, D. E. DORFAN, G. P. DUBOIS, A. L. DUNCAN, K. F. EINSWEILER, B. I. EISENSTEIN, R. FABRIZIO, G. GLADDING, F. GRANCAGNOLO, R. P. HAMILTON, J. HAUSER, C. A. HEUSCH, D. G. HITLIN, L. KÖPKE, W. S. LOCKMAN, U. MALLIK, P. M. MOCKETT, R. F. MOSLEY, A. NAPPI, A. ODIAN, R. PARTRIDGE, J. PERRIER, S. A. PLAETZER, J. D. RICHMAN, J. ROEHRIG, J. J. RUSSELL, H. F. W. SADROZINSKI, M. SCARLATELLA, T. L. SCHALK, R. H. SCHINDLER, A. SEIDEN, C. SIMOPOULOS, J. C. SLEEMAN, A. L. SPADAFORA, I. E. STOCKDALE, J. J. THALER, B. TRIPSAS, W. TOKI, F. VILLA, A. WATTENBERG, A. J. WEINSTEIN, N. WERMES, H. J. WILLUTSKI, D. WISINSKI, W. J. WISNIEWSKI and G. WOLF: SLAC-PUB-3858 (1985); C. BEBEK, K. BERKELMAN, E. BLUCHER, D. G. CASSEL, T. COPIE, R. DESALVO, J. W. DEWIRE, R. EHRLICH, R. S. GALIK, M. G. D. GILCHRIESE, B. GITTELMAN, S. W. GRAY, A. M. HALLING, D. L. HARTILL, B. K. HELTSLEY, S. HOLZNER, M. ITO, J. KANDASWAMY, R. KOWALEWSKI, D. L. KREINICK, Y. KUBOTA, N. B. MISTRY, J. MUELLER, E. NORDBERG, M. OGG, D. PETERSON, D. PERTICONE, M. PISHARODY, K. READ, D. RILEY, A. SILVERMAN, P. C. STEIN and S. STONE: CLNS-86/715, CLEO-86-2 (1986).

[60] For a recent review see, for example, E. H. THORNDIKE: in *Proceedings of the 1985 International Symposium on Lepton and Photon Interactions at High Energies*, edited by M. KONUMA and K. TAKAHASHI (Kyoto University, Kyoto, Japan, 1986), p. 406.

[61] R. RÜCKL: in *Proceedings of the XXII International Conference on High Energy Physics*, edited by A. MEYER and E. WIECZOREK (Akademie der Wissenschaften der DDR, Zeuthen, GDR, 1984), Vol. I, p. 135; M. BAUER and B. STECH: *Phys. Lett. B*, **152**, 380 (1985); B. STECH: in *Proceedings of the Moriond Workshop on Flavour Mixing and CP Violation*, edited by J. TRAN THANH VAN (Editions Frontières, Gif-sur-Yvette, France, 1985), p. 151; A. J. BURAS, J.-M. GERARD and R. RÜCKL: *Nucl. Phys. B*, **268**, 16 (1986); MPI-PAE/PTh 16/86 (1986); A. N. KAMAL and M. D. SCADRON: *Phys. Rev. D*, **32**, 1164 (1985); A. N. KAMAL: *Phys. Rev. D*, **33**, 1344 (1986); I. I. BIGI: *Phys. Lett. B*, **169**, 101 (1986); H.-Y. CHENG: Indiana University preprint, IUHET Nr. 113 (1986).

Compositeness.

R. Petronzio

Theoretical Physics Division, CERN - 1211 Geneva 23, Switzerland

1. – Introduction.

So far, most of the particles which were thought to be elementary turned into composite ones. The closest example is the one of hadrons: both the complexity of hadron spectrum and the simple scaling laws observed in deep inelastic experiments have suggested the existence of hadron constituents—the quarks—bound by a colour force mediated by gluons.

Today, elementary particles according to the « standard model » based on the $SU_3 \times SU_2 \times U_1$ gauge group are quarks, leptons, the electroweak intermediate vector bosons, the gluons and the scalar Higgses [1]. The question is: do we have any experimental evidence for the compositeness of these particles?

The answer is no (modulo, maybe, the recent « anomalous » events observed at the Sp$\bar{\text{p}}$S collider).

However, compositeness might be desirable to reduce the number of free parameters of the standard model; they include Λ_{QCD}, e, g_{W}, $\langle\phi\rangle$, m_{Higgs}, $\{m_{\text{quarks}}\}$, $\{m_{\text{leptons}}\}$, θ_i ($i=1,4$), where θ_i are the four parameters (three angles plus a phase) entering the Kobayashi-Maskawa mass matrix [2] and $\langle\phi\rangle$ is the vacuum expectation value of the Higgs field responsible for the $SU_2 \times U_1$ spontaneous symmetry breaking. Masses and mixing angles all come from the Yukawa couplings of the original Higgs doublet to the initially massless fermions.

Certainly, the number of free parameters is rather large, and, in particular, the occurrence of three « families » of quarks and leptons remains unexplained. If the standard model is not the ultimate theory, at which energy scale « E » is something new expected to occur?

The answer of grand unified theories is « E » $\sim 10^{17}$ GeV [3] but at the price of « unnaturalness ». For, in order to obtain two independent symmetry breakings, one from the unifying group (SU_5, for example) to $SU_3 \times SU_2 \times U_1$ and one from the latter to $SU_3 \times U_1'$, at energy scales so different from each other (10^{17} and 10^2 GeV, respectively), one must perform a very « fine tuning » (with a precision of the order of 10^{-15}!) of the parameters of the Higgs potential responsible for the two breakings.

This unnaturalness can be eliminated by constructing models where supersymmetry keeps some parameters of the scalar potential—the mass terms in particular—zero in a natural way [4]. The presence of a Higgs mass (*i.e.* the $SU_2 \times U_1$ symmetry-breaking scale) is then associated with the breaking of SUSY and of its projection. In this way something new (the presence of the supersymmetric partners of the ordinary fermions) is expected in the W-Z mass range (of course, this includes a few times their mass).

The other possibility is that the Higgs theory is only an effective one. In order to figure out the value of the scale at which the effectiveness will show up, one can repeat an argument given by 'T Hooft in his Cargèse lectures [5], based on a more quantitative analysis of the concept of unnaturalness. Define the scale where something new happens as μ, which is supposed to be much larger than m_H, the Higgs mass. The theory is said natural *up to* μ if, by setting $m_\text{H}/\mu = 0$, some new symmetry is recovered.

In the absence of gauge forces, the symmetry is easily found to be

$$(1.1) \qquad \phi'(x) = \phi(x) + \text{const},$$

where the constant drops when only the kinetic term of the Lagrangian is considered. (Of course, neither ϕ^2 nor ϕ^4 terms are allowed.)

The gauge forces of the standard model break the symmetry of eq. (1.1): indeed, they imply the following symmetry operation:

$$(1.2) \qquad \phi'(x) = \Omega(x)\phi(x),$$

which, if compatible with the one of eq. (1.1), would lead to

$$(1.3) \qquad \phi'(x) = \phi(x) + c(x),$$

which, by choosing $c(x) = -\phi(x)$, would completely annihilate the scalar field. If standard gauge forces break the symmetry of eq. (1.1), the forbidden mass terms will now appear as radiative corrections due to the gauge fields:

$$(1.4) \qquad m_\text{H} \sim \alpha\mu,$$

where α is a generic electroweak coupling constant. Similarly, the ϕ^4 coupling constant, λ, can be expected to be of order α. In a « natural » scenario, one would get

$$(1.5) \qquad m_\text{H} \sim \alpha\mu = \lambda\langle\varphi\rangle \sim \alpha\langle\varphi\rangle,$$

from which follows

$$\mu \sim \langle\varphi\rangle, \qquad m_\text{H} \sim \alpha\mu.$$

Conversely, naturalness will break (*i.e.* something new must happen) when μ is much larger than $\langle\phi\rangle$. For the actual world, this means $\mu \sim$ a few TeV. There, the Higgs theory is expected to show up as an effective one.

My lectures will be devoted to an introductory discussion of compositeness for the Higgses, the W and Z and the quarks and leptons. This list does not contain massless vector bosons, like the gluons, associated to a local gauge symmetry. A theorem [6] states that a theory with a Lorentz-covariant conserved current cannot have massless particles of spin greater than $\frac{1}{2}$ with a non-vanishing value of the associated charge. This implies that massless spin-one particles can exist only if they are explicitly associated to a locally gauge-invariant symmetry (in this case, the current J^μ is not Lorentz-covariant). In other words, if one wants gauge invariance, one has to buy it from the beginning. There is no point in seeking for composite gluons when, in order to obtain them, one must have local colour symmetry already realized (and, therefore, gluons already there).

2. – Composite Higgs.

The nicest idea (for me) to introduce composite Higgses is the one of technicolour [7]. Something new happens at $\mu^2 \sim G_F^{-1}$: a new type of strong interactions coupled to a new type of fermions (technifermions). In fig. 1, one sees

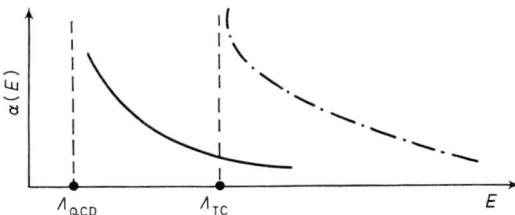

Fig. 1. – The qualitative behaviour of running coupling constants of QCD and QTD.

the behaviour, as a function of the energy scale E, of the running coupling constants of QCD and TC (technicolour). They become large at $E = \Lambda_{\text{QCD}}$ and $E = \Lambda_{\text{TC}}$. Of course, $\Lambda_{\text{TC}} \sim \sqrt{G_F^{-1}} \gg \Lambda_{\text{QCD}}$.

In the simplest scheme, one introduces an SU_2 doublet of left-handed technifermions

(2.1) $$T_{\text{left}} \equiv \begin{pmatrix} u_{\text{left}} \\ d_{\text{left}} \end{pmatrix}$$

and two right-handed singlets

(2.2) $$u_{\text{right}}, \quad d_{\text{right}}.$$

These fermions are the exact replicas of those in an ordinary quark doublet. The fermion's Lagrangian (without mass terms) has an overall global chiral symmetry

$$(2.3) \quad \mathscr{L} = i\bar{\psi}^i \hat{D} \psi^i = i\bar{\psi}^i_L \hat{D} \psi^i_L + i\bar{\psi}^i_R \hat{D} \psi^i_R,$$

where $\psi^i_{L,R} \equiv (1 \pm \gamma^5) \psi^i$ are helicity eigenstates, and the index i refers to the fermion's flavour. The symmetry is

$$(2.4) \quad \begin{cases} \psi^i_L \to U^{ij}_L \psi^j_L, \\ \psi^i_R \to U^{ij}_R \psi^j_R, \end{cases}$$

where U_L, U_R are unitary matrices acting in the flavour space.

For a flavour doublet, like the one in eqs. (2.1) and (2.2), one has the following *global* symmetry: $SU_{2,L} \times SU_{2,R} \times U_1$. The extra U_1 symmetry is broken by the instantons of the techniforce (*).

The technicolour, very much like the ordinary colour, spontaneously breaks the L-R SU_2 symmetry to the vector SU_2. This happens through the formation of a vacuum condensate:

$$(2.5) \quad \langle 0 | \bar{u}u + \bar{d}d | 0 \rangle \neq 0.$$

This is invariant under the (2.4) transformation *if* $U_L = U_R$ (note that $\bar{u}u \equiv \bar{u}_L u_R$).

The breaking of a global symmetry leads to Goldstone bosons carrying the quantum numbers of the broken generators [8]. In full analogy with QCD, one can write

$$(2.6) \quad \langle 0 | J^{\mu,a}_5 | \Pi^b \rangle = F_\pi \zeta^\mu \delta^{ab},$$

where Π^b are the three technipions corresponding to the three generators of the broken $SU_{2,\text{axial}}$. The Kronecker δ^{ab}, eq. (2.6), is the consequence of isospin conservation which holds true after the spontaneous symmetry breaking. These zero-mass particles are coupled to the intermediate vector boson of the *local* $SU_2 \times U_1$ theory $(B_\mu, \vec{W}_\mu)$ and shift their mass.

Let us consider charged W's first: the effect of the vacuum polarization « renormalizes » their propagator:

$$(2.7) \quad \frac{g_{\mu\nu} - q_\mu q_\nu / q^2}{q^2} \xrightarrow{\text{vacuum polarization}} \frac{g_{\mu\nu} - q_\mu q_\nu / q^2}{q^2 [1 + \pi(q^2)]}.$$

(*) You are not meant to understand this statement from my lectures: nevertheless it is true.

The contribution of techniquark currents which couple to thechnipions through eq. (2.6) is (see fig. 2a))

$$(2.8) \qquad \pi(q^2) \xrightarrow[q^2 \to 0]{} \frac{g_2^2 F_\pi^2}{4q^2},$$

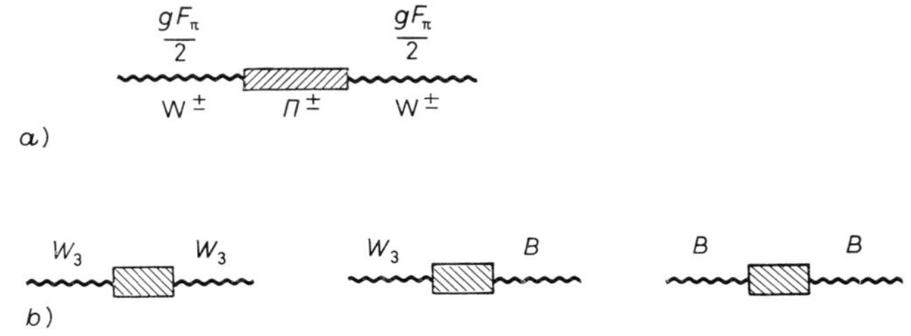

Fig. 2. – a) Diagram giving a mass term to charged W's, b) diagrams producing the mixing mass matrix of W^3 and B.

where $g_2/2$ is the coupling of techniquarks to the charged W's. By inserting (2.8) in (2.7), one gets

$$(2.9) \qquad q^2[1 + \pi(q^2)] \to q^2 + \frac{g_2^2 F_\pi^2}{4}.$$

The W mass has been shifted from zero to a finite value:

$$(2.10) \qquad m_W^2 = \frac{F_\pi^2 g_2^2}{4}.$$

For the case of neutral bosons, one has a 2×2 matrix to diagonalize; the diagrams are the ones of fig. 2b) and the matrix is

$$(2.11) \qquad \begin{pmatrix} g_1^2 & g_1 g_2 \\ g_1 g_2 & g_2^2 \end{pmatrix} \frac{F_\pi^2}{4},$$

where g_1 is the coupling of techniquarks to the U_1 field B_μ.

The eigenvalues are

$$(2.12) \qquad m_\gamma = 0, \quad m_Z = \tfrac{1}{4}(g_1^2 + g_2^2) F_\pi^2.$$

Therefore, the Weinberg relation

$$(2.13) \qquad \frac{m_W}{m_Z} = \cos\theta_W$$

is preserved. It was very essential here that $F_\pi^\pm$ and F_π^0 be the same for the relation (2.13) to be true. The effects of the Higgs potential have been replaced by vacuum condensates. It is instructive to verify that also the quantum numbers of the original Higgses and of technipions are the same.

In the standard model, the Higgses form a complex doublet

$$\varphi = \begin{pmatrix} \varphi^+ \\ \varphi^0 \end{pmatrix} \tag{2.14}$$

and the symmetry breaking is produced by

$$\langle \operatorname{Re} \varphi_0 \rangle = \langle \phi_0 + \phi_0^* \rangle \neq 0 \,. \tag{2.15}$$

The isospin/hypercharge assignments for the Higgses are

$$\tag{2.16} \begin{cases} & \quad\quad T_3 \quad\quad Y \\ \varphi^+ & \quad\quad \tfrac{1}{2} \quad\quad \tfrac{1}{2} \\ \varphi^- & \quad -\tfrac{1}{2} \quad -\tfrac{1}{2} \\ \varphi_0 & \quad -\tfrac{1}{2} \quad\quad \tfrac{1}{2} \\ \bar{\varphi}_0 & \quad\quad \tfrac{1}{2} \quad -\tfrac{1}{2} \end{cases}$$

to be compared with the techni-q techni-$\bar{q}$ states:

$$\tag{2.17} \begin{cases} & \quad\quad T_3 \quad\quad Y \\ \bar{u}_L d_R & \quad -\tfrac{1}{2} \quad -\tfrac{1}{2} \\ \bar{d}_L u_R & \quad\quad \tfrac{1}{2} \quad\quad \tfrac{1}{2} \\ \bar{u}_L u_R & \quad -\tfrac{1}{2} \quad\quad \tfrac{1}{2} \\ \bar{d}\; d_R & \quad\quad \tfrac{1}{2} \quad -\tfrac{1}{2} \end{cases}$$

Therefore, $\operatorname{Im} \phi_0 \equiv \phi_0 - \phi_0^*$ transforms as $\bar{u}u - \bar{d}d$ while $\operatorname{Re} \phi_0 \equiv \phi_0 + \phi_0^*$ goes like $\bar{u}u + \bar{d}d$. The vacuum condensate and the « eaten » technipions have the same quantum numbers of the corresponding Higgses.

The order of magnitude of the technicolour scale can be approximated by comparing F_π, eq. (2.6), to the familiar f_π of QCD:

$$\frac{\Lambda_{\text{QCD}}}{\Lambda_{\text{TC}}} \sim \frac{f_\pi}{F_\pi} = \frac{f_\pi}{\sqrt{4m_W^2/g_2^2}} \sim \frac{1}{2500} \,. \tag{2.18}$$

Numerically, one gets $\Lambda_{\text{TC}} \sim 0.5$ TeV. If more fundamental doublets of techniquarks are present, let us say r, one gets

$$\Lambda_{\text{TC}} \sim \frac{1}{\sqrt{r}} \cdot (0.5 \text{ TeV}) . \tag{2.19}$$

The main problem of the scheme that I have presented so far is how to give masses to ordinary fermions. In the standard model, masses come from the Yukawa coupling of Higgses to fermions: the Higgs vacuum expectation value then turns into a mass term

$$\lambda \varphi \bar{\psi}_L \psi_R \to \lambda \langle \varphi \rangle \bar{\psi}_L \psi_R \equiv m \bar{\psi}_L \psi_R . \tag{2.20}$$

In the case of technicolour, this corresponds to

$$\langle \varphi \rangle \bar{\psi} \psi \to \langle \bar{u}u + \bar{d}d \rangle \bar{\psi} \psi . \tag{2.21}$$

We see that *four-fermion interactions* (between ordinary and technifermions) are needed in order to produce the usual mass terms of the standard model. This implies the existence of new gauge bosons mediating these new interactions: mass terms are then obtained through diagrams like that in fig. 3.

Fig. 3. – A mass term for ordinary fermions induced by a nonzero condensate of technifermions.

Unfortunately, similar diagrams can also give rise to new flavour *non*-conserving neutral currents whose presence is experimentally severely restricted. This has caused endless readjustments of the technicolour scheme: I will not discuss them at all, since none of them seemed to me attractive. Maybe the fermion mass problem is not for the technicolour energy scale, and still Higgses may turn out to be composite of technifermions.

If this is the case, a rich new phenomenology opens up around the technicolour scale. First, a new spectroscopy of technihadrons is expected, a heavy replica of the standard QCD spectroscopy. Second, if not only a « techniquark doublet », but a complete technifamily is introduced (techniquarks + technileptons), one has a chiral symmetry which is larger than the $SU_{2,\text{left}} \times SU_{2,\text{right}} \times U_1$ considered above. The number of fermion species is in this case given by

$$\underbrace{\underset{\underset{\text{flavour}}{\downarrow}}{\overset{\text{colour}}{3*2}}}_{\text{techniquark doublet}} + \underbrace{\overset{\underset{\text{flavour}}{\hookrightarrow}}{2}}_{\text{technilepton doublet}} = 8 . \tag{2.22}$$

This leads to a $SU_{8,\text{left}} \times SU_{8,\text{right}} \times U_1$ chiral symmetry. When the axial symmetry is broken, one obtains Goldstone bosons corresponding to the following currents:

$$(2.23) \qquad J_\mu^{5a} = \bar{F} \gamma_\mu \gamma_5 T^a F,$$

where F is a column vector with eight components grouping techniquarks + + technileptons and T^a is a generator of the SU_8 group.

By denoting with Q the techniquark and with L the technilepton, one can classify the technipions corresponding to the currents, eq. (2.22), under $SU_3 \times \times SU_2$ by introducing a set of Gell-Mann (λ_i) and Pauli (τ) matrices as follows:

$$(2.24) \quad \begin{cases} \theta_a^i = \bar{Q}\gamma_5 \lambda_a \tau^i Q & \sim 8 \otimes 3, \\ \theta_a = \bar{Q}\gamma_5 \lambda_a Q & \sim 8 \otimes 1, \\ T_c^i = \bar{Q}_c \gamma_5 \tau^i L & \sim \bar{3} \otimes 3, \\ \bar{T}_c^i = \bar{L}\gamma_5 \tau^i Q_c & \sim 3 \otimes 3, \\ T_c = \bar{Q}_c \gamma_5 L & \sim \bar{3} \otimes 1, \\ \bar{T}_c = \bar{L}\gamma_5 Q_c & \sim 3 \otimes 1, \\ P^\pm = \bar{Q}\gamma_5 \tau^\pm Q - 3\bar{L}\gamma_5 \tau^\pm L & \sim 1 \otimes 3, \\ P_3 = \bar{Q}\gamma_5 \tau_3 Q - 3\bar{L}\gamma_5 \tau_3 L & \sim 1 \otimes 3, \\ P_0 = \bar{Q}\gamma_5 Q - 3\bar{L}\gamma_5 L & \sim 1 \otimes 1, \\ \pi_i = \bar{Q}\gamma_5 \tau_i Q + \bar{L}\gamma_5 \tau_i L & \sim 1 \otimes 3. \end{cases}$$

On the right-hand side of each state, I have given the corresponding classification under colour $\times$ weak isospin. Only the last states in the list, called π^i, are the ones which will appear as longitudinally polarized W's and Z's: all the others would remain massless if ordinary $SU_3 \times SU_2$ gauge forces were switched off. One can expect that their mass will be in general of order $\alpha_s \Lambda_{\text{TC}}$ or $\alpha_{\text{weak}} \Lambda_{\text{TC}}$ when they have nontrivial transformation properties under colour or weak isospin, respectively. Moreover, the state P^0 is a singlet under $SU_2 \times SU_3$ and will remain massless if noting is done (the same happens for the state P^3 but it is less straightforward to see it). These states (axions) require extra interactions to acquire a mass and are certainly a problem for the simplest technicolour scheme.

The pseudoscalars of eq. (2.23) have interesting decays:

$$(2.25) \quad \begin{cases} \theta_a \to q\bar{q}, & \theta_a^3 \to q\bar{q}, & \theta_a^\pm \to q\bar{q}. \\ \quad gg & g\gamma & gW^\pm \\ \quad g\gamma & gZ & ggW^\pm \\ \quad gZ & & \\ \quad ggg & & \end{cases}$$

Indeed they have two or three jets, jets + leptons (when a Z, W is involved) and jets + photons decay modes. These represent interesting signatures for collider experiments. Also the q$\bar{\mathrm{q}}$ decay mode will presumably be seen as a jet(s) + leptons event: indeed, the fact that the Yukawa-like coupling of pseudoscalars to fermions violates chirality implies that its effective strength should be of the order of m_t/F_π, vanishing when fermion masses are set to zero. Heavy pseudoscalars would then be mainly coupled to heavy fermions which, weakly unstable, would then give rise to semi-leptonic decays.

3. – Composite W, Z.

Technicolour introduces a new, strongly interacting, gauge force which produces the « vacuum condensation » replacing the usual spontaneous-symmetry-breaking mechanism of the Higgs potential.

ABBOTT and FARHI [9] explored the possibility that the Fermi scale is the one where again new strong interactions occur, but based this time on the « weak » isospin gauge group $SU_{2,L}$ itself. As we will see, this picture leads to composite W, Z and fermions.

The model contains (in the simplest version) a doublet of left-handed fermions and right-handed isosinglets. A complex doublet scalar is also present; in formulae:

$$(3.1) \quad \begin{cases} \psi^i_{L,a} \leftarrow \text{flavour} \\ \phantom{\psi^i_{L,a}} \leftarrow SU_{L,2} \\ \psi^i_R \leftarrow \text{flavour} , \\ \varphi_a \leftarrow SU_{L,2} . \end{cases}$$

The strong interactions of the $SU_{L,2}$ gauge bosons lead to composite states for which the « iso-colour » is screened. New left-handed fields are then formed by binding the complex scalar to the original left-handed fermions of (3.1). These states are

$$(3.2) \quad U^i_L \equiv \phi^*_a \psi^i_{L,a}, \qquad D^i_L \equiv \varepsilon_{ab} \phi_b \psi^i_{L,a},$$

where ε_{ab} is a 2×2 antisymmetric matrix with unit elements. The right-handed partners of the spinors, eq. (3.2), are the original fields, eq. (3.1), which were isosinglet from the beginning. Fermion masses are given by the Yukawa couplings of the original complex doublet:

$$(3.3) \quad \overline{U}_R \phi^*_a \psi_{L,a} \to \overline{U}_R U_L .$$

Besides the fermions, also the W's are composite: the $\vec{W}$ triplet is associated to bound states of two scalars as follows:

$$\phi^*_a D^{ab}_\mu \varepsilon_{bc} \phi^*_c , \qquad \phi_a \varepsilon_{ab} D^{bc}_\mu \phi_c , \qquad \phi^*_a \overleftrightarrow{D}^{ab}_\mu \phi_b ,$$

where D_μ^{ab} is the covariant derivative of the $\tilde{S}U_{L,2}$ gauge group. The relation between the electric charge and the hypercharge is now simply

$$Q = Y \qquad (3.4)$$

because, by construction, all low-energy states are isosinglets, i.e. they have $T_3 = 0$. For example, $U_L^i \equiv \phi_a^* \psi_{L,a}^i$ has hypercharge $\frac{1}{2}(\phi^+) + \frac{1}{6}(\psi_{L,a}^i) = \frac{2}{3} = Q$. In this model, there is *no* spontaneous symmetry breaking and the U_1 group of $SU_{L,2} \times U_1$ is directly the electromagnetic gauge group.

Besides the original $\tilde{S}U_{2,\text{left}}$ local invariance, the Lagrangian of the Abbott-Farhi model possesses a global $SU_{2,\text{vector}}$ invariance, provided electromagnetic interactions are switched off and Yukawa couplings of up and down quarks are the same. This extra symmetry can be defined as follows. Group the scalar doublet and its complex conjugate into a 2×2 matrix:

$$\Omega = \begin{pmatrix} \phi_+ & -\phi_0^* \\ \phi_0 & \phi_- \end{pmatrix}. \qquad (3.5)$$

The potential $V(\phi^* \phi) \equiv V[\text{Tr}(\Omega \Omega^+)]$ is then invariant under the unitary transformation

$$\Omega \to U \Omega U^+ . \qquad (3.6)$$

If, correspondingly,

$$\hat{A}_\mu \to U \hat{A}_\mu U^+ \qquad (3.7)$$

and

$$\begin{pmatrix} u \\ D \end{pmatrix} \to U \begin{pmatrix} u \\ D \end{pmatrix}, \qquad (3.8)$$

eqs. (3.6)-(3.8) are a set of transformations which leave the Lagrangian invariant. Only the operation (3.7) is the global version of the usual SU_2 local invariance; eq. (3.8) is a transformation which applies to both left- and right-handed fields and cannot be identified with a global version of the original $SU_{2,L}$. Under the global SU_2, the W's transform as a triplet; this guarantees that, in the absence of electromagnetic interactions, their masses will be the same. The general form of the low-energy effective Lagrangian (current-current type interaction) is

$$\mathcal{L} \sim \tilde{g}^2 \mathbf{J}_\mu \cdot \mathbf{J}^\mu + e^2 J_\mu^{\text{e.m.}} \frac{1}{q^2} J^{\text{e.m.},\mu} + \tilde{g}'^2 \cdot J_\mu^0 J^{0\mu} . \qquad (3.9)$$

The last term corresponds to the exchange of a composite global isosinglet state: Bose symmetry forbids the existence of such a state made out of two scalar, but it is always possible to form it out of two fermion fields. Maybe (*) this impies that its mass is heavier than the one of the isotriplet and then justifies its omission in the effective low-energy interaction.

As remarked before, electromagnetic interactions break the global SU_2 (members of the same doublet have different electric charges!) and the actual mass eigenstates in the neutral sector have to be recomputed after the breaking. A way to parametrize it is to introduce a phenomenological L_{breaking} which, however, should conserve the electromagnetic gauge invariance. This is simply achieved by introducing the electromagnetic field always in the combination entering $F_{\mu\nu}$ [10]. One gets

$$\mathscr{L}_{\text{breaking}} \sim -\tfrac{1}{2}\lambda \left[F^{\mu\nu} W_{0\mu\nu} + W_0^{\mu\nu} F_{\mu\nu}\right], \tag{3.10}$$

where $W_{0\mu\nu}$ is the analogue of the electromagnetic $F_{\mu\nu}$ but made with the neutral W boson. The total Lagrangian of the neutral sector looks like

$$\mathscr{L}^{\text{total}}_{\text{neutral sector}} = -\tfrac{1}{4} \Lambda^{\text{T}}_{\mu\nu} K \Lambda^{\mu\nu} - \Lambda^{\text{T}}_\mu M^2 \Lambda^\mu + \Lambda^{\text{T}}_\mu J^\mu, \tag{3.11}$$

where

$$\begin{cases} \Lambda_{\mu\nu} = \begin{pmatrix} F^{\mu\nu} \\ W_0^{\mu\nu} \end{pmatrix}, \\ \Lambda_\mu = \begin{pmatrix} A^\mu \\ W_0^\mu \end{pmatrix}, \\ J_\mu = \begin{pmatrix} eJ_\mu^{\text{e.m.}} \\ gJ_\mu^{W0} \end{pmatrix}, \end{cases} \tag{3.12}$$

K and M^2 are matrices:

$$M^2 = \begin{pmatrix} 0 & 0 \\ 0 & M^2 \end{pmatrix}, \qquad K = \begin{pmatrix} 1 & \lambda \\ \lambda & 1 \end{pmatrix}. \tag{3.13}$$

The new mass eigenstates can be obtained by looking at the inverse propagator matrix and by finding the values of q^2 at which one has zero eigenvalues:

$$D^{-1}(q^2) = \begin{pmatrix} q^2 & \lambda q^2 \\ \lambda q^2 & q^2 - M_W^2 \end{pmatrix}. \tag{3.14}$$

The product of the eigenvalues is given by the determinant

$$\det[D^{-1}(q^2)] = q^2 [q^2 - M_W^2 - \lambda^2 q^2]$$

(*) Personally, I do not really see why.

which is zero at

(3.15)
$$\begin{cases} q^2 = 0 & \text{(the photon)}, \\ q^2 = \dfrac{M_W^2}{1-\lambda^2} & \text{(the } Z^0\text{)}. \end{cases}$$

Rewritten in terms of the mass eigenstates, the low-energy effective Lagrangian of the neutral sector is

(3.16) $$\mathscr{L} = \frac{1}{2}\left\{ e^2 J_\lambda^{e.m.} \frac{1}{q^2} J^{e.m.,\lambda} + \frac{g^2}{1-\lambda^2}\left[J_\varrho^3 - \frac{e\lambda}{g}J_\varrho^{e.m.}\right]\frac{1}{q^2 - M_Z^2}\left[J_3^\varrho - \frac{e\lambda}{g}J_{e.m.}^\varrho\right]\right\}.$$

Provided one identifies $e\lambda/g \equiv \sin^2\theta_W$, one exactly recovers the form of the low-energy Lagrangian of the standard model.

However, Glashow-Weinberg-Salam theory says also

(3.17) $$M_Z^2 = \frac{M_W^2}{\cos^2\theta_W} \quad and \quad \frac{37.4 \text{ GeV}}{M_W(\text{GeV})} = \sin\theta_W,$$

which would imply

(3.18) $$\lambda = \sin\theta_W \quad and \quad e/g = \sin\theta_W.$$

The last two conditions are not automatic in the Abbott-Farhi model (only the product of the two has already been fixed); they are often referred to as the « unification condition ». If eqs. (3.18) are not both satisfied, one obtains a curve in the M_Z/M_W plane corresponding to the equation

(3.19) $$M_Z^2 = M_W^2 \frac{1}{1 - (M_W(\text{GeV})/37.4 \text{ GeV})^2 \sin^4\theta_W}.$$

Only by fixing $(M_W/37.4 \text{ GeV})^2 = 1/\sin^2\theta_W$, one gets a unique prediction for the M_W and M_Z values.

Is the unification condition understandable?

Intuitively, it has to be related to the *local* (and not global as in the Abbott-Farhi model) SU_2 gauge invariance. Two exercises [11] can make this statement a quantitative one. The first one consists in taking the high-q^2 limit of the current-current interaction, eq. (3.16) ($q^2 \gg M_Z^2$):

(3.20) $$\mathscr{L}_{\text{eff}} = \frac{1}{2q^2}\left\{\frac{e^2}{1-\lambda^2}J_\mu^\gamma J^{\gamma\mu} + \frac{e^2 + g^2 - 2eg}{1-\lambda^2}J_\mu^3 J^{\mu 3} + \frac{2e(e-\lambda g)}{1-\lambda^2}J_\mu^\gamma J^{\mu 3}\right\},$$

where

$$J_\mu^\gamma \equiv J_\mu^{e.m.} - J_\mu^3.$$

If there are no constraints between e/g and λ, one obtains a breaking of the $SU_2 \times U_1$ invariance even at high energies. Instead, the condition $e/g = \lambda$ sets to zero the last term in eq. (3.19) and provides (as in the standard model) an $SU_2 \times U_1$ invariant effective Lagrangian at high energies.

The second example is the calculation of the limit of large S (centre-of-mass total energy squared) for the $\nu + \bar{\nu} \to W^+ W^-$ cross-section: it can be seen that it violates unitarity unless the unification condition is implemented.

From the point of view of the strong-interaction dynamics of the Abbott-Farhi model, the condition $\lambda = \sin^2 \theta_W \sim 0.23$ can be seen as the statement of a very large mixing (compared to the γ-ρ mixing in QCD) between the photon and the Z^0. This phenomenon is often referred to as « strong W dominance »: one in fact can obtain the condition

$$(3.21) \qquad \frac{e}{g} = \lambda,$$

if one assumes that the electromagnetic form factor is entirely dominated by W exchange [12]. The corresponding diagram is in fig. 4. Other states

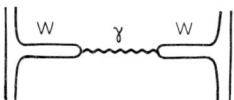

Fig. 4. – The diagram representing a strong W dominance.

can dominate the electromagnetic form factor besides the W and modify the condition eq. (3.21): what does it mean dynamically the fact that those higher excited states are far enough to leave eq. (3.21) unchanged?

Some understanding of this problem can be obtained by applying to this new strong-interaction dynamics the techniques known in QCD as the sum rule approach [13]. I will first explain the main ideas and then develop the analysis for the case of weak composite bosons. Generally speaking, one wants to relate the nonperturbative properties of the *bound states* of the theory (masses, wave function, ...) to the nonperturbative properties of the *vacuum* by introducing vacuum condensates like $\langle F_{\mu\nu} F^{\mu\nu} \rangle$, $\langle \bar{\psi}\psi \rangle$, ...

The method is based on three steps.

A) Choose an operator O_R with the quantum numbers of the resonance R to be studied and form the correlator

$$(3.22) \qquad \Pi(q^2) = \int d^4x \, \exp[iqx] \, \langle 0 | T[O_R(x) O_R^\pm(0)] | 0 \rangle,$$

where T means T-ordered product.

For $\Pi(Q^2)$ one can write an operator product expansion. If, for example, we are in the case of QCD and O_R is a fermion-antifermion local current, one gets the type of diagrams in fig. 5a) and b). The ones in fig. 5a) have a logarithmic

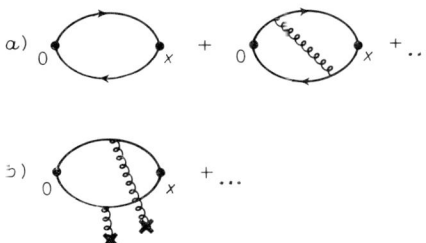

Fig. 5. – a) The perturbative contribution to the correlator, b) nonperturbative contributions to the correlator.

dependence upon q^2 and can be calculated for large q^2 in asymptotic QCD perturbation theory (where the coupling constant is small); the ones of fig. 5b) involve a knowledge of the vacuum expectation values of dimensionful operators and exhibit a power law behaviour with q^2.

B) Make use of the analyticity properties of the theory by writing a dispersion relation (possibly subtracted) for $\bar\Pi(q^2)$

$$(3.23) \qquad \Pi(q^2) \sim \frac{1}{2\pi i} \oint \frac{ds}{s+q^2} [\operatorname{Im} \Pi(s)] .$$

C) Parametrize $\operatorname{Im} \Pi(s)$ as a superposition of (zero width) resonances plus a continuum

$$(3.24) \qquad \operatorname{Im} \Pi(s) \sim \delta(s - m_R^2) f_R + \theta[S - S_0] \cdot \mathrm{const},$$

where the constant in eq. (3.24) is known from perturbation theory, and S_0, the «continuum threshold», is an extra unknown. By plugging C) + B) into the operator product expansion of A), one gets a relation among the vacuum expectation values and the perturbative contribution entering the expansion, on the one hand, and, on the other, the resonance parameters of eq. (3.24).

Physically, the method relies on a duality between resonances and the continuum: the latter realizes an «average» of the resonance contribution to the correlator, eq. (3.22). The average over the q^2 values where the resonance contribution is important is provided by the dispersive integral, eq. (3.23).

The previous machinery has been applied to the case in which the resonances are the W's of the Abbott-Farhi model [14]. A simple choice for the operator O_R is

$$(3.25) \qquad O_R = \tfrac{1}{2} \phi^* \overleftrightarrow{D}_\mu \phi ,$$

having the W_3 quantum numbers.

The operator product expansion of $\Pi(q^2)$ in this case contains « scaling » terms (logarithmically dependent upon q^2) coming from the diagrams like that in fig. 6a) and power law suppressed terms, diagrams 6b) and c). A particular

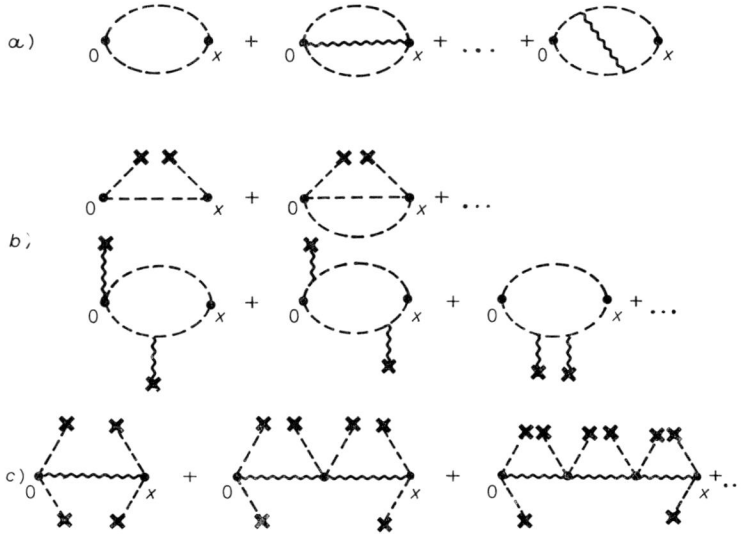

Fig. 6. – Perturbative (a)) and nonperturbative (b), c)) contributions to the correlator dominated by composite W exchange.

role among the latter diagrams is played by the ones in fig. 6c), where the original $SU_{2,L}$ gauge field propagates in the nonperturbative vacuum made by condensates of the scalar field.

The contribution to $\Pi(q^2)$ given by these diagrams is

$$(3.26) \qquad \Pi_{6c}(q^2) \sim \sum_{N=1}^{\infty} \frac{(g^2/2)^N}{(q^2)^N} \langle \phi^* \sigma^i \phi \phi^* \sigma^i \phi \, [\phi^*\phi]^{N-1} \rangle \,.$$

If a very specific factorization property of the matrix element, eq. (3.26), is assumed

$$(3.27) \qquad \langle \phi^*_{\alpha_1} \phi_{\beta_1} \cdots \phi^*_{\alpha_N} \phi_{\beta_N} \rangle = \frac{\langle \phi^*\phi \rangle^N}{(N+1)!} \sum_{\substack{\text{permutations}\\ \beta_i}} \delta_{\alpha_1 \beta_1} \cdots \delta_{\alpha_N \beta_N} \,,$$

then

$$(3.28) \qquad \langle \phi^* \sigma^i \phi \phi^* \sigma^i \phi (\phi^*\phi)^{N-2} \rangle \to \langle \phi^*\phi \rangle^N \,.$$

Note that the sum over permutations compensates for the $1/(N+1)!$ of eq. (3.27). The previous assumption, admittedly *ad hoc*, allows us to resum the series, eq (3.26), which now becomes a geometric series:

$$\Pi(q^2) = g^2 \langle \phi^* \phi \rangle \frac{1}{q^2 - (g^2/2)\langle \phi^* \phi \rangle} \, . \qquad (3.29)$$

From eq. (3.29), one reads off

$$M_{\rm W}^2 = (g^2/2)\langle \phi^* \phi \rangle \, . \qquad (3.30)$$

This relation is very close to the one of the standard model

$$M_{\rm W}^2 = (g^2/2)\langle \phi \rangle^2 \, . \qquad (3.31)$$

The difference is that now the vacuum expectation value is gauge-invariant and there is no spontaneous symmetry breaking. The other outcome of the factorization condition, eq. (3.27), and of the dominance of diagrams 6c) is that the whole nonperturbative part of the dispersion relation is saturated by a single resonance only: the W. Where is the « continuum » contribution? The technical steps to estimate the continuum scale by looking at which q^2 the perturbative (continuum) contribution equals the nonperturbative one proved to be very annoying(*) in the oral version of these lectures, and I decided to omit them in the written version.

The result is

$$\Lambda_{\rm continuum}^2 \sim 40 \, M_{\rm W}^2 \, . \qquad (3.32)$$

The continuum starts at a scale which is several times the W mass scale and confirms the picture of W dominance in the Fermi region. The picture gets, of course, modified when a factorization scheme different from eq. (3.27) is assumed: if, for example, one has instead the following equality

$$\langle \phi^* \sigma_i \phi \phi^* \sigma_i \phi (\phi^* \phi)^{N-2} \rangle = \frac{(N+1)!}{2^N} \langle \phi^* \phi \rangle^N \, , \qquad (3.33)$$

which is true in the large-N (N of SU_N) limit, one gets more than a single W. Still, the mass of the second one M_2 lies sufficiently far away from the lowest-lying state ($M_{\rm W}$)

$$M_2 \sim 3 M_{\rm W} \, . \qquad (3.34)$$

The above discussion is an attempt to find a dynamical translation of the « W dominance » idea: this can be obtained by assuming a specific factoriza-

---

(*) Students, private communication.

tion scheme for the vacuum expectation values of the diagrams where the original gauge bosons propagate as « free » in a nontrivial vacuum which, by screening the W's iso-colour, also provides it a mass.

4. – Composite fermions.

The Abbott-Farhi model not only predicts composite W's but also composite fermions. In this case, the natural question is: why does one get a very light fermion spectrum if the compositeness scale is in the TeV range? According to 't Hooft's concept of naturalness, fermion masses can be « naturally » small compared to 1 TeV because, by setting them to zero, one recovers an extra symmetry for the Lagrangian: the chiral symmetry. The main purpose of this section is to discuss how the chiral dynamics can impose constraints on the admissible states of a theory of composite fermions. These constraints go under the name of 't Hooft anomaly conditions [5].

Before stating these conditions, it is useful to recall a few basic points about chiral invariance. A massless fermion Lagrangian coupled to gauge fields

(4.1) $$\mathscr{L} = i\bar{\psi}\hat{D}\psi$$

can be rewritten in terms of left- and right-handed fields as

(4.2) $$\mathscr{L} = i[\bar{\psi}_L \hat{D}\psi_L + \bar{\psi}_R \hat{D}\psi_R],$$

where

$$\psi_{L,R} \equiv \frac{1 \mp \gamma_5}{2}\psi.$$

By choosing the following basis for the Dirac matrices

(4.3) $$\begin{cases} \gamma_0 = \begin{pmatrix} 0 & 1 \\ 1 & 0 \end{pmatrix}, & \gamma^i = \begin{pmatrix} 0 & \sigma_i \\ -\sigma_i & 0 \end{pmatrix}, \\ \gamma_5 = \begin{pmatrix} -1 & 0 \\ 0 & 1 \end{pmatrix}, \end{cases}$$

where **1** is unity 2×2 matrix and σ_i are the Pauli matrices, the Lagrangian, eq. (4.2), can be written as

(4.4) $$\mathscr{L} = \psi_L^{\dagger}[(-\boldsymbol{\sigma})\cdot(\boldsymbol{p} + g\hat{\boldsymbol{A}}) - g\hat{A}_0]\psi_L + \psi_R^{\dagger}[(\boldsymbol{\sigma})\cdot(\boldsymbol{p} + g\hat{\boldsymbol{A}}) - g\hat{A}_0]\psi_R,$$

where A_μ is the gauge vector field and g the corresponding coupling constant.

The point that I want to recall to you is that one can eliminate the right-handed fields from expression (4.4) in favour of left-handed, charge-conjugated fields. Define

(4.5) $\quad\begin{cases} \tilde{\psi}_L \equiv \psi_R^+ \sigma_2, \\ \tilde{\psi}_L^+ \equiv \sigma_2 \psi_R. \end{cases}$

Then

(4.6) $\quad \int \psi_R^+ [(\boldsymbol{\sigma}) \cdot (\boldsymbol{p} + g\boldsymbol{A} \cdot \boldsymbol{\tau}) - gA_0 \cdot \boldsymbol{\tau}] \psi_R =$
$= \int \tilde{\psi}_L^+ [(-\boldsymbol{\sigma}) \cdot (\boldsymbol{p} + g\boldsymbol{A} \cdot (-\boldsymbol{\tau}^T)) - gA_0 \cdot (-\boldsymbol{\tau}^T)] \tilde{\psi}_L,$

where the product between the vector field A_j and the generators of the gauge group t_j has been made explicit: $\sum_j A_j \tau_j \equiv \boldsymbol{A} \cdot \boldsymbol{\tau}$. The expression on the right-hand side of eq. (4.6) contains $\tilde{\psi}_L$ fields but with generators $-\tau^T$ (T ≡ transposed) instead of τ, transforming then under the conjugate representation. Right-handed fields can then be eliminated in favour of new left-handed ($\tilde{\psi}_L$) fields forming a conjugate representation of one of the original right-handed fields. If all representations (including the conjugate ones) are present, the most general form of a chiral Lagrangian is

(4.7) $\quad \sum_{r,\bar{r}} \sum_{i=1}^{n_r} \psi_{L,r_i}^+ [(-\boldsymbol{\sigma})(\boldsymbol{p} + g\boldsymbol{A} \cdot \tau_{r_i}) - gA_0 \cdot \tau_{r_i}] \psi_{L,r_i},$

where r and $\bar{r}$ are conjugate representations.

The general global symmetry of the above Lagrangian is

(4.8) $\quad \psi_{L,i} \to U_{ij} \psi_{L,j},$

where U is a unitary transformation for each representation r_i of the gauge group and the index i runs over the number of «flavours».

Now comes the anomaly.

In a «handed» theory where both vector and axial currents are present, the axial-current conservation is spoiled by quantum effects. The diagrams responsible for the anomaly in lowest order (and, by topological reason, giving the correct answer to all orders in perturbation theory) are depicted in fig. 7,

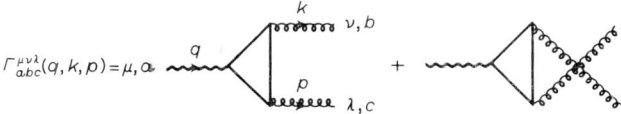

Fig. 7. – The Lorentz/colour structure of the triangle diagram contributing to the anomaly.

where also the notation for the Lorentz and « colour » indices and four-momenta is given. The vertices of the triangle are saturated with two-vector (on the right) and one-axial (on the left) currents. At $p^2 = k^2 = 0$, the general Lorentz decomposition of the effective vertex is

(4.9) $\Gamma_{\mu\nu\lambda} \sim A_1(q^2)\,\varepsilon_{\mu\nu\lambda\alpha}(k-p)^\alpha + A_2(q^2)\,q_\mu\,\varepsilon_{\nu\lambda\alpha\beta}\,k^\alpha p^\beta +$
$+ A_3(q^2)\,[k_\nu\,\varepsilon_{\mu\lambda\alpha\beta}\,k^\alpha p^\beta - p_\lambda\,\varepsilon_{\mu\nu\alpha\beta}\,k^\alpha p^\beta] + A_4(q^2)\,[k_\lambda\,\varepsilon_{\mu\nu\alpha\beta}\,k^\alpha p^\beta - p_\nu\,\varepsilon_{\mu\lambda\alpha\beta}\,k^\alpha p^\beta]\,.$

The colour structure is proportional to

(4.10) $$\text{Tr}\,[t^a\,\{t^b, t^c\}]\,.$$

The anticommutator comes from the Bose symmetry of the two vector currents. (The expression, eq. (4.9), is symmetric in the simultaneous exchange of $\nu \leftrightarrow \lambda$ and $k \leftrightarrow p$.) Note that the presence of the antisymmetric tensor is forced by the odd parity of the effective vertex containing two-vector and one-axial currents.

The contribution to the anomaly comes entirely from the first term, eq. (4.9):

(4.11) $$q_\mu \Gamma_{\mu\nu\lambda} \underset{|q^2|\to 0}{\neq} 0 \propto A_1\,.$$

We are almost ready to state 't Hooft's anomaly conditions: i) Call « meta-colour » the force responsible for the compositeness of fermions, and « meta-flavour » any remaining symmetry (gauged or not). (For example, meta-flavour might be the set of gauge forces of the standard model.) ii) The theory, if the meta-flavour symmetry is « handed », can have anomalies (including the value zero, *i.e.* no anomalies at all), coming from meta-flavour currents.

Then (here come the conditions...), if *chirality is unbroken*,

1) the anomaly must be the same no matter whether in the diagrams of fig. 7 one puts the fundamental or the composite fermion fields,

2) the matching must work independently of the size of the meta-flavour group (persistent mass condition).

The first condition can be pictorially represented by the « equation » in fig. 8. If chirality is broken, the anomaly will be matched in general by the

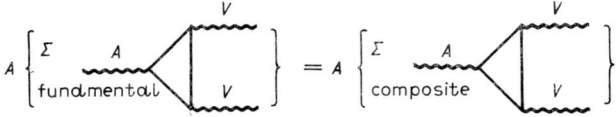

Fig. 8. – A pictorial representation of the anomaly matching condition.

Goldstone bosons associated to the spontaneous symmetry breaking. This can be seen by considering the condition on the effective vertex from vector current conservation [15]:

(4.12) $$k_\nu \Gamma^{\mu\nu\lambda} = 0,$$

which implies

(4.13) $$A_1(q^2) = q^2 \frac{A_4(q^2)}{2}$$

and, by writing a dispersion relation for A_4 in the « q^2 » channel,

(4.14) $$A_1(q^2) = \frac{q^2}{2} \frac{1}{2\pi i} \oint ds \frac{\text{Disc } A_4(s)}{s - q^2}.$$

We know that A_1 is real (because the value of the anomaly is real): then

(4.15) $$\text{Disc } A_1 = q^2 \text{ Disc } A_4 = 0.$$

The second identity, eq. (4.15), and eq. (4.14) cannot be made consistent unless

(4.16) $$\text{Disc } A_4(q^2) = c\delta(q^2),$$

where the coefficient c is fixed by the anomaly. There are two possibilities of having a polelike behaviour in the discontinuity of A_4 in the « q^2 » channel: either one has zero-mass pseudoscalar states coupled to the axial current carrying momentum q (Goldstone bosons), or *zero-mass* fermions are coupled to the current. The first possibility corresponds to a spontaneously broken chiral symmetry, the second to a conserved one: the anomaly always manages to stay the same, but, in the second case, zero-mass fermions (composite or not) are needed.

It was impossible for 'T HOOFT (and for the others) to satisfy the anomaly constraints for SU_N meta-colour groups with $N \geq 3$. This could be taken as a serious indication for the necessity of spontaneous chiral-symmetry breaking in QCD. For $N = 2$, solutions exist and one clear example is provided by the Abbott-Farhi model. Let us rewrite the fermion fields of the model:

(4.17) $$\psi^a_{L,i} \begin{smallmatrix} \leftarrow \text{meta-flavour} \\ \leftarrow \text{meta-colour} \end{smallmatrix}.$$

For three families, the meta-flavour index a ranges from 1 to 12 ((number of colours $+ 1$) * number of families) and the meta-colour one i from 1 to 2.

Right-handed fields are « spectators » of the confinement game of $SU_{2,L}$.

The composite fields, eq. (3.2), have the same meta-flavour index, they do not have any $SU_{2,L}$ index (they are isosinglets), but they come in doublets.

Therefore, as far as the meta-flavour currents and their anomalies are concerned, there is always a quark doublet for each family and the 't Hooft conditions are satisfied.

Note in particular that the matching works independently of the number of families (persistent mass condition). Why does a solution exist in this case? The physical reason is that the isocolour of left-handed fermions could be « screened » by the scalars: the number of states has been conserved, while their symmetry properties have been « dressed » (the $SU_{2,L}$ local doublets become SU_2 global doublets). Can one have screening with gluons? Yes, in models where the meta-colour gauge group is O_n, as discussed by BARBIERI et al. in ref. [16]. I will now explain how they work.

The fundamental fermion fields (preons), denoted by χ and all left-handed, are

$$(4.18) \qquad \chi^i_\alpha ,$$

where i, the meta-colour index, runs from 1 to n (for a O_n meta-colour group) and α, the meta-flavour one, goes from 1 to N (capital n!): the model possesses an SU_N symmetry. They transform as the vector representation on O_n and as the fundamental one of SU_N.

I will now specialize for simplicity to O_3; in this case, the possible composite states are made of

i) one « preon » plus meta-gluons,

$$(4.19) \qquad \chi^i_\alpha F^j_{\mu\nu} F^k_{\mu\nu} \varepsilon^{ijk} , \qquad \chi^i_\alpha F^j_{\mu\nu} F^k_{\varrho\sigma} \varepsilon_{\mu\nu\varrho\sigma} \varepsilon^{ijk} \dots ;$$

ii) three preons

$$(4.20) \qquad \chi^i_\alpha \chi^j_\beta \chi^k_\gamma \varepsilon^{ijk} .$$

These two classes of states transform under meta-flavour as

$$(4.21) \qquad \begin{cases} \text{i)} \ T_\alpha , \\ \text{ii)} \ T_{\alpha\beta\gamma} . \end{cases}$$

The first one is again the fundamental representation of SU_N, the second one is a reducible representation, given that there are no symmetry conditions among the indices $\alpha\beta\gamma$.

For each representation r of $SU_{N,\text{meta-flavour}}$, one can calculate the value of the anomaly and impose

$$(4.22) \qquad A(\text{preons}) = \sum_r A[l(r)] ,$$

where $l(r)$ is the number of composite (left-handed) states in the representation r and $A[l(r)]$ the corresponding value of the anomaly. Table I contains, for each type of composite state, the decomposition under SU_N representation and the corresponding value of the anomaly. The last column contains the labelling of the free parameter (number of states for each representation) which have to be fixed to satisfy the anomaly condition.

TABLE I. – *The value of the anomaly for different SU_N meta-flavour representations for states made of one (χ) or three ($\chi\chi\chi$) preons.*

Composite state	SU_N representation	SU_N anomaly	$l(r)$
χ_α + metaglue	T_α	1 (by definition)	f
$\chi_\alpha \chi_\beta \chi_\gamma$	$T_{[\alpha\beta\gamma]}$	$(N^2 - 9N)/2 + 9$	a
	$T_{\{\alpha\beta\gamma\}}$	$(N^2 + 9N)/2 + 9$	b
	$T_{\{[\alpha\beta]\gamma\}}$	$N^2 - 9$	c
$\chi^\alpha \chi_\beta \chi_\gamma$	$T^\alpha_{[\beta\gamma]}$	$(N^2 - 7N)/2$	d
	$T^\alpha_{\{\beta\gamma\}}$	$(N^2 + 7N)/2$	e

The preon anomaly value (the left-hand side of eq. (4.22)) is

$$A(\text{preons}) = n \tag{4.23}$$

since there are n states in the fundamental representation of O_n (for which the anomaly is 1 by definition).

Before matching the preon anomaly with the composite one, we derive the conditions among the parameters of table I which ensure the independence of the result from N (the size of the meta-flavour group):

$$a = b, \quad d = e, \quad c + a + e = 0. \tag{4.24}$$

By eliminating part of the parameters with eqs. (4.24), the anomaly matching condition, eq. (4.22), reads

$$n = f + 27a + 9e. \tag{4.25}$$

The general solution is the following superposition of states

$$nT_\alpha + l_1[T_{[\alpha\beta\gamma]} \oplus T_{\{\alpha\beta\gamma\}} \oplus T^{\{[\alpha\beta],\gamma\}} \oplus 27T^\alpha] + \\ + l_2[T^\alpha_{[\beta\gamma]} \oplus T^\alpha_{\{\beta\gamma\}} \oplus T^{\{[\alpha\beta]\gamma\}} \oplus 9T^\alpha], \tag{4.26}$$

where the coefficients 27 and 9 inside the square brackets represent the multiplicity of the neighbour representation. The coefficients l_1 and l_2 are arbitrary, for the states inside the square brackets are anomaly free.

For special values of n, one can saturate the anomaly with three preon states only, but the simplest case is the one where the anomaly is entirely matched by the first term in eq. (4.26). This is again, like in the Abbott-Farhi case, a « screening model »: meta-gluons have screened the original preon meta-colour. The anomaly condition then predicts that exactly n massless states must form: the original O_n local symmetry has been dressed into a « family symmetry »; the low-energy spectrum contains in fact n states with identical meta-flavour properties.

The real problem of all these models is a satisfactory answer to the question: how do these massless states acquire a mass? Meta-colour cannot provide a mass, given that chiral symmetry was assumed to remain unbroken under meta-colour: the only possibility is meta-flavour, but no appealing (to me) scheme has been found for the mass generation.

The 't Hooft anomaly conditions represent a sharp constraint imposed on any underlying dynamics and one of the few solid results in the field of composite models.

5. – Phenomenology.

The way the new physics can manifest itself experimentally at energies below the compositeness scale is

  i) by producing new effective interactions among ordinary particles,

  ii) by the appearance of « excited » states which are partners of the observed particles.

The first type of experimental signals can be discussed in the framework of effective Lagrangians. The ordinary standard-model Lagrangian gets modified by form factors which spoil the pointlike character of the original interactions and by the addition of operators of dimensions higher than four. The presence of form factors has been parametrized at PETRA by the factor

(5.1) $$F(q^2) = \frac{1}{1 + q^2/\Lambda^2}$$

which multiplies the strength of pointlike electromagnetic interactions. The bound for Λ which is obtained is

(5.2) $$\Lambda \gtrsim 200 \text{ GeV},$$

a lousy bound.

Higher-dimension operators can have effects at low energy and high energies. An example for low energies [17] is given by the unseen reaction $\mu \to e\gamma$: if both the muon and the electron are made by the same preons, meta-gluon exchanges at distances of order $1/\Lambda_{\text{compositeness}} \equiv 1/\Lambda_c$ like the ones imagined on fig. 9 give rise to a magnetic-type transition of the type

$$(5.3) \qquad \mathscr{L}_{\text{eff}}^{\mu \to e\gamma} = \bar{e}_L \sigma^{\mu\nu} F_{\mu\nu} \mu_R \frac{e}{\Lambda_c} \frac{m_\mu}{\Lambda_c},$$

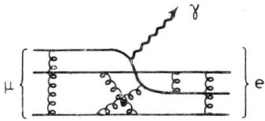

Fig. 9. – Diagram contributing to the transition of $\mu \to e\gamma$ if they were composite.

where e is the electric charge and m_μ the muon mass. The presence of the extra factor m_μ/Λ_c, besides the $1/\Lambda_c$ required by dimensional reasons, comes from an argument of chiral invariance. Indeed, the interaction, eq. (5.3), breaks the chiral invariance and it ought vanish in the limit of massless fermions. Similarly, one obtains corrections to the $g-2$ of the electron (muon) coming from a term

$$(5.4) \qquad \mathscr{L}_{\text{eff}}^{g-2} = \bar{e}_L \sigma^{\mu\nu} F_{\mu\nu} e_R \frac{e}{\Lambda_c} \frac{m_e}{\Lambda_c}.$$

The limits for Λ_c resulting from the experimental bounds for $\mu \to e\gamma$ and $g-2$ are

$$(5.5) \qquad \begin{cases} g-2 \to \Lambda_c \geqslant 800 \text{ GeV}, \\ \mu \to e\gamma \to \Lambda_c \geqslant 200 \text{ TeV}! \end{cases}$$

Obviously, the second limit rules out the possibility of compositeness « nearby » the present energies. If one wants to evade it, one must have i) muons and electrons *not* made by the same preons (this, however, would mean, for me, that the family problem is not going to be solved by compositeness: a very unpleasant case...), or ii) a new dynamical « GIM » mechanism preventing again the flavour-changing neutral currents.

Effective four-fermion interactions generated by meta-colour can interfere with ordinary four-fermion amplitudes at high energies and produce observable effects. The coupling of the new interactions is of the order of $1/\Lambda_c^2$: the best effects will then be obtained when the interference occurs with weak or electromagnetic interactions. The results that I will present are all taken from the review talk given at the Moriond Conference of 1984 by

RÜCKL [18]. The chosen form of the effective Lagrangian is

$$\mathcal{L}_{\text{eff}}^{\text{4 fermions}} \sim \frac{2\pi}{\Lambda_c^2} [J_{V,A}^\mu \cdot J_{\mu V,A}], \tag{5.6}$$

where the current can be of vector or axial type. The result for the bounds for Λ_c are summarized in table II.

TABLE II. – *Limits on the compositeness scale Λ (see text) coming from different reactions with present and future machines.*

Reaction	Limits for Λ, present machines	Limits for Λ, foreseeable machines
$e^+e^- \to e^+e^-$	$\Lambda > 0.75$ TeV (LL, RR)	$\Lambda > 2$ TeV (LL, RR)
	$\Lambda > 1.5$ TeV (VV, AA)	$\Lambda > 5$ TeV (VV, AA) (LEP)
$\nu N \to \nu +$ all	$\Lambda \geqslant 2.5$ TeV (RL)	
$ep \to e$ all		$\Lambda \geqslant 3.4$ TeV (HERA)

Much less sensitivity is obtained with pp ($\bar{p}p$) machines where the interference occurs with strong interactions.

We move to the last topic of my lectures: the signal for excited quarks and leptons. Again, I will only discuss a particular form of the couplings of the new fermions to the ordinary ones [19, 20]:

$$\mathcal{L}_{\text{leptons}} = \frac{gf}{M}\left(\bar{L}\sigma_{\mu\nu}q^\nu \frac{\tau}{2} l_L + \text{h.c.}\right)\vec{W}^\mu + \frac{g'f'}{M}\left(-\frac{1}{2}\bar{L}\sigma_{\mu\nu}q^\nu l_L + \ldots\right)B^\mu, \tag{5.7}$$

where $L \equiv \begin{pmatrix} N \\ E \end{pmatrix}$ is a *vector* doublet of excited leptons and l_L is the ordinary lepton doublet.

$$\mathcal{L}_{\text{quarks}} = \frac{g_s f_s}{M}\left(\bar{q}^* \sigma_{\mu\nu} q^\nu \frac{\lambda^i}{2} q_L\right) G^{\mu,i} + \mathcal{L}_{\text{leptons}}(L \leftrightarrow q^*; l_L \leftrightarrow q_L), \tag{5.8}$$

where $q^* \equiv \begin{pmatrix} U^* \\ D^* \end{pmatrix}$ is an excited quark doublet and $q_L \equiv \begin{pmatrix} u_L \\ d_L \end{pmatrix}$ is the ordinary quark doublet. The coupling constants g, g', g_s are the usual ones of electroweak and gauge bosons $\vec{W}_\mu$, B_μ and of gluons G^i_μ; f, f', f_s are phenomenological parameters. The search for these states at present collider energy is feasible: to show this, I will try to be provocative and convince you that maybe these states *have been* found in the « strange collider events » where a (maybe) anomalous inclusive production of jets and/or leptons is observed. These events can be divided into

   i) leptonic like
      a) $e^+e^-\gamma$, $\mu^+\mu^-\gamma$,
      b) missing energy plus an energetic gamma;
   ii) semi-leptonic:
      a) missing energy + jet(s),
      b) missing energy + electron + jet(s);
   iii) hadronic: jet-jet.

I will briefly discuss, given the hypothesis of excited leptons and quarks,

   i) the signature of their decays,
   ii) the relative normalization for different decay modes,
   iii) the absolute production rate.

First, the excited leptons.
Their decay through the coupling eq. (5.7) is

(5.9) $$E \to e\gamma, \quad N \to \nu\gamma.$$

They can be produced, again by the couplings, eq. (5.9), by the Z in association with a normal lepton:

(5.10) $$\begin{cases} Z^0 \to e^+ E^-, \\ \hookrightarrow e^-\gamma \\ \to \gamma \bar{N} \\ \hookrightarrow \bar{\nu}\gamma \end{cases}$$

Given the total number of Z^0 collected up to now, one expects about 2÷3 events of the $e^+e^-\gamma$ type and one or two of the type $\nu\bar{\nu} + \gamma$, *i.e.* missing energy recoiling against an energetic gamma. The masses of the excited leptons are taken to be of the order of 75 GeV. The amount of events $W \to E\nu$
$\hookrightarrow e\gamma$
$= e\nu\gamma$ can be kept below unity given the freedom in choosing the couplings f and f'.

There is a main problem though: the angular configuration of the events is very unlikely in this interpretation.

Excited-quark decays can give rise to many different signatures:

(5.11)
$$\begin{cases} q^* \to qg \to \text{jet-jet}, \\ \to q\gamma \to \text{jet-}\gamma, \\ \to qW \to \text{jet} + e + \nu, \\ \hookrightarrow e\nu \\ \to qZ \to \text{missing energy} + \text{jet}, \\ \hookrightarrow \nu\bar{\nu} \\ \to qZ \to \text{jet} + \text{jet} + \text{jet}. \\ \hookrightarrow q\bar{q} \end{cases}$$

The main features are the production of leptons associated to *one jet*, the existence of a decay γ + jet and of three-jet events.

The relative normalizations of the different decay channels are summarized in fig. 10 for an initial absolute normalization of hundred jet-jet events. This rate for the presently total collected luminosity of about 130 inverse nanobarns is realistic and can be obtained by choosing $f_s = 1$. The total number

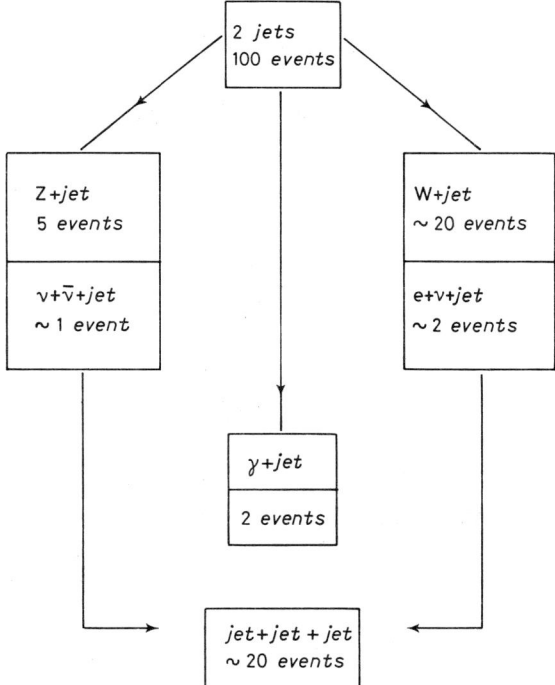

Fig. 10. – The relative normalization of various events at p$\bar{\text{p}}$ collider when excited quarks are produced.

of « strange events » is very small and it is impossible to draw any conclusion from them: moreover, some of them, like the ones of the type ii), have sometimes more jets associated to the missing energy or missing energy plus a lepton signal. The production of many jets and leptons is hard to obtain with the excited-quark production: this could come from the decay on an up-type excited quark into an ordinary top quark which would then give rise to many jets in its decay. However, the mechanism would certainly imply a dangerous rate of flavour-changing neutral currents. The best signature of excited quarks is the respectable production rate of events with a signe photon plus a jet, which have not (yet?) been seen by the UA's experiments.

Many of the signatures of excited-quark decays are actually shared by the pseudo-Goldstone bosons of technicolour, eq. (2.24). Given their preference to couple to heavy quarks, they could more easily produce multijet events: the production rate would be comparable to the one of excited quarks in the case of gluon-gluon fusion into the coloured octet states in the first two lines of eqs. (2.24).

6. – Concluding remarks.

Composite Higgses are certainly appealing theoretically and can be accommodated with the technicolour idea. Composite W's and Z look improbable if the standard-model mass formulae will turn out to be exact, *including* the calculated radiative corrections. For composite fermions, there are no objections to a compositeness scale Λ_c of the order of $(1 \div 2)$ TeV, with the exception of the limits coming from $\mu \to e\gamma$ which would naively be like $\Lambda_c \gtrsim 200$ TeV. Excited fermions could produce a very rich and distinct phenomenology already at present CERN collider energies, if their mass would be below the 200 GeV.

Compositeness is a way of answering the question: what happens at energies just above the Fermi scale? I mentioned in the introduction that another way is to introduce supersymmetry. Maybe both will occur at the same time: the advantages of a picture of supersymmetric compositeness will be the subject of Buchmüller's lectures.

REFERENCES

[1] S. L. GLASHOW: *Nucl. Phys.*, **22**, 579 (1961); S. WEINBERG: *Phys. Rev. Lett*, **19**, 1264 (1967); A. SALAM: in *Proceedings of the VIII Nobel Symposium, Aspenäsrgarden, 1968*, edited by N. SVARTHOLM (Almqvist and Wiksells, Stockholm, 1968), p. 367; C. BOUCHIAT, J. ILIOPOULOS and PH. MEYER: *Phys. Lett. B*, **38**, 519 (1972); S. L. GLASHOW, J. ILIOPOULOS and L. MAIANI: *Phys. Rev. D*, **2**, 1285 (1970). For a review of the present status of the standard model, see P. LANGACKER: to be published.

[2] M. KOBAYASHI and T. MASKAWA: *Prog. Theor. Phys.*, **49**, 652 (1973).
[3] For GUT reviews, see P. LANGACKER: *Phys. Rep. C*, **72**, 185 (1981); J. ELLIS: in *Gauge Theories in High Energy Physics, Les Houches, Session XXXVII, 1981*, edited by M. K. GAILLARD and R. STORA (North-Holland Publ. Co., Amsterdam, 1983), p. 161; D. V. NANOPOULOS: rapporteur talk at *The XXII International Conference on High Energy Physics, Leipzig, July 19-25, 1984*.
[4] See, for example, J. ELLIS: *Supersymmetry. Spectroscopy of the future? or of the present?*, CERN preprint TH. 4017 (1984).
[5] G. 'T HOOFT: in *Recent Developments in Gauge Theories, Cargèse 1979*, edited by G. 'T HOOFT *et al.* (Plenum Press, New York, N. Y., London, 1980).
[6] S. WEINBERG and E. WITTEN: *Phys. Lett. B*, **96**, 59 (1980).
[7] « Technicolour » is a trade mark; about the question of the exact spelling, see ref. [19] of *Phys. Rep.*, **74**, 278 (1981) by E. FARHI and L. SUSSKIND. For an introduction to the physics, read the remaining part of the quoted reference.
[8] J. GOLDSTONE: *Nuovo Cimento*, **19**, 15 (1961).
[9] L. F. ABBOTT and E. FARHI: *Phys. Lett. B*, **101**, 69 (1981); *Nucl. Phys. B*, **189**, 547 (1981).
[10] See, for example, the lectures by J. J. SAKURAI at the XVII Rencontre de Moriond, Les Arcs (March 1982).
[11] P. Q. HUNG and J. J. SAKURAI: *Nucl. Phys. B*, **143**, 81 (1978).
[12] See, for example, D. SCHILDKNECHT: in *Proceedings of the VIII International Workshop on Weak Interactions and Neutrinos, Javea, Spain, 1982*.
[13] M. A. SHIFMAN, A. I. VAINSHTEIN and V. I. ZAKHAROV: *Nucl. Phys. B*, **147**, 385, 448, 519 (1979).
[14] S. NARISON: CERN preprint TH. 3871 (1984); H. G. DOSCH, M. KREMER and M. G. SCHMIDT: Heidelberg preprint (1984) (I have followed in detail the latter reference).
[15] Y. FRISHMAN, A. SCHWIMMER, T. BANKS and S. YANKIELOWICZ: *Nucl. Phys. B*, **177**, 157 (1981).
[16] R. BARBIERI, L. MAIANI and R. PETRONZIO: *Phys. Lett. B*, **96**, 63 (1980).
[17] M. PESKIN: in *Proceedings of the International Symposium on Lepton and Photon Interactions at High Energies* (1981).
[18] For a review, see R. RÜCKL: CERN preprint TH. 3897 (1984).
[19] N. CABIBBO, L. MAIANI and Y. SRIVASTAVA: *Phys. Lett. B*, **139**, 459 (1984).
[20] A. DE RÚJULA, L. MAIANI and R. PETRONZIO: *Phys. Lett. B*, **140**, 253 (1984).

Some Aspects of Supersymmetric Composite Models of Quarks and Leptons.

W. BUCHMÜLLER

CERN - Geneva, Switzerland

1. – Are quarks and leptons composite?

The topic of supersymmetric preon models is very speculative. There is neither experimental evidence for supersymmetry or quark-lepton substructure [1] nor does a satisfactory theoretical model exist. Yet the considerable amount of recent work [2] on this subject is not without motivation: it is based on the belief that the Higgs sector of the standard model is only a low-energy effective Lagrangian and the experience that the dynamical understanding of a mass spectrum generally involves more fundamental constituents. Indeed, focusing on the family replication and the quark-lepton mass spectrum, it seems difficult to escape the problem of quark-lepton substructure. If quarks and leptons are composite, however, their structure must be very different from the bound states we know. Contrary to atoms, nuclei and hadrons, quarks and leptons are very pointlike, *i.e.* their size $r_{q,l}$ is much smaller than their Compton wavelength:

$$\xi_{q,l} = \left(\frac{1}{m_{q,l}}\right)\frac{1}{r_{q,l}} \gg 1 \ . \tag{1}$$

This inequality represents the main dynamical problem of composite quarks and leptons. 'T HOOFT [3] has shown that unbroken chiral symmetries imply the existence of massless composite fermions. Furthermore, it appears that the most interesting models which satisfy 't Hooft's consistency conditions require spin-0 preons in addition to spin-$\frac{1}{2}$ preons. Fundamental scalars, however, are « unnaturally » light unless they are part of a supersymmetric theory. Thus one is led to supersymmetric theories as the most promising candidates for a theory of quark-lepton substructure. Indeed, as we will see in the next section, supersymmetric confining gauge theories lead almost unavoidably to light composite fermions.

An important issue in the context of composite models is the nature of the weak interactions. It is conceivable that the W vector bosons are also composite and that the substructure scale is related to the Fermi scale, *i.e.* $1/r_{q,1} \sim G_F^{-\frac{1}{2}} \sim 300$ GeV. It is believed [1, 2] that composite W-bosons are consistent with the neutral-current phenomenology as well as the successful mass predictions of the standard model if the heavy bound states predicted by the preon theory have masses of the order of 1 TeV.

In these lectures we will discuss a supersymmetric toy model in which the left-handed particles of one family and the W-bosons are bound states, and we will use it to illustrate some techniques which are important in the context of supersymmetric composite models. In sect. **2** we discuss the idea of quasi-Goldstone fermions and the structure of their residual interactions. Section **3** deals with the coset space $U_6/SU_2 \times U_4$, and in sect. **4** a particular preon model is described which realizes this symmetry breaking. Section **5** contains some remarks on how the U_6 toy model may be extended to a more realistic theory and some constraints on models with composite W-bosons are listed.

2. – Quasi-Goldstone fermions.

The only known bound states, which are light compared to their inverse size, are the pions for which one has $\xi_\pi \sim 2.0$ [4] (cf. eq. (1)). They arise as pseudo-Goldstone bosons from the spontaneous breaking of chiral invariance. As the Goldstone mechanism plays a crucial role in supersymmetric preon models, let us briefly recall some features of dynamical symmetry breaking in QCD. In the case of two flavours $u_{L,R}$ and $d_{L,R}$ the QCD Lagrangian possesses the (approximate) global symmetry

(2) $$G = SU_{2,L} \times SU_{2,R} \times U_{1,V},$$

where the two SU_2 subgroups are generated by the charges

$$Q_{L,R}^A = \tfrac{1}{2}(T^A \mp X^A),$$

with

(3) $$\begin{cases} T^A = \int d^3x\, \bar{q}(x)\gamma_0 \frac{1}{2}\tau^A q(x), \\ X^A = \int d^3x\, \bar{q}(x)\gamma_0\gamma_5 \frac{1}{2}\tau^A q(x), \end{cases} \qquad q = \begin{pmatrix} u \\ d \end{pmatrix}.$$

The formation of the vacuum expectation values

(4) $$\langle 0|\bar{u}u|0\rangle = \langle 0|\bar{d}d|0\rangle \sim \Lambda_{QCD}^3$$

breaks the chiral symmetry G dynamically to the diagonal subgroup $H = SU_{2,V} \times U_{1,V}$. The spontaneous breaking of the axial generators X^A leads to an isotriplet of (almost) massless (pseudo) Goldstone bosons, the pions,

$$(5) \qquad \pi^A(x) \sim \bar{q}(x) \gamma_5 \tfrac{1}{2} \tau^A q(x) \,.$$

It is a crucial feature of the Goldstone phenomenon that the interactions of the Goldstone bosons at energies small compared to Λ_{QCD} are determined entirely by the coset space G/H [5] and do not depend on details of the dynamics of the underlying theory. In order to obtain the effective low-energy Lagrangian, one constructs a representation of the full group G acting on the pion fields where the broken generators X^A are realized nonlinearly,

$$(6) \qquad \begin{cases} \dfrac{1}{i}[T^A, \pi^B] = \varepsilon_{ABC} \pi^C \,, \\[6pt] \dfrac{1}{i}[X^A, \pi^B] = 2f_\pi \delta^{AB} + \dots \,, \end{cases}$$

and $f_\pi \sim \Lambda_{\text{QCD}}$ is the pion decay constant:

$$(7) \qquad \langle 0|\bar{q}(0) \gamma_\mu \gamma_5 \tfrac{1}{2} \tau^A q(0)|\pi^B(p)\rangle = ip_\mu \delta^{AB} f_\pi \,.$$

The effective Lagrangian describing the $\pi\pi$ interaction is now obtained by demanding that its variation with respect to T^A and X^A is a total derivative. This yields the result [5]

$$(8) \qquad \mathscr{L}_{\text{eff}} = \frac{1}{2} \partial^\mu \pi^A \partial_\mu \pi^A - \frac{1}{f_\pi^2} \pi^A \pi^A \partial^\mu \pi^B \partial_\mu \pi^B + O(\pi^6) \,,$$

which incorporates the low-energy theorems of current algebra.

It is expected that in general also in supersymmetric (SUSY) confining gauge theories dynamical symmetry breaking will take place. SUSY gauge theories are built from chiral superfields $\varphi_i = (\chi_i, \eta_{Li})$, containing complex scalars χ_i and left-handed Weyl fermions η_{Li}, and vector superfields $V^I = (\lambda_L^I, V_\mu^I)$, containing Weyl fermions λ_L^I and vector bosons V_μ^I. Vacuum expectation values of scalar fields

$$(9) \qquad \langle 0|\chi_i \chi_j|0\rangle \sim \Lambda_{\text{HC}}^2 \,,$$

which are of the order of the hypercolour scale Λ_{HC}, can break the symmetry G of the Lagrangian to a subgroup H. Due to supersymmetry, the resulting Goldstone bosons have to be part of chiral superfields $\varphi_i = (\varphi_i, \psi_{Li})$. Applying Weinberg's method to the supersymmetric case an effective low-energy

Lagrangian for the Goldstone superfields can be constructed [6] which has the generic form

$$(10) \quad \mathscr{L}_{\text{eff}} = -\frac{1}{f^2} \bar{\varphi}_i \bar{\varphi}_j \varphi_i \varphi_j \big|_{\theta\theta\bar{\theta}\bar{\theta}} + \ldots = $$

$$= -\frac{1}{f^2} \varphi_i^* \varphi_i \partial^\mu \varphi_j^* \partial_\mu \varphi_j - \frac{1}{2f^2} \bar{\psi}_{\text{L}i} \gamma^\mu \psi_{\text{L}i} \bar{\psi}_{\text{L}j} \gamma_\mu \psi_{\text{L}j} + \ldots, \quad f \sim \Lambda_{\text{HC}}.$$

The almost unavoidable appearance of massless composite fermions in supersymmetric theories leads naturally to the conjecture [6, 7] that quarks and leptons may be identified as « quasi-Goldstone fermions », *i.e.* as superpartners of Goldstone bosons. In such a scenario (cf. fig. 1), light composite

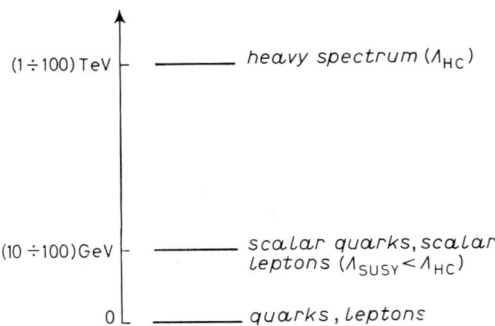

Fig. 1. – Possible mass spectrum of scalar and fermionic components of Goldstone supermultiplets in supersymmetric preon models.

Goldstone supermultiplets arise from a spontaneous symmetry breaking at a large mass scale (*e.g.* $(1 \div 100)$ TeV). Mass splittings within the Goldstone multiplets which render the quasi-Goldstone fermions lighter than their scalar superpartners can occur as a consequence of soft SUSY breaking, explicit breaking of the original global symmetry—for instance, through gauge interactions—and unbroken chiral symmetries [8]. Some interesting explicit examples have been discussed by LERCHE and LÜST [9].

An important feature of the effective Lagrangian eq. (10) is the current-current self-interaction of the fermions. This suggests that the weak interactions of quarks and leptons may be residual interactions which are mediated by the exchange of composite W-bosons, a possibility which we will pursue in the following sections. The current-current form of the residual fermion interactions in eq. (10) is a special property of quasi-Goldstone fermions. If quarks and leptons are (pseudo) Goldstone fermions arising from the spontaneous breakdown of an extended supersymmetry, as suggested by BARDEEN and VIŠNJIĆ [10], their residual interactions will involve derivatives, in accord with the low-energy theorems for Goldstone particles,

$$\mathcal{L}_{\text{eff}}^{3F} = -\frac{1}{4f^4} \bar{\psi}_i \gamma^\mu \partial^\nu \psi_i \bar{\psi}_j \gamma_\nu \partial_\mu \psi_j + \dots, \tag{11}$$

and can, therefore, not be identified as weak interactions.

3. – The coset space $U_6/SU_2 \times U_4$.

In models with composite W-bosons weak isospin has to be introduced as a global symmetry. Furthermore one has to ensure that quarks and leptons couple universally to the composite W-bosons. This can be achieved by imposing a global Pati-Salam SU_4 invariance [11] with lepton number as fourth colour. In the case of residual weak interactions parity violation must be due to a different bound-state structure of left- and right-handed fermions. It is, therefore, natural to consider in first approximation only the left-handed fields as composite and to treat, as in the Abbott-Farhi model [12], right-handed fields as elementary spectators.

The left-handed fermions u_L^α, d_L^α, ν_L and e_L^- of one family transform with respect to $SU_2 \times SU_4$ as

$$\begin{pmatrix} u_L^\alpha & \nu_L \\ d_L^\alpha & e_L^- \end{pmatrix} \equiv (\psi_L^{ia}) \sim (2, 4). \tag{12}$$

It is easy to see that one can obtain the ψ_L^{ia} as quasi-Goldstone fermions from the spontaneous symmetry breaking $U_6 \to U_2 \times U_4$ [8, 13-16]:

$$U_{(6)}: \left(\begin{array}{c|c} U_2 & (2, 4^*) \\ \hline (2^*, 4) & U_4 \end{array} \right). \tag{13}$$

The 16 Goldstone bosons transform as $(2, 4^*) + (2^*, 4)$ and can be embedded into 8 chiral multiplets $\varphi_i^a \equiv (\varphi_i^a, \psi_{iL}^a)$. The assignment of Goldstone bosons to chiral superfields is not unique. The only restriction is that the coset space G/H be embedded in a Kähler manifold [17]. The above choice of 8 chiral multiplets is clearly the minimal one and it is possible because $U_6/U_2 \times U_4$ is a Grassmann manifold, *i.e.* a special Kähler manifold. One can also associate, however, the 16 Goldstone bosons with 16 chiral superfields [13]. It is a dynamical question which case is realized. Another important point concerns the two U_1 factors in $H = U_2 \times U_4$. One or both of them may also be spontaneously broken. This leads to one or two additional neutral chiral multiplets and, as we will see, changes the low-energy effective Lagrangian of the Goldstone multiplets φ_i^a.

It turns out that the relevant coset space is $U_6/SU_2 \times U_4$. The broken U_1 factor yields one neutral Goldstone superfield φ, the « novino ». The effective Lagrangian for the superfields φ_i^a and φ can be constructed in a straightforward manner following Weinberg's method [5]. The first step is the construction of a nonlinear realization of the U_6 algebra. The generators T_β^α ($\alpha, \beta = 1, ..., 6$) which satisfy the commutation relation

(14) $$[T_\beta^\alpha, T_\delta^\gamma] = \delta_\delta^\alpha T_\beta^\gamma - \delta_\beta^\gamma T_\delta^\alpha$$

are split into the unbroken generators ($\alpha = (i, a); i = 1, 2; a = 3, ..., 6$)

(15) $$L_j^i = T_j^i - \tfrac{1}{2} \delta_j^i T_k^k, \qquad L_b^a = T_b^a,$$

which belong to the unbroken subgroup $SU_2 \times U_4$, and the broken generators

(16) $$X_i^a = T_i^a, \quad X_a^i = T_a^i, \quad X = \frac{1}{\sqrt{2}} T_i^i.$$

The Goldstone superfields φ_i^a and φ transform linearly with respect to $H = SU_2 \times U_4$:

(17) $$[L_j^i, \varphi_k^a] = \delta_k^i \varphi_j^a - \tfrac{1}{2} \delta_j^i \varphi_k^a, \qquad [L_b^a, \varphi_i^c] = - \delta_b^c \varphi_i^a.$$

For the broken generators in G/H a nonlinear realization has to be constructed $\left[A, B, ... = \binom{a}{i}, \binom{i}{a}, \binom{i}{i}\right]$:

(18) $$[X_A, \varphi_B] = F_{AB}(\varphi),$$

where $F_{AB}(\varphi)$ satisfies the « Jacobi identities » [5] $\left(P, Q = \binom{i}{j}, \binom{a}{b}\right)$

(19) $$\begin{cases} [X_A, [X_B, \varphi_C]] - [X_B, [X_A, \varphi_C]] = [[X_A, X_B], \varphi_C], \\ [X_A, [L_P, \varphi_B]] - [L_P, [X_A, \varphi_B]] = [[X_A, L_P], \varphi_B], \\ [L_P, [L_Q, \varphi_A]] - [L_Q, [L_P, \varphi_A]] = [[L_P, L_Q], \varphi_A]. \end{cases}$$

Equations (19) ensure that the functions $F_{AB}(\varphi)$ define a representation of the U_6 Lie algebra. One can check by explicit calculation that an exact solution of eqs. (19), which is unique up to redefinition of the Goldstone superfields, is given by [16]

$$
(20) \quad \begin{cases} \dfrac{1}{i}[X_i^a, \varphi_j^b] = \dfrac{1}{f_2} \varphi_j^a \varphi_i^b, & \dfrac{1}{i}[X_a^i, \varphi_j^b] = f_2 \delta_j^i \delta_a^b, \\[2mm] \dfrac{1}{i}[X_i^a, \varphi] = \dfrac{i}{\sqrt{2}} \varphi_i^a, & \dfrac{1}{i}[X_a^i, \varphi] = 0, \\[2mm] \dfrac{1}{i}[Y, \varphi_i^a] = -\dfrac{i}{\sqrt{2}} \varphi_i^a, & \dfrac{1}{i}[Y, \varphi] = f_1. \end{cases}
$$

ZUMINO has shown [17] that in general the Lagrangian of a supersymmetric σ-model takes the form

$$(21) \qquad \mathscr{L} = K(\bar{\varphi}_A, \varphi_A)|_{\theta\theta\bar\theta\bar\theta},$$

where K is the Kähler potential of the associated Kähler manifold. Therefore, the effective Lagrangian for Goldstone superfields φ_i^a and φ can be constructed by making an $SU_2 \times U_4$ invariant ansatz for K and by demanding that a variation of K yields a sum of chiral and antichiral superfields:

$$(22) \quad [X_A, K(\bar{\varphi}_a^i, \varphi_j; \varphi_i^a, \varphi)] = h_A^{(0)} + h_{AB}^{(1)} \varphi_B - h_{ABC}^{(2)} \varphi_B \varphi_C + \ldots$$
$$\ldots + \bar{h}_{AB}^{(1)} \bar\varphi_B + \bar{h}_{ABC}^{(2)} \bar\varphi_B \bar\varphi_C + \ldots.$$

Equations (21) and (22) imply that the variation of the Lagrangian is a total derivative and that the action is consequently invariant under U_6 transformations. Using the described procedure, one finds [16]

$$(23) \quad K = \bar\varphi_a^i \varphi_i^a + \bar\varphi\varphi - \dfrac{1}{2f_2^2} \bar\varphi_a^i \bar\varphi_b^j \varphi_j^a \varphi_i^b + \dfrac{1}{4f_1^2} \bar\varphi_a^i \bar\varphi_b^j \varphi_i^a \varphi_j^b + $$
$$+ F\bar\varphi_a^i \bar\varphi \varphi_i^a \varphi + G \bar\varphi \bar\varphi \varphi \varphi + \ldots,$$

where F and G are unconstrained parameters. This arbitrariness of the Kähler potential is related to the presence of one «quasi-Goldstone boson» in φ and reflects the different ways in which the odd-dimensional coset space $U_6/SU_2 \times U_4$ can be embedded in a Kähler manifold.

Of particular interest are the quark-lepton residual interactions which are contained in eq. (23). Using the identity

$$(24) \qquad \bar\tau_j^i \bar\tau_l^k = 2(\delta_l^i \delta_j^k - \tfrac{1}{2}\delta_j^i \delta_l^k),$$

one obtains [16]

$$(25) \quad \mathscr{L}_{\text{eff}}^{(1,1)} = -\dfrac{1}{2f_2^2}\left[\bar\psi_{La}^i \gamma_\mu \left(\dfrac{1}{2}\bar\tau\right)_i^j \psi_{Lj}^a\right]^2 - \dfrac{f_2^2 - f_1^2}{2f_2^2}\left[\dfrac{1}{2}\bar\psi_{La}^i \gamma_\mu \psi_{Li}^a\right]^2 + \ldots.$$

Equation (25) contains the phenomenologically wanted isovector exchange term as well as an isoscalar exchange term whose presence is a familiar problem of models with composite W-bosons [1, 2]. Indeed, in the limit $v_1 \to 0$, *i.e.* in the absence of the novino, the result is unacceptable because isovector and isoscalar contributions are of equal magnitude. In the case $v_1 \approx v_2$, however, which one may expect in a preon theory with a single scale, the isoscalar exchange term is suppressed. We are thus led to the unexpected result that in supersymmetric preon models with composite W-bosons there is a direct relation between the weak interactions of quarks and leptons and the existence of a new neutral Goldstone superfield, the novino.

In this section we have shown how one can construct the effective Lagrangian for Goldstone superfields in a direct pedestrian way. More elegant methods which are crucial if one wants to construct the σ-models beyond the quartic terms can be found in the literature [18]. The very interesting subject of gauged supersymmetric σ-models [19] is beyond the scope of these lectures.

4. – A U_6 model.

The simplest supersymmetric « preon model » with global U_6 invariance is a SU_2 gauge theory with 6 doublets of chiral superfields $\chi_\alpha^p = (\tilde{\chi}_\alpha^p, \eta_{L\alpha}^p)$, where $\alpha = 1, ..., 6$ denotes the U_6 flavour index and $p = 1, 2$ is the SU_2 hypercolour index. The interaction Lagrangian for the chiral multiplets χ_α^p and the gauge vector multiplets $V = \frac{1}{2} \tau^I V^I$, $V^I = (\lambda_L^I, V_\mu^I)$, is given by [20]

$$(26) \qquad \mathscr{L} = \int d^4\theta \, \bar{\chi}_p^\alpha \, (\exp [2gV])_q^p \, \chi_\alpha^q \, .$$

The simplest gauge-invariant composite operators are the bilinear chiral superfields

$$(27) \qquad \Phi_{\alpha\beta} = \varepsilon_{pq} \chi_\alpha^p \chi_\beta^q$$

and the vector superfields

$$(28) \qquad J_\beta^\alpha = \bar{\chi}_p^\alpha \, (\exp [2gV])_p^q \, \chi_\beta^q \, ,$$

which, except for the anomalous U_1 factor J_α^α, represent the conserved currents of the theory:

$$(29) \qquad D^2 J_\beta^\alpha = \bar{D}^2 J_\beta^\alpha = 0 \, .$$

This U_6 « preon model » corresponds to SUSY QCD with two colours and three flavours. As the fundamental representation of SU_2 is pseudoreal, the global symmetry is U_6 rather than $U_3 \times U_3$. Vacuum expectation values of the operators $\Phi_{\alpha\beta}$ and J_β^α can break the U_6 symmetry down to $(SU_2)^3$ ($i = 1, 2$; $a_1 = 3, 4$; $a_2 = 5, 6$):

$$(30) \quad \begin{cases} \langle \Phi_{ij} \rangle = v_1^2 \varepsilon_{ij}, & \langle \Phi_{a_1 b_1} \rangle = v_2^2 \varepsilon_{a_1 b_1}, \\ \langle \Phi_{a_2 b_2} \rangle = v_3^2 \varepsilon_{a_2 b_2}, & \langle J_j^i \rangle = \tilde{v}_1^2 \delta_j^i, \\ \langle J_{b_1}^{a_1} \rangle = \tilde{v}_2^2 \delta_{b_1}^{a_1}, & \langle J_{b_2}^{a_2} \rangle = \tilde{v}_3^2 \delta_{b_2}^{a_2}. \end{cases}$$

Constraints [21] on the possible values of $v_1^2, ..., \tilde{v}_3^2$ can be obtained by breaking the U_6 invariance explicitly to $(SU_2)^3$ through the superpotential

$$(31) \quad g_m = -\tfrac{1}{2} m_1 \varepsilon^{ij} \Phi_{ij} - \tfrac{1}{2} m_2 \varepsilon^{a_1 b_1} \Phi_{a_1 b_1} - \tfrac{1}{2} m_3 \varepsilon^{a_2 b_2} \Phi_{a_2 b_2}$$

and considering the chiral limit $m_i \to 0$. The effective-Lagrangian approach [22], analyticity arguments [23] and instanton calculations [24] lead to the following mass dependence of the chiral condensates:

$$(32) \quad v_i^2 \sim \frac{1}{m_i} (m_1 m_2 m_3)^{\frac{1}{2}}.$$

If one demands that in the chiral limit all vacuum expectation values are finite, only two cases are possible: either the full U_6 symmetry remains unbroken ($v_1^2 = v_2^2 = v_3^2 = 0$) or the unbroken subgroup is $SU_2 \times U_4$ ($v_1^2 \neq 0$, $v_2^2 = v_3^2 = 0$). Constraints on the vector condensates follow from the SUSY Dashen formulae [25, 26]

$$(33) \quad \begin{cases} f_A (M^2)_{AB} f_B = \tfrac{1}{2} \langle 0 | [X_A, [X_B, \mathscr{L}_F]] | 0 \rangle, \\ f_A M_{AB} f_B = \langle 0 | [X_A, [X_B, \mathscr{L}_S]] | 0 \rangle, \end{cases}$$

where X_A are the broken charges, f_A the related decay constants and M_{AB} the pseudo-Goldstone boson mass matrix. $\mathscr{L}_F$ and $\mathscr{L}_S$ are the fermionic and scalar terms in the symmetry-breaking part of the Lagrangian:

$$(34) \quad \begin{cases} \mathscr{L}_F = \dfrac{1}{2} \dfrac{\partial^2 g_m}{\partial \chi_\alpha^p \partial \chi_\beta^q} \eta_\alpha^p \eta_\beta^q + \text{c.c.}, \\ \mathscr{L}_S = \dfrac{1}{2} \dfrac{\partial^2 g_m}{\partial \chi_\alpha^p \partial \chi_\beta^q} \tilde{\chi}_\alpha^p \tilde{\chi}_\beta^q. \end{cases}$$

Demanding that the decay constants f_1 and f_2 of the Goldstone superfields φ and φ_i^a remain finite in the chiral limit, one finds [21] that the vector condensates follow the pattern of the chiral condensates: either the U_6 symmetry remains unbroken ($\tilde{v}_1^2 = \tilde{v}_2^2 = \tilde{v}_3^2 = 0$) or the unbroken subgroup is $SU_2 \times U_4$ ($\tilde{v}_1^2 \neq 0$, $\tilde{v}_2^2 = \tilde{v}_3^2 = 0$). The size of the condensates is expected to be large, i.e. $\alpha(v_1^2) \sim$ $\sim \alpha(\tilde{v}_1^2) \sim 1$ ($\alpha = g^2/4\pi$).

The analysis of the bilinear condensates suggests that the model may have a phase in which the full U_6 symmetry is unbroken. It is a remarkable feature of the model that this possibility is indeed compatible [27] with 't Hooft's anomaly conditions [3]. The symmetry of the classical Lagrangian is $G_{cl} = U_6 \times U_{1,R}$, where the R symmetry acts differently on the scalar and the fermionic component of the superfield χ_α^p:

$$(35) \qquad \tilde{\chi}_\alpha^p \to \tilde{\chi}_\alpha^p, \qquad \eta_{L\alpha}^p \to \exp[-i\gamma]\,\eta_{L\alpha}^p.$$

Instanton effects reduce the classical invariance G_{cl} to $G_{qu} = SU_6 \times U_{1,X}$ with [16]

$$(36) \qquad X = T_\alpha^\alpha + 3R.$$

The preons $\eta_{L\beta}^\alpha$, the gauginos λ_{Lq}^p and the composite fermions $\psi_{L\alpha\beta}$, which are contained in the chiral superfields $\Phi_{\alpha\beta}$, transform with respect to $SU_6 \times U_{1,X}$ as follows:

$$(37) \qquad (\eta_{L\alpha}^p) \sim 2(\overline{6})_{-2}, \qquad (\lambda_{Lq}^p) \sim 3(1)_3, \qquad (\psi_{L\alpha\beta}) \sim (\overline{15})_{-1}.$$

At the preon level one obtains for the triangle anomalies (in units of $(6)_1$)

$$(38) \quad [SU_6]^3 : -2, \quad X[SU_6]^2 : 2(-2) = -4, \quad X^3 : 12(-2)^3 + 3 \cdot 3^3 = -15.$$

In terms of the composite fermions, the anomalies read

$$(39) \quad \begin{cases} [SU_6]^3 : K(\overline{\square\!\square}) = -(6-4) = -2, \\ X[SU_6]^2 : (-1)\,C(\overline{\square\!\square}) = (-1)(6-2) = -4, \\ X^3 : 15(-1)^3 = -15, \end{cases}$$

where the quantities K and C in eq. (39) are defined [3] through

$$(40) \quad \begin{cases} \operatorname{tr}[\lambda^a(R)\,\lambda^b(R)] = C(R)\,\operatorname{tr}[\lambda^a(\square)\,\lambda^b(\square)], \\ \operatorname{tr}[\{\lambda^a(R), \lambda^b(R)\}\,\lambda^c(R)] = K(R)\,\operatorname{tr}[\{\lambda^a(\square), \lambda^b(\square)\}\,\lambda^c(\square)]. \end{cases}$$

The matching of the anomalies for all $SU_6 \times U_{1,X}$ currents does not necessarily imply that a phase with unbroken SU_6 symmetry exists. In the « Higgs phase », where the SU_2 gauge invariance is spontaneously broken through large vacuum expectation values of the fundamental scalar fields $\tilde{\chi}_\alpha^p$, i.e.

$\alpha(\langle\tilde{\chi}\rangle^2) \ll 1$, the unbroken global symmetry is $SU_2 \times U_4$ [16]. Up to U_6 transformations, the only solution to the D-term constraint

$$\langle \tilde{\chi}_p^{*\alpha}\rangle (\tau^I)_q^p \langle \tilde{\chi}_\alpha^q\rangle = 0, \qquad I = 1, ..., 3, \tag{41}$$

reads [28]

$$\langle \tilde{\chi}_i^p\rangle = v\delta_i^p, \qquad \langle \tilde{\chi}_a^p\rangle = 0. \tag{42}$$

This corresponds to the result obtained in the «confining phase» (cf. eq. (30)) with $v_1^2 = \tilde{v}_1^2 = r^2$. According to the idea of complementarity [29], there is no phase transition between the «Higgs phase» and the «confining phase». This suggests that also in the strong-coupling regime the unbroken global symmetry is $SU_2 \times U_4$ rather than U_6.

An important question concerns the existence of heavy vector bosons in the theory. In analogy with QCD one expects naively that the currents J_β^α create 36 massive vector supermultiplets from the vacuum. The suppression of the isoscalar exchange term in the effective Lagrangian eq. (25) suggests [16] that among these 36 vector bosons the isotriplet of W-bosons plays a special role. The additional 33 vector bosons should be either more heavily or weakly coupled than the W-bosons. Such a qualitative difference can only be caused by the scalar condensates. A means to study the effect of scalar condensates on the physical spectrum of a theory are the SVZ sum rules [30]. In the standard manner, the 2-point function of the currents J_β^α in the Euclidean region can be evaluated using dispersion relations and an ansatz for the physical spectrum, on one side, and in perturbation theory (including power corrections), on the other side. A straightforward supergraph calculation (cf. fig. 2) yields the result [21]

$$(43a) \quad i\int d^4x \exp[iqx] \langle T(J_\beta^\alpha(x,\theta,\bar{\theta}) J_\delta^\gamma(0,0,0))\rangle =$$
$$= \tfrac{1}{2} P_T \delta^4(\theta) q^2 \Pi_{\beta\delta}^{\alpha\gamma}(q^2) - \tfrac{1}{2}(P_1 - P_2)\tilde{\delta}^4(\theta)\delta_\delta^\alpha \delta_\beta^\gamma (\langle\bar{\chi}^\alpha \chi_\alpha\rangle - \langle\bar{\chi}^\beta \chi_\beta\rangle) -$$
$$- \tfrac{1}{2}\delta^4(\theta)\delta_\delta^\alpha \delta_\beta^\gamma (\langle\bar{\chi}^\alpha \chi_\alpha\rangle + \langle\bar{\chi}^\beta \chi_\beta\rangle),$$

with

$$(43b) \quad \Pi_{\beta\delta}^{\alpha\gamma}(q^2) = \delta_\delta^\alpha \delta_\beta^\gamma \left[-\frac{1}{8\pi^2}\ln\frac{q^2}{\Lambda^2} + \frac{1}{q^2}(\langle\bar{\chi}^\alpha\chi_\alpha\rangle + \langle\bar{\chi}^\beta\chi_\beta\rangle)\right] -$$
$$- 2\frac{2\pi\alpha}{(q^2)^2}\langle(\bar{\chi}^\alpha\tau^A\chi_\beta)(\bar{\chi}^\gamma\tau^A\chi_\delta)\rangle + 2\frac{(2\pi\alpha)^2}{(q^2)^3}\langle(\bar{\chi}^\alpha\tau^A\chi_\beta)(\bar{\chi}^\gamma\tau^A\chi_\delta)(\bar{\chi}\chi)\rangle + ...,$$

$$\alpha = \frac{g^2}{4\pi}, \quad \bar{\chi}\chi \equiv \sum_\varepsilon \bar{\chi}^\varepsilon \chi_\varepsilon.$$

The polarization tensor $\Pi_{\beta\delta}^{\alpha\gamma}(q^2)$ has 5 irreducible components with respect to the unbroken $SU_2 \times U_4$ symmetry: $(3,1)$, $(1,1)$, $(2,4)$, $(1,15)$ and $(1,1)'$. In

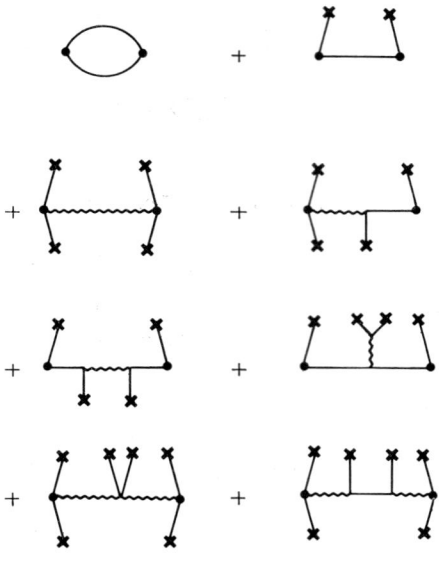

Fig. 2. – Supergraphs contributing to the two-point function of the U_6 currents J^α_β (cf. eq. (43)). Full lines denote chiral superfield propagators, wavy lines vector superfield propagators. Crosses indicate scalar condensates.

the channels (1, 1) and (2, 4), the Goldstone superfields φ and φ_{ia} contribute in addition to the vector superfields. Using the results for the bilinear condensates (30) and factorization for higher condensates, one finds large power corrections in the (3, 1), (1, 1) and (2, 4) channels which are directly related to the mass of the composite W-bosons and the decay constants f_1 and f_2 [21]:

$$M_W^2 \approx 2\pi\alpha_W \langle \bar\chi\chi \rangle, \tag{44a}$$

$$f_1^2 \approx f_2^2 \approx \langle \bar\chi\chi \rangle, \quad \left|\frac{f_1^2 - f_2^2}{f_2^2}\right| \ll 1. \tag{44b}$$

The sum rule in the (3, 1) channel, which yields the standard-model mass formula (44a), is identical with a sum rule [31] which has previously been derived in the context of the Abbott-Farhi model.

The result eq. (44) shows clearly the importance of scalar condensates for the vector-boson mass spectrum. The W mass is directly related to the size of the condensate and, as a consequence of eqs. (25) and (44b), the additional 33 vector bosons must be substantially heavier than the W-bosons, if they exist at all. Given a spectrum of resonances and the condensates of the underlying theory sum rules constrain the resonance parameters. In general, it is not possible to prove the existence or absence of certain states by means of sum rules. The qualitative features of the sum rules [21] obtained from eq. (43) indicate, however, that the physical states in the strong-coupling regime may be completely identical with the ones in the Higgs phase and that an additional strongly interacting spectrum of heavy states may not exist. This would mean that the intuition derived from ordinary QCD is totally misleading and that the spectrum of physical states can essentially be read off from the classical Lagrangian, a question which clearly deserves further studies. Even if such radical conclusions are unjustified, there remains at least a striking similarity in the vacuum structure between the Higgs phase and the confining phase.

5. – Towards a realistic preon model.

In the previous sections we have discussed the idea of quasi-Goldstone fermions and we have illustrated some techniques, which are used in the context of supersymmetric composite models, by means of a U_6 toy model. This « preon model » yields the left-handed sector of one family and the W-bosons as bound states. The right-handed part of one family can be incorporated by extending the group U_6 to $U_6 \times U_6$ [8, 13, 15, 32]. The preons now consist of two sextets of chiral superfields which are usually chosen to transform as $N + N^*$ with respect to the hypercolour group SU_N. A multiplicity of families, which are labelled by means of a discrete or a broken U_1 symmetry, could emerge in such models as a consequence of the dimension of the hypercolour group [33], but no fully satisfactory example has been found so far. In $U_6 \times U_6$ models with composite W-bosons one has to understand the origin of parity violation. It is conceivable that, like in ordinary left-right symmetric models, parity is broken spontaneously, yet a detailed mechanism has not yet been proposed. It is also possible, although not particularly attractive, to break parity explicitly by choosing $SU_2 \times SU_2'$ as hypercolour group with different coupling constants for the two SU_2 subgroups [32].

In general, one expects in models with composite W-bosons and global $SU_2 \times SU_4$ invariance additional composite vector bosons which transform as (1, 15) [34, 35]. The effect of these V-bosons on low-energy weak interactions within one family can be estimated and one finds a lower mass bound of ~ 500 GeV [34]. Much more stringent bounds are obtained if families are included, especially from the process $K_L^0 \to \mu^\pm e^\mp$ [11, 36]. The exchange of the leptoquark [11] among the $\overline{V}$-bosons yields the effective four-fermion Lagrangian (cf. fig. 3)

$$(45) \qquad L_{V_3} = -\frac{g_V^2}{2 M_{V_3}^2} (\bar{\mu}_L \gamma^\mu s_L^\alpha \bar{d}_{L\alpha} \gamma_\mu e_L + \text{c.c.}),$$

where the coupling constant g_v satisfies the bound [34] $g_v^2/4\pi > \alpha_s(M_v^2) \sim 0.1$. From the experimental bound [37]

$$\frac{\Gamma(K_L^0 \to \mu^\pm e^\mp)}{\Gamma(K^- \to \mu^- \bar{\nu}_\mu)} < 2 \cdot 10^{-6}$$

one obtains

(46) $$\frac{M_{V_3}}{M_W} > \left[\frac{\alpha_s^2}{\alpha_W^2 \sin^2 \theta_C} \frac{\Gamma(K^+ \to \mu^+ \nu_\mu)}{\Gamma(K_L^0 \to \mu^\pm e^\mp)}\right]^{\frac{1}{4}} > 0.8 \cdot 10^2 \,.$$

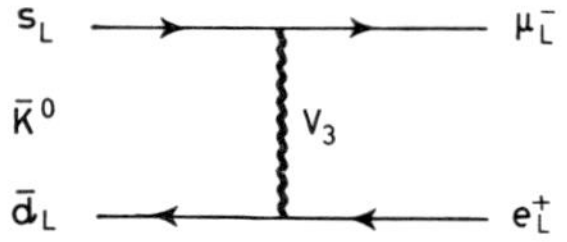

Fig. 3. – Contribution of the leptoquark V_3 to the decay $\bar{K}^0 \to \mu^- e^+$.

In models with a single scale one is thus faced with the problem of explaining dynamically vector-boson masses which differ by a factor of about 100! The bound eq. (46) is based on the assumption that the V-bosons couple universally to different families. This is indeed very likely to be a consequence of the same mechanism which enforces a universal coupling of the W-bosons. On the other side, without a « standard composite model » providing a convincing explanation for families, one can always hope for some mechanism which circumvents the bound eq. (46). In supersymmetric theories, for instance, the V-bosons may simply not exist, as we saw in the previous section.

Within SUSY composite models, the weak interactions may also be treated as fundamental gauge interactions. Such models [8, 13, 32] are based on the coset space $U_6 \times U_6/U_4 \times U_4 \times SU_{2,D}$, and the W-bosons acquire mass from the spontaneous symmetry breaking $SU_2 \times SU_2 \to SU_{2,D}$. Thus the confining hypercolour group acts like technicolour [38] and quarks and leptons are superpartners of the pseudo-Goldstone bosons familiar from extended technicolour theories. As in models with composite W-bosons, the substructure scale is related to the Fermi scale. A potential problem for these technicolour type models are the residual interactions [32] due to compositeness which now compete with the weak gauge interactions. An interesting possibility is that the two scales are separated as a result of soft SUSY breaking terms in the preon Lagrangian [39].

Such soft SUSY breaking terms are needed in any case in order to obtain a realistic boson and fermion mass spectrum [9, 13, 39]. In addition, explicit breaking of the original global symmetry is required in order to generate a mass for the Goldstone bosons. Part of this explicit breaking is provided by the colour and electromagnetic gauge interactions, but presumably this will not be sufficient. In general, some neutral fermions and bosons, such as the novino, are likely to remain massless.

Supersymmetric gauge theories are interesting candidates for a theory of quark-lepton substructure because they provide naturally light composite fermions. During the last two years, we have become more familiar with various technical aspects of such theories, but the main challenge of composite models, with or without supersymmetry, remains to find the solution of the family problem.

* * *

It is a pleasure to thank R. D. PECCEI, M. G. SCHMIDT and T. YANAGIDA for collaboration on the novino model.

REFERENCES

[1] For a general review, see R. PETRONZIO: this volume, p. 76.
[2] For a recent review, see R. D. PECCEI: in *Proceedings of the IV Topical Workshop on Proton-Antiproton Collider Physics, Bern*, edited by H. HÄNNI and J. SCHACHER (1984), p. 483.
[3] G. 'T HOOFT: in *Recent Developments in Gauge Theories, Cargèse 1979*, edited by G. 'T HOOFT et al. (Plenum, New York, N. Y., London, 1980).
[4] O. DUMBRAJS, R. KOCH, H. PILKUHN, G. C. OADES, H. BEHRENS, J. J. DE SWART and P. KROLL: *Nucl. Phys. B*, **216**, 304 (1983).
[5] S. WEINBERG: *Phys. Rev.*, **166**, 1568 (1968). For a recent review, see M. E. PESKIN: preprint SLAC-PUB-3021 (1982).
[6] W. BUCHMÜLLER, R. D. PECCEI and T. YANAGIDA: *Phys. Lett. B*, **124**, 67 (1983).
[7] W. BUCHMÜLLER, S. T. LOVE, R. D. PECCEI and T. YANAGIDA: *Phys. Lett. B*, **115**, 233 (1982).
[8] R. BARBIERI, A. MASIERO and G. VENEZIANO: *Phys. Lett. B*, **128**, 493 (1983).
[9] W. LERCHE and D. LÜST: *Nucl. Phys. B*, **244**, 157 (1984).
[10] W. A. BARDEEN and V. VIŠNJIĆ: *Nucl. Phys. B*, **194**, 422 (1982).
[11] J. C. PATI and A. SALAM: *Phys. Rev. D*, **10**, 275 (1974).
[12] L. F. ABBOTT and E. FARHI: *Phys. Lett. B*, **101**, 69 (1981).
[13] W. BUCHMÜLLER, R. D. PECCEI and T. YANAGIDA: *Nucl. Phys. B*, **227**, 503 (1983).
[14] F. BORDI, R. CASALBUONI, D. DOMINICI and R. GATTO: *Phys. Lett. B*, **127**, 419 (1983).
[15] O. W. GREENBERG, R. N. MOHAPATRA and M. YASUE: *Phys. Lett. B*, **128**, 65 (1983).
[16] W. BUCHMÜLLER, R. D. PECCEI and T. YANAGIDA: *Nucl. Phys. B*, **231**, 53 (1984).
[17] B. ZUMINO: *Phys. Lett. B*, **87**, 203 (1979).
[18] C. K. LEE and H. S. SHARATCHANDRA: preprint MPI-PAE/PTh 54/83 (1983); W. LERCHE: *Nucl. Phys. B*, **238**, 582 (1984); T. EUGO, I. OJIMA and T. YANAGIDA: *Phys. Lett. B*, **135**, 402 (1984); M. BANDO, T. KURAMOTO, T. MASKAWA and S. UEHARA: *Phys. Lett. B*, **138**, 94 (1984).

[19] C. L. ONG: *Phys. Rev. D*, **27**, 911 (1983); J. BAGGER and E. WITTEN: *Phys. Lett. B*, **118**, 103 (1982); A. BURAS and W. SŁOMINSKI: *Nucl. Phys. B*, **223**, 157 (1983).
[20] We use the conventions of J. WESS and J. BAGGER: *Supersymmetry and Supergravity* (Princeton University Press, Princeton, N. J., 1983).
[21] W. BUCHMÜLLER and M. G. SCHMIDT: *Nucl. Phys. B*, **258**, 230 (1985).
[22] T. R. TAYLOR, G. VENEZIANO and S. YANKIELOWICZ: *Nucl. Phys. B*, **231**, 493 (1983).
[23] G. VENEZIANO: *Phys. Lett. B*, **124**, 357 (1983).
[24] G. C. ROSSI and G. VENEZIANO: *Phys. Lett. B*, **138**, 195 (1984); M. G. SCHMIDT: *Phys. Lett. B*, **141**, 236 (1984).
[25] G. VENEZIANO: *Phys. Lett. B*, **128**, 199 (1983).
[26] G. M. SHORE: *Nucl. Phys. B*, **231**, 139 (1984); T. E. CLARK and S. T. LOVE: *Nucl. Phys. B*, **232**, 306 (1984); W. LERCHE: *Nucl. Phys. B*, **246**, 475 (1984).
[27] J. M. GERARD: private communication.
[28] F. BUCCELLA, J. P. DERENDINGER, S. FERRARA and C. A. SAVOY: *Phys. Lett. B*, **115**, 375 (1982).
[29] E. FRADKIN and S. SHENKER: *Phys. Rev. D*, **19**, 3682 (1979); S. DIMOPOULOS, S. RABY and L. SUSSKIND: *Nucl. Phys. B*, **173**, 208 (1980).
[30] M. A. SHIFMAN, A. I. VAINSHTEIN and V. T. ZAKHAROV: *Nucl. Phys. B*, **147**, 385, 448 (1979).
[31] H. G. DOSCH, M. KREMER and M. G. SCHMIDT: *Phys. Lett. B*, **137**, 88 (1984); *Z. Phys. C*, **26**, 569 (1985).
[32] W. BUCHMÜLLER, R. D. PECCEI and T. YANAGIDA: *Nucl. Phys. B*, **244**, 186 (1984).
[33] O. W. GREENBERG, R. N. MOHAPATRA and M. YASUE: *Phys. Rev. Lett.*, **51**, 1737 (1983).
[34] W. BUCHMÜLLER: *Phys. Lett. B*, **145**, 151 (1984).
[35] B. SCHREMPP and F. SCHREMPP: preprint DESY 84-055 (1984).
[36] I thank J. L. ROSNER for pointing out the importance of this process which has recently also been emphasized by J. C. PATI: Maryland University preprint 84-146 (1984); see also R. N. CAHN and H. HARARI: *Nucl. Phys. B*, **176**, 135 (1980).
[37] *Rev. Mod. Phys.*, **36**, S12 (1984).
[38] L. SUSSKIND: *Phys. Rev. D*, **20**, 2619 (1979); S. WEINBERG: *Phys. Rev. D*, **19**, 1277 (1979).
[39] A. MASIERO and G. VENEZIANO: *Nucl. Phys. B*, **248**, 593 (1985).

Neutrino Masses 1984.

M. RONCADELLI

Dipartimento di Fisica Nucleare e Teorica - Pavia, Italia (*)
Istituto Nazionale di Fisica Nucleare - Sezione di Pavia, Italia

Introduction.

The subject of neutrino masses is both theoretically fashinating and experimentally challenging. There has been a lot of activity in this research field over the last few years. Many workshops have been organized on this subject and a lot of review papers are presently available [1, 2]. The aim of the present lectures is to provide an *elementary* introduction to this topic, in a manner which should be as much *pedagogical* as possible. The selection of the arguments and the way to present them reflect both this hope and the wish not to make these lectures exact replicas of the many others given on the same subject elsewhere. And they will obviously reflect the author's personal point of view about neutrino masses.

The paper is organized as follows. Section **1** contains a careful discussion of « spinology » starting from first principles: this allows for a better understanding of Majorana masses as contrasted to the familiar Dirac masses. In sect. **2** we discuss the physical difference between Dirac and Majorana masses and present the « classic » theoretical arguments in favour of nonvanishing Majorana neutrino masses. Next, we consider in detail the general case in which both Dirac and Majorana mass terms are present. Then we turn our attention to the four most interesting scenarios which arise as particular cases. Finally, we briefly discuss the neutrinoless double beta-decay. In sect. **3** we review the present experimental status about neutrino masses by considering the three relevant experiments: neutrino oscillation, neutrinoless double beta-decay and tritium beta-decay. A discussion of the results is offered. Section **4** contains the description of a very recent model for naturally light Dirac neutrinos, a possibility suggested but not compelled by experiment. Finally, we have collected in the appendix some technical details about « spinology ».

(*) Postal address: Via A. Bassi 6, I-27100 Pavia, Italy.

1. – Generalities on fermion masses.

Let us first consider the Lie algebra of the Lorentz group $O_{3,1}$, which has the well-known structure

(1.1) $$\begin{cases} [J_i, J_j] = i\varepsilon_{ijk} J_k, \\ [J_i, K_j] = i\varepsilon_{ijk} K_k, \\ [K_i, K_j] = -i\varepsilon_{ijk} J_k, \end{cases}$$

where J_i and K_i ($i = 1, 2, 3$) are the Hermitian generators of space rotations and Lorentz boosts, respectively, and ε_{ijk} is the totally antisymmetric Levi-Civita tensor. The minus sign in the last of eqs. (1.1) arises from the non-compact nature of $O_{3,1}$. By introducing the non-Hermitian operators

(1.2) $$N_i \equiv \tfrac{1}{2}(J_i + iK_i)$$

eqs. (1.1) can be rewritten as

(1.3) $$\begin{cases} [N_i, N_j^\dagger] = 0, \\ [N_i, N_j] = i\varepsilon_{ijk} N_k, \\ [N_i^\dagger, N_j^\dagger] = i\varepsilon_{ijk} N_k^\dagger. \end{cases}$$

This shows that the Lie algebra of $O_{3,1}$ is isomorphic to the Lie algebra of $SU_2 \times SU_2$.

This is a good thing, for it allows us to classify the representations of $O_{3,1}$ in terms of the well-known representations of SU_2. The latters are labelled by the eigenvalues of the Casimir operator, which take integer or half-integer values. Consequently, the representations of $O_{3,1}$ can be labelled as (m, n), where m and n are the eigenvalues of the Casimirs of the two SU_2, $\sum N_i N_i$ and $\sum N_i^\dagger N_i^\dagger$, respectively. Moreover, since $J_i = N_i + N_i^\dagger$, it follows that the spin s of the representation (m, n) is given by $s = m + n$.

Since we want to discuss here spin–one-half fermions, they should necessarily belong to the representations $(\tfrac{1}{2}, 0)$ or $(0, \tfrac{1}{2})$ of $O_{3,1}$. It is clear from eqs. (1.2) and (1.3) that they are not independent, but rather complex conjugate of each other.

We would like to stress a very remarkable fact: any other representation of the Lorentz group can be constructed by multiplication of these two building blocks.

A two-component *left*-handed Weyl spinor $\psi_L \equiv \psi_\alpha$ ($\alpha = 1, 2$) transforms according to the $(\tfrac{1}{2}, 0)$ representation.

Likewise, a two-component *right*-handed Weyl spinor $\psi_R \equiv \bar{\psi}^{\dot\alpha}$ ($\dot\alpha = 1, 2$) transforms according to the $(0, \frac{1}{2})$ representation. Naively, one might expect $(\psi_\alpha)^* = \bar{\psi}^{\dot\alpha}$. This is false. Instead, one has $(\psi_\alpha)^* = \bar{\psi}_{\dot\alpha}$ with $\bar{\psi}_{\dot\alpha} \neq \bar{\psi}^{\dot\alpha}$. The latter fact is a consequence of the nontrivial structure of the charge conjugation operator (see appendix). The Van der Waerden notation we are employing makes it evident that ψ_L is a doublet under the first SU_2 and a singlet under the second SU_2, and *vice versa* for ψ_R. Now the $(\frac{1}{2}, 0)$ and $(0, \frac{1}{2})$ representations can be assembled to form again a spin–one-half representation of $O_{3,1}$: $(\frac{1}{2}, 0) \oplus (0, \frac{1}{2})$. Actually, there are *two* distinct possibilities of doing this, depending on whether the spinors in $(\frac{1}{2}, 0)$ and in $(0, \frac{1}{2})$ are either *independent* or « complex conjugate » of each other. The first case yields a four-component *Dirac* spinor:

$$(1.4) \qquad \psi_D = \begin{pmatrix} \varphi_\alpha \\ \bar{\chi}^{\dot\alpha} \end{pmatrix},$$

whereas the second possibility gives a four-component *Majorana* spinor:

$$(1.5) \qquad \psi_M = \begin{pmatrix} \chi_\alpha \\ \bar{\chi}^{\dot\alpha} \end{pmatrix}.$$

As is evident from eq. (1.5), there is a one-to-one correspondence between two-component Weyl spinors and four-component Majorana spinors.

Needless to say, in ψ_D all four complex components are independent, while in ψ_M only two of them are independent. This reflects the fact that a Majorana spinor is self-conjugate, in the sense that $(\psi_M)^c = \psi_M$ (see appendix). *Physically*, this means the particles described by a Majorana spinor coincide with their antiparticles. Therefore, a Majorana particle cannot carry any *conserved* fermion number, like electric charge, colour, baryon or lepton number, etc. This important point will be emphasized later on.

A fermion mass term is *defined* as a real Lorentz-invariant fermion bilinear not involving derivatives.

A *Dirac* mass term is then

$$(1.6) \qquad \bar{\psi}_D \psi_D = (\chi^\alpha, \bar{\varphi}_{\dot\alpha}) \begin{pmatrix} \varphi_\alpha \\ \bar{\chi}^{\dot\alpha} \end{pmatrix} = (\chi^\alpha, 0) \begin{pmatrix} \varphi_\alpha \\ 0 \end{pmatrix} + (0, \bar{\varphi}_{\dot\alpha}) \begin{pmatrix} 0 \\ \bar{\chi}^{\dot\alpha} \end{pmatrix} = \bar{\psi}_R \psi_L + \bar{\psi}_L \psi_R .$$

It is straightforward to check that an equivalent form of the Dirac mass term is

$$(1.7) \qquad \bar{\psi}_D \psi_D = - \chi_L^T C^{-1} \varphi_L + \text{h.c.} = \overline{(\chi_L)^c} \varphi_L + \text{h.c.} ,$$

where one should bear in mind that $\varphi_L = \psi_L$ and $\chi_L = C\gamma^0(\psi_R)^*$.

A *Majorana* mass term is instead

$$(1.8) \qquad \bar{\psi}_M \psi_M = (\chi^\alpha, \bar{\chi}_{\dot\alpha}) \begin{pmatrix} \chi_\alpha \\ \bar{\chi}^{\dot\alpha} \end{pmatrix} = - \chi_L^T C^{-1} \chi_L + \text{h.c.} = \overline{(\psi_L)^c} \psi_L + \text{h.c.}$$

As a final remark, we observe that $\overline{(\psi_R)^c} \psi_R$ is the Hermitian conjugate of

$\overline{(\psi_L)^c}\psi_L$ for a Majorana spinor but *not* for a Dirac spinor, since in the latter case ψ_L and ψ_R are independent objects.

2. – Varieties of neutrino masses.

At this point, the reader might probably ask: what is the *physical* difference between Dirac and Majorana masses? To illuminate this issue, consider a global, vectorial fermion number transformation $U_{1,F}$ (physically: electric charge, baryon or lepton number, etc.):

(2.1) $$\psi_L \to \exp[i\theta]\psi_L, \qquad \psi_R \to \exp[i\theta]\psi_R.$$

It is immediately clear that the Dirac mass term (1.6) is $U_{1,F}$ invariant, whereas the Majorana mass term (1.8) is *not*, and in fact violates $U_{1,F}$ by *two* units. This circumstance has an obvious physical consequence. Since nobody wants to break electric charge, all electrically charged fermions can only have Dirac masses: Majorana masses are forbidden by electric-charge conservation! Yet another question arises: what about neutral fermions? They are clearly allowed to have also Majorana masses if we are to accept that they do not carry any conserved fermion number. As far as we presently know, the only elementary neutral fermions are the neutrinos. Historically, a conserved lepton number has been attributed to them, since all observed reactions involving neutrinos conserve lepton number. This implies that they can only have Dirac masses, which—as we saw—necessarily involve *independent* ν_L and ν_R. Yet, only ν_L participates in the weak interactions, and so—again historically—it has been assumed that ν_R simply does not exist. This is true also in the $SU_2 \times U_1$ standard model. Then neutrinos must necessarily be massless. The moral is, therefore, that *lepton number conservation plus the absence of ν_R imply massless neutrinos*.

The development in particle physics over the last decade has been characterized by the so-called *gauge dogma*: *only gauge symmetries are good symmetries*. As a consequence, lepton number can well be violated. Then neutrinos are allowed to have Majorana masses, according to the prevailing attitude among particle physicists: *everything is allowed that is not explicitly forbidden*, as GLASHOW likes to put it. Moreover, in some extensions of the $SU_2 \times U_1$ standard model, like $SU_{2,L} \times SU_{2,R} \times U_{1,B-L}$ [3] or in some grand unified theories like SO_{10} [4], both baryon and lepton numbers are actually broken, and neutrinos do have both Dirac and Majorana masses.

Actually, Majorana masses for the neutrinos have been favoured over Dirac masses in the last five years. Why? Experimentally, we know that [5]

(2.2) $$m_{\nu_e} < 46 \text{ eV},$$

(2.3) $$m_{\nu_\mu} < 0.5 \text{ MeV},$$

(2.4) $$m_{\nu_\tau} < 164 \text{ MeV},$$

and so the real main question becomes: why are neutrinos so light with respect to the charged fermions? Nobody has, of course, any answer, but the general feeling has been that serious models for neutrino masses should *naturally* account for this fact.

It is a trivial game to get Dirac neutrino masses in the $SU_2 \times U_1$ standard model. It is enough to add an independent ν_R for each fermion family and fine-tune the corresponding Yukawa coupling to accont for the neutrino-charged lepton mass difference. This strategy looks, however, quite ugly, and this is why people turned to Majorana neutrino masses. And this attitude yields immediately a *bonum*. The correlation between neutrino masses and lepton number violation naturally explains why no lepton number violation has been detected so far, as a consequence of the smallness of neutrino masses.

We proceed now to first describe the most general neutrino mass matrix (for simplicity, we confine the discussion to the one-family case), and we will then consider various attractive situations which arise as particular cases. Given two *independent* Weyl spinors ψ_L, ψ_R, the most general mass Lagrangian we can construct following the prescriptions of the preceding section is

$$(2.5) \quad \mathscr{L}_{\text{mass}} = D[\bar{\psi}_L \psi_R + \bar{\psi}_R \psi_L] + A[\overline{(\psi_L)^c}\psi_L + \bar{\psi}_L(\psi_L)^c] + B[\overline{(\psi_R)^c}\psi_R + \bar{\psi}_R(\psi_R)^c].$$

By assembling ψ_L, ψ_R into two independent Majorana spinors χ, ξ:

$$(2.6) \quad \chi \equiv \psi_L + (\psi_L)^c = \chi^c,$$

$$(2.7) \quad \xi \equiv \psi_R + (\psi_R)^c = \xi^c,$$

we rewrite eq. (2.5) as

$$(2.8) \quad \mathscr{L}_{\text{mass}} = \frac{1}{2} D(\bar{\chi}\xi + \bar{\xi}\chi) + A\bar{\chi}\chi + B\bar{\xi}\xi = (\bar{\chi}, \bar{\xi}) \begin{pmatrix} A & D/2 \\ D/2 & B \end{pmatrix} \begin{pmatrix} \chi \\ \xi \end{pmatrix}.$$

The eigenvalues of this mass matrix are

$$(2.9) \quad m_{1,2} = 1/2 \{(A + B) \pm [(A - B)^2 + D^2]^{\frac{1}{2}}\}$$

and for the eigenvectors we find

$$(2.10) \quad \begin{cases} \zeta_1 = \chi \cos\theta - \xi \sin\theta, \\ \zeta_2 = \chi \sin\theta + \xi \cos\theta, \end{cases}$$

with $\theta = \frac{1}{2} \text{arctg} [D/(A - B)]$. Clearly ζ_1, ζ_2 are *Majorana spinors* (since χ and ξ are self-conjugate), and so neutrinos are Majorana particles in the most general case of eq. (2.5).

Let us consider now some interesting particular cases.

1) $A \neq 0$, $B = D = 0$. Neutrinos are obviously *Majorana* particles. This situation corresponds to *no* introduction of ν_R, and is in a sense the *minimal* extension of the $SU_2 \times U_1$ standard model. Neutrino masses arise upon spontaneous symmetry breaking from the following invariant Lagrangian:

(2.11) $$\mathscr{L}'_Y = h' \overline{(L)^c} \sigma_i \Phi_i L + \text{h.c.},$$

where $L^T \equiv (\nu_L, e_L)$, σ_i are the Pauli matrices and Φ is a complex isotriplet Higgs field. The many-family generalization is trivial. If Φ does not carry lepton number, this symmetry is *explicitly* broken [6]. Then the magnitude of A is totally arbitrary, and so no understanding emerges about the smallness of neutrino masses. An attractive possibility is that the origin of lepton number violation is *just* due to nonvanishing neutrino masses, so as to recover lepton number conservation in the limit of vanishing neutrino masses—like, *e.g.*, in the high-temperature regime in the early Universe—a peculiar feature of the $SU_2 \times U_1$ standard model. This philosophy *demands* that Φ should carry (minus) two units of lepton number, which is, therefore, *spontaneously* broken in order to generate $A \neq 0$. Correspondingly, a physical Goldstone boson—named *majoron*—comes into play. Then an astrophysical bound exists, which is *independent* of neutrino masses and requires $\langle \Phi \rangle < 10^5$ eV [7], so that *a fortiori* $A < 10^5$ eV. This provides a rationale for the neutrino-charged lepton mass splitting. Such a scenario is implemented in the *triplet majoron model* [8], which leads to various new predictions for particle physics, astrophysics and cosmology [9]. Finally, a very speculative but fashinating possibility is that neutrino masses of this kind arise as a low-energy remnant of gravitational interactions. Then they would be in the 10^{-5} eV range, and so naturally small [10].

2) $A \ll D \ll B$. Neutrinos are again *Majorana* particles. This is the well-known « see-saw » mechanism [11]. Take first $A = 0$. Then eq. (2.9) yields

(2.12) $$m_1 \simeq D^2/B,$$

(2.13) $$m_2 \simeq B.$$

Since D is expected to be in the range of some charged-fermion mass in the same family, we see that a large B naturally implies a small neutrino mass m_1. Certainly, one would like to correlate B to some hypothetical new physical scale, and various proposals have been made, connecting B to the grand-unification scale [12], the scale of spontaneous parity violation [13], the scale of spontaneous lepton number violation [14] or the scale associated with horizontal interactions [15]. Not to make our discussion exceedingly long, we consider this mechanism within the $SU_2 \times U_1$ standard model. The Dirac mass term in (2.5) arises upon spontaneous symmetry breaking of the gauge group from the invariant Lagrangian:

(2.14) $$\mathscr{L}_Y = h\bar{L}\varphi e_R + h'\bar{L}\tilde{\varphi}\nu_R + \text{h.c.},$$

where φ is the usual Higgs doublet and $\tilde{\varphi}$ its conjugate. Consequently, such a term is chirally protected by $SU_2 \times U_1$, and so $D < \langle\varphi\rangle$. Since ν_R is a gauge singlet, B can be whatever, and the «survival hypothesis» [16] strongly suggests that it should be very large. This is why m_1 is naturally small. It should be noted that A cannot really vanish because of radiative corrections, but, since one estimates $A \sim D^2/B$, the above scenario is not essentially altered [17]. In this way we have a naturally light Majorana neutrino with a very heavy and unobservable partner.

3) $A = B \ll D$. This case is referred to as *pseudo-Dirac* neutrinos [18], although they are *Majorana* particles. It is a logical possibility which is interesting for neutrinoless double beta-decay (see below). Unfortunately, neutrinos are *not* naturally light in this scheme, and so we do not discuss it any further.

4) $D \neq 0$, $A = B = 0$. Since the Majorana mass terms in eq. (2.5) are killed by definition, the two Majorana spinors χ, ξ pair up to form a *Dirac* neutrino.

In the multigenerational extensions of the above-discussed schemes, there are in general interfamily mixings which break the separate lepton numbers, exactly the same way as it occurs in the quark sector. This fact gives rise to the important phenomenon of *neutrino oscillations* [19] (see below).

Finally, we would like to discuss the *neutrinoless double beta-decay* $(\beta\beta)_0$ [20], a unique process which arises as a consequence of lepton number violation: just the case of Majorana neutrinos. Basically, it consists of the following nuclear decay: $(Z, A) \to (Z+2) + 2e^-$ where the neutrino legs originating from the decay of the two neutrons are *virtual*. Why this is possible can be understood intuitively. In fact, both neutrino legs carry minus one unit of outgoing lepton number. To tie the two legs together, a Lorentz-invariant neutrino bilinear not involving derivatives is needed, which violates lepton number by two units: this is just a Majorana mass term! This fact makes it also clear why lepton number conservation forbids the $(\beta\beta)_0$ process [21]. Quantitatively, the corresponding amplitude $A\,[(\beta\beta)_0]$ is the sum of the contributions from all the neutrino mass eigenstates ν_i with mass m_i which couple to the electron. One finds [22]

(2.15) $$A[(\beta\beta)_0] = \sum_i \eta(\nu_i)|U_{ei}|^2 m_i \tilde{A},$$

where $\eta(\nu_i)$ is the CP parity of ν_i, U_{ei} its mixing parameters with the electron, and $\tilde{A}$ is a quantity independent of m_i and related to the nuclear matrix elements.

3. – A glimpse to experiment.

It is now time to discuss the three classic phenomena in which nonvanishing neutrino masses can show up and hopefully be detected. We will pay attention to what has been already learned this way.

i) *Neutrino oscillations*. A detailed discussion would be exceedingly long, so we give only a qualitative account, which is, however, sufficient for our purposes [19]. The essential point is that, in any experiment, neutrinos are produced and detected as *current* eigenstates, which have well-defined momentum but *not* well-defined energy, as a consequence of the nonvanishing off-diagonal elements in the (neutrino) mass matrix. Actually, any current eigenstate is a linear superposition of the various neutrino mass eigenstates. Now, if the latters are nondegenerate, each of them will have a different time evolution. Therefore, the coefficients of the decomposition of the current eigenstates into mass eigenstates will change with the time. This simple fact has the following implication. Suppose that at time t_1 only, *e.g.*, ν_μ are produced in experiment. Then, at a later time t_2 the neutrino beam will consist not only of ν_μ, but of ν_e and ν_τ as well. Also the latter can, therefore, be detected. Since the phenomenon of neutrino oscillations necessarily requires at least one neutrino type with nonvanishing mass, experiments of this kind can yield a positive evidence for neutrino masses. Although all neutrino oscillation experiments led to negative results in the past, there is a very recent claim in favour of a nonvanishing neutrino mass [23] in the form

$$(3.1) \qquad \Delta m^2 = 0.2 \text{ (eV)}^2, \quad \sin^2 2\theta = 0.2,$$

where Δm^2 is the difference of two squared neutrino masses and θ is the mixing angle, under the assumption that only two kinds of neutrinos participate in the oscillation. Obviously, this result requires further experimental confirmation before being taken seriously.

ii) *Neutrinoless double beta-decay* $(\beta\beta)_0$. We have already discussed this process from a theoretical point of view. Experimentally, there is the hope to observe it by looking at nuclei for which ordinary β-decay is energetically forbidden or at least strongly suppressed by angular-momentum conservation. Clearly, also the lepton-number-conservaing two-neutrino mode $(\beta\beta)_2$ is then allowed, but is expected to be strongly suppressed with respect to the $(\beta\beta)_0$ mode by the four-body phase-space. There are basically two alternative ways to search for $(\beta\beta)_0$ decay. One is the geological method, which is based on the following strategy. It is well known that during the Earth's history, some nuclei undergo β-decay. One looks for anomalously large abundances of those isotopes that are the products of *double* β-decay. Since the lifetimes corresponding to the $(\beta\beta)_0$ and $(\beta\beta)_2$ modes can be predicted and are generally different, from the age of the rock in which the daughter nuclei $(A, Z+2)$ are trapped one can determine which of the two decay modes is the leading one. An anomalously short lifetime for the decay ^{128}Te $\to$ ^{128}Xe was found [24] and interpreted [25] as the $(\beta\beta)_0$ decay mode induced by Majorana neutrinos with

$$\text{(3.2)} \qquad \sum_i \eta(\nu_i)|U_{ei}|^2 m_i \simeq 30 \text{ eV}.$$

However, such a claim has been recently disproved [26]. At present, the geological analysis yields the following upper bound [26]:

$$\text{(3.3)} \qquad \sum_i \eta(\nu_i)|U_{ei}|^2 m_i \leqslant 6 \text{ eV}.$$

It goes without saying that the geological experiments are beset by many uncertainties, and so their results should be taken with great caution.

Quite recently, laboratory experiments performed on the decay ^{76}Ge $\to$ ^{76}Se have reached a high sensitivity, thus providing interesting and very reliable results. No positive evidence exists by now in favour of $(\beta\beta)_0$ decay, and the following upper bound has been obtained [27]:

$$\text{(3.4)} \qquad \sum_i \eta(\nu_i)|U_{ei}|^2 m_i \leqslant 3.7 \text{ eV}.$$

The only uncertainty in eq. (3.4) comes from the actual evaluation of the nuclear matrix elements.

   iii) *Tritium beta-decay*. The electron energy spectrum in nuclear β-decay is affected by finite $\bar{\nu}_e$ mass. By denoting by $N(E)$ and P the electron energy spectrum and its momentum, respectively, consider the Kurie plot of $[N(E)/PE]^{\frac{1}{2}}$ vs. E. The case of vanishing m_{ν_e} gives a straight line with negative slope, which intercepts the E axis at $\Delta \equiv$ neutron mass minus proton kinetic energy. The effect of $m_{\nu_e} \neq 0$ is to cause the line to curve downward near the endpoint with infinite slope. The point where it intercepts the E axis is then $\Delta - m_{\nu_e}$, which, therefore, yields the value of m_{ν_e} [28]. The tritium β-decay ^{3}H $\to$ $\to$ ^{3}He $+ e^- + \bar{\nu}_e$ turns out to be the most suitable decay from an experimental point of view, to look for finite neutrino mass effects. In 1980 the ITEP group at Moscow reported the evidence for $14 \text{ eV} < m_{\nu_e} < 46 \text{ eV}$ [29]. This experiment has been re-done by the same group, who now finds [30]

$$\text{(3.5)} \qquad m_{\nu_e} \simeq 33 \text{ eV}.$$

Unfortunately, the tritium used in this experiment is contained in the molecular environment of valine ($C_5H_{11}NO_{26}$), so that eq. (3.5) has to be taken with some caution. Various experiments are underway to confirm or disprove this result, including the same kind of decay using, however, free tritium.

iv) *Moral.* Although there is by now no firm evidence in favour of nonvanishing neutrino masses, one can be positive enough to take the results (3.1), (3.4) and (3.5) at face value. The first question that comes to mind is whether or not they are compatible.

One should remember that eq. (3.4) applies only to Majorana masses, whereas eqs. (3.1) and (3.5) have a general validity. It should be perfectly clear that eqs. (3.1), (3.4) and (3.5) can well be compatible for suitable values of m_i, U_{ei} and $\eta(\nu_i)$. This is an important point, since it has been claimed sometimes that the $(\beta\beta)_0$ results cannot be reconciled with eq. (3.5), thereby ruling out the possibility of Majorana neutrino masses. In fact, a look at eq. (3.4) makes it obvious that this is *not* the case.

The question whether one is fully satisfied in this way is of a completely different nature. One might wish to have eqs. (3.4) and (3.5) *automatically* compatible, and/or understand the above results as a first hint toward the situation in which a future improved knowledge will make eqs. (3.4) and (3.5) incompatible. Then one can consider *pseudo-Dirac* neutrinos, which have opposite CP parity, so that $\eta(\nu_i) = 0$ in eq. (3.4). Unfortunately, as we discussed in the preceding section, they are not naturally light. So the only possibly left open along the avenue of *naturalness* is Dirac neutrinos.

We stressed that the theoretical motivation to consider Majorana instead of Dirac neutrinos was to have them naturally light. As it happens often in the development of science, a fruitful initial assumption can lead to achievements which finally end up in showing that the initial motivation was wrong. This can also be the case here. In the next section we will describe a model for naturally light Dirac neutrinos, which provides an elegant solution to eqs. (3.4) and (3.5) in a *natural* manner.

Are neutrino really massive? Are they Dirac or Majorana particles? Only time will tell.

4. – A concrete proposal.

With the motivation discussed at the end of last section, we consider the $SU_2 \times U_1$ standard model and enlarge the spectrum of its physical states by only introducing gauge group *singlets* [31]. Specifically, for each fermion family, we add the « right-handed neutrino » D_R (note the change of notation with respect to sect. **2**) along with a pair of independent Weyl spinors S_L, S_R which carry lepton number plus one. Moreover, we add a complex Higgs field Φ carrying no lepton number. For simplicity, we discuss the one-family case only. Now the most general gauge and lepton number invariant Lagrangian reads

$$(4.1) \quad \mathscr{L}'_Y = h_1 \bar{L}\tilde{\varphi}e_R + h'_1 \bar{L}\varphi D_R + h_2 \bar{L}\varphi S_R + h'_2 \Phi \bar{S}_L S_R + \\ + h_3 \Phi \bar{S}_L D_R + M \bar{S}_L S_R + M' \bar{S}_L D_R + \text{h.c.},$$

where $L^T \equiv (D_L, e_L)$ and φ are the usual lepton and Higgs doublets. To get *naturally* light Dirac neutrinos we impose a new *symmetry* in order to constrain the otherwise arbitrary neutrino mass matrix arising from L'_Y. We take the following continuous global symmetry [32]:

(4.2) $$U_{1,D}: \begin{cases} D_R \to \exp[i\theta] D_R, \\ \Phi \to \exp[-i\theta] \Phi, \end{cases}$$

with all the other fields invariant. The requirement that L'_Y should be $U_{1,D}$ invariant implies $h'_1 = h'_2 = M' = 0$. As a consequence, L'_Y reduces to

(4.3) $$\mathscr{L}_Y = h_1 \bar{L} \tilde{\varphi} e_R + h_2 \bar{L} \varphi S_R + h_3 \Phi \bar{S}_L D_R + M \bar{S}_L S_R + \text{h.c.}$$

Upon spontaneous symmetry breaking, L_Y yields the following neutrino mass matrix:

(4.4) $$\mathscr{M} = \begin{array}{c|cccc} & D_L & S_L & D_R & S_R \\ \hline \bar{D}_L & 0 & 0 & 0 & m_1 \\ \bar{S}_L & 0 & 0 & m_2 & M \\ \bar{D}_R & 0 & m_2 & 0 & 0 \\ \bar{S}_R & m_1 & M & 0 & 0 \end{array}$$

with $m_1 = h_1 \langle \varphi \rangle$, $M_2 = h_3 \langle \Phi \rangle$. By denoting by Λ the natural cut-off of the theory, we expect

(4.5) $$m_1 \lesssim G_F^{-\frac{1}{2}} \approx 250 \text{ GeV}, \quad m_2 \ll \Lambda, \quad M \approx \Lambda,$$

since m_1 is chirally protected by $SU_2 \times U_1$, m_2 breaks $U_{1,D}$, while M does not break any symmetry (survival hypothesis). It is straightforward to see that the four Weyl spinors D_L, D_R, S_L, S_R pair up into two Dirac spinors which are eigenstates of $\mathscr{M}$:

(4.6) $$\nu \simeq D_L + D_R - \frac{m_1}{M} S_L - \frac{m_2}{M} S_R, \quad N \simeq S_L + S_R + \frac{m_1}{M} D_L + \frac{m_2}{M} D_R,$$

corresponding to the eigenvalues

(4.7) $$m_\nu \simeq \frac{m_1 m_2}{M}, \quad m_N \simeq M.$$

We see that eqs. (4.5) imply $m_\nu \ll G_F^{-\frac{1}{2}}$, so that neutrinos are *naturally* light. Notice that we are working at the leading order dictated by eqs. (4.5).

Before proceding further, we would like to stress a delicate point. For the electron family, eqs. (4.7) should yield $m_{\nu_e}/m_e \simeq 10^{-6} \ldots 10^{-5}$, which is such a tiny ratio that might well be totally upset by radiative corrections. By carefully looking at the one-loop level, one finds only one possible contribution to m_ν, which is less than $m_1 m_2/M$. So no problem arises.

Clearly a Goldstone boson—named *Diron*—is present in the theory, arising from the spontaneous breakdown of $U_{1,D}$. A detailed analysis shows that the couplings of the Diron both to charged fermions and to neutrinos are extremely small, thereby making it totally harmless. The smallness of the latter couplings is of particular importance once cosmology is considered. It is well known that neutrinos stable on the age of the Universe should weigh less than 100 eV [33]. Now, in a many-family extension of the present model, one might get ν_μ or ν_τ heavier than 100 eV, and no cosmological trouble would arise if they could decay sufficiently fast into ν_e + Diron. This is, however, *not* the case, as a consequence of the tiny neutrino-Diron coupling. Therefore, the model under consideration predicts that *all* neutrinos should weigh less than 100 eV, in order to be consistent with cosmology. The reader can check that eqs. (4.5) and (4.7) actually lead to such a numerology, since Λ should be take on the order of the Planck mass $\simeq 10^{19}$ GeV.

* * *

I should like to thank Prof. N. CABIBBO for his kind invitation to lecture at this School and Dr. G. MARTINELLI for his very efficient organization.

APPENDIX

It is assumed throughout that all spinors are anticommuting objects, in accord with the connection between spin and statistics.

As we mentioned in the text,

(A.1) $\qquad\qquad\qquad \psi_\alpha \sim (\tfrac{1}{2}, 0) \qquad\qquad (\alpha = 1, 2),$

(A.2) $\qquad\qquad\qquad \bar{\psi}^{\dot{\alpha}} \sim (0, \tfrac{1}{2}) \qquad\qquad (\dot{\alpha} = 1, 2),$

under $O_{3,1}$. The indices can be raised and lowered as follows:

(A.3) $\qquad\qquad \psi^\alpha \equiv \varepsilon^{\alpha\beta} \psi_\beta, \qquad \bar{\psi}_{\dot{\alpha}} \equiv \varepsilon_{\dot{\alpha}\dot{\beta}} \bar{\psi}^{\dot{\beta}},$

where $\varepsilon^{\alpha\beta}$, $\varepsilon_{\dot{\alpha}\dot{\beta}}$ are the totally antisymmetric tensors with $\varepsilon^{12} = \varepsilon_{21} = +1$. The multiplication conventions are

(A.4) $\qquad\qquad \psi\varphi \equiv \varepsilon^{\alpha\beta} \psi_\beta \varphi_\alpha, \qquad \bar{\psi}\bar{\varphi} \equiv \varepsilon_{\dot{\alpha}\dot{\beta}} \bar{\psi}^{\dot{\beta}} \bar{\varphi}^{\dot{\alpha}}.$

It can be shown that the SU_2 dotted and undotted indices are covariantly saturated in eq. (A.4), and so $\psi\varphi$ and $\bar{\psi}\bar{\varphi}$ are actually Lorentz-invariant expressions. Finally $(\psi\varphi)^\dagger \equiv \bar{\varphi}\bar{\psi}$. We work in the Weyl basis for the gamma-matrices, which is defined as

(A.5) $$\gamma^\mu \equiv \begin{pmatrix} 0 & \sigma^\mu \\ \bar{\sigma}^\mu & 0 \end{pmatrix} \qquad (\mu = 0, 1, 2, 3)$$

in terms of the Pauli matrices σ^i ($i = 1, 2, 3$) and $\bar{\sigma}^0 = \sigma^0$, $\bar{\sigma}^i = -\sigma^i$.

Consequently, we have $\{\gamma^\mu, \gamma^\nu\} = 2\eta^{\mu\nu}$, $\eta^{\mu\nu} \equiv (+, -, -, -)$. The γ_5 matrix is

(A.6) $$\gamma_5 \equiv -i\gamma^0\gamma^1\gamma^2\gamma^3 = \begin{pmatrix} 1 & 0 \\ 0 & 1 \end{pmatrix}$$

and the charge conjugation matrix C satisfying the relation $C^{-1}\gamma^\mu C = -\gamma^{\mu\mathrm{T}}$ is taken to be

(A.7) $$C = -i\gamma^2\gamma^0 = \begin{pmatrix} \varepsilon_{\alpha\beta} & 0 \\ 0 & \varepsilon^{\dot{\alpha}\dot{\beta}} \end{pmatrix}.$$

Correspondingly, eq. (A.3) can be rewritten as

(A.8) $$\psi^\alpha = (C^{-1})^{\alpha\beta}\psi_\beta, \qquad \bar{\psi}_{\dot{\alpha}} = (C^{-1})_{\dot{\alpha}\dot{\beta}}\bar{\psi}^{\dot{\beta}}.$$

The left- and right-handed projectors for a four-component spinor ψ are defined as

(A.9) $$P_\mathrm{L} \equiv \tfrac{1}{2}(1 + \gamma_5), \qquad P_\mathrm{R} \equiv \tfrac{1}{2}(1 - \gamma_5).$$

By applying these operators to a Dirac and a Majorana spinor, we have, respectively,

(A.10) $$P_\mathrm{L}\psi_\mathrm{D} = \frac{1}{2}(1+\gamma_5)\begin{pmatrix}\varphi_\alpha \\ \bar{\chi}^{\dot{\alpha}}\end{pmatrix} = \begin{pmatrix}\varphi_\alpha \\ 0\end{pmatrix}, \qquad P_\mathrm{R}\psi_\mathrm{D} = \frac{1}{2}(1-\gamma_5)\begin{pmatrix}\varphi_\alpha \\ \bar{\chi}^{\dot{\alpha}}\end{pmatrix} = \begin{pmatrix}0 \\ \bar{\chi}^{\dot{\alpha}}\end{pmatrix},$$

and

(A.11) $$P_\mathrm{L}\psi_\mathrm{M} = \frac{1}{2}(1+\gamma_5)\begin{pmatrix}\chi_\alpha \\ \bar{\chi}^{\dot{\alpha}}\end{pmatrix} = \begin{pmatrix}\chi_\alpha \\ 0\end{pmatrix}, \qquad P_\mathrm{R}\psi_\mathrm{M} = \frac{1}{2}(1-\gamma_5)\begin{pmatrix}\chi_\alpha \\ \bar{\chi}^{\dot{\alpha}}\end{pmatrix} = \begin{pmatrix}0 \\ \bar{\chi}^{\dot{\alpha}}\end{pmatrix},$$

where eqs. (1.4) and (1.5) have been used. Equations (A.10) and (A.11) show an important difference between a Dirac and a Majorana spinor: in the former case the left- and right-handed projections are *independent*, which is *not* any longer true in the latter case.

Given a four-component spinor ψ, we denote by $\bar{\psi}$ the one which transforms oppositely to ψ under $O_{3,1}$. Explicitly, this means

(A.12) $$\bar{\psi} \equiv \psi^\dagger \gamma^0.$$

Now, the charge conjugated of ψ is defined as

(A.13) $$\psi^c \equiv C(\bar{\psi})^\mathrm{T}$$

with C given by eq. (A.7). Inserting ψ_D and ψ_M as given by eqs. (1.4) and (1.5) into eq. (A.12), we get

(A.14) $$\bar{\psi}_D = (\chi^\alpha, \bar{\varphi}_{\dot\alpha}),$$

(A.15) $$\bar{\psi}_M = (\chi^\alpha, \bar{\chi}_{\dot\alpha}).$$

Finally, by inserting again ψ_D and ψ_M into eq. (A.13), it follows

(A.16) $$(\psi_D)^c = \begin{pmatrix} \chi_\alpha \\ \bar{\varphi}^{\dot\alpha} \end{pmatrix},$$

(A.17) $$(\psi_M)^c = \begin{pmatrix} \chi_\alpha \\ \bar{\chi}^{\dot\alpha} \end{pmatrix}.$$

Equations (A.16) and (A.17) show the most important difference between a Dirac and a Majorana spinor: in the former case φ_α gets interchanged with χ_α under charge conjugation, whereas a Majorana spinor is *self-conjugate*.

REFERENCES

[1] S. M. BILENKY and B. PONTECORVO: *Phys. Rep.*, **41**, 225 (1978); P. H. FRAMPTON and P. VOGEL: *Phys. Rep.*, **82**, 339 (1982); L. WOLFENSTEIN: CMU-HEP-84-18 preprint, to appear in the *Proceedings of Neutrino 84, June 1984, Dortmund (W. Germany)*; any many others.

[2] A. D. DOLGOV and YA. B. ZEL'DOVICH: *Rev. Mod. Phys.*, **53**, 1 (1981); P. H. FRAMPTON and P. VOGEL: *Phys. Rep.*, **82**, 339 (1982); G. STEIGMAN: *Annu. Rev. Nucl. Part. Sci.*, **29**, 313 (1979); J. BERNSTEIN: CERN Yellow Report 84-06; E. W. KOLB: Fermilab-84/90-A preprint, to appear in the *Proceedings of Neutrino 84, June 1684, Dortmund (W. Germany)*; and many others.

[3] R. N. MOHAPATRA and G. SENJANOVIC: *Phys. Rev. Lett.*, **44**, 912 (1980); *Phys. Rev. D*, **23**, 165 (1981). For an early version of the model, see J. C. PATI and A. SALAM: *Phys. Rev. D*, **10**, 275 (1974); R. N. MOHAPATRA and J. C. PATI: *Phys. Rev. D*, **11**, 566, 2558 (1975); G. SENJANOVIC and R. N. MOHAPATRA: *Phys. Rev. D*, **12**, 1502 (1975).

[4] H. GEORGI: in *Particles and Fields-1974, Proceedings of the Meeting of the APS Division of Particles and Fields, Williamsburg, Virginia*, edited by C. E. CARLSON (AIP, New York, N. Y., 1975); H. FRITZSCH and P. MINKOWSKI: *Ann. Phys. (N. Y.)*, **93**, 193 (1975); H. GEORGI and D. V. NANOPOULOS: *Nucl. Phys. B*, **155**, 59 (1979).

[5] PARTICLE DATA GROUP: *Rev. Mod. Phys.*, **56**, No. 2, Part 2 (1984).

[6] T. P. CHENG and L. F. LI: *Phys. Rev. D*, **22**, 2860 (1980).

[7] H. GEORGI, S. L. GLASHOW and S. NUSSINOV: *Nucl. Phys. B*, **193**, 297 (1981); M. FUKUGITA, S. WATAMURA and M. YOSHIMURA: *Phys. Rev. Lett.*, **48**, 1522 (1982); *Phys. Rev. D*, **26**, 1840 (1982); J. ELLIS and K. OLIVE: *Nucl. Phys. B*, **223**, 252 (1983).

[8] G. B. GELMINI and M. RONCADELLI: *Phys. Lett. B*, **99**, 411 (1981); H. GEORGI, S. L. GLASHOW and S. NUSSINOV: *Nucl. Phys. B*, **193**, 297 (1981).

[9] G. B. GELMINI and M. RONCADELLI: *Phys. Lett. B*, **99**, 411 (1981); H. GEORGI, S. L. GLASHOW and S. NUSSINOV: *Nucl. Phys. B*, **193**, 297 (1981); G. B. GELMINI, S. NUSSINOV and M. RONCADELLI: *Nucl. Phys. B*, **209**, 157 (1982); S. NUSSINOV and M. RONCADELLI: *Phys. Lett. B*, **122**, 387 (1983); V. BARGER, W. Y. KEUNG and S. PAKVASA: *Phys. Rev. D*, **25**, 907 (1982); V. BARGER, H. BOER, W. Y. KEUNG and R. J. N. PHILLIPS: *Phys. Rev. D*, **26**, 218 (1982); T. GOLDMAN, E. W. KOLB and G. J. STEPHENSON: *Phys. Rev. D*, **26**, 2503 (1982); J. ELLIS and J. HAGELIN: *Nucl. Phys. B*, **217**, 189 (1983); E. W. KOLB, D. L. TUBBS and D. A. DICUS: *Astrophys. J. Lett.*, **225**, L57 (1982); D. A. DICUS, E. W. KOLB and D. L. TUBBS: *Nucl. Phys. B*, **223**, 532 (1983); M. FUKUGITA, S. WATAMURA and M. YOSHIMURA: *Phys. Rev. Lett.*, **48**, 1522 (1982); *Phys. Rev. D*, **26**, 1840 (1982); J. ELLIS and K. A. OLIVE: *Nucl. Phys. B*, **223**, 252 (1983); P. LANGACKER, G. SEGRÉ and S. SONI: *Phys. Rev. D*, **26**, 3425 (1982); A. SANTAMARIA, A. PICH and J. BERNABEU: Valencia preprints (1984). Extension of the model have been discussed in J. SCHECHTER and J. W. F. VALLE: *Phys. Rev. D*, **25**, 774 (1982); R. BARBIERI and R. N. MOHAPATRA: *Z. Phys. C*, **11**, 175 (1981); R. N. MOHAPATRA and G. SENJANOVIC: *Phys. Rev. Lett.*, **49**, 7 (1982).

[10] R. BARBIERI, J. ELLIS and M. K. GAILLARD: *Phys. Lett. B*, **90**, 249 (1980).

[11] T. YANAGIDA: KEK lectures (1979); M. GELL-MANN, P. RAMOND and R. SLANSKY: in *Supergravity*, edited by D. Z. FREEDMAN and P. VAN NIEUWENHUIZEN (North-Holland, Amsterdam, 1979).

[12] M. GELL-MANN, P. RAMOND and R. SLANSKY: in *Supergravity*, edited by D. Z. FREEDMAN and P. VAN NIEUWENHUIZEN (North-Holland, Amsterdam, 1979); R. BARBIERI, D. V. NANOPOULOS, G. MORCHIO and F. STROCCHI: *Phys. Lett. B*, **90**, 91 (1980); E. WITTEN: *Phys. Lett. B*, **91**, 81 (1980).

[13] R. N. MOHAPATRA and G. SENJANOVIC: *Phys. Rev. Lett.*, **44**, 912 (1980); *Phys. Rev. D*, **23**, 165 (1981).

[14] Y. CHIKASHIGE, R. N. MOHAPATRA and R. D. PECCEI: *Phys. Lett. B*, **98**, 265 (1981); *Phys. Rev. Lett.*, **45**, 1926 (1980).

[15] Y. CHIKASHIGE, G. B. GELMINI, R. D. PECCEI and M. RONCADELLI: *Phys. Lett. B*, **94**, 492 (1980); T. YANAGIDA: *Prog. Theor. Phys.*, **64**, 1103 (1980); F. WILCZEK: *Phys. Rev. Lett.*, **49**, 1549 (1982); G. B. GELMINI, S. NUSSINOV and T. YANAGIDA: *Nucl. Phys. B*, **219**, 31 (1983).

[16] H. GEORGI: *Nucl. Phys. B*, **156**, 126 (1979).

[17] R. BARBIERI, D. V. NANOPOULOS, G. MORCHIO and F. STROCCHI: *Phys. Lett. B*, **90**, 91 (1980); M. MAGG and C. WETTERICH: *Phys. Lett. B*, **94**, 61 (1980).

[18] L. WOLFENSTEIN: *Nucl. Phys. B*, **186**, 147 (1981); S. T. PETCOV: *Phys. Lett. B*, **110**, 245 (1982); J. W. F. VALLE: *Phys. Rev. D*, **27**, 1672 (1983); C. N. LEUNG and S. T. PETCOV: *Phys. Lett. B*, **125**, 461 (1983); M. DOI, M. KENMOKU, T. KOTANI, H. NISHIURA and E. TAKASUGI: *Prog. Theor. Phys.*, **70**, 1331 (1983).

[19] B. PONTECORVO: *Sov. Phys. JETP*, **6**, 429 (1958); **7**, 172 (1958); **26**, 984 (1968). For a review, see S. M. BILENKY and B. PONTECORVO: *Phys. Rep.*, **41**, 225 (1978).

[20] For a review, see D. BRYMAN and C. PICCIOTTO: *Rev. Mod. Phys.*, **50**, 11 (1978); H. PRIMAKOFF and S. P. ROSEN: *Annu. Rev. Nucl. Sci.*, **31**, 145 (1981); W. C. HAXTON and G. J. STEPHENSON: *Double beta-decay*, to appear in *Prog. Part. Nucl. Phys.*

[21] The $(\beta\beta)_0$ decay can also be induced by right-handed weak currents. We assume throughout that this possible contribution is negligible compared to the possible lepton-number-violating contribution. Actually, both of them are operative in inducing $0^+ \to 0^+$ transitions between the ground states of father and daughter nuclei, while the $0^+ \to 2^+$ decay to the first excited state of $(A, Z+2)$ can only be induced by right-handed weak currents. Therefore, a discrimination between

the two mechanisms is experimentally possible by considering the energy and angular distribution of the single outgoing electrons.

[22] B. KAYSER and A. S. GOLDHABER: *Phys. Rev. D*, **28**, 2341 (1983); B. KAYSER: *Phys. Rev. D*, **30**, 1023 (1984).
[23] J. F. CAVAIGNAC *et al.*: Lapp exp. 84-03 and ISN-84-11 preprints.
[24] E. W. HENNECKE, O. K. MANUEL and D. D. SABU: *Phys. Rev. C*, **11**, 1378 (1975); E. W. HENNECKE: *Phys. Rev. C*, **17**, 1168 (1978).
[25] M. DOI, T. KOTANI, H. NISHIURA, K. OKUDA and E. TAKASUGI: *Phys. Lett. B*, **103**, 219 (1981); *Prog. Theor. Phys.*, **66**, 1765 (1981).
[26] T. KIRSTEN, H. RICHTER and E. JESSBERGER: *Phys. Rev. Lett.*, **50**, 474 (1983); *Z. Phys. C*, **16**, 189 (1983).
[27] E. BELLOTTI, O. CREMONESI, E. FIORINI, C. LIGUORI, A. PULLIA, P. SVERZELLATI and L. ZANOTTI: *New limits on double beta decay of* Ge, Milano preprint (1984), to appear in *Phys. Lett. B*,
[28] If ν_e is not a mass eigenstate, the mixing effects will imply the existence of kinks from various superimposed endpoints in the Kurie plot.
[29] V. LUBIMOV, E. G. NOVIKOV, V. Z. NOZIK, E. F. TRETYAKOV and V. S. KOSIK: *Phys. Lett. B*, **94**, 266 (1980).
[30] S. BORIS *et al.*: *Proceedings of the International Europhysics Conference on High Energy Physics, Brighton, July 1983.*
[31] The model described in this section is due to D. WYLER and M. RONCADELLI: *Phys. Lett. B*, **133**, 325 (1983). A very similar model has been proposed by P. ROY and O. SHANKER: *Phys. Rev. Lett.*, **52**, 713 (1984).
[32] Although a discrete symmetry could be sufficient, it would lead to the cosmological «domain wall» problem.
[33] S. S. GERSTEIN and YA. B. ZEL'DOVICH: *JETP Lett.*, **4**, 120 (1966); R. COWSIK and J. MCCLELLAND: *Phys. Rev. Lett.*, **29**, 669 (1972).

Dark Matter, Galaxies and Large-Scale Structure in the Universe.

J. R. PRIMACK (*)

Stanford Linear Accelerator Center, Stanford University - Stanford, CA 94305

The standard theory of cosmology is the hot big bang, according to which the early universe was hot, dense, very nearly homogeneous, and expanding adiabatically according to the laws of general relativity. This theory nicely accounts for the cosmic background radiation, and accurately predicts the abundances of the lightest nuclides. It is probably even true, as far as it goes; at least, I will assume so here. But as a fundamental theory of cosmology, the standard theory is seriously incomplete. One way of putting this is to say that it describes the middle of the story, but leaves us guessing about both the beginning and the end.

Galaxies and large-scale structure—clusters of galaxies, superclusters and voids—are the grandest structures visible in the Universe, but their origins are not yet understood. Moreover, there is compelling observational evidence that most of the mass detected gravitationally in galaxies and clusters is dark —that is, visible neither in absorption nor emission of any frequency of electromagnetic radiation.

Explaining the rich variety and correlations of galaxy and cluster morphology will require filling in much more of the history of the Universe:

Beginnings, in order to understand the origin of the fluctuations which eventually collapse gravitationally to form galaxies and large-scale structure. This is a mystery in the standard expansionary universe, because the matter which comprises a typical galaxy, for example, first came into causal contact about a year after the big bang. It is very hard to see how galaxy-size fluctuations could have formed after that, but even harder to see how they could have formed earlier.

(*) Permanent address: Santa Cruz Institute for Particle Physics, University of California, Santa Cruz, CA 95034.

Denouement, since even given appropriate initial fluctuations, we are far from understanding the evolution of clusters and galaxies, or even the origins of stars and the stellar initial mass function.

And the mysterious *dark matter* is probably the key to unravelling the plot since it appears to be gravitationally dominant on all scales larger than the cores of galaxies. The dark matter is, therefore, crucial for understanding the evolution and present structure of galaxies, clusters, superclusters and voids.

Most reviews of cosmology have until recently concentrated on explaining the hot big bang, especially primordial nucleosynthesis. With the advent of grand unified theories (GUTs) in particle physics, and especially the lovely idea of cosmic inflation, it has also become possible to give an account of the very early universe which is at least coherent, if not yet very well grounded observationally.

The present lectures take a different approach, emphasizing the period *after* the first three minutes, during which the Universe expands by a factor of $\sim 10^8$ to its present size, and all the observed structures form. This is now an area undergoing intense development in astrophysics, both observationally and theoretically. It is probably now ripe for major progress. It is not impossible that the present decade will see the construction at last of a fundamental theory of cosmology, with perhaps profound implications for particle physics as well.

Although I will concentrate in these lectures on the development of galaxies and large-scale structure in the relatively « recent » universe, I can hardly avoid retelling some of the earlier parts of the story. Primordial nucleosynthesis will be important in this context primarily as a source of information on the amount of ordinary (« baryonic ») matter in the Universe; GUT baryosynthesis, for its implication that the primordial fluctuations were probably adiabatic; and inflation, for the constant-curvature (« Zel'dovich ») spectrum of fluctuations and a plausible solution to the problem of generating these large-scale fluctuations without violating causality. I will be especially concerned with evidence and arguments bearing on the astrophysical properties of the dark matter, which can also help to constrain possible particle physics candidates. The list of these now includes ~ 30 eV neutrinos, very massive right-handed neutrinos, other heavy stable particles such as photinos, massive unstable neutrinos or their decay products, very light « invisible » axions, u-d-s symmetric « quark nuggets » and primordial black holes. One of these hypothetical species may be the dominant form of matter in the Universe—or perhaps it is something no one has even thought of yet!

I will begin by discussing the basic astronomical data on the distribution of matter in the Universe: galaxies, clusters, superclusters and voids, and the strong evidence that all the visible matter on galaxy scales and larger is moving in the vast potential wells of the gravitationally dominant dark matter. If this is so, the inevitable question is how these enormous ghostly structures formed.

To prepare to discuss the answers that have been proposed, I will need to review the theory of gravity, not merely standard general-relativistic cosmology, but also the theory of the growth and collapse of fluctuations in an expanding universe. Learning the basic theory of gravitational collapse—including « virialization » by « violent relaxation »—was a revelation to me, and it has been my experience that it is not generally appreciated outside astrophysics. Under the rubric of gravity theory, I will also discuss briefly the idea of cosmic inflation and its implications for the origin of fluctuations. And I will discuss even more briefly some recent suggestions of modified gravity, with a r^{-1} force law at large distances, as an alternative to dark matter.

Next comes the most conventional part of these lectures, describing the standard hot big bang: decoupling, nucleosynthesis, recombination. This provides the essential background for the three astrophysical arguments that the dark matter is probably not baryonic: excluding various possible forms of baryonic dark matter in galaxy halos, bounding the abundance of baryonic matter using the observed deuterium abundance, and bounding the magnitude of adiabatic fluctuations at recombination from the observational upper limits on fluctuations in the cosmic background radiation. (I will also point out explicitly the loopholes in each of these arguments.)

Finally I take up the key question: what is the dark matter that the Universe is mostly made of? From the viewpoint of astrophysics, it is useful to categorize the dark matter as *hot, warm,* or *cold*, depending on its thermal velocity compared to the Hubble flow (expansion). Hot dark matter, such as ~ 30 eV neutrinos, is still relativistic when galaxy-size masses ($\sim 10^{12} M_\odot$, where $M_\odot = 2.0 \cdot 10^{33}$ g is the mass of the Sun) are first encompassed within the horizon. Warm dark matter is just becoming nonrelativistic then. Cold dark matter, such as axions or massive photinos, is nonrelativistic when even globular-cluster masses ($\sim 10^6 M_\odot$) come within the horizon. As a consequence, fluctuations on galaxy scales are wiped out with hot dark matter but preserved with warm, and all cosmologically relevant fluctuations survive in a universe dominated by cold dark matter.

The first possibility for nonbaryonic dark matter that was examined in detail was massive neutrinos, assumed to have mass ~ 30 eV—both because that mass corresponds to closure density, and because the Moscow tritium β-decay experiment continues to provide evidence that the electron neutrino has that mass. Although this picture leads to superclusters and voids of the size seen, superclusters are the first structures to collapse in this theory since smaller-size fluctuations do not survive. The theory founders on this point, however, since galaxies are almost certainly older than superclusters. A related problem is that galaxy and cluster formation is sufficiently complicated in the neutrino picture that no theory of it has yet been worked out.

A currently popular possibility is that the dark matter is cold. I have been one of those who have been studying the consequences of this picture. Its virtues include an account of galaxy and cluster formation that appears —at least to me and my co-workers—to be very attractive. Its defects are less clear, perhaps at least partly because it has not yet been subjected to enough critical scrutiny. Some recent work suggests that the size of the large-scale structure in a cold-dark-matter universe will come out right only if the density is not more than about half the critical density, but this is contrary to prejudice, the inflationary hypothesis and the latest upper bounds on small-angle fluctuations in the microwave background radiation. Another problem with hot as well as cold dark matter is understanding the strong correlations in the locations of rich (*i.e.* populous) clusters of galaxies across tremendous distances, large even compared to the scale of superclusters.

These lectures end with a survey of new ideas for solving these problems, new sources of observational data which may differentiate more clearly between the various possibilities for the dark matter, and finally some possible broader implications of the picture that is emerging from particle physics and cosmology of the structure of the Universe on both the smallest and largest scales.

1. – Matter.

1˙1. *Sizes.* – This section is mainly about the distribution of matter in the Universe on galaxy and larger scales, and the evidence that most of the mass is dark. But I think it may be useful to provide a little orientation about sizes and distances before getting into details.

Figure 1.1 attempts to illustrate the relative distances and sizes of various objects in the Universe. I also find it helpful in grasping astronomical distances to make analogies to ordinary-size objects. For example, if the Sun is a grain of sand (1 mm), the orbit of the Earth is 10 cm and that of Pluto is 4 m. The nearest star is 30 km away and the center of the Galaxy is five times the distance to the Moon.

There are $\sim 10^{11}$ stars in our Galaxy, and $\sim 10^{11}$ galaxies in the visible Universe—a star in the Milky Way for every grain of sand it would take to fill a large lecture room, and then a galaxy for every star. There are more stars than all the grains of sand in all the beaches of the Earth.

If our Galaxy is the size of a half dollar (3 cm), the nearest big galaxy is almost 1 m away, and the Virgo cluster of galaxies, located near the center of the local supercluster, is 10 m away. The most distant quasars are more than a kilometer away.

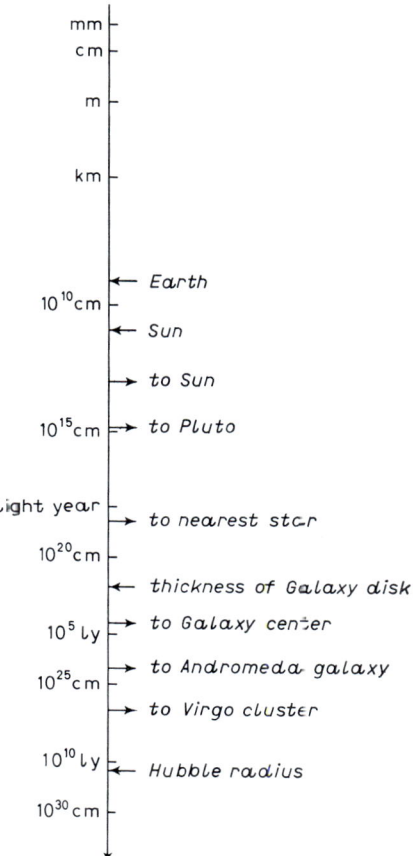

Fig. 1.1. – Sizes and distances.

Table I lists the values of the most important physical constants used in these lectures. Astronomers measure distance in parsec (pc). The Sun is about 8 kpc from the center of the Milky Way galaxy, which is about halfway to the edge of the visible Galaxy. As we will see, the Milky Way's dark halo extends considerably farther.

The distance to distant galaxies is deduced from their red-shifts using Hubble's constant $H_0 = 100h$ km s^{-1} Mpc^{-1}, the value of which remains uncertain by about a factor of two: $\frac{1}{2} \leqslant h \leqslant 1$. Consequently, the parameter h appears in many formulae where the distance matters.

1'2. *Galaxies*. – The nearest large galaxy to ours is the great galaxy in the constellation Andromeda. It was first recorded on an astronomical map by ABD-AL-RAHMAN AL SUFI in 964 A.D., and first drawn—as an elliptical nebula (Latin for cloud)—in an engraving of the Andromeda constellation by

Table I.

parsec	pc	$= 3.09 \cdot 10^{18}$ cm $= 3.26$ light-years (LY)
Newton's constant	G	$= 6.67 \cdot 10^{-8}$ dyn cm^2 g^{-2}
Hubble parameter	H	$= 100h$ km s^{-1} Mpc^{-1}, $\tfrac{1}{2} \leqslant h \leqslant 1$
Hubble time	H^{-1}	$= h^{-1} 9.78 \cdot 10^9$ y
Hubble radius	R_H	$= cH^{-1} = 3.00 h^{-1}$ Gpc
critical density	ϱ_c	$= 3H^2/8\pi G = 1.9 \cdot 10^{-29} h^2$ g cm^{-3} = $= 11 h^2$ keV cm^{-3} $= 2.8 \cdot 10^{11} h^2 M_\odot$ Mpc^{-3}
speed of light	c	$= 3.00 \cdot 10^{10}$ cm s^{-1} $= 306$ Mpc Gy^{-1}
solar mass	$M_\odot$	$= 2.00 \cdot 10^{33}$ g
solar luminosity	$L_\odot$	$= 3.83 \cdot 10^{33}$ erg s^{-1}
Planck's constant	$\hbar$	$= 1.06 \cdot 10^{-27}$ erg s $= 6.58 \cdot 10^{-16}$ eV s
Planck mass	M_{Pl}	$= (\hbar c/G)^{\frac{1}{2}} = 2.18 \cdot 10^{-5}$ g $= 1.22 \cdot 10^{19}$ GeV
proton mass	m_p	$= 1.67 \cdot 10^{-24}$ g $= 0.938$ GeV/c^2
Boltzmann constant	k_B	$= 1.38 \cdot 10^{-16}$ erg K^{-1} $= (1.16 \cdot 10^4)^{-1}$ eV K^{-1}
sidereal year	y	$= 3.155\,815 \cdot 10^7$ s
radian		$= 57°.2958 = 3437'.75 = 206\,265''$

Bouillaud in 1667 [1]. Messier included it in his catalogue of nebulas as number 31. It was not until 1923, however, that Hubble first recognized the true nature of M31.

Like our own galaxy, M31 is a typical giant spiral. It is perhaps twice as massive as the Milky Way, with a mass in stars of about $4 \cdot 10^{11} M_\odot$. Its radius is about 25 kpc. It is located about 0.7 Mpc from us, and its velocity along the line of sight (measured by the Doppler shift) is 270 km s^{-1} toward us.

There are about thirty other galaxies known in our local group of galaxies, but all are much smaller than these two giants. M33, the only other spiral galaxy, has perhaps a tenth the mass of the Milky Way. M32, the largest elliptical galaxy in the local group, is considerably less massive. Both M32 and M33 are fairly close to M31. The largest galaxies in the immediate vicinity of the Milky Way are two irregular galaxies, known as the Large and Small Magellanic Clouds. In addition, seven dwarf spheroidal galaxies have been found near our Galaxy: Draco, Ursa Minor, Carina and Sculptor within 100 kpc, and Fornax and Leo I and II at about twice that distance. (They are named after the constellations in which they lie.) Fornax, the most massive of them, has a mass in stars of only about $2 \cdot 10^7 M_\odot$. Partly because of the fact that their masses are so tiny (for galaxies), these dwarf spheroidals may give us important clues about the origin of galaxies and the composition of the dark matter, as I will discuss later on.

Figure 1.2 is the traditional Hubble «tuning-fork» diagram of galaxy types, from ellipticals, through lenticular (S0) and spiral galaxies (with and without central bars), to irregulars. This progression of galaxy morphologies

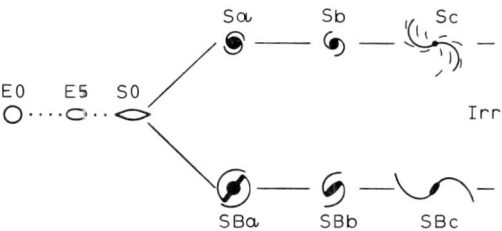

Fig. 1.2. – Hubble's classification of galaxy types [2]. Elliptical (E) galaxies have a spheroidal appearance with no visible disk, while lenticular (S0) and spiral (S) galaxies have a disk in addition to a spheroidal central bulge or nucleus. S0's have no spiral arms, and relatively little dust and ionized hydrogen in the disk; these are more prominent in spiral galaxies. A large minority of S galaxies have « bars » across their nuclei, with the spiral arms beginning from the ends of the bar rather than winding out directly from the nucleus. Some irregular (Irr) galaxies resemble disk galaxies but with less symmetry; others are even more irregular than that.

corresponds to decreasing prominence of the spheroidal component and increasing importance of disk. HUBBLE thought it possible that his classification was evolutionary, and although this is no longer believed the sequence Sa - Sb - Sc - Sd is called by astronomers the progression from « early » to « late » spiral types. The disk-to-bulge luminosity ratio increases from ~ 1 for Sa to ~ 10 for Sd. Late spiral types also have more gas and young, blue stars. Roughly 10 % of all bright galaxies are ellipticals, 20 % are S0, 65 % are spirals, and 5 % are irregulars, with higher fractions of S0 and E in regions of higher galaxy number density—another important clue to galaxy origins.

Spiral galaxies. Spiral galaxies have three visible components: the disk with its spiral arms, the nucleus or bulge and the stellar halo or corona. In addition, spiral galaxies generally appear to possess extensive dark-matter halos.

Although the spiral arms are the distinguishing feature of spiral galaxies, there is less to them than meets the eye. The spiral arms are bright because of the short-lived luminous supergiant stars and emission nebulae they contain, but the number density of long-lived stars like the Sun is not much different in the arms than in the interarm regions of the disk. Following the work of C. C. Lin and Frank Shu, it is now thought that the arms are the result of density waves travelling around the galaxy: the passage of the disk matter through such a wave triggers the process of star formation. Incidentally, the spiral arms curve backward, opposite to the direction of rotation of the galaxy. Thus a spiral galaxy rotates like a pinwheel, the spiral-arm density waves rotating in the same direction but slower than the stars and gas in the disk.

The disk is remarkably thin. In our Galaxy, it is a few hundred parsec thick at the radius of the Sun (about 8 kpc). Perhaps $(10 \div 20)$ % of the mass in the disk is in gas (mostly hydrogen and helium) and dust (composed of what astronomers call « metals »: elements more massive than helium). The disk becomes thicker and more gaseous at large radii, and in some galaxies the outer edge of the disk is warped. Spiral galaxies are generally surrounded by a diffuse envelope of neutral atomic hydrogen (H I, observed with radio telescopes in 21 cm emission), sometimes extending to several times the optical radius [3].

The stars of a spiral galaxy were classified by BAADE in 1944 into two broad categories, Population I and II. The relatively young, metal-rich stars of the disk are Population I. The older, lower-metallicity stars are Population II; these are found mainly in the nucleus and stellar halo, including the globular clusters. Globular clusters are dense spherical assemblages of stars, having typically $\sim 10^6$ stars within a radius of a few pc. There are about 200 globular clusters in the Milky Way. Thus only a tiny fraction of the $\sim 10^{11}$ stars in the Galaxy are in globular clusters. The total number of stars in the diffuse stellar halo is also a tiny fraction of the total. The stellar halo and about half the globular clusters are distributed roughly spherically. The other half of the globular clusters are associated with the disk. Most of the Population II stars lie in the spheroidal bulge which occupies the center of the galaxy, with radius ~ 4 kpc and very little gas, dust, or young stars. The majority of the stars in a galaxy like ours are Population I stars in the disk.

The surface brightness (luminosity per unit area) distribution as a function of radius in the disk component of typical S and S0 galaxies is of the form

$$(1.1) \qquad I_D(r) = I_0 \exp[-\alpha r].$$

The corresponding disk luminosity is $L_D = 2\pi I_0 \alpha^{-2}$, half of which is emitted within the effective radius $r_e = 1.67 \alpha^{-1}$. For example, the Milky Way has an effective radius $r_e \approx 5$ kpc, a total (disk plus bulge) luminosity $L = L_D + L_B \approx 1.6 \cdot 10^{10} L_\odot$, a disk-to-bulge ratio $L_D/L_B \approx 2$, and is classified as an Sb or Sc galaxy [4].

The most important source of information about the dynamics of a galaxy is Doppler shift measurements of the line-of-sight velocities of its components. By 1979, the evidence had become overwhelming that the rotation velocity of spiral galaxies remains roughly constant from a few kpc to the largest radii at which observations are possible [5]. This is surprising, since if the mass were mainly associated with the stars, which are centrally concentrated, then the velocity in the outer regions would fall as $v \propto r^{-\frac{1}{2}}$, like that of the planets in the solar system. Figure 1.3 shows rotation curves for many spiral galaxies, obtained both from 21 cm observations and from measurements of velocities of the clouds of ionized gas surrounding hot blue stars. (Because these gas clouds emit most of their light in a few spectral lines, their velocities can be measured in a fraction of the exposure time required for stellar measurements [6].) By simple Newtonian arguments, a constant rotation velocity

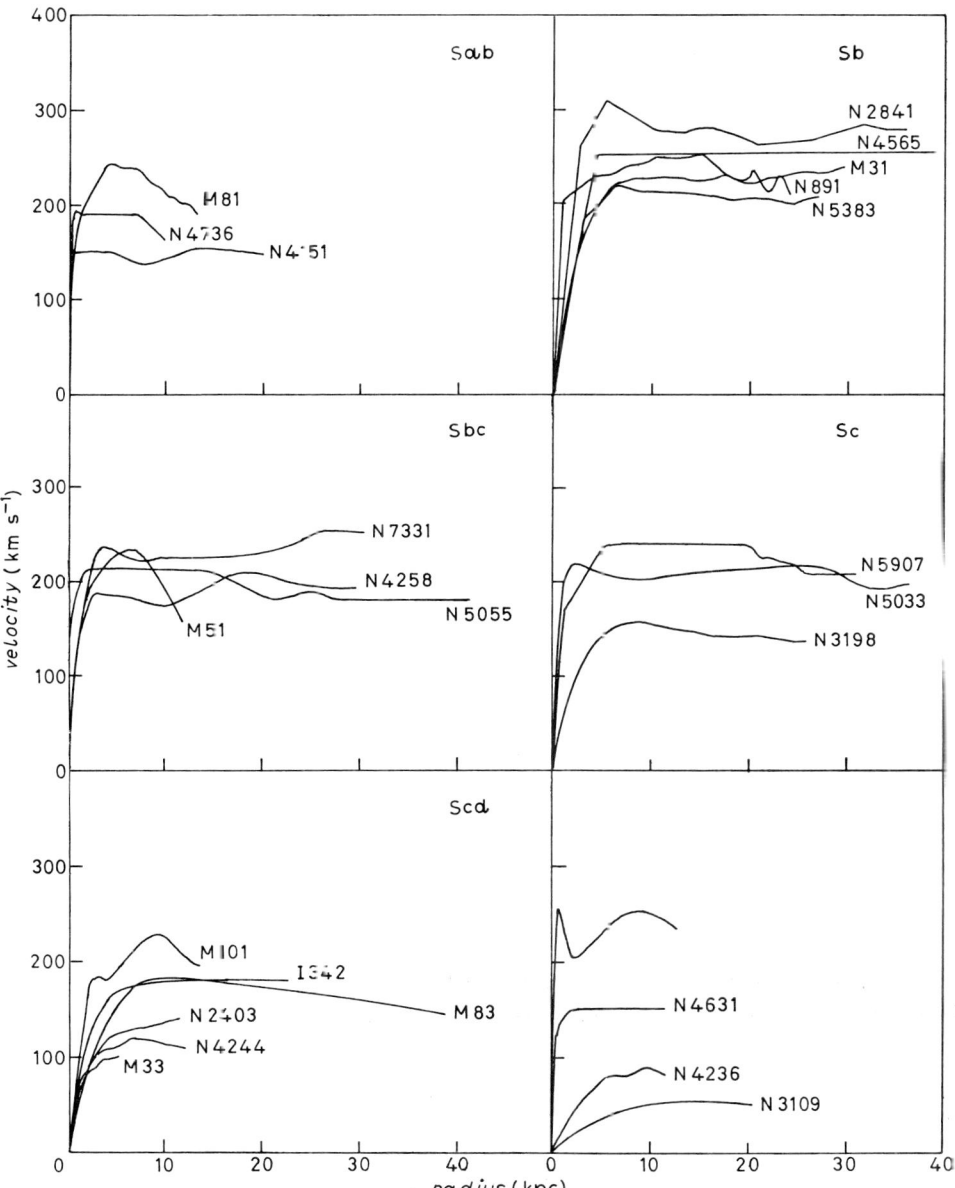

Fig. 1.3a). – Rotation curves for many spiral galaxies, obtained from 21 cm observations. (Source: ref. [7], reprinted in ref. [5]. Reproduced by permission.)

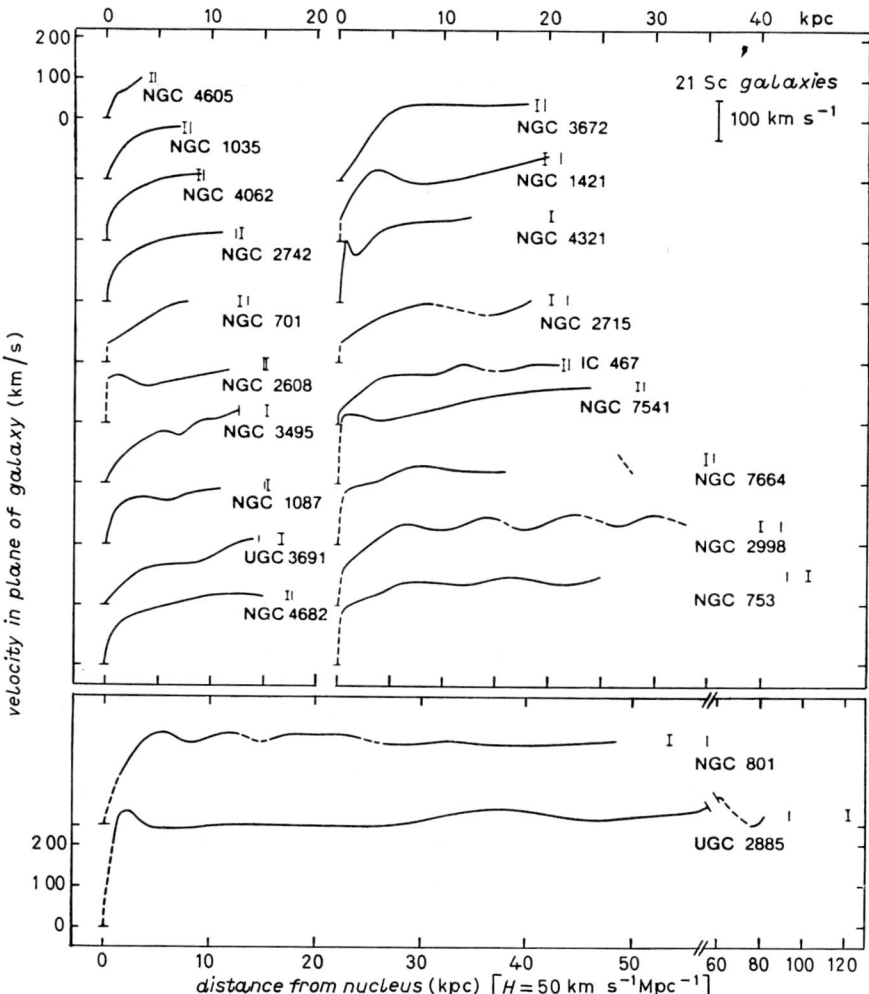

Fig. 1.3b). – Rotation curves for many spiral galaxies, obtained from optical measurements (see text). (Source: ref. [8]. Reproduced by permission.)

$v_{\rm rot}$ implies that the mass $M(r)$ within radius r grows linearly with radius:

(1.2) $$M(r) = (v_{\rm rot}^2/G)\, r\,.$$

Correspondingly, the mass density falls as r^{-2}. Since the luminosity falls exponentially with radius, the mass-to-light and total-to-luminous-mass ratios M/L and $M/M_{\rm lum}$ grow with radius. From the fact that the rotation velocity is constant to several times the effective radius, it follows that the mass associated with the dark halos of these galaxies is at least several times that of all the visible matter.

Actually, the existence of massive dark-matter halos was not entirely a surprise: at least two pieces of evidence had pointed toward it. Since the mid-1930's the astronomer Fritz ZWICKY had been emphasizing that there is much more mass detected dynamically in great clusters of galaxies than can be attributed to the stars in their galaxies [6, 9]. And in 1973 it was pointed out [10] that a self-gravitating disk is unstable toward collapse to a rotating bar—indeed, this bar instability probably is responsible for the fact that roughly a third of spiral galaxies have central bars—but that the disk can be stabilized by a roughly spherical halo containing comparable mass at the same radius. More recent detailed studies of galactic disks have confirmed that most of the dark matter cannot be in the disk [11]. The existence of warps in the outer parts of disk galaxies is also evidence that the dark halo is roughly spherical, since such warps would be smeared out in a nonspherical halo.

How large is the total mass associated with a typical spiral galaxy? Since according to the above equation the mass grows linearly with radius, one can equivalently ask, how large is the halo? We can set lower limits of $r_{halo} \geq$ ≥ 70 kpc, and correspondingly $M/M_{lum} \geq 10$ and $M \geq 2 \cdot 10^{12} M_\odot$ for our own galaxy from studies of its satellites (see fig. 1.4). This mass is comparable to that suggested by studies of the dynamics of the local group of galaxies [12]. As I will discuss shortly, the evidence from studies of the dynamics of all assemblages of galaxies, from small groups to rich (*i.e.* very populous) clusters, is consistent with $M/M_{lum} \approx 10$. The only significant evidence to the contrary of which I am aware is a recent paper reporting the results of a new technique for measuring galaxy mass based on the distortion of the images of background galaxies by gravitational deflection of their light by foreground galaxies [13].

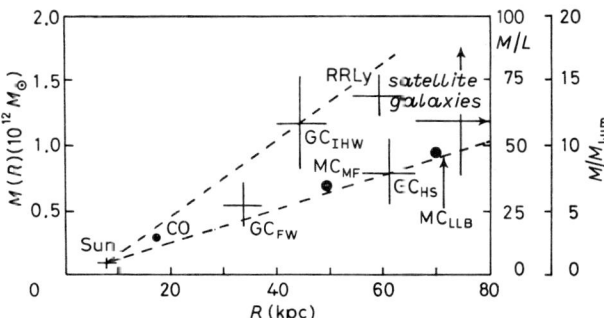

Fig. 1.4. – Mass of the Milky Way galaxy interior to radius R, deduced from the dynamics of its components and satellites: carbon monoxide clouds, globular clusters [14-16], the Magellanic Clouds [17, 18], a distant RR Lyrae star (assumed to be bound) [12] and satellite galaxies. (Adapted from ref. [20].)

Elliptical galaxies. Elliptical galaxies are spherical or ellipsoidal stellar systems consisting almost entirely of old stars. They contain very little dust and show no evidence of spiral arms. The larger ellipticals contain many globular clusters. In all these respects, they resemble the nucleus and stellar halo components of spiral galaxies.

Elliptical galaxies are classified in several ways. One is by ellipticity, with the integer n in En designating $10(a-b)/a$, where a and b are the projected major and minor axes. Ellipticals have projected axial ratios b/a between 1.0 (E0) and 0.3 (E7). It was once widely believed that elliptical galaxies are oblate spheroids flattened by rotation, but in the past few years rotation curves and velocity dispersion data have shown that some ellipticals, especially the larger ones, rotate much too slowly to account for their flattening. There is evidence that some ellipticals, again especially the larger ones, are actually triaxial, and that their flattening is due to velocity anisotropies. It is not yet known whether these are more nearly oblate or prolate [21].

Ellipticals vary very widely in mass, from dwarf spheroidals to supergiant galaxies. The latter are the largest known galaxies, with extensive (~ 100 kpc) amorphous stellar envelopes and masses as much as an order of magnitude larger than that of M31. Called cD galaxies, they are usually found in the cores of rich, regular clusters of galaxies; and they are often flattened, the major axis aligned with that of the cluster. Roughly a third of all cD galaxies have multiple nuclei, which suggests that they formed through mergers. At the other end of the size scale, there are probably more dwarf ellipticals (dE) than any other type of galaxies in the Universe—as is true in our local group of galaxies. Or perhaps the most populous galaxy species is dwarf irregulars [22]. In any case, dwarf galaxies represent only a small fraction of the stars and mass in the Universe since they are so small.

The projected distribution of surface brightness in E galaxies is well fitted by the de Vaucouleurs formula

$$(1.3) \qquad I(r) = I_e \exp\left[-7.67((r/r_e)^{\frac{1}{4}} - 1)\right],$$

where the « effective radius » r_e is the radius enclosing half of the total light. $I(r)$ falls off more slowly than r^{-2} for $r < r_e$ and more rapidly than that for $r > r_e$. The same formula fits the bulges of S0 and S galaxies.

The total luminosity L of an elliptical galaxy is observed to be related to its stellar velocity dispersion σ by the formula $L \approx L_\star (\sigma/220 \text{ km s}^{-1})^\gamma$, where $L_\star = 10^{10} L_\odot$ and $\gamma = 4 \pm 1$. This is the Faber-Jackson relation [23]. There is an analogous relation $L_H \propto v_{\text{rot}}^4$ between the total infra-red luminosity and rotation velocity of spirals, called the infra-red Tully-Fisher relation [24]. A similar relation has been found to hold between the total luminosity in the blue spectral band and the rotation velocity for a sample of spiral galaxies [25].

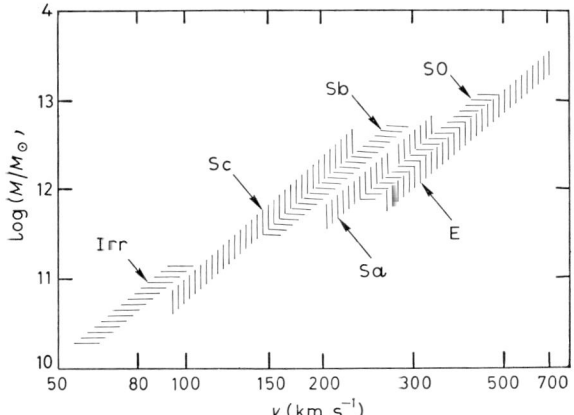

Fig. 1.5. – Total mass (assumed to be $10 \times M_{\text{lum}}$) vs. velocity (rotation velocity for disks, velocity dispersion for spheroids) for various galaxy types. Slopes are roughly consistent with $L \propto v^4$. (From fig. 4 of ref. [26]; reprinted here as fig. 4.4.)

These empirical relations, displayed in fig. 1.5, are important in providing both cosmic yardsticks and insights into the formation and dynamics of galaxies.

Luminosity distribution. The galaxy luminosity function is defined such that $\varphi(L)\,\mathrm{d}L$ is the number density of galaxies having total luminosity in the interval $(L, L + \mathrm{d}L)$. The available data are fitted by Schechter's convenient function [21]

$$(1.4) \qquad \varphi(L) = \frac{\varphi_\star}{L_\star} \left(\frac{L}{L_\star}\right)^\alpha \exp\left[-L/L_\star\right],$$

where [21]

$$(1.5) \qquad \begin{cases} \alpha = -1.29 \pm 0.11, \\ \varphi_\star = (1.3 \pm 0.3) \cdot 10^{-2} h^3 \,\mathrm{Mpc}^{-3}, \\ L_\star = 1.1 \cdot 10^{10} h^{-2} L_\odot. \end{cases}$$

This is sketched in fig. 1.6. Actually, $\varphi(L)$ must fall off more rapidly than the function (1.4) at small L, since the mean space density of galaxies corresponding to (1.4),

$$(1.6) \qquad \langle n \rangle = \varphi_\star \Gamma(\alpha + 1),$$

diverges if $\alpha < -1$. The shape of the luminosity function for $L < 0.005 L_\star$ is uncertain, but the number of small nearby galaxies is indeed less than predicted by (1.4). The shapes of the luminosity functions for the different morphological types of galaxies differ at the faint end, dwarf E and I galaxies

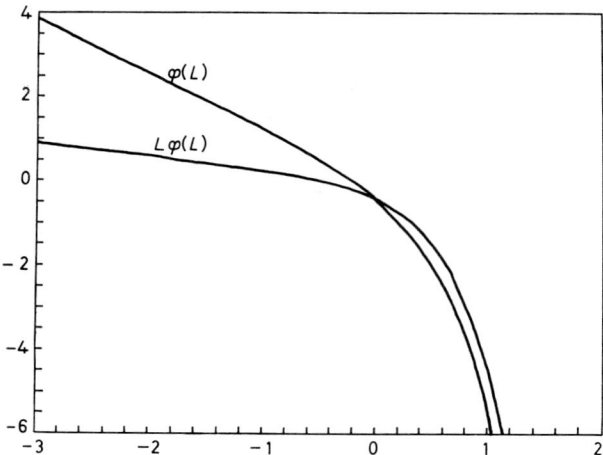

Fig. 1.6. – Schechter luminosity function. Axes are $\log \varphi(L)$ and $L\varphi(L)$ vs. $\log L/L_\star$.

being more numerous than dwarf S and S0, but the luminosity functions have similar shapes at the bright end.

The mean luminosity density corresponding to (1.4) is perfectly finite:

$$(1.7) \qquad \langle \mathscr{L} \rangle = \varphi_\star L_\star \Gamma(\alpha + 2) \approx 1.8 \cdot 10^8 h L_\odot \text{ Mpc}^{-3}.$$

The majority of galaxies are faint, but most of the light comes from those that are of luminosity $\geqslant L_\star$. With (1.7) we can evaluate the mean mass-to-light ratio of the Universe:

$$(1.8) \qquad M/L = \Omega \varrho_c / \langle \mathscr{L} \rangle \approx 1500 \Omega h (M_\odot/L_\odot),$$

where ϱ_c is the critical density for closure (see table I and sect. 2) and Ω is the average density of the Universe in units of ϱ_c. Typically,

$$M/L \approx 14 h (M_\odot/L_\odot)$$

in the centers of galaxies [21]; thus $\Omega(\text{galaxy cores}) \approx 10^{-2}$, with perhaps another factor of two including the entire visible mass in galaxies. If the total galactic mass, including that of the halo, is about ten times greater (*i.e.* $M/M_{\text{lum}} \approx 10$), as discussed above, then $\Omega \approx 0.2$ and the Universe is open.

Interpretations. Although it is perhaps premature to sketch a theoretical framework for understanding the basic facts about galaxies, both in the context of these lectures and given the available astronomical data, I think it is nevertheless useful to do so at this point. The great advantage of keeping a tentative theory in mind as one thinks about data is that it helps in organizing and remembering the facts. If it is a good theory, it will also call attention to particularly important facts—especially those that may contradict it!

The basic picture of galaxy formation that I have in mind is that galaxies collapsed gravitationally from initially rather homogeneous mixtures of dark matter and ordinary matter (in about the ratio 10:1). As I will explain in the next section, the result of virialization by violent relaxation in gravitational collapse is a roughly isothermal halo, with density falling as r^{-2}, as required to produce the observed constant-velocity rotation curves. The ordinary matter continued to radiate away its kinetic energy and sink toward the center, eventually forming the visible stars. This process is called dissipational collapse. Meanwhile the dark matter retained its post-virialization velocity and density distribution, and it forms the galactic halos. We do not know what the dark matter is, but its key property, in addition to being invisible, is that it is dissipationless. Probably both properties are a consequence of its lack of significant interaction with electromagnetic radiation, perhaps because the dark matter is composed of neutral elementary particles.

In this picture, the disk in disk galaxies formed when the dissipational collapse of the baryonic matter was halted by angular-momentum conservation. (The symmetrical configuration of minimum kinetic energy for given angular momentum is a disk.) Galactic spheroids resulted when dissipational collapse was halted by some other process, presumably star formation. (A collection of virialized gravitating mass points is dissipationless.) Evidently, spheroids result from matter which had either (or both) higher initial density or smaller initial angular momentum than that which formed disks [26, 27].

It follows that all galaxies should be surrounded by massive dark-matter halos. I have already discussed the strong evidence that this is true for spiral galaxies. Although relevant observations are more difficult for other galaxy types, the data available are consistent with the ubiquity of massive halos [5, 28-31].

A useful way of visualizing galaxies is sketched in fig. 1.7, where density is plotted vs. distance from the center of our Galaxy, looking toward M31. In the central region of a typical large galaxy the density is high—perhaps even infinite at the very center if there is a black hole there. This is surrounded by a region of rapidly falling baryonic-matter density, so that there are comparable total amounts of ordinary and dark matter enclosed within a few effective radii (r_e). The density of the dark-matter halo (dashed line) declines $\propto r^{-2}$ out at least to $\sim 10^2$ kpc. If it continues to follow a r^{-2} law between the galaxies (dotted line), then the average density is approximately that required for closure, i.e. $\Omega \approx 1$. Jim PEEBLES calls this the «alpine model». On the other hand, if the dark-matter density falls off rapidly beyond $\sim 10^2$ kpc («crayon model»), then, as I mentioned before, $\Omega \approx 0.2$ and the Universe is open.

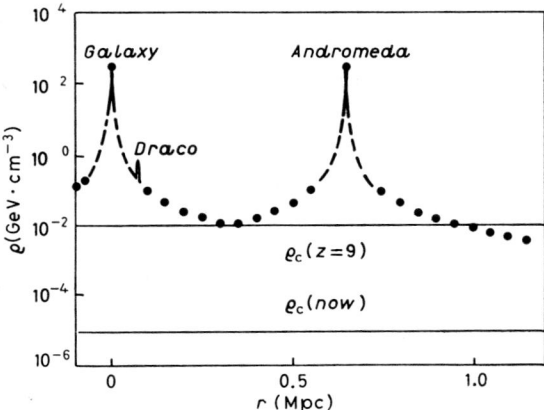

Fig. 1.7. – Schematic plot of mass density *vs.* distance from center of the Milky Way. (See text for explanation. From ref. [32].)

The horizontal lines in fig. 1.7 represent critical density today and at the earlier epoch when the Universe had expanded only 1/10 as much, *i.e.* when $R = 0.1$. The expansion factor R, defined to equal unity now, is given in terms of the red-shift $z \equiv \delta\lambda/\lambda$ by $R = (1 + z)^{-1}$. In the next section I will remind you of the relationship between z or R and the time t since the big bang, and also explain why the fact that the halo at 100 kpc is roughly an order of magnitude more dense than the higher of the two light horizontal lines suggests that the galaxy interior to that collapsed gravitationally before z of 10.

1`3. *Groups and clusters.* – Half or more of all galaxies are members of groups or clusters. « Groups » of galaxies are systems containing at most a few tens of bright galaxies, while « clusters » are richer (*i.e.* more populous) systems. They are identified as density enhancements, either in surface number density of galaxies on the sky, or in red-shift-space volume density. It is thought that most of them are also gravitationally bound structures, especially those of high density.

After a particular variety of astronomical object has been discovered, it has usually proved very valuable to make a catalogue of such objects, in order to study them systematically. There are two great catalogues of clusters of galaxies, Abell's catalogue of 2712 rich clusters [33] and the more extensive Zwicky catalogue [34], which lists and classifies poorer (*i.e.* less populous) clusters as well. Both catalogues are based on the Palomar Sky Survey plates, and so are limited to the northern sky.

Reliable identification of groups of galaxies requires red-shift data, which have only recently become available for large numbers of galaxies. The best catalogue of groups is that recently compiled by GELLER and HUCHRA [35], obtained by applying a group-finding algorithm to the NB whole-sky catalogue [36] of the 1312 galaxies brighter than $m_B = 13.2$ (*), and to the Harvard-Smithsonian Center for Astrophysics («CfA») survey of the northern sky

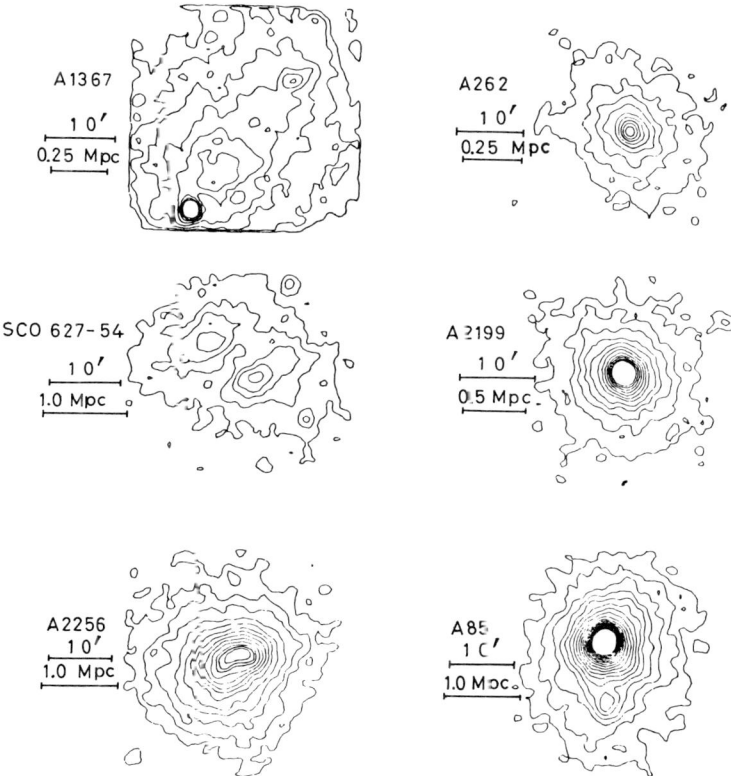

Fig. 1.8. – X-ray isodensity contours for six rich clusters of galaxies. The clusters on the left represent those having large X-ray core radii and no central, dominant galaxy. Those on the right have smaller X-ray core radii and contain cD galaxies. The clusters at the top of the figure are less dynamically evolved than those at the bottom. (From (0.5÷3.0) keV imaging proportional counter images from the Einstein satellite Reproduced, with permission, from ref. [37]; © 1982 by Annual Reviews, Inc.)

(*) The notation m_B represents apparent magnitude in the B blue spectral band. Apparent magnitude is related to the measured flux S by $m = A - 2.5 \log S$, where the constant A depends on the spectral band; thus a galaxy of $m = 12$ appears to be 100 times brighter than one of $m = 17$. The naked eye can see to $m \approx 6.5$; a six inch (15 cm) telescope, to $m = 13$; and the Palomar 5 m telescope, to $m = 24$ (photographically). Astronomical traditions can be long lived. The magnitude scale was adopted in the 19th century to agree approximately with the brightness classification given in the catalogue of 850 stars complied by HIPPARCHUS in the second century B.C., whose 6th magnitude stars are about 100 times fainter than those of 1st magnitude.

(2396 galaxies, complete to $m_B = 14.5$ about 20% of the sky) [36]. They found 92 groups in the former catalogue and 176 in the latter; about 60% of all the galaxies in the catalogues are assigned to groups.

There are several classification schemes for clusters, but a simple one that overlaps with the others is «regular» vs. «irregular» [37, 38]. Regular clusters have a smooth and symmetric structure, with high central galaxy density ($\geqslant 10^3$ per Mpc^3), a small fraction of spiral galaxies ($\leqslant 20\%$), high velocity dispersion (~ 1000 km s^{-1}) and a high X-ray luminosity ($> 10^{44}$ erg s^{-1}) from hot gas (of temperature $T \geqslant 6$ keV). Examples include A85 and A2256 (the bottom two X-ray images in fig. 1.8), A496 (upper right in fig. 1.9) and the Coma cluster. (This cluster, designated A1656—i.e. No. 1656 in Abell's catalogue—, is the nearest rich cluster, at about $45h^{-1}$ Mpc. As usual, it is named after the constellation in which it lies on the sky, Coma Berenices—Berenice's Hair.)

Only about a quarter of all rich clusters are regular. Irregular clusters have a rather lumpy structure, lower central galaxy density, a somewhat higher spiral fraction ($\geqslant 40\%$) than regular clusters, lower velocity dispersion, lower X-ray luminosity and cooler gas $((1 \div 2)$ keV). Examples include A262,

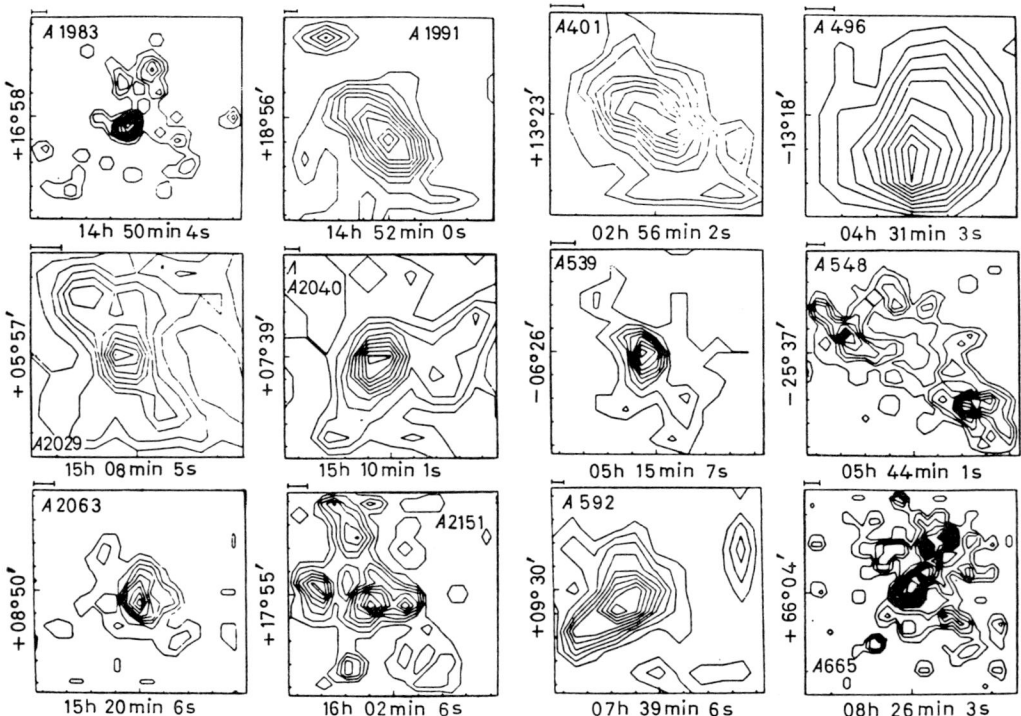

Fig. 1.9. – Contours of galaxy density in rich clusters. (Reproduced, with permission, from ref. [39], as reproduced in ref. [40].)

A1367 and the Virgo cluster. In addition, there are intermediate cases, exemplified by the middle two images in fig. 1.8 and many of the clusters in fig. 1.9. Many of these are elongated and have prominent subclusters.

Another distinction that is especially apparent on the X-ray images is between those clusters with a central, dominant galaxy (*e.g.*, the three clusters on the right half of fig. 1.8) and those without a cD (left half of fig. 1.8). In their cores, cD galaxies look like giant ellipticals, except that some have multiple nuclei. This core is surrounded by a very extensive stellar and gaseous envelope, with optical surface brightness decreasing much more slowly than the de Vaucouleurs (eq. (1.3)) profile of a typical elliptical at large distances, and with extended, centrally peaked X-ray emission from the hot gas.

There is no sharp dividing line between « groups » and « clusters », and a substantial overlap of physical characteristics between these two categories [41]. Most groups are loose, but there are compact groups with galaxy densities comparable to those in the cores of rich clusters. Some groups even contain small cD galaxies.

Alan DRESSLER first demonstrated that in rich clusters there is a well-defined relationship, shown in fig. 1.10, between the local number density of galaxies and the local fraction of each galaxy Hubble type [42]. The local density was computed using the 10 nearest (projected) neighbors of each galaxy. The fractions of E and S0 galaxies increase, and the fraction of S+I decreases, smoothly and monotonically as the local galaxy density increases. This relation between population and density holds for individual clusters as well as, on

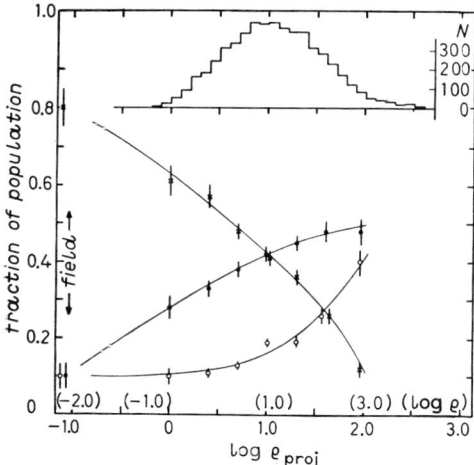

Fig. 1.10. – Fractions of elliptical (o), lenticular (•) and spiral plus irregular galaxies (×) as a function of the logarithm of the projected density, in galaxies per Mpc^2. Also shown is an estimate of the space density, in galaxies per Mpc^3. The upper histogram shows the number distribution of the galaxies in the sample ($\sim$ 6000 galaxies in 55 rich clusters) over the bins of projected density. (Reproduced, with permission, from ref. [42].)

the average, from cluster to cluster. In particular, it holds for both regular and irregular clusters. And it has recently been shown to hold for groups as well as clusters [43].

Interpretations. As I will discuss in the next section, the process of dissipationless gravitational collapse produces a smooth, centrally concentrated distribution of matter. The obvious interpretation of the difference between regular and irregular clusters of galaxies is, therefore, that the former have undergone collapse, while the latter have not yet done so [44]. If they are indeed in virial equilibrium, regular clusters' large velocity dispersions are strong evidence for a large quantity of dark matter to provide the required gravitational binding energy. Although the mass-to-light ratio (M/L) implied for rich clusters is about a factor of 6 larger than that for galaxies (including their massive dark halos), the ratio of total to luminous mass (M/M_{lum}), which is physically more meaningful (*), is about the same for both.

The analysis by GELLER and HUCHRA [35] of groups and clusters in the CfA catalogue finds that they have approximately constant M/L. An earlier study [45] which claimed to find a trend of increasing M/L with increasing size of the cluster is now known to have been misled by a flaw in the cluster-finding algorithm.

The data on M/L and M/M_{lum} are plotted in fig. 1.11. It is apparent that the data are consistent with roughly constant M/M_{lum} across the entire range of masses from dwarf spheroidal galaxies (using the dynamical mass estimates for them) to the cores of rich clusters. The most straightforward interpretation of this constancy is that there is about an order of magnitude more dark than luminous matter in the Universe.

cD galaxies are thought to form through galactic cannibalism as (gravitational) dynamical friction causes cluster galaxies to spiral into the centrally located giant, where they are disrupted by tidal forces [37, 46]. The fact that many cD galaxies have multiple nuclei is evidently direct evidence for galactic cannibalism. Computer simulations of the evolution of groups and clusters have shown that mergers and tidal stripping are most rapid in small groups, including those that form in the early stages of the collapse of larger clusters, and that it is possible to understand the origin of cD galaxies in this way if cluster galaxies initially possess massive dark halos which only later become smeared out as the cluster relaxes [47].

(*) The old red stars of the E and S0 galaxies in regular rich clusters are less luminous per unit mass than the younger and bluer stars of S galaxies, which are not as prevalent in rich clusters, and the X-ray observations [37] show that there is at least as much mass in the hot gas in the cores of rich clusters as there is in galaxies. M_{lum} compensates for intrinsic luminosity differences of different galaxy types and includes the mass in hot gas; that is why it is physically more meaningful than L. For more details, see ref. [26], especially table I.

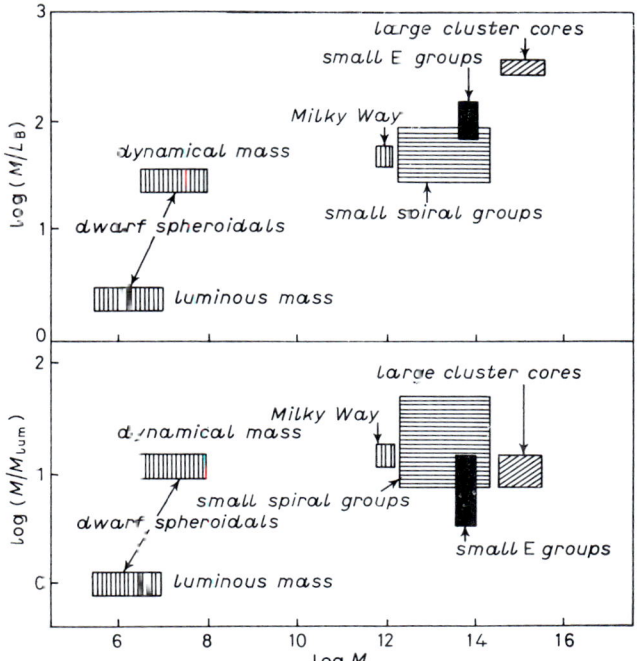

Fig. 1.11. – Mass-to-light ratio, M/L_B, and total-to-luminous mass, M/M_{lum}, for structures of various sizes in the Universe. Although M/L_B increases systematically with mass, the more physically meaningful ratio M/M_{lum} appears to be constant on all scales within the errors. If the velocity dispersion data for the dwarf spheroidal galaxies are interpreted to imply heavy halos, the upper estimates in the figure result. The lower estimates follow from assuming that all the mass is visible. The upper estimates are probably more realistic. (Reproduced from ref. [26]. The data come from table I of this reference.)

Finally, regarding Dressler's correlation between galaxy type and number density, the key question is whether it is caused by heredity (*i.e.* factors present when galaxies formed) or environment (evolutionary effects after galaxy formation, such as galaxy mergers or stripping of gas from spirals to form S0s). There is evidence that both heredity and environment are important [41, 42, 48]. I will return to all of these questions in later sections; they are crucial to unraveling the mystery of the origin of galaxies and clusters.

1'4. *Superclusters and voids.* – Thirty years ago, astronomers knew that rich clusters consist mostly of E and S0 galaxies, and that the majority of galaxies are spirals and lie outside these clusters in relative isolation in the « field ». But they did not yet know about superclusters and voids [49].

Gerard DE VAUCOULEURS was the first to define and describe the Local Supercluster, the vast aggregation of several thousand galaxies of which our own Local Group, containing the Milky Way, is an outlying member. The Local Supercluster is centered on the Virgo cluster, about $15h^{-1}$ Mpc away from us. It has recently been mapped in some detail by TULLY [50], who finds that it consists of a fairly thin disk component containing about 60% of the luminous galaxies and a halo component with 40%, and that almost all the luminous galaxies of the halo are associated with a few clusters leaving most of the volume off the disk empty.

Although there was some recognition that there are other superclusters in addition to our own on the basis of (two-dimensional) sky surveys, we have only begun to see the large-scale structure of the Universe clearly with the advent of large-scale red-shift surveys. The limitation of these surveys is that, while thousands of galaxy positions can be read off of a single photographic plate, spectral red-shifts must be obtained one by one. Roughly 10^4 of them are presently available, including deep surveys of a few percent of the sky (« drilling holes in space »: measuring red-shifts for all galaxies brighter than a faint limiting magnitude in a small angular region) and shallower surveys covering larger angular area (the prime examples being the NB and CfA catalogues). The data are growing rapidly: the doubling time for the number of galaxy red-shifts available is presently about three years. Technological advances, including image tubes and CCD (charge-coupled device) detectors that allow modern astronomers to record as much information in an exposure of a few minutes as their predecessors could in an entire night, have helped to make this possible.

Figure 1.12 shows an example of the results of these surveys. The top portion shows the positions of bright galaxies in a region of the sky in the direction of the constellation Perseus. A chain of galaxies is apparent—the clearest such « filament » known. The lower portion of the figure, in which the galaxy positions are plotted in a red-shift-angle « wedge » diagram, shows that these galaxies are concentrated at a particular distance, about $50h^{-1}$ Mpc; thus they really do lie in a filamentary band across the sky. Equally striking in this figure is the fact that most of the wedge diagram is empty. Such voids in the galaxy distribution are apparent on all diagrams of this sort. Galaxies are concentrated in flattened or filamentary superclusters, leaving most of the volume of the Universe virtually devoid of bright galaxies [40].

All nearby Abell clusters are now known to belong to superclusters. For example, Coma and A1367 are connected by a bridge of galaxies, including several large groups. The whole structure stretches at least 20 degrees across the sky, corresponding to a length of $\sim 30h^{-1}$ Mpc; some astronomers argue that it is even larger. What is really needed now are catalogues of superclusters and voids, so that their statistical properties can be learned. Astronomers will be able to obtain enough red-shift data in five to ten years for this to be possible. The largest void discovered to date is the « great void in Boötes » lying between two large superclusters, the Hercules supercluster on the near

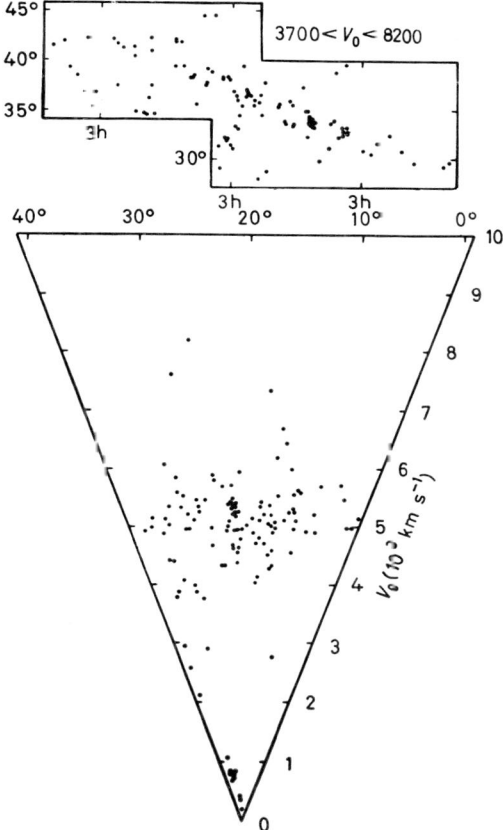

Fig. 1.12. - Perseus supercluster. (Top) Map of the surveyed region of the sky (co-ordinates are declination vs. right ascension), showing galaxies with $3700 \text{ km s}^{-1} < V_0 < 8200 \text{ km s}^{-1}$. (Bottom) Wedge diagram of all galaxies in the survey with $V_0 < 10000 \text{ km s}^{-1}$. The position variable is each galaxy's projected position on the major axis of the Perseus supercluster filament with arbitrarily chosen zero point. The foreground galaxies are clustered, although one galaxy is found in an otherwise void region. At red-shifts higher than the supercluster's, very few galaxies are found. (Reproduced, with permission, from ref. [51].)

side and the great Corona Borealis supercluster, which contains 15 Abell clusters, on the far side. The Boötes void is perhaps $60h^{-1}$ Mpc across, and the density of bright galaxies in it is probably less than a tenth, and almost certainly less than a quarter, of the average density [52].

Any data regarding correlations of galaxy and cluster properties across the vast distances spanned by superclusters and voids are potentially important in indicating how they may have formed. Probably the most interesting datum of this sort obtained thus far is Binggeli's observation that the position angles of all nearby, elongated Abell clusters lie within 45° of the direction to the nearest cluster, provided the clusters are separated by less than $\sim 15h^{-1}$ Mpc. He found a similar, though less significant, correlation on larger scales, and also a correlation between the position angle of the brightest cluster galaxy's major axis and the direction to the nearest cluster [53]. Similar, but substantially weaker, correlations were found in a recent analysis of a larger sample of clusters [54]. In a similar vein, the analysis of local flattening of the galaxy distribution and its correlation across space may help to clarify the nature of superclustering [55, 56].

Interpretations. The cores of rich clusters and compact groups represent enhancements of 10^4 or more over the average galaxy number density. They are certainly bound and relaxed structures. On the other hand, the galaxy density enhancement represented by the Local Supercluster is much smaller, perhaps a factor of three [57]. The *peculiar velocity* (deviation from uniform Hubble flow $v = H_0 r$) of galaxies in superclusters is typically $\lesssim 10^3$ km s^{-1}. A velocity of 10^3 km s^{-1} is equivalent to a Mpc/Gy. Thus, while galaxies in rich cluster cores have had plenty of time since the big bang to cross from one side to the other, probably several times, the vast majority of galaxies have hardly had time to move more than a small fraction of the distance across their local superclusters. For example, the component of the Local Group's peculiar velocity in the direction of the Virgo cluster is $(200 \div 400)$ km s^{-1} (measured both via the dipole anisotropy of the cosmic background radiation and with respect to an ensemble of moderately distant galaxies [57, 58]), but the LG is nevertheless still expanding away from the Virgo cluster with a velocity of ~ 1000 km s^{-1}. Thus the Local Supercluster has not yet had time to collapse, certainly not in its longer dimensions across the disk, and it is perhaps not even gravitationally bound.

It is precisely because of their unrelaxed state that superclusters are so valuable to cosmologists: gravity has not yet had time to mix them up, so their structure may reflect in a rather simple way the nature of the primordial conditions that gave rise to them.

The big question is, which came first, superclusters (and voids), or galaxies? One popular view, *hierarchical clustering*, has it that galaxies formed more or less at random locations, and were subsequently gathered up into clusters of ever increasing size, culminating in the vast superclusters whose dimensions we are only now beginning to appreciate. This view has long been championed by Jim PEEBLES of Princeton, among others. A competing «top down» view, long advocated by the Russian astrophysicist Yakov ZEL'DOVICH and his colleagues [59, 60], among others, contends that it is the superclusters that formed first, subsequently fragmenting into smaller objects which then formed galaxies. On the face of it, there are several outstanding pieces of evidence that superclusters were primary: the very existence of superclusters and large voids is pretty direct evidence that galaxy formation could not have occurred at random locations in the early universe, and the Binggeli correlation discussed above is easy to understand as a reflection of superclustering preceding the formation of clusters, if not also galaxies. However, there are serious problems with this view, too. As I will explain in more detail in sect. **3**, galaxies appear to be much older structures than superclusters. In addition, in the « top down » scenario it is hard to understand the observed clustering substructure [61] as well as the structure of individual galaxies.

2. – Gravity.

Gravity is the subject of this second section. In it I will try to introduce the basic ideas necessary to understand the effects of gravity both on the evolution of the entire Universe and on the growth and collapse of the fluctuations that presumably formed galaxies and all larger-scale structures. I will also briefly explain how the hypothesis of cosmic inflation can account for the origin of these large-scale fluctuations without violating causality.

I will assume here that Einstein's general theory of relativity (GR) accurately describes gravity. Although it is important to appreciate that there is no observational confirmation of this on extragalactic scales, the tests of GR on smaller scales are becoming increasingly precise, especially with the discovery of pulsars in binary star systems [62]. There are two other reasons most cosmologists believe in GR: it is conceptually so beautifully simple that it is hard to believe it could be wrong, and anyway it has no serious theoretical competition. Nevertheless, since a straightforward interpretation of the available data in the context of this standard theory of gravity leads to the disquieting conclusion that most of the matter in the Universe is dark, there have been suggestions that perhaps our theory of gravity is inadequate on large scales. I will mention them briefly at the end of this section.

2`1. *Cosmology*. – The « cosmological principle » is logically independent of our theory of gravity, so it is appropriate to state it before discussing GR further. But before I can state it, some definitions are necessary:

A *comoving observer* is at rest and unaccelerated with respect to nearby material (in practice, with respect to the center of mass of galaxies within, say, 50 Mpc).

The Universe is *homogeneous* if all comoving observers see identical properties.

The Universe is *isotropic* if all comoving observers see no preferred direction.

The *cosmological principle* asserts that the Universe is homogeneous and isotropic on large scales. (It is not difficult to see that isotropy actually implies homogeneity, but the counterexample of a cylinder shows that the reverse is not true.) In reality, the matter distribution in the Universe is in our common experience exceedingly inhomogeneous on small scales, and increasingly homogeneous on scales approaching the entire horizon. The cosmological principle is in practice the assumption that for cosmological purposes we can neglect this inhomogeneity. The great advantage of assuming homogeneity is that our own neighborhood becomes representative of the whole Universe, and the range of cosmological models to be considered is also enormously reduced.

The cosmological principle implies the existence of a universal cosmic time, since all observers see the same sequence of events with which to synchronize their clocks. (This assumption is sometimes explicitly included in the statement of the cosmological principle; see, *e.g.*, ref. [63], p. 203.) In particular, they can all start their clocks with the big bang.

Astronomers observe that the red-shift

$$(2.1) \qquad z \equiv \frac{\lambda - \lambda_0}{\lambda_0}$$

of distant galaxies is proportional to their distance. We assume, for lack of any viable alternative explanation, that this red-shift is a Doppler shift: the Universe is expanding. The cosmological principle then implies (see, for example, ref. [64], subsect. 4'3) that the expansion is homogeneous:

$$(2.2) \qquad r = R(t)\,r_0\,,$$

which immediately implies Hubble's law:

$$(2.3) \qquad v = \dot{r} = \dot{R}R^{-1}r = Hr\,.$$

Here r_0 is the present distance of some distant galaxy (the subscript 0 in cosmology denotes the present era), r is its distance as a function of time and v is its velocity, and $R(t)$ is the scale factor of the expansion (scaled to be unity at the present: $R(t_0) = 1$). The scale factor is related to the red-shift by $R = (1 + z)^{-1}$. Hubble's « constant » $H(t)$ (constant in space, but a function of time except in an empty universe) is

$$(2.4) \qquad H(t) = \dot{R}R^{-1}\,.$$

Finally, it can be shown [63, 65] that the most general metric satisfying the cosmological principle is the Robertson-Walker metric

$$(2.5) \qquad \mathrm{d}s^2 = c^2\,\mathrm{d}t^2 - R(t)^2\left[\frac{\mathrm{d}r^2}{1 - kr^2} + r^2(\sin^2\theta\,\mathrm{d}\varphi^2 + \mathrm{d}\theta^2)\right],$$

where the curvature constant k, by a suitable choice of units for r, has the value 1, 0, or -1, depending on whether the Universe is closed, flat, or open, respectively. For $k = 1$ the spatial universe can be regarded as the surface of a sphere of radius $R(t)$ in four-dimensional Euclidean space; and although for $k = 0$ or -1 no such simple geometric interpretation is possible, $R(t)$ still sets the scale of the geometry of space.

2˙2. *General relativity*. – Formally, GR consists of the assumption of the equivalence principle (or the principle of general covariance [65]) together with Einstein's field equations

$$(2.6) \qquad R^{\mu\nu} - \frac{1}{2} R g^{\mu\nu} = -\frac{8\pi G}{c^4} T^{\mu\nu} - \Lambda g^{\mu\nu}.$$

The equivalence principle implies that space-time is locally Minkowskian and globally (pseudo-) Riemannian, and the field equations specify precisely how space-time responds to its contents. The essential physical idea underlying GR is that space-time is not just an arena, but rather an active participant in the dynamics. Fortunately, there are several excellent introductions to GR for cosmologists [63, 65, 66].

It will not be necessary to discuss the details of GR here, but I think it may be useful to spend a little time on the concept of horizons, since in my experience this is one of the things that most confuse newcomers to cosmology —in particular, the apparent contradiction between Hubble's law and the speed of light as a speed limit.

I find it helpful to picture the behavior of space-time near horizons using the somewhat artificial concept of a static point, which is fixed in space. Figure 2.1a) shows a number of static points located at various distances from a black-hole singularity. Imagine that each static point emits a pulse of light; the light circles in the figure show schematically the positions of the wave fronts a moment later. Far from the black hole, space-time is flat and the light circle is centered on the static point. But closer to the black hole, the light is increasingly dragged toward the singularity, as if space itself were flowing into the black hole. As HARRISON amusingly puts it [67], the event horizon, located at the Schwarzschild radius, « is the country of the Red Queen where one must move as fast as possible in order to remain on the same spot ». At the horizon, the light circle lies on the static point and no light can escape outward. Inside the horizon space effectively flows inward faster than light, and outward-moving light cannot even reach the horizon. It is important to understand that special relativity remains locally valid except at the singularity itself, and light always moves at the speed of light c with respect to freely falling observers.

A Hubble sphere in the expanding universe is like a Schwarzschild event

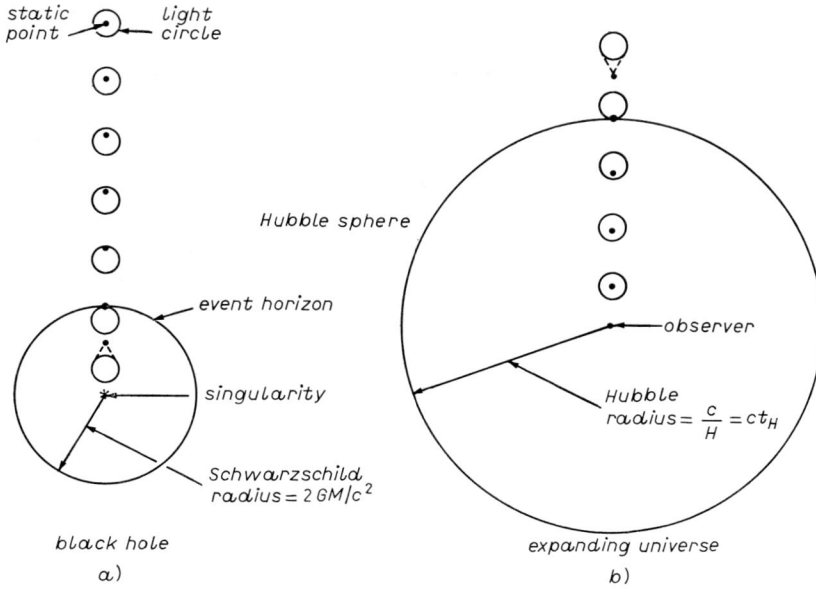

Fig. 2.1. – (Left) Schwarzschild and (right) Hubble spheres. (Left figure adapted from ref. [67], fig. 9.7.)

horizon turned inside out. As fig. 2.1b) shows, the light circles are centered on their static points well inside the Hubble sphere, but dragged increasingly outward at larger radii. At the Hubble sphere, the light circle lies on the static point and no light can escape inward. And beyond the Hubble sphere, space effectively flows outward faster than the speed of light. But the galaxies in that space are not moving at all (except for their small peculiar motions); it is the expansion of space that is carrying them away from us. The recession velocity in Hubble's law (2.3) is thus not an ordinary (local) velocity. The picture of the Hubble expansion as arising from galaxies flying apart in an underlying Euclidean space is only mildly misleading locally, but completely untenable on the scale of the Hubble radius. It is space itself that is expanding. This idea of space-time as an active participant in the dynamics of the Universe is also crucial for understanding the inflationary universe.

Comoving co-ordinates are co-ordinates with respect to which comoving observers are at rest. A comoving co-ordinate system expands with the Hubble expansion. It is convenient to specify linear dimensions in comoving co-ordinates scaled to the present, as in eq. (2.2). For example, if I say that two objects were 1 Mpc apart in comoving co-ordinates at a red-shift of $z = 9$, their actual distance then was 0.1 Mpc.

In a nonempty universe with vanishing cosmological constant, the case first studied in detail by the Russian cosmologist Alexander FRIEDMANN in

1922-1924, gravitational attraction ensures that the expansion rate is always decreasing. As a result, the Hubble radius

$$(2.7) \qquad R_{\mathrm{H}}(t) = cH(t)^{-1}$$

is increasing. The Hubble radius of a Friedmann universe expands even in comoving co-ordinates. Our backward light-cone encompasses more of the Universe as time goes on.

I will conclude these preliminary reflections on horizons in the Universe with fig. 2.2. In this figure mass is plotted against linear size. The upper left portion of the graph is the region excluded by gravity: the heavy diagonal line is the Schwarzschild radius $R_{\mathrm{s}} = 2GMc^{-2}$. An object of mass M having radius $< R_{\mathrm{s}}(M)$ lies inside its horizon and has effectively no size at all. There is reason to believe that such black holes are formed in the gravitational collapse of stars, and that massive black holes power quasars and other active galactic nuclei. There is no known way to make black holes of substellar mass

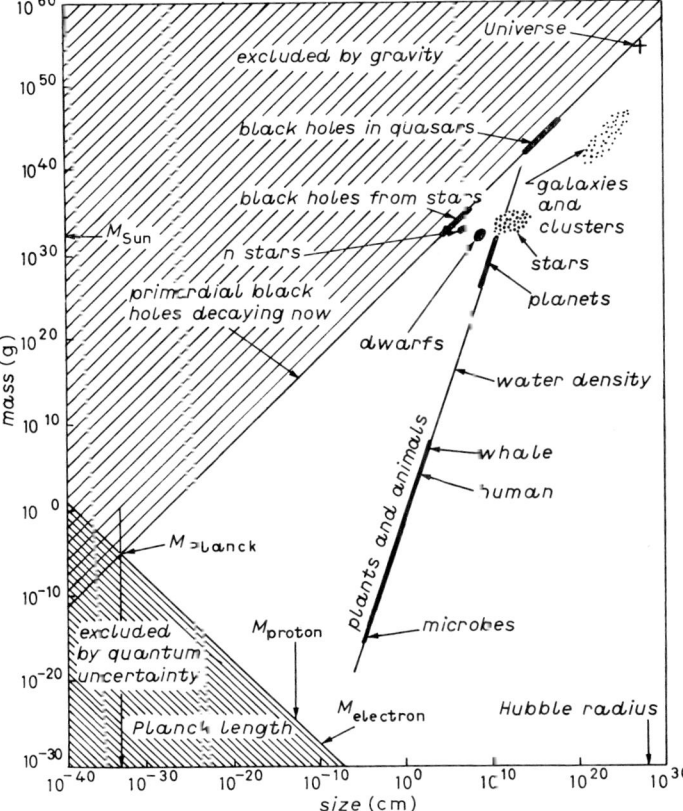

Fig. 2.2. – Mass and linear size.

except perhaps in the early universe; any lighter than 10^{15} g will already have decayed by now with the emission of Hawking radiation.

Gravity is more important, the closer an object is to the Schwarzschild line. Gravity is, of course, important for planets, stars, galaxies, clusters and the Universe as a whole; it is relatively unimportant for objects that are small or have low density.

The Heisenberg uncertainty principle excludes the shaded region in the lower left corner of fig. 2.2: trying to look in smaller and smaller regions requires larger and larger amounts of energy. Combining the constraints of gravity and quantum mechanics, there is a smallest length, the Planck length $\lambda_{Pl} = (G\hbar/c^3)^{\frac{1}{2}} = 2 \cdot 10^{-33}$ cm, and a characteristic mass of a quantum black hole, the Planck mass $M_{Pl} = (\hbar c/G)^{\frac{1}{2}} = 2 \cdot 10^{-5}$ g (see table I). To understand the origins of the big bang before the Planck time $t_{Pl} = \lambda_{Pl}/c$ will require a quantum theory of gravity.

A universe of vanishing curvature $k = 0$ has critical density; the mass enclosed by the Hubble sphere lies on the heavy diagonal line in fig. 2.2. A closed (open) universe with $k = 1$ ($k = -1$) lies above (below) this line. Presently available data indicate that the Universe is actually within about an order of magnitude of critical density, as indicated by the cross in the upper left corner of the figure.

2˙3. *Friedmann universes.* – Einstein's equations (2.6) for a homogeneous and isotropic fluid of density ϱ and pressure p are

$$(2.8) \qquad \frac{\dot{R}^2}{R^2} + \frac{kc^2}{R^2} = \frac{8\pi}{3} G\varrho + \frac{\Lambda c^2}{3}$$

for the 00 component, and

$$(2.9) \qquad \frac{2\ddot{R}}{R} + \frac{\dot{R}^2}{R^2} + \frac{kc^2}{R^2} = -\frac{8\pi}{c^2} Gp + \Lambda c^2$$

for the ii components [68]. Multiplying (2.8) by R^3, differentiating and comparing with (2.9) gives the equation of continuity

$$(2.10) \qquad \frac{d}{dR}(\varrho R^3) = -3pR^2 c^{-2} .$$

Given an equation of state $p = p(\varrho)$, this equation can be integrated to determine $\varrho(R)$; then (2.8) can be integrated to determine $R(t)$.

Consider, for example, the case of vanishing pressure $p = 0$, which is presumably an excellent approximation for the present Universe since the contribution of radiation and massless neutrinos (both having $p = \varrho c^2/3$) to the mass-energy density is at the present epoch much less than that of nonrelativistic matter (for which p is negligible). Equation (2.10) reduces to

$$(4\pi/3)\varrho R^3 = M = \text{const}, \tag{2.11}$$

and (2.8) yields *Friedmann's equation*

$$\dot{R}^2 = \frac{2GM}{R} + \frac{\Lambda c^2 R^2}{3} - kc^2. \tag{2.12}$$

This can be integrated in general in terms of elliptic functions, and for $\Lambda = 0$ in terms of elementary functions (see below).

Notice the analogy with Newtonian physics. Applying energy conservation to a self-gravitating sphere gives (2.12) with $k/2$ as the net energy (kinetic minus potential) per unit mass, and $\Lambda = 0$. The cosmological constant can be given a pseudo-Newtonian interpretation as a Klein-Gordon modification of the Poisson equation [63]:

$$\nabla^2 \varphi + \Lambda \varphi = -4\pi G \varrho. \tag{2.13}$$

For the time being, let us set $\Lambda = 0$. (I will discuss the case of a non-vanishing cosmological constant in sect. 4.) Solving the Friedmann equation for k at the present time (since k is a constant, any time will do),

$$kc^2 = R_0^2 \left(\frac{8\pi}{3} G \varrho_0 - H_0^2 \right). \tag{2.14}$$

Thus the Universe is flat ($k = 0$) if its density equals the *critical density*

$$\varrho_{c,0} \equiv \frac{3H_0^2}{8\pi G}. \tag{2.15}$$

It is convenient to specify the density in units of critical density via the *density parameter*

$$\Omega \equiv \varrho/\varrho_c. \tag{2.16}$$

It is also conventional to introduce the *deceleration parameter*

$$q_0 \equiv -R_0 \ddot{R}_0 / \dot{R}_0^2. \tag{2.17}$$

It follows that, if $\Lambda = 0$ and the Universe is dominated today by nonrelativistic matter, $q_0 = \Omega_0/2$.

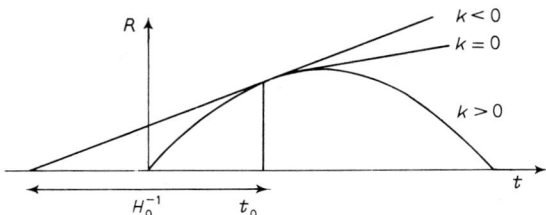

Fig. 2.3. – Sketch of solutions of Einstein's equations for the scale factor $R(t)$ in open ($k < 0$), flat ($k = 0$) and closed ($k > 0$) Friedmann universe models. Hubble's constant H_0 is the slope $\dot R$, and deceleration due to gravity acting since the big bang (at $t = 0$ in the figure) implies that the actual age of the Universe t_0 is less than the Hubble time H_0^{-1}; the precise relationship is given by eq. (2.18).

The results obtained by integrating the Friedmann equation for positive-, vanishing- and negative-curvature universes are sketched in fig. 2.3 and summarized below. In each case, the time since the big bang is given by the expression

$$(2.18) \qquad t_0 = H_0^{-1} f(\Omega).$$

The function $f(\Omega)$ is graphed in fig. 2.4. It is a monotonically decreasing function, with $f(0) = 1$.

Open, $k = -1$, $\Omega_0 < 1$

$$(2.19) \qquad \begin{cases} R(\eta) = GM(\cosh \eta - 1), \\ t = GM(\sinh \eta - \eta), \\ f(\Omega_0) = \dfrac{1}{1 - \Omega_0} - \dfrac{\Omega_0}{2(1 - \Omega_0)^{\frac{3}{2}}} \cosh^{-1}\left(\dfrac{2}{\Omega_0} - 1\right). \end{cases}$$

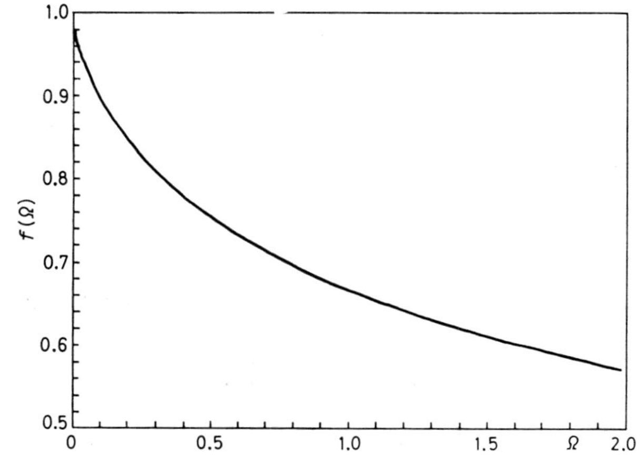

Fig. 2.4. – The function $f(\Omega) = t_0 H_0$ for a Friedmann universe ($\Lambda = 0$).

Flat, $k = 0$, $\Omega_0 = 1$ (*Einstein-de Sitter universe*)

(2.20) $$R(t) = (9GM/2)^{\frac{1}{3}} t^{\frac{2}{3}}, \quad f(1) = \tfrac{2}{3}.$$

Closed, $k = +1$, $\Omega_0 > 1$ (*Friedmann-Einstein universe*)

(2.21) $$\left|\begin{array}{l} R(\eta) = GM(1 - \cos\eta), \\ t = GM(\eta - \sin\eta), \\ f(\Omega_0) = \dfrac{\Omega_0}{2(\Omega_0 - 1)^{\frac{3}{2}}} \cos^{-1}\left(\dfrac{2}{\Omega_0} - 1\right) - \dfrac{1}{\Omega_0 - 1}. \end{array}\right.$$

Figure 2.5 shows how Ω_0 is related to H_0 in these Friedmann models, for various values of t_0.

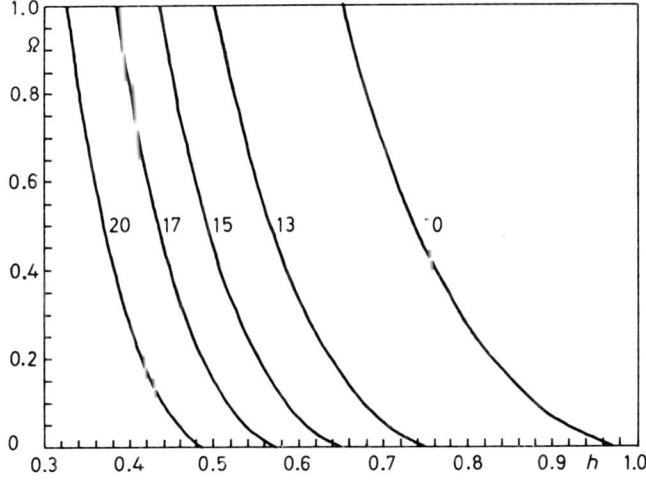

Fig. 2.5. – Relationship of cosmological density parameter Ω_0 and Hubble's parameter $H_0 \equiv 100h$ km s^{-1} Mpc^{-1} for various values of t_0 (in Gy).

2`4. Comparison with observations.

Age of the Universe t_0. Observational evidence bearing on the age of the Universe and other fundamental cosmological parameters was reviewed at the 1983 ESO-CERN conference by SANDAGE [69]. The best lower limits for t_0 come from studies of the stellar populations of globular clusters (GCs). SANDAGE concludes that a conservative lower limit on the age of GCs is (16 ± 3) Gy, which is then a lower limit on t_0. SANDAGE goes on to assume *a*) that the apparent cut-off in quasar red-shifts at $z \sim 4$ implies that galaxy formation ended at that epoch, about 2 Gy, and *b*) that the stars in the oldest GCs studied formed at that epoch; thus he estimates $t_0 \approx (18 \pm 3)$ Gy. I prefer simply to conclude that $t_0 > (16 \pm 3)$ Gy. Figure 2.5 shows that $t_0 > 13$ Gy implies that $H_0 \leqslant 76$ km s^{-1} Mpc^{-1} even for Ω very small, and that $H_0 \leqslant 50$ km s^{-1} Mpc^{-1} for $\Omega = 1$. (Figure 4.5 gives the analogous constraints for the case of a flat universe with nonvanishing cosmological constant.)

Hubble's parameter H_0. Hubble's parameter $H_0 \equiv 100h$ km s^{-1} Mpc^{-1} has in recent years been measured in two basic ways: *a)* using type-I supernovae as «standard candles» and *b)* using the Tully-Fisher relation between the rotation velocity and luminosity of spiral galaxies. Both methods depend on measuring the distance to nearby calibrating galaxies. SANDAGE has long contended that $h \approx 0.5$, and he concludes [69] that using both methods the latest data are consistent with $h = 0.50 \pm 0.07$. DE VAUCOULEURS has long contended that $h \approx 1$, and he has recently argued that the data still support this value [70]. Another method for determining H_0 has recently been proposed which, like *a)*, uses type-I supernovae, but which avoids the uncertainties of the «distance ladder» by calculating the absolute luminosity of type-I supernovae from first principles (using a very plausible but as yet unproved physical model). The result obtained is that h lies between 0.38 and 0.71, with a best estimate of 0.58 [71].

Cosmological density parameter Ω. In the first section I summarized the evidence on the mass associated with galaxies from luminosity and dynamical mass measurements: Ω (luminous) $\approx 0.01 \div 0.02$ and Ω (dark halos) $\approx 0.1 \div 0.2$. Here I will discuss several other observations that are relevant to cosmological mass estimates: galaxy position and velocity correlation functions, the infall velocity of the Local Group toward the Virgo cluster, the dynamics of other superclusters, constraints on the density of diffuse neutral and ionized hydrogen, and attempts to measure the deceleration parameter.

Galaxy correlation functions. PEEBLES and his collaborators have analyzed the available data on the angular positions of $\sim 10^6$ galaxies in terms of low-order correlation functions [72]. More recently, red-shift data from both the CfA survey [73] and a deeper red-shift survey [74] have also given estimates of the relative peculiar velocity between pairs of galaxies as a function of their separation, which in turn can be used to estimate Ω.

The galaxy two-point correlation function $\xi(r)$ (also called the autocorrelation or autocovariance function) is defined by

(2.22) $$\delta P = \bar{n}^2 [1 + \xi(r_{12})] \delta V_1 \delta V_2,$$

where δP is the joint probability of finding galaxies in volumes δV_1 and δV_2 separated by distance r_{12}, and $\bar{n}$ is the average number density of galaxies. Equivalently, the probability of finding a galaxy in δV at distance r, given one at the origin, is

(2.23) $$\delta P(1|2) = \bar{n}[1 + \xi(r)]\delta V.$$

The three-point correlation function is defined analogously to (2.22):

(2.24) $$\delta P = \bar{n}^3[1 + \xi(r_{12}) + \xi(r_{23}) + \xi(r_{13}) + \zeta(r_{12}, r_{23}, r_{13})]\delta V_1 \delta V_2 \delta V_3.$$

The corresponding triangle geometry is sketched in fig. 2.6.

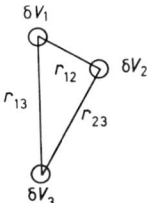

Fig. 2.6. – Geometry for the three-point correlation function.

The two-point correlation function has a remarkably simple form: it is approximately a power law over the interval $0.1 h^{-1}$ Mpc $\lesssim r \lesssim 10 h^{-1}$ Mpc:

(2.25) $$\xi(r) = (r/r_0)^{-\gamma},$$

with [73] index

$$\gamma = 1.77 \pm 0.04$$

and correlation length

$$r_0 = (5.4 \pm 0.3) h^{-1} \text{Mpc},$$

and $\xi \ll 1$ (and rather uncertain) for $r \gtrsim 10 h^{-1}$ Mpc. Values of $\xi(r)$ determined from the CfA data are plotted in fig. 2.7. The three-point function is found [75] to be well approximated by a symmetric sum of products of two-point functions:

(2.26) $$\zeta(r_{12}, r_{23}, r_{31}) = Q[\xi(r_{12})\xi(r_{23}) + \xi(r_{23})\xi(r_{31}) + \xi(r_{31})\xi(r_{12})],$$

with $Q \approx 1$.

In order to use these data to estimate the average mass density $\bar{\varrho}$, it is assumed that the galaxy distribution accurately traces the mass distribution. However, it is known that rich clusters are more strongly correlated than galaxies, with $\xi_{cc}(r) \approx 10 \xi_{gg}(r)$, as shown in fig. 2.8 [76-78]. Thus rich clusters and galaxies cannot both be good tracers of the mass distribution; perhaps neither is. For the time being I will blithely ignore this cautionary aside. (It will come up again.)

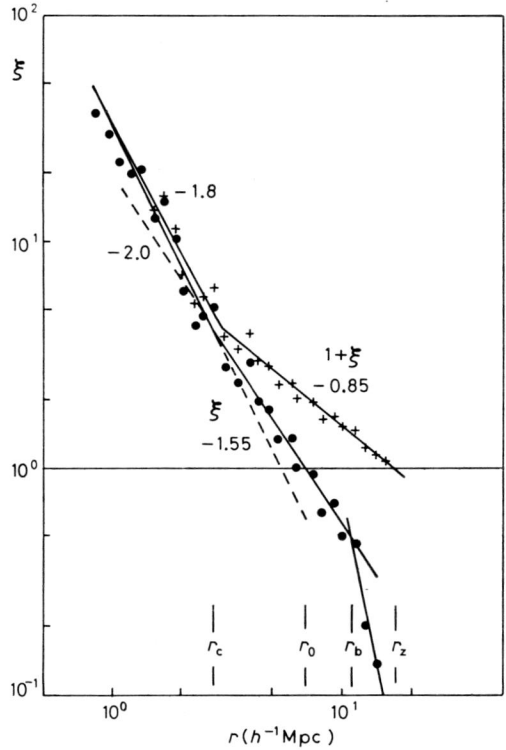

Fig. 2.7. – Two-point galaxy correlation function $\xi(r)$ calculated from the CfA data [73], as replotted and fitted by DEKEL and AARSETH. Dots (•) are ξ and pluses (+) are $1 + \xi$. (Reproduced, with permission, from ref. [79], which discusses a possible physical interpretation of the fits and the corresponding length scales r_c, r_0, r_b, r_z.)

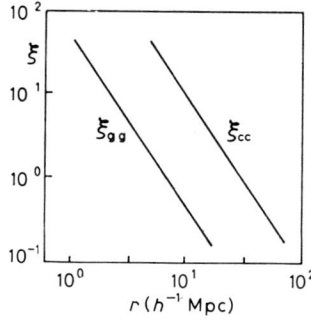

Fig. 2.8. – The cluster-cluster correlation function has the same slope as the galaxy-galaxy correlation function, but is roughly an order of magnitude larger.

Let us assume then that the mass is distributed like the galaxies, with $\delta\varrho/\varrho \sim \xi(r)$ [21]. The mass of a typical bound clump of size $r \ll r_0$ is $M \sim \bar{\varrho}\xi(r)r^3$, so by the virial theorem the internal velocity should be

$$(2.27) \qquad v^2 \sim GM/r \propto r^{2-\gamma}.$$

PEEBLES [80] has shown more precisely that

$$(2.28) \qquad \langle v_{12}^2(r) \rangle = \frac{6G\bar{\varrho}}{\xi(r)} \int_r^\infty \frac{\mathrm{d}r}{r} \int \mathrm{d}^3 z \, \frac{\mathbf{r}\cdot\mathbf{z}}{z^3} \zeta(\mathbf{r},\mathbf{z},|\mathbf{r}-\mathbf{z}|),$$

which is called the *cosmic virial theorem*. Using (2.25) and (2.26), (2.28) implies that

$$(2.29) \qquad \Omega \approx Q^{-1} \left(\frac{\langle v_{12}^2 \rangle^{\frac{1}{2}}}{800 \text{ km s}^{-1}} \right)^2 \left(\frac{1 h^{-1} \text{ Mpc}}{r} \right)^{2-\gamma}.$$

The red-shift survey data [73, 74] give

$$(2.30) \qquad \langle v_{12}^2 \rangle^{\frac{1}{2}} = (300 \pm 50) \text{ km s}^{-1},$$

for $r \sim 1 h^{-1}$ Mpc, with a weak dependence on r_{12} consistent with (2.27). Applying the cosmic virial theorem with [21, 73, 81] $Q = 0.7 \div 1.3$ gives $\Omega \approx 0.1 \div 0.2$.

DAVIS and PEEBLES [73] discuss two other methods of extracting estimates of Ω from red-shift data. One is based on the Irvine-Layzer *cosmic energy equation*, which relates the single-galaxy one-dimensional velocity dispersion $\bar{v}_\mathrm{p}$ to the potential energy stored in fluctuations. With reasonable approximations, this yields $\Omega \approx (\bar{v}_\mathrm{p}/660 \text{ km s}^{-1})^2 \approx 0.2$. The second method is based on the assumption that the mass clustering on scales $\leqslant 1 h^{-1}$ Mpc is statistically stable, neither expanding with the Hubble expansion nor collapsing. This leads to the expression

$$(2.31) \qquad \sigma^2(r) \approx 4.13 Q (H_0 r)^2 \xi(r) \Omega,$$

where $\sigma(r)$ is the pair-weighted one-dimensional relative velocity dispersion, given approximately by [73]

$$(2.32) \qquad \sigma(r) = \sigma_0 (h r_{\text{Mpc}})^{0.13 \pm 0.04}$$

with $\sigma_0 = (340 \pm 40)$ km s^{-1}. Then

$$(2.33) \qquad \Omega = Q^{-1} (\sigma_0 / 900 \text{ km s}^{-1})^2 = 0.20(1.5^{\pm 1}).$$

All these methods give estimates of the mass on scales of order 1 Mpc. It is perhaps significant that they all agree that $\Omega(\sim 1 \text{ Mpc}) \approx 0.1 \div 0.3$, and that this agrees with the estimates of the amount of dark matter around galaxies and clusters discussed in sect. 1. But all of these estimates are insensitive to a possible component of dark matter that is not clustered on small scales but instead distributed rather uniformly.

Infall toward Virgo. The best method presently available for estimating the cosmic density on scales of ~ 10 Mpc uses the infall velocity v_V of the Local Group toward the center of the Local Supercluster, which is located in the Virgo cluster at a distance of about $10h^{-1}$ Mpc. In linear perturbation theory, the infall peculiar velocity at a radius R resulting from a mass excess δM distributed spherically within that radius is given by [57]

$$(2.34) \qquad v_V = \frac{2G\delta M f(\Omega)}{3H_0 R^2 \Omega},$$

where $f(\Omega) \approx \Omega^{0.6}$ (see ref. [72], eq. (14.8)). Assuming that the distribution of bright galaxies N traces the mass distribution on large scales so that

$$(2.35) \qquad \frac{\delta M}{M} \approx \frac{\delta N}{N} \equiv \delta,$$

and using eq. (2.15), it follows that

$$(2.36) \qquad \Omega^{0.6} = 3v_V/v_H \delta,$$

where v_H is the unperturbed Hubble velocity. SANDAGE [69, 82] takes $v_V = (200 \pm 50)$ km s^{-1}, $v_H = 1170$ km s^{-1} and $\delta = 2.8 \pm 0.5$, which implies $\Omega \approx 0.06$. DAVIS and PEEBLES [57] argue that the predominance of the evidence, especially the agreement between the velocity of the Local Group with respect to the cosmic background radiation and with respect to distant galaxies [58], suggests rather that $v_V = (400 \pm 60)$ km s^{-1}, and they take $\delta = 2.2 \pm 0.3$. With the linear spherical approximation (2.36), this gives $\Omega \approx 0.2$. As I discussed in the first section, the Local Supercluster is not very spherical. DAVIS and PEEBLES [57] obtain $\Omega = 0.35 \pm 0.15$ in a nonlinear spheroidal model. However, even if the mass distribution is well represented by the galaxy distribution, if it is aspherical, then v_V can reflect the effect of mass outside R [83], adding further uncertainty to the determination of $\Omega(\sim 10 \text{ Mpc})$. Perhaps in a few years the galaxy flow in the neighborhood of the Local Supercluster will be better understood through the interplay of theory and observation (especially needed are reliable distance indicators independent of red-shift) [70].

Dynamics of superclusters. The dynamics of other superclusters can also be used to estimate the value of Ω, especially as better data become available on the peculiar velocities of the galaxies and clusters within them. As an example of this approach, HARMS *et al.* [84] used a spherical model to estimate the density required to account for the observed velocity dispersion of the galaxy clusters in the supercluster 1451 + 22. The observed velocity dispersion is about half that expected from unperturbed Hubble flow. Their model gave an average density within this supercluster between 1.01 and $1.99\varrho_{c,0}$. They estimate that the space density of galaxies is enhanced in this supercluster by a factor between about 17 and 71. Making the crucial assumption that the mass density is enhanced by the same factor, it follows that $0.014 \leqslant \Omega \leqslant 0.12$. Relaxing the assumption of spherical symmetry would allow $\Omega \leqslant 0.3$. Similar results are obtained for other superclusters [85].

Density of hydrogen. You may wonder how much of the dark matter could be ordinary hydrogen. GUNN and PETERSON pointed out that absorption of quasar light by intervening atomic hydrogen (HI) would cause an absorption trough on the short-wavelength side of the Lyman-α line at 1216 Å, as sketched in fig. 2.9. The Ly α line is conveniently red-shifted into the visible range for quasars with $z > 1.5$, and the absence of such an absorption trough implies that

(2.37) $$\Omega(\text{H I}) < 3 \cdot 10^{-7} h^{-1}$$

with a similar result for molecular hydrogen

(2.38) $$\Omega(\text{H}_2) < 5 \cdot 10^{-5} h^{-1}.$$

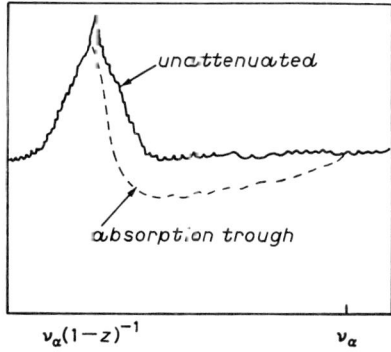 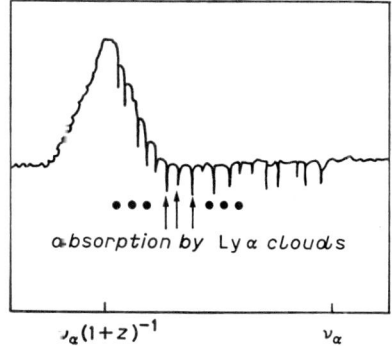

Fig. 2.9. – Schematic sketches of quasar spectra, showing (left) the unattenuated red-shifted Ly α emission feature and the absorption trough that would be seen if there were significant absorption by intervening hydrogen, and (right) the actual sort of spectra seen, with discrete absorption lines from small intervening « Ly α clouds » of neutral hydrogen. (From unpublished manuscript by George BLUMENTHAL.)

Although there is no absorption trough, there are many discrete absorption lines in quasar spectra caused by small «Ly α clouds» of neutral hydrogen (this interpretation is confirmed by the presence of Ly β absorption as well). These Ly α clouds are important as cosmological tracers (more on that later), but their total mass is less than that of the luminous parts of galaxies.

What about ionized hydrogen? $\Omega(\text{H II}) \ll 1$ from nonobservation of radiation, except possibly for plasma at a temperature of $\sim 3 \cdot 10^8$ K. The observed X-ray background in the range $3 \text{ keV} < h\nu \lesssim 50 \text{ keV}$ could be produced by nearly a closure density of ionized hydrogen at this temperature—but an enormous amount of energy would be required to heat so much gas to so high a temperature, and another explanation would still be required for the X-ray background above ~ 60 keV. Moreover, as I will explain in the next section, the standard theory of hot-big-bang nucleosynthesis produces the observed abundances of deuterium, ^{3}He and ^{4}He only if the primordial baryon abundance Ω_b lies between $0.01h^{-2}$ and $0.035h^{-2} \lesssim 0.14$ [86]. The upper limit is (barely) consistent with all the dark matter being baryonic, but I will discuss other arguments against this in the next section.

Deceleration parameter q_0. A way of determining Ω on very large scales is to measure the deceleration parameter q_0, given by eq. (2.17). If the cosmological constant vanishes, then $q_0 = 2\Omega$. Although q_0 can in principle be measured by determining the deviation of very distant objects from Hubble's law, in practice it has been impossible to determine their distances very accurately. The traditional approach, based on the assumed constant luminosity of the brightest galaxies in each rich cluster, is fraught with uncertainties—in particular, the effects of evolution (time variation in absolute luminosity, caused, for example, by the aging of the stellar populations) and sampling bias (near and distant samples may not be comparable). Nevertheless a recent review [87] obtains an upper limit $q_0 \lesssim 1$ from radio galaxies observed in the near IR having red-shifts in the range ~ 0.5 to ~ 1. Alternative approaches are unfortunately also problematic. Since quasars have by far the highest observed red-shifts ($z \lesssim 3.8$), they would provide an ideal sample for determining q_0 if some feature of their spectra could be used to determine their intrinsic luminosity. A recent study, exploiting an observed correlation between the strength of the C IV (triply ionized carbon) 1550 Å emission line and the luminosity of the underlying continuum in flat-radio-spectrum quasars, finds $q_0 = 1^{+1.5}_{-0.7}$ assuming no evolution [88]. This result may suffer from possible selection and evolution effects [89], however, and it is based entirely on an empirical correlation whose origin is not well understood.

To summarize, the accurate measurement of the cosmological density parameter Ω is difficult, but it probably lies in the range $0.1 \lesssim \Omega \lesssim 2$. Large Ω, such as the Einstein-de Sitter value $\Omega = 1$, is excluded unless mass density is distributed considerably more broadly than luminosity.

2'5. Growth and collapse of fluctuations. – The continuity or energy conservation equation (2.10) can be integrated, given an equation of state $p = p(\varrho)$, to determine $\varrho(R)$. Then the Einstein equation (2.8) can be integrated to give $R(t)$. Consider the equation of state $p = w\varrho$, where w is a constant. Integrating (2.10) gives

$$\varrho \propto R^{-3(1+w)}, \tag{2.39}$$

and then integrating (2.8) in the approximation that $k = 0$, which is always valid at early times (*), gives

$$R \propto t^{2/3(1+w)}. \tag{2.40}$$

There are two standard cases:

radiation dominated

$$w = \tfrac{1}{3}, \quad \varrho \propto R^{-4}, \quad R \propto t^{\tfrac{1}{2}}; \tag{2.41}$$

matter dominated

$$w = 0, \quad \varrho \propto R^{-3}, \quad R \propto t^{\tfrac{2}{3}}. \tag{2.42}$$

The crossover between these two regimes occurs at $R = R_{eq}$, when relativistic particles (photons and N_ν species of two-component neutrinos of negligible mass) and nonrelativistic particles (ordinary and dark matter) make equal contributions to ϱ:

$$R_{eq} = \frac{4\sigma T_0^4(1+\gamma)}{\Omega \varrho_{c,0} c} = \frac{4.05 \cdot 10^{-5}}{\Omega h^2} \frac{1+\gamma}{1.681} \theta^4. \tag{2.43}$$

Here the scale factor R has been normalized so that $R_0 \equiv R(t_0) = 1$; γ is the ratio of neutrino to photon energy densities (discussed further in sect. **3**),

$$\gamma \equiv \frac{\varrho_{\nu,0}}{\varrho_{\gamma,0}} = \frac{7}{8}\left(\frac{4}{11}\right)^{\tfrac{4}{3}} N_\nu (= 0.681 \text{ for } N_\nu = 3); \tag{2.44}$$

σ is the Stefan-Boltzmann constant and $\theta \equiv T_0/2.7$ K. The contribution of relativistic particles to the cosmological density is very small today in the standard model; for example, the contribution of photons is $\Omega_{\gamma,0} = 3.0 \cdot 10^{-5} h^{-2} \theta^4$.

(*) The curvature term, which is $\propto R^{-2}$, is possibly important today. But in the early universe it is always much smaller than the density term, which is $\propto R^{-3}$ (matter dominated) or $\propto R^{-4}$ (radiation dominated).

It is also possible to obtain a simple expression for $t(R)$ that is valid in both radiation- and matter-dominated eras, for the case of a flat universe (*i.e.* $k = 0$). Simply integrate the Einstein equation (2.8) with

(2.45) $$\varrho = \varrho_{\text{rel}} + \varrho_{\text{nonrel}} \approx \varrho_{c,0}\Omega_0(R_{\text{eq}}R^{-4} + R^{-3}).$$

The result is

(2.46) $$t = \tfrac{2}{3}H_0^{-1}\Omega_0^{-\frac{1}{2}}R_{\text{eq}}[(R - 2R_{\text{eq}})(R + R_{\text{eq}})^{\frac{1}{2}} + 2R_{\text{eq}}^{\frac{3}{2}}],$$

with the following limiting behaviors:

(2.47) $$\begin{cases} R \ll R_{\text{eq}}: & t \approx \tfrac{1}{2}H_0^{-1}\Omega_0^{-\frac{1}{2}}R_{\text{eq}}^{-\frac{1}{2}}R^2; \\ R = R_{\text{eq}}: & t_{\text{eq}} = 0.3905 H_0^{-1}\Omega_0^{-\frac{1}{2}}R_{\text{eq}}^{\frac{3}{2}}; \\ R \gg R_{\text{eq}}: & t \approx \tfrac{2}{3}H_0^{-1}\Omega_0^{-\frac{1}{2}}R^{\frac{3}{2}}. \end{cases}$$

It is now easy to calculate the mass M_{H} of nonrelativistic matter encompassed by the horizon $ct(R)$ as a function of scale factor R:

(2.48) $$M_{\text{H}} = \tfrac{4}{3}\pi c^3 t^3 \frac{\varrho_{c,0}\Omega_0}{R^3} = \frac{2.41 \cdot 10^{15} M_\odot}{\Omega_0^2 h^4}\left[\frac{(y-2)(y+1)^{\frac{1}{2}} + 2}{y}\right]^3,$$

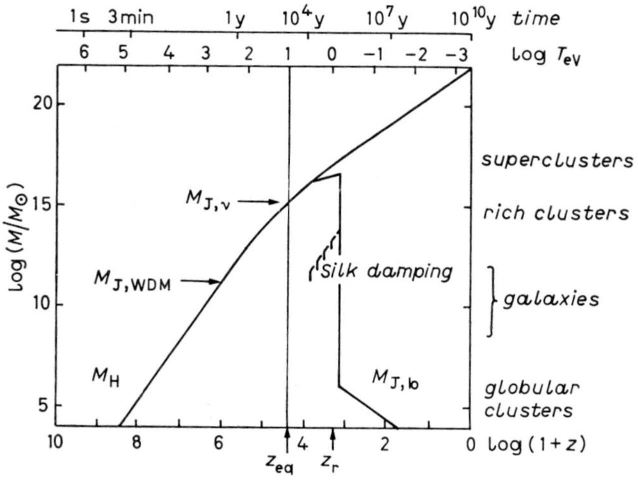

Fig. 2.10. – Mass M_{H} (heavy solid line) of nonrelativistic matter encompassed by the horizon as a function of red-shift. The Jeans mass M_J is approximately equal to M_{H} until the era (z_{eq}) when nonrelativistic matter begins to dominate gravitationally; at recombination (z_r), M_J drops sharply (light solid line). Photon diffusion (Silk damping) erases adiabatic fluctuations in radiation plus ordinary matter that cross into the region below the dashed line. Fluctuations in a universe dominated by hot or warm dark matter are damped by free streaming until the temperature drops to the mass of the dark-matter particles and they become nonrelativistic; the corresponding (free-streaming) Jeans masses are indicated. (See sect. **3**.)

where $y \equiv R/R_{eq}$. The behavior of $M_{\rm H}$ is sketched in fig. 2.10 (heavy solid curve).

Top hat model. It is now time to consider the evolution of small fluctuations in the density. In the linear regime $\delta \equiv \delta\varrho/\varrho \ll 1$, the growth rate is independent of shape. It is simplest to consider a spherical («top hat») fluctuation, say a region of radius $R(1+a)$ with uniform density $\bar{\varrho}(1+\delta)$ in a background of density $\bar{\varrho}$ (see fig. 2.11).

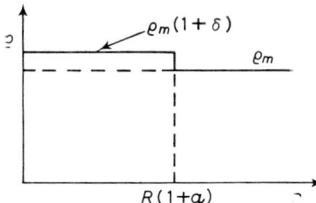

Fig. 2.11. – «Top hat» fluctuation: constant density $\varrho_{\rm m}(1+\delta)$ inside a spherical volume of radius $R(1+a)$ in a universe of background density $\varrho_{\rm m} = \bar{\varrho}$.

Consider first the growth of fluctuations in a matter-dominated universe ($p = 0$). Conservation of mass implies

$$\bar{\varrho}(1+\delta)R^3(1+a)^3 = \text{const}, \tag{2.49}$$

or

$$\delta = -3a. \tag{2.50}$$

Now it is necessary to bring in gravity:

$$\ddot{R} = -\frac{4\pi G}{3}\left(\varrho + \frac{3p}{c^2}\right)R. \tag{2.51}$$

(This equation follows by subtracting (2.8) from (2.9). Alternatively, it is the 00 component of Einstein's equations in the form $R_{\mu\nu} = -(8\pi G/c^4) \cdot (T_{\mu\nu} - \frac{1}{2}g_{\mu\nu}T^\lambda_\lambda)$ applied to the Robertson-Walker metric.) Applying (2.51) to the background and to the fluctuation, we obtain

$$\ddot{R}(1+a) + 2\dot{R}\dot{a} + R\ddot{a} = -(4\pi G/3)\bar{\varrho}R(1+a+\delta),$$

or

$$\ddot{\delta} + 2(\dot{R}/R)\dot{\delta} = 4\pi G\bar{\varrho}\delta. \tag{2.52}$$

Substituting $\dot{R}/R = \frac{2}{3}t^{-1}$, valid for a flat ($k = 0$) matter-dominated universe, and trying $\delta = t^\alpha$, one finds $(\alpha + 1)(\alpha - \frac{2}{3}) = 0$. The general solution of (2.52) is thus

(2.53) $$\delta = At^{\frac{2}{3}} + Bt^{-1}.$$

Notice that the amplitude of the fluctuation in the growing mode has the same rate of growth as the scale factor R in the matter-dominated universe.

An analogous calculation for a radiation-dominated universe gives

(2.54) $$\delta = At + Bt^{-1}.$$

This time the growing mode for the amplitude grows as the square of the scale factor (i.e. $\delta \propto R^2$) in the radiation-dominated universe. The solution (2.54) is actually relevant only on scales larger than the horizon, since, once the fluctuations come within the horizon, the radiation and baryons start to oscillate and the neutrinos freely stream away. (I will discuss this further in sect. **3**.) One must be careful in discussing behavior on scales larger than the horizon, since the freedom to choose co-ordinates or gauge can complicate the physical interpretation. In these lectures I am using «time-orthogonal» co-ordinates and the «synchronous gauge» formalism [65, 72, 90]. (Bardeen's gauge-invariant formalism is an attractive alternative [91].) Indeed it may seem paradoxical even to consider fluctuations larger than the horizon—but it is necessary to do so, since all cosmologically interesting fluctuations are larger than the horizon at early times. What we are doing effectively is comparing the growth rates of universes differing slightly in density. The region of slightly higher density (the fluctuation) expands slightly more slowly; consequently, the density contrast δ between it and the background grows with time. (Birkhoff's theorem [66] permits us to ignore the Universe outside our spherically symmetric fluctuation.)

Since cosmological curvature is at most marginally important at the present epoch, it was negligible during the radiation-dominated era and at least the beginning of the matter-dominated era. But for $k = -1$, i.e. $\Omega < 1$, the growth of δ slows for $R/R_0 \gtrsim \Omega_0$, as gravity becomes less important and the Universe begins to expand freely. To discuss this case, it is convenient to introduce the variable

(2.55) $$x \equiv \Omega^{-1}(t) - 1 = (\Omega_0^{-1} - 1) R(t)/R_0.$$

(Note that $\Omega(t) \to 1$ at early times.) The general solution in the matter-dominated era is then [92]

(2.56) $$\delta = \tilde{A} D_1(t) + \tilde{B} D_2(t),$$

where the growing solution is

(2.57) $$D_1 = 1 + \frac{3}{x} + \frac{3(1+x)^{\frac{1}{2}}}{x^{\frac{3}{2}}} \ln[(1+x)^{\frac{1}{2}} - x^{\frac{1}{2}}]$$

and the decaying solution is

(2.58) $$D_2 = (1+x)^{\frac{1}{2}}/x^{\frac{3}{2}}.$$

These agree with the Einstein-de Sitter results (2.53) at early times ($t \ll t_0$, $x \ll 1$). For late times ($t \gg t_0$, $x \gg 1$) the solutions approach

(2.59) $$D_1 = 1, \quad D_2 = x^{-1};$$

in this limit the Universe is expanding freely and the amplitude of fluctuations stops growing.

Spherical collapse. At early times, an overdense fluctuation expands with the Hubble flow. Eventually, however, it reaches a maximum radius, and then «turns around» and begins to contract, just like a small piece of a positive-curvature Robertson-Walker universe. Continuing the analogy, one might suppose that it would collapse to a point—but, of course, it does not; « violent relaxation » rapidly brings it into virial equilibrium at a radius about half the maximum radius. Since the fluctuation is now well inside the horizon and there are no relativistic velocities, the Newtonian approximation is valid.

Figure 2.12 summarizes the collapse process with sketches of the radius, density and density contrast as a function of scale factor R. This paragraph and the next are devoted to filling in the details in this figure.

I will start by deriving an expression for the maximum radius r_m, and the time t_m at which it is reached, for a spherical «top hat» fluctuation. As above, let the density in the fluctuation equal $\bar{\varrho}(1+\delta)$, but let the radius be $r = r_i(R/R_i) + \xi$, where $\xi = 0$ at the initial time t_i. The initial time t_i is arbitrary, except that I will assume that it is in the matter-dominated era, that $\delta_i \ll 1$, and that the fluctuation is described by the growing mode $\delta \propto t^{\frac{2}{3}}$.

Conservation of mass ($= \varrho r^3$) implies that the initial velocity at the edge of the spherical fluctuation is

(2.60) $$v_i = H_i r_i + \dot{\xi}_i = H_i r_i - r_i \dot{\delta}_i/3,$$

so the corresponding kinetic energy per unit mass is

(2.61) $$K_i = \tfrac{1}{2} H_i^2 r_i^2 - \tfrac{1}{3} H_i r_i^2 \dot{\delta}_i = \tfrac{1}{2} H_i^2 r_i^2 (1 - \tfrac{2}{3}\delta_i).$$

Since the potential energy per unit mass at the edge of the sphere is

(2.62) $$W_i = -\tfrac{1}{2} H_i^2 r_i^2 \Omega_i (1 + \delta_i),$$

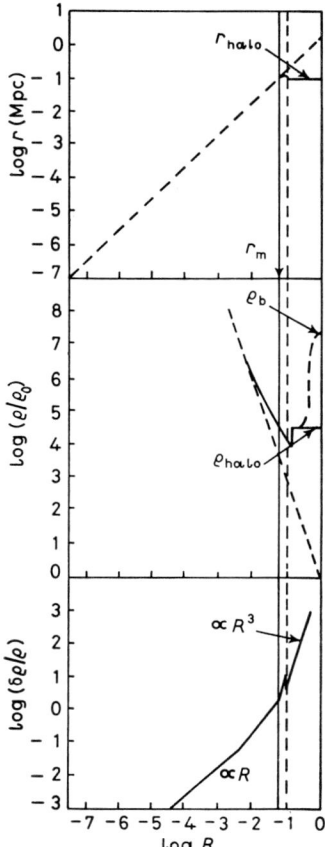

Fig. 2.12. – Schematic sketches of radius, density and density contrast of an overdense fluctuation. It initially expands with the Hubble expansion, reaches a maximum radius (solid vertical line), and undergoes violent relaxation during collapse (dashed vertical line), which results in the dissipationless matter forming a stable halo. Meanwhile the ordinary matter (ϱ_b) continues to dissipate kinetic energy and contract, thereby becoming more tightly bound, until dissipation is halted by star or disk formation.

the total energy per unit mass is

$$(2.63) \quad E = K_i + W_i = \frac{W_i}{1+\delta_i}\left[(1+\delta_i) - \frac{1}{\Omega_i}(1-\tfrac{2}{3}\delta_i)\right].$$

Maximum expansion corresponds to $K_m = 0$, so $E = W_m = (r_i/r_m)W_i$ and

$$(2.64) \quad \frac{r_m}{r_i} = \frac{W_i}{E} \approx \frac{1+\delta_i}{1-\Omega_i^{-1}+\tfrac{5}{3}\delta_i}.$$

This result, derived by BLUMENTHAL and me [93], differs slightly from that given in PEEBLES ([72], § 19) because I here assume a purely growing mode for δ_i and allow a nonzero deviation of the expansion velocity from pure Hubble flow at t_i. It can be rewritten in terms of Ω_0 using the fact that

$$\Omega_i = \Omega_0 \frac{1+z}{1+\Omega_0 z_i} ; \tag{2.65}$$

namely,

$$\frac{r_m}{r_i} \approx \frac{(1+z_i)(1+\delta_i)}{1 - \Omega_0^{-1} + \frac{5}{3}(1+z_i)\delta_i} . \tag{2.66}$$

The corresponding time can be calculated from standard Newtonian expressions. The force law

$$\ddot{r} = -GM/r^2$$

implies that

$$\dot{r}^2 = \frac{2GM}{r}\left(1 - \frac{r}{r_{m}}\right),$$

which can be integrated giving

$$t_m = \left(\frac{\pi^2 r_m^2}{8GM}\right)^{\frac{1}{2}} . \tag{2.67}$$

The density in the top hat is then

$$\varrho_m = \frac{3\pi}{32 G t_m^2} ; \tag{2.68}$$

since the background density in the Einstein-de Sitter ($k = 0$) approximation is $\bar{\varrho} = (6\pi G t^2)^{-1}$,

$$\varrho_m/\bar{\varrho} = 9\pi^2/16 = 5.6 , \tag{2.69}$$

and the density contrast is $\delta_m = 4.6$ at maximum expansion.

Violent relaxation. Figure 2.13 shows the result of a computer « N-body simulation » [94] of the late stages of dissipationless gravitational collapse of a top hat mass distribution: the bodies fall together (a)) into a dense « crunch » (b)), from which they emerge into a centrally condensed distribution (c)) that remains remarkably stable thereafter. The process occurs rapidly, in a time on the order of the gravitational dynamical time

$$\tau = (G\varrho)^{-\frac{1}{2}} . \tag{2.70}$$

It is called « violent relaxation ».

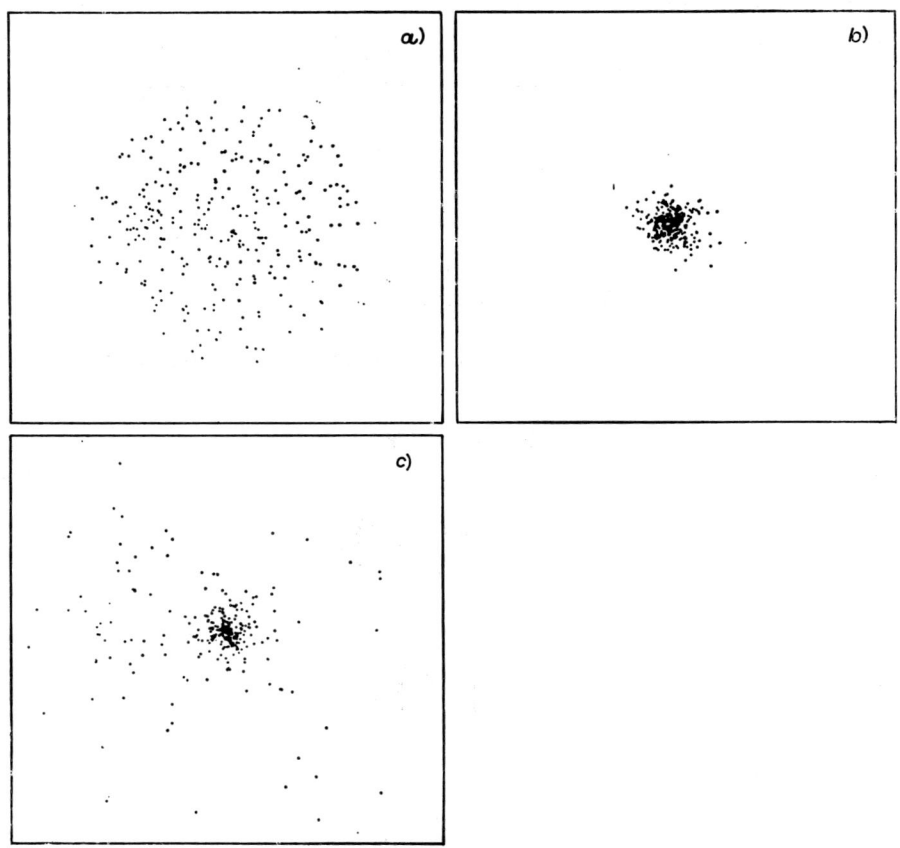

Fig. 2.13. – N-body simulation of gravitational collapse. (Reproduced, with permission, from ref. [94].)

The bound particles in the final configuration (c) accurately satisfy the virial theorem $\langle K \rangle = -\frac{1}{2} \langle W \rangle$. The potential energy varies inversely as the radius, $W = A/r$, so the radius after virialization r_v is given by

$$\frac{A}{r_m} = E = \tfrac{1}{2}\langle W \rangle = \frac{A}{2r_v}, \tag{2.71}$$

which implies that $r_v = \frac{1}{2} r_m$. As fig. 2.13 illustrates, the radius shrinks only by roughly a factor of 2 in the collapse. (Actually, this «radius» is effectively defined by the last equality in (2.71); since the mass is redistributed in the collapse, it is somewhat arbitrary.)

LYNDEN-BELL [95] and SHU [96] have shown by statistical methods that the distribution resulting from dissipationless violent relaxation via chaotic changes of the collective gravitational field, with the total mass much greater than that of any component particles, is to an excellent approximation a Maxwell-Boltzmann distribution, but with components of different masses having the same velocity dispersion and not the same «temperature». In other words, the distribution is a Maxwellian in the velocities, independent of the mass. Such a distribution is nevertheless called an «isothermal sphere» [97]. As I discussed in sect. 1, constancy of the velocity implies that the total mass increases linearly with radius, or equivalently that the density falls as r^{-2}, outside the central core; this is roughly what is found in computer simulations. Of course, this can only be true for intermediate values of r, since the total mass is finite. Another way of saying this is that high-energy orbits with periods longer than the collapse time cannot be very fully populated [21]. Thus the density falls faster than r^{-2} at large r.

In any case, the simple model of a spherical top hat initial distribution is rather unrealistic in at least two respects: it is likely that the initial density distribution is smoothly peaked rather than a step function, and moreover somewhat aspherical. The outer parts of the initial dark-matter density fluctuation will collapse later, perhaps resulting in a large constant-velocity halo with density falling roughly as r^{-2} to considerable distances [98]. Asphericity is amplified in the collapse, and the most probable result is that the collapse will actually occur in one direction first: «pancake collapse» [99]. This can happen even if the bulk of the matter in the fluctuation is not even bound, so that the expansion continues in the perpendicular directions; this is a popular model for the origin of superclusters [100-102]. In the case of protogalaxies, the subsequent violent relaxation of a flattened intermediate configuration produces an ellipsoidal rather than a spherical virialized distribution [103]; perhaps this is the typical shape of galactic halos.

A key feature of the dark matter is that it is dissipationless, whereas ordinary (baryonic) matter can convert its kinetic energy into radiation and thereby cool via bremsstrahlung (also called by astrophysicists «free-free scattering»), Ly α and Ly β radiation, and excitation of molecular and metallic energy levels. If the ordinary matter and dark matter are initially well mixed (which is a plausible initial condition before violent relaxation, at least in the cold-DM picture, as I will discuss in sect. 4), then dissipation during the «crunch» and afterward will cause the baryonic matter to sink to the center. The baryonic matter can radiate away energy but not angular momentum. If the dissipative collapse is halted by angular momentum, a disk will result. If it is halted by star formation (stars have negligible collision cross-sections), then a spheroidal system will result. These are, of course, the two elements of galaxy structure.

Presumably the processes just discussed occur on a variety of scales. If, as usually assumed, smaller-mass fluctuations have higher amplitudes, then they will turn around and virialize within larger-mass fluctuations, which subsequently themselves virialize, and so on until the present. The virialization of the next larger scale of the clustering hierarchy will tend to disrupt the smaller-scale structures within it. The crucial question for galaxy formation in this gravitational collapse picture is: What sets the mass scale of galaxies? (Recall that most of the mass in galaxies is in big galaxies whose mass is within an order of magnitude of that of the Milky Way.) At least two factors must be considered: the initial fluctuation spectrum and its modification as the Universe evolves and the rate of dissipation compared to gravitational collapse on different scales.

I will return to this in sect. 4. But first, in order to begin to discuss the fluctuation spectrum, I must ask where the fluctuations themselves came from.

2`6. *Inflation and the origin of fluctuations.* – The basic idea of inflation is that before the Universe entered the present adiabatically expanding Friedmann era, it underwent a period of de Sitter exponential expansion of the scale factor, termed *inflation* [104].

Then de Sitter cosmology corresponds to the solution of Friedmann's equation in an empty universe (*i.e.* with $\varrho = 0$ or, in (2.12), $M = 0$) with vanishing curvature ($k = 0$) and positive cosmological constant ($\Lambda > 0$). The solution is

(2.72) $$R = R_0 \exp[Ht],$$

with constant Hubble parameter

(2.73) $$H = (\Lambda/3)^{\frac{1}{2}}.$$

There are analogous solutions for $k = +1$ and $k = -1$ with $R \propto \cosh Ht$ and $R \propto \sinh Ht$, respectively. The scale factor expands exponentially because the positive cosmological constant corresponds effectively to a negative pressure. de Sitter space is discussed in textbooks on general relativity (for example, ref. [63, 105]) mainly for its geometrical interest. Until recently, the chief significance of the de Sitter solution (2.72) in cosmology was that it is a kind of limit to which all indefinitely expanding models with $\Lambda > 0$ must tend, since, as $R \to \infty$, the cosmological constant term ultimately dominates the right-hand side of the Friedmann equation (2.12).

As GUTH [104] emphasized, the de Sitter solution might also have been important in the very early universe because the vacuum energy that plays such an important role in spontaneously broken gauge theories also acts as an effective cosmological constant. A period of de Sitter inflation preceding ordinary radiation-dominated Friedmann expansion could explain several features of the observed universe that otherwise appear to require very special initial conditions: the horizon, smoothness, flatness, rotation and monopole problems. (A number of other people independently appreciated the power of an initial de Sitter period to generate desirable initial conditions for a subsequent Friedmann era [106, 107]. A paper by KAZANAS [108] is apparently the first published discussion of this in the context of grand unified theories.)

I will illustrate how inflation can help with the horizon problem. At recombination (p$^+$+e$^-$ → H), which occurs at $R/R_0 \approx 10^{-3}$, the mass encompassed by the horizon was $M_H \approx 10^{18} M_\odot$, compared to $M_{H,0} \approx 10^{22} M_\odot$ today. Equivalently, the angular size today of the causally connected regions at recombination is only $\Delta\theta \sim 3°$. Yet the fluctuation in temperature of the cosmic background radiation from different regions is so small that only an upper limit is presently available: $\Delta T/T < 10^{-4}$. How could regions far out of causal contact have come to temperatures which are so precisely equal? This is the « horizon problem ». With inflation, it is no problem because the entire observable universe initially lay inside a single causally connected region that subsequently inflated to a gigantic scale.

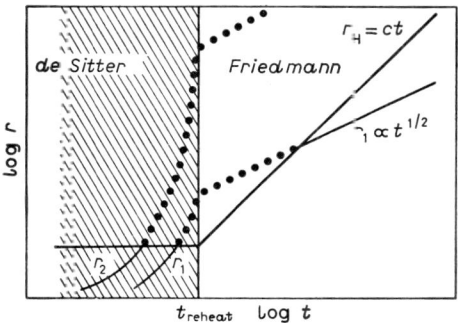

Fig. 2.14. – During the inflationary (de Sitter) expansion, regions once causally connected expand much faster than light (dotted lines). After reheating, the Universe becomes dominated by radiation rather than the effective cosmological constant associated with the Higgs self-energy; and as the horizon expands faster than the Hubble expansion ($R \propto t^{\frac{1}{2}}$ in a radiation-dominated universe), regions again become causally connected.

This is illustrated in fig. 2.14. The Hubble parameter $\dot{R}/R$ is constant in size during the de Sitter era; then, after reheating, the horizon size, of course, just grows linearly with time. A region of size r_1, initially smaller than the de Sitter horizon, inflates to a size much larger than the de Sitter horizon and is no longer causally connected (dots). After reheating, r_1 expands with the scale factor ($\propto t^{\frac{1}{2}}$ in the radiation-dominated Friedmann era) and eventually crosses back inside the horizon. The curve labeled r_2 shows the similar fate of a larger region. The region encompassed by the present horizon presumably all lay within a region like this that started smaller than the de Sitter horizon.

In inflationary models, the dynamics of the very early universe is typically controlled by the self-energy of the Higgs field associated with the breaking

of a grand unified theory (GUT) into the standard 3-2-1 model: GUT →
→ $SU_{3,\text{color}} \otimes [SU_2 \otimes U_1]_{\text{electroweak}}$. This occurs when the cosmological temperature drops to the unification scale $T_{\text{GUT}} \sim 10^{14}$ GeV at about 10^{-35} s after the big bang. GUTH [104, 109] initially considered a scheme in which inflation occurs while the Universe is trapped in an unstable state (with the GUT unbroken) on the wrong side of a maximum in the Higgs potential. This turns out not to work: the transition from a de Sitter to a Friedmann universe never finishes [110]. The solution in the « new inflation » scheme [111] is for inflation to occur *after* barrier penetration (if any). It is necessary that the Higgs potential be nearly flat (*i.e.* decrease very slowly with increasing Higgs field) for the inflationary period to last long enough. This nearly flat part of the Higgs potential must then be followed by a very steep minimum, in order that the energy contained in the Higgs potential be rapidly shared with the other degrees of freedom (« reheating »).

In turns out to be necessary to inflate by a factor $\geqslant e^{66}$ in order to solve the flatness problem, *i.e.* that $\Omega_0 \sim 1$. (With $H^{-1} \sim 10^{-34}$ s during the de Sitter phase, this implies that the inflationary period needs to last for only a relatively small time $\tau \geqslant 10^{-32}$ s.) The « flatness problem » is essentially the question why the Universe did not become curvature dominated long ago. Neglecting the cosmological constant on the assumption that it is unimportant after the inflationary epoch, the Friedmann equation can be written

$$(2.74) \qquad \left(\frac{\dot{R}}{R}\right)^2 = \frac{8\pi G}{3} \frac{\pi^2}{30} g(T) T^4 - \frac{kT^2}{(RT)^2},$$

where the first term on the right-hand side is the contribution of the energy density in relativistic particles and $g(T)$ is the effective number of degrees of freedom (discussed in detail in sect. **3**). The second term on the right-hand side is the curvature term. Since $RT \approx$ const for adiabatic expansion, it is clear that, as the temperature T drops, the curvature term becomes increasingly important. The quantity $K \equiv k/(RT)^2$ is a dimensionless measure of the curvature [112]. Today, $|K| = |\Omega - 1| H_0^2/T_0^2 \leqslant 2 \cdot 10^{-58}$. Unless the curvature exactly vanishes, the most « natural » value for K is perhaps $K \sim 1$. Since inflation increases R by a tremendous factor $\exp[H\tau]$ at essentially constant T (after reheating), it increases RT by the same tremendous factor and thereby decreases the curvature by that factor squared. Setting $\exp[-2H\tau] \leqslant 2 \cdot 10^{-58}$ gives the needed amount of inflation: $H\tau \geqslant 66$. This much inflation turns out to be enough to take care of the other cosmological problems mentioned above as well [113].

Of course, this is only the minimum amount of inflation needed; the actual inflation might have been much greater. Indeed it is frequently argued that, since the amount of inflation is a tremendously sensitive function of the initial value of the Higgs field (for example), it is extremely likely that there was much more inflation than the minimum necessary to account for the fact that Ω_0 is of order unity [114]. It then follows that the curvature constant is probably vanishingly small after inflation, which implies (in the absence of a cosmological constant today) that $\Omega_0 = 1$ to a very high degree of accuracy. However, in view of our lack of knowledge of the true dynamics of the inflationary epoch (assuming that there really was one), it is at least conceivable that increasing inflation becomes increasingly less likely [26]. This could happen, for example, through the effects of a « compensating field » of the sort proposed by ABBOTT [115] to explain why the cosmological constant is so small today. Then we might live in a part of the Universe that happened to inflate only enough to make $\Omega_0 \approx 0.2$.

Thus far, I have sketched how inflation stretches, flattens and smooths out the Universe, thus greatly increasing the domain of initial conditions that could correspond to the Universe that we observe today. But inflation also can explain the origin of the fluctuation necessary in the gravitational-instability picture of galaxy and cluster formation. Recall that the very existence of these fluctuations is a problem in the standard big-bang picture, since these fluctuations are much larger than the horizon at early times (see fig. 2.10). How could they have arisen?

The answer in the inflationary-universe scenario is that they arise from quantum fluctuations in the scalar field φ whose vacuum energy drives inflation. The scalar fluctuations $\delta\varphi$ during the de Sitter phase are of the order of the Hawking temperature $H/2\pi$. Because of these fluctuations, there is a time spread $\Delta t \approx \delta\varphi/\dot{\varphi}$ during which different regions of the same size complete the transition to the Friedmann phase. The result is that the density fluctuations when a region of a particular size re-enters the horizon are equal to [116]

$$(2.75) \qquad \delta_H \equiv \left(\frac{\delta\varrho}{\varrho}\right)_H = 2^{\frac{3}{2}} \Delta t \,.$$

The time spread Δt can be estimated from the equation of motion of φ (the free Klein-Gordon equation in an expanding universe)

$$(2.76) \qquad \ddot{\varphi} + 3H\dot{\varphi} = - (\partial V/\partial\varphi) \,.$$

Neglecting the $\ddot{\varphi}$ term, since the scalar potential V must be very flat in order for enough inflation to occur, $\dot{\varphi}$ and hence δ_H will be essentially constant. These are fluctuations of all the contents of the Universe, so they are adiabatic fluctuations.

Thus *inflation predicts the constant-curvature spectrum*

$$(2.77) \qquad \delta_H = \text{const} \,.$$

Some time ago HARRISON [117], ZEL'DOVICH [118] and others had emphasized that this is the only scale-invariant (*i.e.* power law) fluctuation spectrum that avoids trouble at both large and small scales. If

(2.78) $$\delta_H \propto M_H^{-\alpha},$$

then, if $-\alpha$ is too large, the Universe will be less homogeneous on large than small scales, contrary to observation; and if α is too large, fluctuations on sufficiently small scales will enter the horizon with $\delta_H \gg 1$ and collapse to black holes [119, 120]; thus $\alpha \approx 0$.

Inflation predicts more: it allows the calculation of the value of the constant δ_H in terms of the properties of the scalar potential $V(\varphi)$. Indeed, this has proved to be embarrassing, at least initially, since the Coleman-Weinberg potential, the first potential studied in the context of the new inflation scenario, results in $\delta_H \sim 10^2$ [116], some six orders of magnitude too large. But this does not seem to be an insurmountable difficulty. A prescription for a suitable potential has been given [121], and particle physics models that are more or less satisfactory have been constructed [122].

Thus inflation at present appears to be a plausible solution to the problem of providing reasonable cosmological initial conditions (although it sheds no light at all on the fundamental question why the cosmological constant is so small now). In particular, it predicts the constant-curvature fluctuation spectrum $\delta_H = \text{const}$, at the price of also predicting that the Universe is essentially flat. Inflation is not the only way to get the constant-curvature spectrum, however; there is also the possibility of cosmic strings [123]. Discussing cosmic strings would take us rather far afield. I just want to note here that, even though they have the same spectrum, the fluctuations generated by the motion of relativistic strings are rather different from those arising from quantum fluctuations of an essentially free field. In particular, the latter are Gaussian [124].

2`7. *Is the gravitational force $\propto r^{-1}$ at large r?* – In concluding this section on gravity and cosmology, I return to the question whether our conventional theory of gravity is trustworthy on large scales. The reason for raising this question is that interpreting modern observations within the context of the standard theory leads to the conclusion that at least 90% of the matter in the Universe is dark. Moreover, there is no observational confirmation that the gravitational force falls as r^{-2} on galactic and extragalactic scales.

TOHLINE [125] pointed out that a modified gravitational force law, with the gravitational acceleration given by

(2.79) $$a = \frac{GM_{\text{lum}}}{r^2}\left(1 + \frac{r}{d}\right),$$

could be an alternative to dark-matter galactic halos as an explanation of the constant-velocity rotation curves of fig. 1.3. (I have written the mass in (2.79) as M_lum to emphasize that there is not supposed to be any dark matter.) Indeed, (2.79) implies

$$v^2 = \frac{GM_\text{lum}}{d} = \text{const}$$

for $r \gg d$. The trouble is that, with the distance scale d where the force shifts from r^{-2} to r^{-1} taken to be a physical constant, the same for all galaxies, this implies that $M_\text{lum} \propto v^2$, whereas observationally $M_\text{lum} \propto L \propto v^4$, as I mentioned in sect. **1** (« Tully-Fisher law »).

MILGROM [126] proposed an alternative idea, that the separation between the classical and modified regimes is determined by the value of the gravitational acceleration a rather than the distance scale r. Specifically, MILGROM proposed that

(2.80) $$\begin{cases} a = GM_\text{lum}\, r^{-2}, & a \gg a_0, \\ a^2 = GM_\text{lum}\, r^{-2} a_0, & a \ll a_0, \end{cases}$$

where the value of the critical acceleration $a_0 \approx 8 \cdot 10^8 h^2$ cm s^{-1} (where h is the Hubble parameter) is determined for large spiral galaxies with $M_\text{lum} \sim 10^{11} M_\odot$. (This value for a_0 happens to be numerically approximately equal to cH_0.) Equation (2.80) implies that

(2.81) $$v^4 = a_0 GM_\text{lum}$$

for $a \ll a_0$, which is now consistent with the Tully-Fisher law $M_\text{lum} \propto v^4$. However, there is a problem with (2.80): data for the largest elliptical galaxies still require the existence of large amounts of dark matter [127]. For example, the cD galaxy in the Abell cluster A2029 has $M_\text{lum} \geqslant 1.5 \cdot 10^{13} M_\odot$, which implies that the gravitational acceleration is given by the usual expression (*i.e.* that $a > a_0$) for $r \leqslant 100$ kpc. But the observed increasing velocity dispersion over this region implies that the mass-to-light ratio increases by a factor ~ 10 in this region. Similarly, data on M87, the giant elliptical in the Virgo cluster, imply that $a > a_0$ out to about 84 kpc, where $M/L \approx 50$. Since $M/L \approx 10$ for the nucleus, here again M/L rises dramatically with r in the supposed « classical » region.

Thus the proposed modifications of gravity are not entirely satisfactory empirically. They are also entirely *ad hoc*. Indeed, it would doubtless be difficult if not impossible to fit a r^{-1} force law into the larger framework of either cosmology or theoretical physics [128]. For example, all one needs to assume in order to get the weak-field limit of general relativity is that gravitation is carried by a massless spin-two particle (the graviton); masslessness implies the standard r^{-2} force, and spin two implies coupling to the energy-momentum tensor [129]. It is not at all clear what sort of particle physics could lead to a force law like (2.80). In the absence of an intrinsically attractive and plausible theory of gravity which leads to a r^{-1} force law at large distances, it seems to me to be preferable by far to take dark matter seriously. Moreover, dark matter is quite consistent with modern ideas in particle physics; indeed, there is an abundance of plausible particle candidates.

3. – Dark matter.

3`1. *The hot big bang.* – The main subject of this section is the properties of the cosmological dark matter (DM). Since the various arguments regarding the properties of the DM depend in several ways on our theories regarding the evolution of the Universe, I begin by reminding you about the standard theory. For more details, you may want to consult some standard references [65, 130-133]. Table II and fig. 3.1 summarize several major signposts in cosmic

TABLE II.

Time	Temperature	
$t_{QCD} \approx 10^{-4}$ s	$\sim 10^2$ MeV	π and μ annihilation; color confinement
$t_{\nu d} \approx 1$ s	1 MeV	neutrino decoupling
$t_e \approx 4$ s	0.5 MeV	e annihilation
$t_D \approx 3$ min	0.1 MeV	D bottleneck, He synthesis
$t_{eq} \approx 3 \cdot 10^4$ y	2 eV	nonrelativistic-matter domination
$t_{rec} \approx 4 \cdot 10^5$ y	0.3 eV	atomic H formation (« recombination »)
$t_0 \approx 15$ Gy	$3 \cdot 10^{-4}$ eV	present epoch

history according to the standard hot-big-bang theory. (The numerical entries in the table are estimates for orientation; precise values depend on cosmological parameters.)

The hadronic era comes to an end at $T_{QCD} \sim 10^2$ MeV, with quantum chromodynamic confinement of colored hadrons (quarks and gluons) into ordinary baryons and mesons, with only the slight excess of baryons over antibaryons surviving after π annihilation. There is still some uncertainty regarding the physics of this era, with such exotic possibilities as a substantial fraction of the baryonic matter ending up as *u-d-s* symmetric « quark nuggets »—a possible candidate for the dark matter [134]. Thereafter, the basic physical processes are thought to be well understood, with the major uncertainties being the fluctuation spectrum and the nature of the dark matter. Shortly after νs decouple thermally (1 in fig. 3.1), e^+e^- annihilation decreases ϱ_m and heats the photon gas (2), which thereafter resumes its adiabatic decrease in temperature $\propto R^{-1}$. Since there are $\sim 10^9$ photons per baryon, there are many

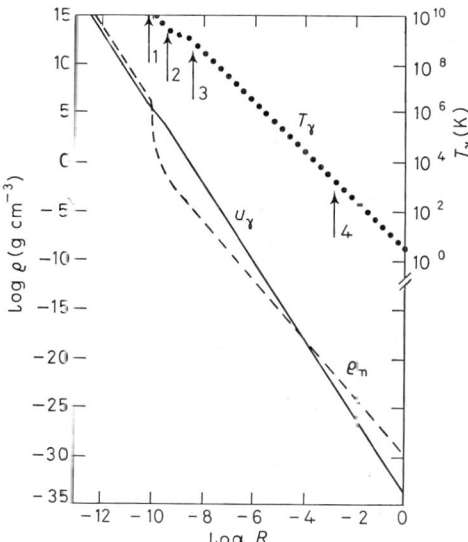

Fig. 3.1. – Radiation temperature (T_γ) and density (u_γ) and nonrelativistic-matter density (ϱ_m) from neutrino decoupling (1) to the present, according to the standard hot-big-bang picture. (See text.)

photons available in the high-energy tail of the distribution to photodissociate D as soon as it forms, so the temperature must decrease still further before primordial nucleosynthesis can begin. Finally the « deuterium bottleneck » is passed (3), and then most of the D is quickly bound into ^{4}He. Galaxy-size masses are first encompassed by the horizon when the scale factor $R \sim 10^7$ (see fig. 2.10). Then fluctuations in the neutrino density are damped by « free streaming »; and fluctuations in the fluid composed of radiation and ionized hydrogen and helium do not grow—rather, they oscillate and (for adiabatic fluctuations) eventually are damped by photon diffusion (« Silk damping »). The density of nonrelativistic matter falls more slowly ($\propto R^{-3}$) than that of relativistic particles (photons and light neutrinos) ($\propto R^{-4}$). They are equal at R_{eq} (see eq. (2.43)), and nonrelativistic matter dominates thereafter in the standard model. Fluctuations in the dark matter can then begin to grow $\propto R$. Finally, at $R_{rec} \approx 10^{-3}$ (4 in fig. 3.1), hydrogen atoms form and the Universe becomes transparent. No longer tied to the radiation by Compton drag, ordinary matter can begin to form the large astronomical objects we see today: globular clusters, galaxies, clusters and superclusters.

In the next subsection, I will consider this story in a little more detail, and point out problems that arise if the nonrelativistic matter is only baryonic. I will begin by discussing evidence that the dark halos of galaxies are not composed of baryonic matter. If the dark matter is not baryonic, what is it? The rest of this section and part of the next will be concerned with a classification of dark-matter candidates by their key astrophysical properties, and an outline of their consequences for the formation of structure in the Universe [135].

3˙2. *The dark matter is probably not baryonic.* – There are three arguments that the DM is not « baryonic », that is, that it is *not* made of protons, neutrons and electrons as all ordinary matter is. As Richard FEYNMAN has said in other contexts, *one* argument would suffice if it were convincing. The three arguments are based on 1) excluding various possible forms of baryonic dark matter in galaxy halos; 2) bounding the abundance of baryonic matter using the observed abundance of light elements, especially deuterium; and 3) bounding the magnitude of adiabatic fluctuations at recombination from the observational upper limits on fluctuations in the cosmic background radiation. All three arguments have loopholes, and I will point them out. Nevertheless, taken together these arguments persuade me, and perhaps will persuade you, that we must take very seriously the possibility that most of the matter in the Universe is not composed of atoms.

Excluding baryonic models. If the dark matter in galaxy halos is baryonic, then it must be gaseous, or agglomerations of atoms held together by chemical or gravitational forces. But arguments can be given against all of these possibilities [11, 136]. The dark matter in galaxy halos cannot be *gas* (it would have to be hot to be pressure supported, and would radiate X-rays that are not seen), nor frozen hydrogen *snowballs* (they would sublimate), nor *dust grains* (their « metals », elements of atomic number $\geqslant 3$, would have prevented formation of the observed low-metallicity Population II stars), nor isolated *jupiters* (how to make so many hydrogen balls too small to initiate nuclear burning without making a few large enough to do so?), nor *collapsed stars* (where is the matter they must have ejected in collapsing?).

The weakest argument is probably that which attempts to exclude « jupiters »: arguments of the form « how could it be that way? » are rarely entirely convincing.

Deuterium abundance. In the early universe, almost all the neutrons remaining after the deuterium bottleneck are synthesized into ^{4}He. (Formation of nuclides with $A \geqslant 6$ is inhibited by the absence of any stable nuclide with $A = 5$.) The fraction remaining in D and ^{3}He is calculated [86] to be a rapidly decreasing function of η, the ratio of baryon to photon number densities, as shown in fig. 3.2. The presently observed D abundance (compared, by number, to H) is $(1 \div 4) \cdot 10^{-5}$. Since D is readily consumed but not produced in stars, 10^{-5} is also a lower limit on the primordial D abundance. This, in turn, implies an *upper* limit $\eta \leqslant 10^{-9}$ or

$$\Omega_b = 0.0035 h^{-2} \theta^3 \eta_{10} \leqslant 0.035 h^{-2} \theta^3 , \tag{3.1}$$

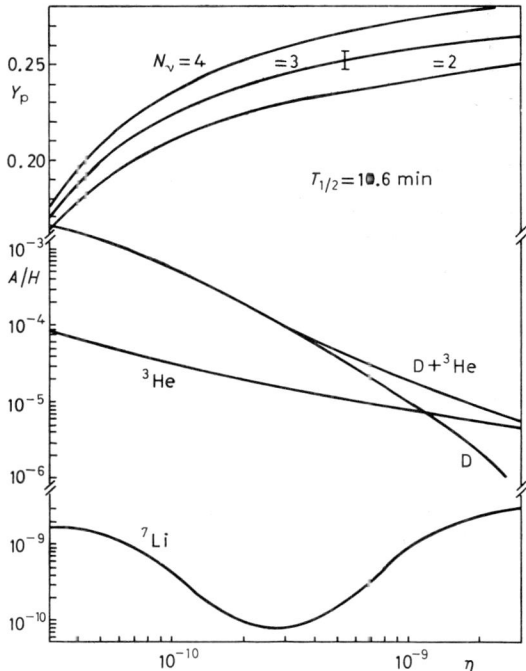

Fig. 3.2. – Calculated primordial abundances, by number relative to hydrogen, of light isotopes D, ^{3}He and ^{7}Li, and by mass of ^{4}He (Y_p) for $N_\nu = 2$, 3 and 4 light ν species (the error bar shows the effect of $\Delta\tau = \pm 0.2$ min uncertainty in the neutron lifetime). The primordial abundances inferred from observations are $Y_p = 0.23 \div 0.25$, $D/H \geqslant 1 \cdot 10^{-5}$, $(D+^3He)/H \leqslant 10^{-4}$, $^7Li/H = (1.1 \pm 0.4) \cdot 10^{-10}$. These are consistent with the predictions for $\eta = (4 \div 7) \cdot 10^{-10}$ and $N_\nu \leqslant 4$. (Reproduced, with permission, from ref. [86].)

where Ω_b is the ratio of the present average baryon density ϱ_b to the critical density $\varrho_{c,0}$, $\theta \equiv T_0/2.7$ K and $\eta_{10} \equiv \eta/10^{-10}$. If use is also made of the somewhat more uncertain upper limit on primordial ^{7}Li abundance [137] of $^7Li/H \leqslant$ $\leqslant 3 \cdot 10^{-10}$ (including an additional factor of two for good measure), then stronger upper limits are obtained: $\eta_{10} \leqslant 7$ and $\Omega_b \leqslant 0.025 h^{-2} \theta^3$. Even with the Hubble parameter at its lower limit $h = 0.5$, this corresponds to $\Omega_b \leqslant 0.1 \theta^3$.

As discussed in sect. **1** and **2**, the observational limits on Ω are $0.1 \leqslant \Omega \leqslant 2$. Therefore, in a baryon-dominated universe ($\Omega \approx \Omega_b$), these bounds are consistent only with the lower limit on Ω, and then only for the Hubble parameter at its lower limit. An Einstein-de Sitter or inflationary ($\Omega = 1$) or closed ($\Omega > 1$) universe cannot be baryonic.

Galaxy and cluster formation. In the gravitational-collapse model for the formation of large-scale structure in the Universe, discussed in sect. **2**, structure forms when fluctuations $\delta \equiv \delta\varrho/\varrho$ grow to nonlinearity ($\delta \geqslant 1$), when they cease to expand with the Hubble flow, and subsequently collapse and virialize. As we will see, the problem is to understand how fluctuations of galaxy and cluster size can grow to nonlinearity by the present without violating the observational bounds on small-angle fluctuations in the cosmic background radiation.

As before, consider a universe with no nonbaryonic dark matter. Before recombination, ionized matter is locked to radiation by Compton scattering (this is called « Compton drag »), and it is appropriate to treat radiation plus baryonic matter as an ideal fluid. Fluctuations in this fluid of wavelength $\lambda = 2\pi R/k$ in the expanding universe obey the wave equation [138]

$$(3.2) \qquad \ddot{\delta}_k + 2 \frac{\dot{R}}{R} \dot{\delta}_k = 4\pi G \varrho \delta_k \left[1 - \frac{\pi k_B T}{\mu \lambda^2 G \varrho}\right],$$

where $\mu \approx 0.78$ is the mean mass per particle of the primordial fluid of ionized hydrogen and helium. Growth of the amplitude of fluctuations occurs only when the right-hand side of (3.2) is positive, i.e. for λ satisfying the Jeans criterion

$$(3.3) \qquad \lambda > \lambda_J = \left(\frac{\pi k_B T}{\mu G \varrho}\right)^{\frac{1}{2}}.$$

That such a condition (first derived for a nonexpanding universe by James JEANS in 1902) should arise is not surprising: the effects of gravity grow with the physical size of the fluctuation, and there is a minimum size above which gravity overwhelms pressure. Equivalently, the Jeans criterion is that the sound-crossing time t_{sound} should exceed the free-fall time t_{dyn}:

$$(3.4) \qquad t_{sound} \approx \lambda/\sqrt{k_B T/\mu} > t_{dyn} \approx (G\varrho)^{-\frac{1}{2}}.$$

The value of the corresponding Jeans mass

$$(3.5) \qquad M_J = (4\pi/3) \varrho (\lambda/2)^3$$

was sketched in fig. 2.10 and is also included in fig. 3.3. In the radiation-dominated era, the Jeans mass M_J is comparable to the mass inside the horizon M_H; between matter domination and recombination, M_J levels out at $\sim 10^{17} M_\odot$; and at recombination M_J drops to about $10^6 M_\odot$ and decreases thereafter as the matter temperature drops.

Now consider what happens when a fluctuation of galaxy size comes within the horizon. (That is, consider a horizontal line across fig. 3.3 at $M \sim 10^{11} M_\odot$.) Since the mass is below the Jeans mass, the fluctuation amplitude cannot grow. Instead, the radiation and ionized-matter fluid oscillates as an acoustic wave.

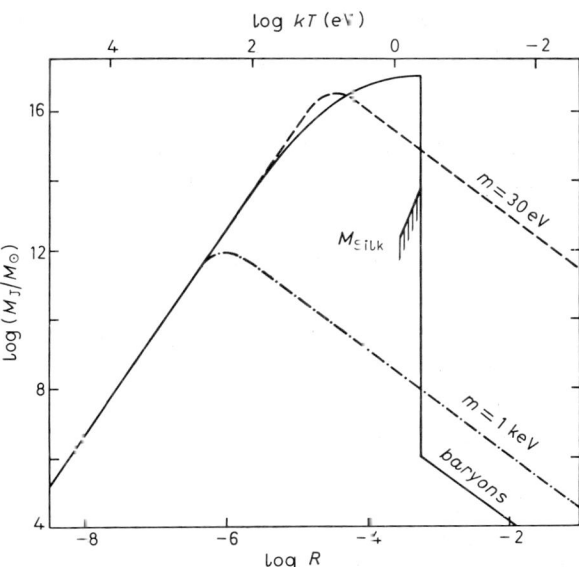

Fig. 3.3. – Jeans mass vs. scale factor for a universe dominated by baryons (solid line), neutrinos ($m = 30$ eV) and warm dark matter ($m = 1$ keV). (From ref. [139].)

I will further assume that fluctuations in baryonic-matter density and in radiation density are correlated: $\delta_r = \frac{4}{3}\delta_b$. These are called adiabatic fluctuations, since the entropy per baryon is constant. These are the sort of fluctuations predicted in most grand unified models, in which $\eta \equiv n_b/n_\gamma$ is a constant determined by the microphysics [140].

It is important to appreciate that photon diffusion damps small adiabatic fluctuations (Silk damping). The photon mean free path is $l_T = (\sigma_T n_e)^{-1}$, where σ_T is the Thomson cross-section and n_e is the electron density. Thus the time to diffuse (random walk) a distance d is $\tau(d) = d^2/cl_T$. Setting this equal to the Hubble time gives the Silk-damping length $d_S \approx (l_T ct)^{\frac{1}{2}}$, with the corresponding Silk mass $M_S = (4\pi/3)\varrho_m d_S^3$. The Silk-damping mass is sketched in fig. 2.10 and 3.3. M_S grows until recombination, and there is strong damping of any fluctuations that lie in the hatched region below the Silk-damping line.

Accurate treatment of Silk damping requires numerical calculations [141]. The result is that there is more than e^{-1} damping of fluctuations with mass smaller than [21]

(3.6) $$M_S \approx 1.3 \cdot 10^{12}(\Omega h^2)^{-\frac{5}{4}} M_\odot.$$

Evaluating this for $\Omega = 0.1$ (consistent with the primordial nucleosynthesis bound) and $h = 1$ gives $M_S = 4 \cdot 10^{13}$; using $h = \frac{1}{2}$ gives $M_S = 3 \cdot 10^{14}$. These masses correspond to clusters of galaxies. Thus ordinary galaxies ($M_b \leqslant$ $\leqslant 10^{11 \div 12} M_\odot$) can form only after the collapse of larger-scale perturbations.

These would be most likely to collapse first in one dimension («pancake» collapse), especially because of the Silk damping of smaller-scale fluctuations [142, 143].

That galaxies form in this indirect manner is a complication of this scenario, but it is not necessarily an argument against it. The serious problem is to get large enough fluctuations. Fluctuations δ grow linearly with the scale factor

$$\delta \propto R = (1+z)^{-1} = T_0/T \tag{3.7}$$

once the Universe becomes matter dominated, but fluctuations smaller than the Jeans mass ($\sim 10^{17} M_\odot$) are prevented from doing so until recombination. Moreover, growth of δ slows when an open universe goes into free expansion, when $z \lesssim \Omega^{-1}$. Thus in a baryonic universe, δ grows only between the epoch of hydrogen recombination ($z_r \approx 10^3$) and $z = \Omega^{-1} \gtrsim 10$ (when free expansion begins). It follows that there is at most a factor of $\sim 10^2$ growth (see fig. 3.4). In order to form galaxies by the present, it is necessary that at recombination $\delta T/T = \frac{1}{3} \delta\varrho/\varrho \gtrsim 3 \cdot 10^{-3}$ for $M \gtrsim M_s$, which corresponds to fluctuations on observable angular scales of a few arc minutes today. Such temperature fluctuations are more than an order of magnitude larger than present observational upper limits [144].

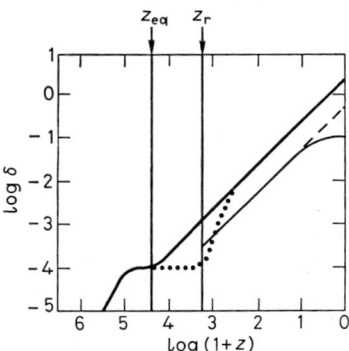

Fig. 3.4. – Overdensity $\delta = \delta\varrho/\varrho$ of a fluctuation vs. red-shift. The light solid line represents the growth of a baryonic fluctuation in a universe with no dark matter. With the constraint $\Omega_b \lesssim 0.1$ from big-bang nucleosynthesis, growth of the fluctuation amplitude δ slows when $z \lesssim 0.1$ and there is only a factor of $\sim 10^2$ growth from recombination (z_r) until the present. Even without this constraint (dashed line), the factor of 10^3 growth of δ since recombination requires too large an amplitude of adiabatic fluctuations at recombination to be consistent with $\Delta T/T$ constraints. This problem is avoided with dark matter (DM): δ_{DM} (heavy solid line) grows $\propto (1+z)^{-1}$ after dark matter becomes gravitationally dominant at z_{eq}; but Compton drag prevents baryonic fluctuations (dotted curve) from growing until after recombination, when they rapidly grow to match δ_{DM}.

The main potential loophole in this argument is the assumption of adiabatic perturbations. It is true that the orthogonal mode, perturbations in baryonic density which are uncorrelated with radiation (called isothermal perturbations), does not arise naturally in most currently fashionable particle physics theories where baryon number is generated in the decay of massive grand-unified-theory bosons («GUT baryosynthesis»), since in such theories $\eta \equiv n_b/n_\gamma$ is determined by the underlying particle physics and should not vary from point to point in space. Nevertheless, there may be ways of generating isothermal density fluctuations on scales much larger than the horizon during GUT baryosynthesis [145]. And galaxies originating as isothermal fluctuations do avoid both Silk damping and contradiction with present $\delta T/T$ limits.

A second loophole is the possibility that matter was reionized at some $z_i \gtrsim 10$, by hypothetical very early sources of UV photons [146]. Then the fluctuations in $\delta T/T$ at recombination associated with baryonic proto-pancakes could be washed out by rescattering. Suppose that the Universe were entirely reionized at a red-shift z_i and remained ionized thereafter. The optical depth is then

$$(3.8) \qquad \tau(z_1) = \int_0^{l_1} \sigma_T n_e \, dl = \frac{c}{H_0} \sigma_T n_{b,0} I = 0.063 \, \Omega_{b,0} h I \,,$$

where

$$(3.9) \qquad \sigma_T = \frac{8\pi e^4}{3 m_e^2} = 0.6652 \cdot 10^{-24} \, \text{cm}^2$$

is the Thomson cross-section, and

$$(3.10) \qquad I = \int_0^{z_1} \frac{(1+z)^3 \, dz}{(1+z)^2 (1 - \Omega_0 z)^{\frac{1}{2}}} =$$
$$= \frac{2}{3\Omega_0^2} \{[(1+\Omega_0 z_1)^{\frac{3}{2}} - 1] - \varepsilon(1 - \Omega_0)[(1+\Omega_0 z)^{\frac{1}{2}} - 1]\} \,.$$

Setting $\tau(z_i)$ equal to unity, it follows that, to be effective in washing out small-angle fluctuations in the cosmic background radiation, any reionization must have occurred before $z_i \approx 20$ for the maximal values $\Omega_0 = \Omega_{b,0} \approx 0.1$ and $h = 1$. Since this is earlier than the period of galaxy formation according to most theories, especially for adiabatic fluctuations where Silk damping prevents galaxy formation until after the collapse of cluster-size pancakes, it is difficult to see how enough matter could have been converted to radiation to cause reionization at such early times.

Despite the loopholes in each of the three arguments against a universe with no nonbaryonic dark matter, I find all the arguments together to be rather persuasive, even if not entirely compelling. If it is indeed true that the bulk of the mass in the Universe is not baryonic, that is yet another blow to anthropocentricity: not only is man not the center of the Universe physically (Copernicus) or biologically (Darwin), it now appears that we and all that we see are not even made of the predominant variety of matter in the Universe!

3'3. *Three types of* DM *particles: hot, warm and cold.* – If the dark matter is not baryonic, what *is* it? I will consider here the physical and astrophysical implications of three classes of elementary-particle DM candidates, which are called hot, warm and cold [147].

Hot DM refers to particles, such as neutrinos, that were still in thermal equilibrium after the most recent phase transition in the hot early universe, the QCD confinement transition, which presumably took place at $T_{QCD} \approx$ $\approx 10^2$ MeV. Hot DM particles have a cosmological number density roughly comparable to that of the microwave background photons, which implies an upper bound to their mass of a few tens of eV. As I shall discuss shortly, this implies that free streaming destroys any fluctuations smaller than supercluster size, $\sim 10^{15} M_\odot$.

Warm DM particles interact much more weakly than neutrinos. They decouple (*i.e.* their mean free path first exceeds the horizon size) at $T \gg T_{QCD}$, and are not heated by the subsequent annihilation of hadronic species. Consequently their number density is roughly an order of magnitude lower, and their mass an order of magnitude higher, than hot DM particles. Fluctuations as small as large galaxy halos, $\geqslant 10^{11} M_\odot$, could then survive free streaming. PAGELS and I initially suggested that, in theories of local supersymmetry broken at $\sim 10^6$ GeV, gravitinos could be DM of the warm variety [148]. Other candidates are also possible, as I will discuss.

Cold DM consists of particles for which free streaming is of no cosmological importance. Two different sorts of cold DM consisting of elementary particles have been proposed, a cold Bose condensate such as axions, and heavy remnants of annihilation or decay such as heavy stable neutrinos. A perennial candidate, primordial black holes, is beginning to be constrained by analysis and observations [149-151]. Finally, I have already mentioned another sort of superdense objects that would behave astrophysically as cold DM: quark nuggets. As we will see, a universe dominated by cold DM looks remarkably like the one astronomers actually observe.

It is of course also possible that the dark matter is NOTA—none of the above! Maybe the dark matter is a mixture, for example « jupiters » plus neutrinos [152] or « jupiters » plus cold dark matter [153]. Some models even include unstable DM that decays into relativistic particles [154-160].

3'4. *Galaxy formation with hot* DM. – The standard hot-DM candidate is massive neutrinos [161-164], although other, more exotic, theoretical possibilities have been suggested, such as a « majoron » [165] of nonzero mass which is lighter than the lightest neutrino species, and into which all neutrinos decay [166]. For definiteness, I will discuss neutrinos.

Mass constraints. Left-handed neutrinos of mass $\leqslant 1$ MeV remain in thermal equilibrium until the temperature drops to $T_{\nu d}$, at which point their mean free path first exceeds the horizon size (see fig. 3.5) and they essentially cease

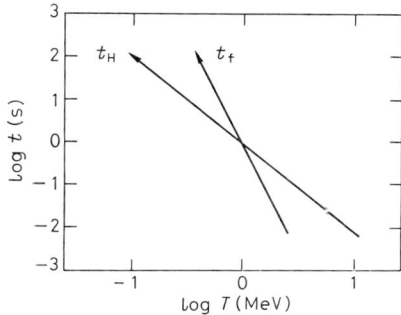

Fig. 3.5. – Decoupling of neutrinos occurs when their mean free time t_f exceeds the Hubble time t_H.

interacting thereafter, except gravitationally [65]. Their mean free path is, in natural units ($\hbar = c = 1$), $\lambda_\nu \sim [\sigma_\nu n_{e^\pm}]^{-1} \sim [(G_F^2 T^2)(T^3)]^{-1}$, where $G_F \approx$ $\approx 10^{-5}$ (GeV)$^{-2}$ is the Fermi constant that measures the strength of the weak interactions. The horizon size is $\lambda_h \sim (G\varrho)^{-\frac{1}{2}} \sim M_{Pl} T^{-2}$, where the Planck mass $M_{Pl} \equiv G^{-\frac{1}{2}} = 1.22 \cdot 10^{19}$ GeV. Thus $\lambda_h/\lambda_\nu \sim (T/T_{\nu d})^3$, with the neutrino decoupling temperature

(3.11) $$T_{\nu d} \sim M_{Pl}^{-\frac{1}{3}} G_F^{-\frac{2}{3}} \sim 1 \text{ MeV}.$$

After T drops below $\frac{1}{2}$ MeV, $e^+ e^-$ annihilation ceases to be balanced by pair creation, and the entropy of the $e^+ e^-$ pairs heats the photons. Above 1 MeV, the number density n_{ν_i} of each left-handed neutrino species (counting both ν_i and $\bar{\nu}_i$) is equal to that of the photons, n_γ, times the factor $\frac{3}{4}$ from Fermi vs. Bose statistics; but then $e^+ e^-$ annihilation increases the photon number density relative to that of the neutrinos by a factor of $\frac{11}{4}$ (*). Thus today, for each

---

(*) Since the argument giving the 11/4 factor is simple, and since the idea of warm DM is based on a generalization of it, I will sketch it here. The key ingredient is that the entropy in interacting particles in a comoving volume is conserved during ordinary Hubble expansion, even during a process such as electron-positron annihilation, so long as it occurs in equilibrium. (This fact should be intuitively obvious, since the process is reversible, and anyway it is easily derived (see, *e.g.*, ref. [65], § 15.6).) That is, $S_I = g_I(T) N_\gamma(T) = $ const, where $N_\gamma = n_\gamma V$ is the number of photons in a given

species,

(3.12) $$n_{\nu,0} = \frac{3}{4}\frac{4}{11} n_{\gamma,0} = 109\theta^3 \text{ cm}^{-3},$$

where, as above, $\theta \equiv T_0/2.7$ K. Since the present cosmological density is

(3.13) $$\bar{\varrho} = \Omega\varrho_c = 11\Omega h^2 \text{ keV cm}^{-3},$$

it follows that

(3.14) $$\sum_i m_{\nu_i} < \bar{\varrho}/n_{\nu,0} \leqslant 100\Omega h^2 \theta^{-3} \text{ eV},$$

where the sum runs over all neutrino species with $M_{\nu_i} \leqslant 1$ MeV. (Heavier neutrinos will be discussed shortly.) Observational data imply that Ωh^2 is less than unity. Thus if one species of neutrino is substantially more massive than the others and dominates the cosmological mass density, as for definiteness I will assume for the rest of this section, then a reasonable estimate for its mass is $m_\nu \sim 30$ eV.

At present the experimental evidence for nonzero neutrino mass is apparently not entirely convincing. Although one group has reported [162] that $14 \text{ eV} < m_{\nu_e} < 40$ eV and that $m_{\nu_e} > 20$ eV at the 95% confidence level [167] from tritium β-decay endpoint data, the experiment is extraordinarily difficult and according to some independent authorities [168] their data are consistent with $m_{\nu_e} = 0$. Several sensitive experiments are in progress using alternative methods. The so far unsuccessful attempts to detect neutrino oscillations also give only upper limits on neutrino masses times (essentially unknown) mixing parameters.

In deriving eq. (3.14), I have been assuming that all the neutrino species are light enough to still be relativistic at decoupling, *i.e.* lighter than one MeV. The bound (3.14) shows that they must then be much lighter than that. In the alternative case that a neutrino species is nonrelativistic at decoupling, it has been shown [169] that its mass must then exceed several GeV, which is not true of the known neutrinos (ν_e, ν_μ and ν_τ). (One might at first think that the Boltzmann factor would sufficiently suppress the number density of neutrinos weighing a few tens of MeV to allow compatibility with the present density of the Universe. It is the fact that they « freeze out » of equilibrium

---

comoving volume V, and $g_I = (g_B + \frac{7}{8} g_F)_I$ is the effective number of helicity states in interacting particles (with the factor of $\frac{7}{8}$ reflecting the difference in energy density for fermions *vs.* bosons). Just above the temperature of electron-positron annihilation, $g_I = g_\gamma + \frac{7}{8} \times g_e = 2 + \frac{7}{8} \times 4 = \frac{11}{2}$; while, below it, $g_I = g_\gamma = 2$. Thus, as a result of the entropy of the electrons and positrons being dumped into the photon gas at annihilation, the photon number density is thereafter increased relative to that of the neutrinos by a factor of $\frac{11}{4}$.

well before the temperature drops to their mass that leads to the higher mass limit.) I have also been assuming that the neutrino chemical potential is negligible, i.e. that $|n_\nu - n_{\bar\nu}| \ll n_\gamma$. This is very plausible, since the net baryon number density $n_b - n_{\bar b} \lesssim 10^{-9} n_\gamma$, but if it is not true the consequences can be rather dramatic [170].

Free streaming. The most salient feature of hot DM is the erasure of small fluctuations by free streaming. Thus even collisionless particles effectively exhibit a Jeans mass. It is easy to see that the minimum mass of a surviving fluctuation is of order M_{Pl}^3/m_ν^2 [163, 171].

Let us suppose that some process in the very early universe—for example, thermal fluctuations subsequently vastly inflated, in the inflationary scenario [172]—gave rise to adiabatic fluctuations on all scales. Neutrinos of nonzero mass m_ν stream relativistically from decoupling until the temperature drops to m_ν, during which time they traverse a distance $d_\nu = R_H(T = m_\nu) \sim$
$\sim M_{Pl} m_\nu^{-2}$. In order to survive this free streaming, a neutrino fluctuation must be larger in linear dimension than d_ν. Correspondingly, the minimum mass in neutrinos of a surviving fluctuation is $M_{J,\nu} \sim d_\nu^3 m_\nu n_\nu(T = m_\nu) \sim$
$\sim d_\nu^3 m_\nu^4 \sim M_{Pl}^3 m_\nu^{-2}$. By analogy with Jeans' calculation of the minimum mass of an ordinary fluid perturbation for which gravity can overcome pressure, this is referred to as the (free-streaming) Jeans mass. (See fig. 3.3.) A more careful calculation [163, 173] gives

$$(3.15) \qquad d_\nu = 41(m_\nu/30 \text{ eV})^{-1}(1+z)^{-1} \text{ Mpc},$$

that is, $d_\nu = 41(m_\nu/30 \text{ eV})^{-1}$ Mpc in comoving co-ordinates, and correspondingly

$$(3.16) \qquad M_{J,\nu} = 1.77 M_{Pl}^3 m_\nu^{-2} = 3.2 \cdot 10^{15}(m_\nu/30 \text{ eV})^{-2} M_\odot,$$

which is the mass scale of superclusters. Objects of this size are the first to form in a ν-dominated universe, and smaller-scale structures such as galaxies can form only after the initial collapse of supercluster-size fluctuations.

The limits on small-angle $\delta T/T$ fluctuations are compatible with this picture. When a fluctuation of total mass $\sim 10^{15} M_\odot$ enters the horizon at $z \sim 10^4$, the density contrast δ_{RB} of the radiation plus baryons ceases growing and instead starts oscillating as an acoustic wave (as usual), while that of the massive neutrinos δ_ν continues to grow linearly with the scale factor $R = (1+z)^{-1}$ since the Compton drag that prevents growth of δ_{RB} does not affect the neutrinos. By recombination, at $z_r \sim 10^3$, $\delta_{RB}/\delta_\nu \lesssim 10^{-1}$, with possible additional suppression of δ_{RB} by Silk damping. (See fig. 3.4.) Thus the hot-DM scheme with adiabatic primordial fluctuations predicts small-angle fluctuations in the microwave background radiation somewhat below current observational upper limits. Similar considerations apply in the warm and cold DM schemes [174, 175].

In numerical simulations of dissipationless gravitational clustering starting with a fluctuation spectrum appropriately peaked at $\lambda \sim d_\nu$ (reflecting damping by free streaming below that size and less time for growth of the fluctuation amplitude above it (cf. fig. 4.2)), the regions of high density form a network of filaments, with the highest densities occurring at the intersections and with voids in between [176-178]. The similarity of these features to those seen in observations is cited as evidence in favor of this model [179].

Potential problems with ν *DM.* A number of potential problems with the neutrino-dominated universe have emerged in recent studies, however.

1) From studies both of nonlinear [79, 178] clustering ($\lambda \leqslant 10$ Mpc) and of streaming velocities [180] in the linear regime ($\lambda > 10$ Mpc), it follows that supercluster collapse must have occurred recently: $z_{sc} \leqslant 0.5$ is indicated [180], and in any case $z_{sc} < 2$ [178]. (See fig. 3.6.) However, the best limits on galaxy ages coming from globular clusters and other stellar populations indicate that galaxy formation took place before $z = 3$ [181]. Moreover, if QSOs are associated with galaxies, as is suggested by the detection of galactic luminosity around nearby QSOs and the apparent association of more distant QSOs with galaxy clusters, the abundance of QSOs at $z > 2$ is also inconsistent with the « top-down » neutrino-dominated scheme in which superclusters form first: $z_{sc} > z_{galaxies}$.

2) Numerical simulations of the nonlinear « pancake » collapse taking into account dissipation of the baryonic matter show that at least 85% of the baryons are so heated by the associated shock that they remain unable to condense, attract neutrino halos, and eventually form galaxies [182]. This could be a problem for the hot-DM scheme for two reasons. With the primordial nucleosynthesis constraint $\Omega_b \leqslant 0.1$, there may be difficulty having enough baryonic matter condense to form the luminosity that we actually observe. And, where are the X-rays from the shock-heated pancakes?

3) The neutrino picture predicts [183] that there should be a factor of ~ 5 increase in M/M_b between large galaxies ($M \sim 10^{12} M_\odot$) and large clusters ($M \geqslant 10^{14} M_\odot$), since the larger clusters, with their higher escape velocities, are able to trap a considerably larger fraction of the neutrinos. As I discussed in sect. 1 (see especially fig. 1.11), although there is some indication that M/L increases with M, the ratio of total to luminous mass M/M_{lum} is probably a better indicator of the value of M/M_b, and it is roughly the same for galaxies with large halos and for rich clusters.

4) Both theoretical arguments regarding the dwarf spheroidal (dS) satellite galaxies of the Milky Way [184] and data on Draco, Carina and Ursa Minor [185, 186] imply that dark matter dominates the gravitational potential

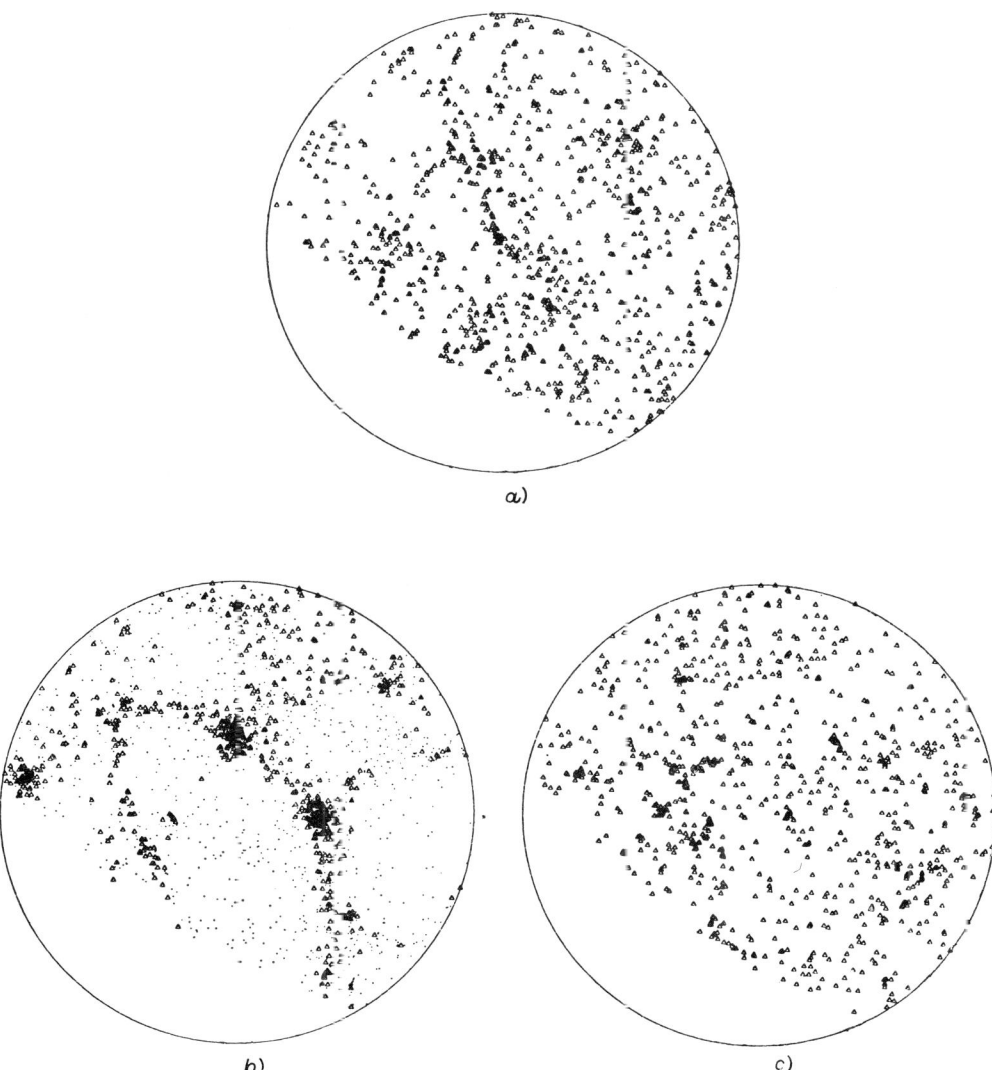

Fig. 3.6. – a) CfA galaxies shown on equal-area plot of the sky (outer circle corresponds to galactic latitude $+40°$, lower blank region below declination $0°$ is blocked by our Galaxy's disk). «Galaxy» locations in N-body simulations, plotted as in a), for Universe dominated by b) neutrinos and c) cold dark matter. The neutrino simulation assumes $\Omega = 1$, $h = 0.54$, and that galaxy formation began at $z = 2.5$; the triangles represent galaxies, while the dots are points in whose neighborhood the matter has not yet collapsed. The cold-dark-matter simulation assumes $\Omega = 0.2$ and $h = 1.1$. (Reproduced, with permission, from ref. [187]).

of these dS galaxies. The phase-space constraint (*) then sets a lower limit [186] $m_\nu > 500$ eV, which is completely incompatible with the cosmological constraint eq. (3.14). (Note that, even for neutrinos as the DM in large spiral galaxies, the phase-space constraint implies $m_\nu > 30$ eV.)

These problems, while serious, may not be fatal for the hypothesis that neutrinos are the dark matter. It is possible that galaxy density does not closely correlate with the density of dark matter—for example, because the first generation of luminous objects heats nearby matter, thereby increasing the baryon Jeans mass and suppressing galaxy formation. This could complicate the comparison of nonlinear N-body simulations with the data. Also, if dark-matter halos of large clusters are much larger in extent than those of individual galaxies and small groups, then virial estimates would underestimate mass on large scales and the data could be consistent with M/M_lum increasing with M. But it is hard to avoid the constraint on z_ec from streaming velocities in the linear regime [180] except by assuming that the Local Group's velocity is abnormally low. And the only explanation for the high M/L of dS galaxies in a neutrino-dominated universe is the rather *ad hoc* assumption that the dark matter in such objects is baryons rather than neutrinos. Of course, the evidence for massive dark halos in dS galaxies is not yet solid.

(*) The phase-space constraint [188] follows from a theorem in classical mechanics to the effect that the maximum 6-dimensional phase-space density cannot increase as a system of collisionless particles evolves. At early times, before density inhomogeneities become nonlinear, the neutrino phase-space density is given by the Fermi-Dirac distribution

$$n_\nu(p) = \frac{g_\nu}{h^3} \left[1 + \exp\left[\frac{pc}{kT_\nu(z)}\right] \right]^{-1},$$

where here h is Planck's constant and $g_\nu = 2$ for each species of left-handed ν plus $\bar\nu$. (Since momentum and temperature both scale as red-shift z as the Universe expands, this distribution remains valid after neutrinos drop out of thermal equilibrium at ~ 1 MeV, and even into the nonrelativistic regime $T_\nu < m_\nu$.) The standard version of the phase-space constraint follows from demanding that the central phase-space density $9[2(2\pi)^{\frac{3}{2}} Gr_c^2 \sigma m_\nu^4]^{-1}$ of the halo, assumed to be an isothermal sphere of core radius r_c and one-dimensional velocity dispersion σ, not exceed the maximum value of the initial phase-space density $n_\nu(0) = g_\nu/2h^3$. The result is

$$m_\nu > (120 \text{ eV}) \left(\frac{100 \text{ km s}^{-1}}{\sigma}\right)^{\frac{1}{4}} \left(\frac{1 \text{ kpc}}{r_c}\right)^{\frac{1}{2}} \left(\frac{g_\nu}{2}\right)^{-\frac{1}{4}}.$$

The lower limits on m_ν quoted in the text use this result. A more conservative phase-space constraint has recently been obtained [189] assuming (perhaps unrealistically) that the neutrinos are in the most compact distribution possible, rather than an isothermal sphere. Assuming that for the dS galaxy Draco $M > 10^8 M_\odot$ and $r_c < 2$ kpc, it follows that $m_\nu > 127$ eV. This is still in trouble with the cosmological upper bound on m_ν.

In summary, the evidence against hot DM is rather impressive. At very least, it indicates that a neutrino-dominated universe must be rather more complicated than theorists have yet envisaged.

3'5. *Galaxy formation with warm* DM. – Suppose the dark matter consists of an elementary particle species X that interacts much more weakly than neutrinos. The X's decouple thermally at a temperature $T_{xd} \gg T_{vd}$ and their number density is not thereafter affected by particle annihilation at temperatures below T_{xd}. With the standard assumption of conservation of entropy in comoving volumes, the X number density today $n_{x,0}$ and mass m_x can be calculated in terms of $g_I(T)$, the effective number of helicity states of interacting particles evaluated at T_{xd} [131]. These are plotted in fig. 3.7, assuming

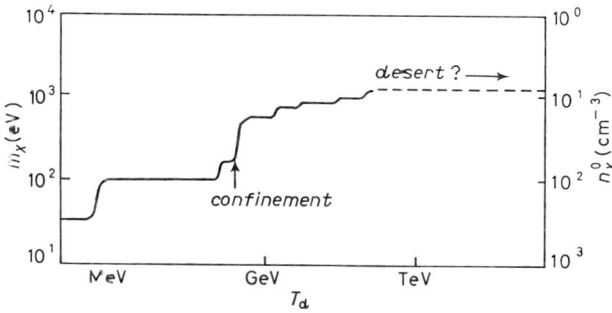

Fig. 3.7. – Mass m_x and present number density of warm-dark-matter particles X, assuming the standard particle physics model with no entropy generation after T_{GUT}. The mass (left axis) scales as Ωh^2.

the « standard model » of particle physics. The simplest grand unified theories predict $g_I(T) \approx 100$ for T between 10^2 GeV and $T_{tot} \sim 10^{14}$ GeV, with possibly a factor of two increase in g_I beginning near 10^2 GeV due to $N = 1$ supersymmetry partner particles. Then for T_{xd} in the enormous range from ~ 1 GeV to $\sim T_{GUT}$, $n_x^0 \sim 5 g_x$ cm^{-3} and correspondingly $m_x = 2\Omega h^2 g_x^{-1}$ keV [190], where g_x is the number of X helicity states. Because of free streaming, such « warm » DM particles of mass $m_x \sim 1$ keV will cluster on a scale $\sim M_{Pl}^3 m_x^{-2} \sim$ $\sim 10^{12} M_\odot$, the scale of large galaxies such as our own.

Candidates form warm DM. What might be the identity of the warm DM particles X? PAGELS and I [148] suggested that they might be the $\pm \frac{1}{2}$ helicity states of the gravitino $\tilde{G}$, the spin-$\frac{3}{2}$ supersymmetric partner of the graviton G. The gravitino mass is related to the scale of supersymmetry breaking m_{SUSY} by $m_{\tilde{G}} = (4\pi/3)^{\frac{1}{2}} m_{SUSY}^2 M_{Pl}^{-1}$, so $m_{\tilde{G}} \sim 1$ keV corresponds to $m_{SUSY} \sim 10^6$ GeV. This now appears to be phenomenologically dubious, and $N = 1$ supersymmetry models with $m_{SUSY} \sim 10^{11}$ GeV and $m_{\tilde{G}} \sim 10^2$ GeV are currently popular [191]. In such models, the photino $\tilde{\gamma}$, the spin-$\frac{1}{2}$ supersymmetric partner of the photon, is probably the lightest particle that is odd under the supersymmetric reflection symmetry R, and hence stable. But in supersymmetric GUT models there is a relation between the mass of the gluino $\tilde{g}$, the supersymmetric partner of the gluon (the gauge boson of QCD), and that of the photino: $m_{\tilde{\gamma}} \sim \frac{1}{7} m_{\tilde{g}}$, and there is a phenomenological lower bound on the mass of the gluino $m_{\tilde{g}} \geqslant 3$ GeV [192]. The requirement that the photinos almost all annihilate, so that they do not contribute too much mass density, also implies that $m_{\tilde{\gamma}} \geqslant 0.5$ GeV [193], and they thus become a candidate for cold rather than warm dark matter.

A hypothetical right-handed neutrino ν_R could be the warm DM particle [194], since if right-handed weak interactions exist they must be much weaker than the ordinary left-handed weak interactions, so $T_{\nu_R d} \gg T_{\nu d}$ as required. But particle physics provides no good reason why any ν_R should be light.

Thus there is at present no obvious warm-DM candidate elementary particle in contrast to the hot and cold DM cases. But our ignorance about the physics above the ordinary weak-interaction scale hardly allows us to preclude the existence of very weakly interacting light particles, so we will consider the warm-DM case, mindful of Hamlet's prophetic admonition

> There are more things in heaven and earth, Horatio,
> Than are dreamt of in your philosophy.

Fluctuation spectrum. The spectrum of fluctuations δ_ν at late times in the hot-DM model is controlled mainly by free streaming; $\delta_\nu(M)$ is peaked at $\sim M_{J,\nu}$, eq. (3.16), for any reasonable primordial fluctuation spectrum. This is *not* the case for warm or cold DM.

The primordial fluctuation spectrum can be characterized by the amplitude of fluctuations just as they enter the horizon (see sect. **2**). It is expected that no mass scale is singled out, so the spectrum is just a power law

$$(3.17) \qquad \delta_{\text{DM,H}} = \varkappa \left(\frac{M_{\text{DM}}}{M_0} \right)^{-\alpha}.$$

Furthermore, to avoid too much power on large or small mass scales requires $\alpha \approx 0$ [117, 118]; and to form galaxies and large-scale structure by the present epoch without violating the upper limits on both small [144] and large [195] scale (quadrupole) angular variations in the microwave background radiation requires $\varkappa \sim 10^{-4}$. Equation (3.17) corresponds in Fourier transform space to $|\delta_k|^2 = k^n$ with $n = 6\alpha + 1$. The case $\alpha = 0$ ($n = 1$) is commonly referred to as the constant-curvature or (Harrison-) Zel'dovich spectrum. As I discussed in sect. **2**, inflationary models predict adiabatic fluctuations with approximately the Zel'dovich spectrum.

The important difference between the fluctuation spectra δ_{DM} at late times in the hot and warm DM cases is that $\delta_{DM,warm}$ has power over an increased range of masses, roughly from 10^{11} to $10^{16} M_\odot$. As for the hot case, the lower limit, $M_x \sim M_{Pl}^3 m_x^{-2}$, arises from the damping of smaller-scale fluctuations by free streaming. In the hot case, the DM particles become nonrelativistic at essentially the same time as they become gravitationally dominant, because their number density is nearly the same as that of the photons. But in the warm case, the x particles become nonrelativistic and thus essentially stop free streaming at $T \approx m_x$, well *before* they begin to dominate gravitationally at $T_{eq} \approx 6\Omega h^2$ eV. (As usual, the subscript «eq» refers to the epoch when the energy density of massless particles equals that of massive ones.) During the interval between $T \approx m_x$ and $T \approx T_{eq}$, growth of δ_{DM} is inhibited by the «stagspansion» phenomenon [135] (the generalization to adiabatic fluctuations of the «Meszaros effect» [196]), which we will discuss in detail in the section on cold DM. Thus the spectrum δ_{DM} is relatively flat between M_x and

$$(3.18) \qquad M_{eq} = \frac{4\pi}{3} \left(\frac{ct_{eq}}{1+z_{eq}} \right)^3 \varrho_{c,0} \Omega_0 = 2.2 \cdot 10^{15} (\Omega_0 h^2)^{-2} M_\odot .$$

Fluctuations with masses larger than M_{eq} enter the horizon at $z < z_{eq}$, and thereafter δ_{DM} grows linearly with $R = (1+z)^{-1}$ until nonlinear gravitational effects become important when $\delta_{DM} \sim 1$. Since for $\alpha = 0$ fluctuations of all sizes enter the horizon with the same root-mean-square amplitude, and those with larger M enter the horizon later in the matter-dominated era and subsequently have less time to grow, the fluctuation spectrum at the present time falls with M for $M > M_{eq}$: $\delta_{DM,0} \propto M^{-\frac{2}{3}}$. For a power law primordial spectrum of arbitrary index α,

$$(3.19) \qquad \delta_{DM,0} \propto M^{-\alpha-\frac{2}{3}} = M^{-(n+3)/6}, \qquad M > M_{eq} .$$

This is true for hot, warm, or cold DM. In each case, after recombination at $z_r \approx 10^3$ the baryons «fall in» to the dominant DM fluctuations on all scales larger than the baryon Jeans mass, and by $z = 100$, $\delta_b = \delta_{DM}$ [197].

In the simplest approximation, neglecting all growth during the «stagspansion» era, the fluctuation spectrum for $M_x < M < M_{eq}$ is just $\delta_{DM,0}(M) \propto$
$\propto M^{-(n-1)/6} = M^{-(n_{eff}+3)/6}$, where $n_{eff} = n - 4$, *i.e.* the spectrum is flattened by a factor of $M^{\frac{2}{3}}$ compared to the primordial spectrum. The small amount of growth that does occur during the «stagspansion» era slightly increases the fluctuation strength on smaller mass scales. Detailed calculations of these spectra are now available [173, 198]. For $\alpha \leq 0$, $\delta_x(M)$ has a fairly broad peak at $M \sim M_x$. Consequently, objects of this mass—galaxies and small groups—are the first to form, and larger-scale structures—clusters and superclusters—form later as $\delta_x(M)$ grows toward unity on successively larger mass scales.

Potential problems with warm DM. The warm-DM hypothesis is probably consistent with the observed features of typical large galaxies, whose formation would probably follow roughly the «core condensation in heavy halos» scenario [199, 200]. The potentially serious problems with warm DM are on scales both larger and smaller than M_x. On large scales, the question is whether the model can account for the observed network of filamentary superclusters enclosing large voids. The most productive approach to this question has employed sophisticated N-body simulations with $N \sim 3 \cdot 10^4$ in order to model the large mass range that is relevant [201]. The N-body results suggest that warm and cold DM can reproduce the observed large-scale structure, although to get good agreement it may be necessary to assume that galaxies do not accurately trace the DM distribution. I will discuss this further in the next section.

On small scales, the preliminary indications that dwarf spheroidal galaxies have large DM halos [184, 185] pose problems nearly as serious for warm as for hot DM. Unlike hot DM, warm DM is (barely) consistent with the phase-space constraint [185, 186]. But since free streaming of warm DM washes out fluctuations δ_x for $M \lesssim 10^{11} M_\odot$, dwarf galaxies with $M \sim 10^7 M_\odot$ can form in this picture only via fragmentation following the collapse of structures of mass $\sim M_x$, much as ordinary galaxies form from supercluster fragmentation in the hot-DM picture. The problem here is that dS galaxies, with their small escape velocities $\lesssim 10$ km s^{-1}, would not be expected to bind more than a small fraction of the X particles, whose typical velocity must be $\sim 10^2$ km s^{-1} ($\sim$ rotation velocity of spirals). Thus we expect M/M_{lum} for dS galaxies to be much smaller than for large galaxies—but the indications are that they are comparable [184-186]. Understanding dwarf galaxies may well be crucial for unravelling the mystery of the identity of the DM [200].

4. – Cold dark matter.

To summarize the terminology introduced in the previous section, the dark matter (DM) that appears to be gravitationally dominant on all astronomical scales larger than the cores of galaxies [5] can be classified, on the basis of its characteristic free-streaming damping mass M_D, as hot ($M_D \sim 10^{15} M_\odot$), warm ($M_D \sim 10^{11} M_\odot$), or cold ($M_D < 10^8 M_\odot$). For the case of cold DM, the main subject of this section, the shape of the DM fluctuation spectrum is determined by *a*) the primordial spectrum (on scales larger than the horizon), which is usually assumed to have a power spectrum of the form $|\delta_k|^2 \propto k^n$ (inflationary models predict the «Zel'dovich spectrum» $n=1$), and *b*) «stagspansion» [135], the stagnation of the growth of DM fluctuations that enter the horizon while the Universe is still radiation dominated, which flattens the fluctuation spectrum for $M \lesssim 10^{15} M_\odot$ [202-204].

An attractive feature of the cold-dark-matter hypothesis is its considerable predictive power: the post-recombination fluctuation spectrum is calculable, and it in turn governs the formation of galaxies and clusters. As I will discuss in this section, good agreement with the galaxy and cluster data is obtained in the cold-DM model for a Zel'dovich spectrum of primordial fluctuations, and the model also appears to be reasonably consistent with the observed large-scale clustering, including superclusters and voids [26].

4`1. *Cold*-DM *candidates.* – Besides the evidence summarized in sect. 3 against hot and warm DM, a further reason to consider cold DM is the existence of several plausible physical candidates, including axions of mass $\sim 10^{-5}$ eV [205, 206]; a heavy, weakly interacting, stable particle, such as the photino, with a mass $\geqslant \frac{1}{2}$ GeV [193]; and primordial black holes [207] with 10^{17} g $\leqslant m_{\text{PBH}} \leqslant$ $\leqslant M_\odot$ [26]. Still another exotic cold-DM candidate has recently been proposed by WITTEN: « nuggets » of *u-d-s* symmetric quark matter [134]. There is thus no shortage of cold-DM candidate particles—although there is admittedly no direct evidence that any of them actually exists. In the remainder of this section, I will briefly discuss the rationale for each of these cold-DM candidates.

First, the axion. Quantum chromodynamics (QCD) with quarks of non-zero mass violates *CP* and *T* due to nonperturbative instanton effects. This leads to a neutron electric-dipole moment that is many orders of magnitude larger than the experimental upper limit, unless an otherwise undetermined complex phase θ_{QCD} is arbitrarily chosen to be extremely small. PECCEI and QUINN [208] have proposed the simplest and probably the most appealing way to avoid this problem, by postulating an otherwise unsuspected symmetry that is spontaneously broken when an associated pseudoscalar field—the *axion* [209]—gets a nonzero vacuum expectation value $\langle \varphi_a \rangle \sim f_a \exp[i\theta]$. This occurs when $T \sim f_a$. Later, when the QCD interactions become strong at $T \sim \Lambda_{\text{QCD}} \sim 10^2$ MeV, instanton effects generate a mass for the axion $m_a =$ $= m_\pi f_\pi / f_a = 10^{-5}$ eV $(10^{12} \text{ GeV}/f_a)$. Thereafter, the axion contribution to the energy density is [210] $\varrho_a = 3 m_a T^3 f_a^2 (M_{\text{Pl}} \Lambda_{\text{QCD}})^{-1}$. (A coherent state of axions behaves cosmologically like pressureless dust, despite the fact that $m_a \ll$ $\ll T_{\text{QCD}}$ [211].) The requirement $\varrho_a^0 < \varrho_0 \Omega$ implies that $f_a \leqslant 10^{12}$ GeV, and $m_a \geqslant 10^{-5}$ eV [210]. The longevity of helium-burning stars implies [212] that $f_a > 10^9$ GeV, $m_a < 10^{-2}$ eV. Thus, if the hypothetical axion exists, it is probably important cosmologically, and for $m_a \sim 10^{-5}$ eV gravitationally dominant. (The mass range $10^{9 \div 12}$ GeV, in which f_a must lie, is also currently popular with particle theorists as the scale of supersymmetry [191] or family symmetry breaking, the latter possibility connected with the axion [213].) If axions comprise the dark halo of our galaxy, laboratory experiments have recently been proposed that could detect them [206].

A quite different sort of cold-DM elementary particle arises naturally in supersymmetry. The lightest supersymmetric partner (LSP) particle is stable because of R-parity, a reflection symmetry equivalent to [214]

(4.1) $$R = (-1)^{3(B-L)}(-1)^F,$$

where B, L and F are baryon, lepton and fermion numbers, respectively. R-parity is, therefore, an exact symmetry in any theory in which $B - L$ and F are conserved. Under R-parity, all ordinary particle fields are unchanged and all superpartner fields change sign.

In many supersymmetric theories, the LSP is a photino of mass $m_{\tilde{\gamma}} \gtrsim \frac{1}{2}$ GeV, this lower limit corresponding to cosmological critical density and dependent on the theoretical parameters controlling the $\tilde{\gamma}$ mass and interactions [193]. (Cf. also subsect. 3`6 above.) The $\tilde{\gamma}$'s almost all annihilate at high temperatures, leaving behind a small remnant that, because $m_{\tilde{\gamma}}$ is large, can contribute a critical density today. Remarkably, there is actually one piece of observational evidence that the Milky Way's DM halo is composed of photinos: the calculated [215] annihilation rate $\tilde{\gamma}\tilde{\gamma} \to p\bar{p}$ for $m_{\tilde{\gamma}} \approx (3 \div 10)$ GeV leads to a flux of low-energy $((0.6 \div 1.2)$ GeV) antiprotons comparable to the observed primary cosmic-ray flux, and no plausible alternative explanation is known for $\bar{p}$'s below the ~ 2 GeV kinematic threshold for secondary production. This theory can be checked by additional observations of the spectrum and anisotropy of cosmic-ray antiprotons, positrons and gamma-rays.

Another possibility for the LSP is a sneutrino $\tilde{\nu}$ (a scalar partner of a neutrino). The main difference from the photino as LSP is that the $\tilde{\nu}$ annihilation rate can be much larger (semi-weak rather than weak because it is mediated by $\tilde{Z}$, a fermion, with amplitude proportional to m_Z^{-1} rather than m_Z^{-2}), so that there is no lower limit on the $\tilde{\nu}$ mass from the cosmological density [216]. However, the assumption that $\tilde{\nu}$ is the LSP implies that $m_{\tilde{\nu}} \lesssim 2$ GeV if $\varrho_{\tilde{\nu},0} \sim \varrho_{c,0}$.

The next cold-DM candidate to be considered may seem rather contrived: a particle, such as a ν_R, that decouples while still relativistic but whose number density relative to the photons is subsequently diluted by entropy generated in a first-order phase transition such as the Weinberg-Salam $SU_2 \otimes U_1 \to U_1$ symmetry breaking [190]. (Recall that the m_x bound in fig. 3.7, corresponding to warm DM, assumes no generation of entropy.) More than a factor $\sim 10^3$ entropy increase would overdilute $\eta = n_b/n_\gamma$, if we assume η was initially generated by GUT baryosynthesis; correspondingly, $m_x \lesssim 1$ MeV and $M_x \gtrsim 10^5 M_\odot$.

Finally, there is the possibility that the DM consists of objects more massive than any stable elementary particle. Two cases that have been investigated are « quark nuggets » and black holes.

A « quark nugget » is a hypothetical chunk of quark matter with roughly equal numbers of u, d and s quarks. Because the Fermi energy is shared among three species, u-d-s symmetric quark matter is more stable than u-d quark matter, and perhaps absolutely stable, as long as the strange-quark mass

$m_s \lesssim 200$ MeV [134, 217]. Note that, even if an ordinary nucleus of atomic weight A is more massive than an amount of u-d-s quark matter with equal baryon number, the decay rate of the ordinary nucleus is extremely small since it is proportional to $\sim A$ powers of the weak-interaction constant G_F. At the time of the QCD confinement phase transition in the early universe, however, there were almost as many s quarks as u and d quarks, since m_s is comparable to T_{QCD}. So if the bubbles of the high-temperature (quark plasma) phase can radiate the annihilation energy of the excess $q\bar{q}$ pairs primarily by bulk neutrino emission, then these bubbles can become u-d-s quark nuggets. But as WITTEN [134] admits, « There is a rich element of wishful thinking here, since this picture assumes neutrino losses are the main way for the high-temperature phase to lose energy, while in fact neutrino losses and surface evaporation appear comparable ».

How large might quark nuggets be? WITTEN argues that, when bubbles of the high-temperature phase begin to shrink, their radius is of order $R_1 \approx$ $\approx (1 \div 10^4)$ cm. The lower limit arises under the assumption that the bubbles grow by coalescence of smaller bubbles of initial size $R_0 \ll R_1$. In this case, the effective bubble size is independent of R_0 and is determined by surface tension, with the result that $R_1 \approx M_{Pl}^{\frac{3}{5}}/T_{QCD}^{\frac{8}{5}} \approx (1 \div 10)$ cm. The upper limit, $R_1 \approx 10^4$ cm, applies if R_0 is as large as it could possibly be (a little smaller than the horizon length M_{Pl}/T_{QCD} at the QCD transition). The linear dimensions of the bubbles then shrink by a factor of $\eta^{\frac{1}{3}} \approx 10^{-3}$, where, as usual, η is the baryon-to-photon ratio. The result is that a quark nugget is expected to have a radius of $(10^{-3} \div 10)$ cm, and correspondingly a mass $m_{qn} \sim (10^6 \div 10^{18})$ g.

It follows from the discussion in sect. 1 that the density of dark matter in our part of the Galaxy is

(4.2) $$\varrho_{DM} \sim \frac{v^2}{4\pi G r^2} \approx 10^{-24} \text{ g cm}^{-3},$$

where $v \approx 220$ km s^{-1} is the circular velocity and $r \approx 8$ kpc the galactic radius of the Sun. The flux intercepted by the Earth is thus $\sim 10^9$ g per year. Given the rather large uncertainty in m_{qn}, this flux corresponds to anywhere from 10^3 light quark nuggets striking the Earth per year, to one heavy one in the Earth's lifetime. Such dense projectiles would pass right through the Earth. DE RUJULA and GLASHOW [218] call any aggregate of stable nuclear matter intermediate in size between nuclei and neutron stars that happen to hit the Earth a « nuclearite ». If they exist in the mass range discussed, they can be detected by looking for linear « astroblems » (the debris of their passage through rock) or linear seismic sources.

Primordial black holes of typical mass $M_{BH} \lesssim M_\odot$ could have formed if there were fluctuations of large amplitude when that mass scale first entered the horizon, or black holes of mass $M_{BH} \gtrsim M_\odot$ could have formed from collapse of ordinary matter after recombination. A rather wide mass range is allowed,

(4.3) $$10^{-16} M_\odot \lesssim M_{BH} \lesssim 10^6 M_\odot,$$

the lower limit implied by the nonobservation of γ-rays from black-hole decay by Hawking radiation, and the upper limit required to avoid disruption of galactic disks and star clusters [207, 219]. Stronger but less certain upper limits are $M_{BH} \lesssim 10^3 M_\odot$ from nonobservation of accretion onto BHs as they plunge through the galactic disk [28, 220], $M_{BH} \lesssim 10^2 M_\odot$ from dwarf spheroidal halos [186], and $M_{BH} \lesssim 10^{-2} M_\odot$ from the nonobservation of the focusing of quasar cores [221].

You will doubtless have noticed that, while there is no direct evidence against any of these candidates for cold DM, there are no convincing arguments in their favor either. Actually, it is not clear that we have a good basis to judge the plausibility of any DM candidates, since in no case except possibly « quark nuggets » is there a fundamental explanation—or, even better, a prediction—for the ratio $\omega \equiv \varrho_{DM\,0}/\varrho_{lum,0}$, which is thought to lie in the range $10 \leqslant \omega \leqslant 10^2$, the lower limit corresponding to an open universe and the upper limit to $\Omega \approx 1$. Two fundamental questions about the Universe which the fruitful marriage of particle physics and cosmology has yet to address successfully are the value of ω and of the cosmological constant Λ.

4'2. *Galaxy and cluster formation with cold* DM.

« *Stagspansion* ». I will follow the current conventional wisdom and assume that the primordial fluctuations were adiabatic. In the standard formalism, fluctuations $\delta \equiv \delta\varrho/\varrho$ grow as $\delta \sim R^2$ on scales larger than the horizon (cf. subsect. 2'5), where $R = (1+z)^{-1}$ is the scale factor normalized to 1 at the present. When a fluctuation enters the horizon in the radiation-dominated era, the photons (together with the charged particles) oscillate as an acoustic wave (cf. discussion of Jeans mass in subsect. 3'1), and the noninteracting neutrinos freely stream away (they are still relativistic, since in the cold-DM case their masses are $\ll$ 30 eV). As a result, the main driving terms for the growth of δ_{DM} disappear and the growth accordingly stagnates as the Universe continues to expand (« stagspansion ») until matter dominates (see fig. 4.1). Matter domination first occurs at $z = z_{eq}$, where (cf. eq. (2.43))

(4.4) $$z_{eq} = 4.2 \cdot 10^4 h^2 \Omega (1 + 0.68 N_\nu)^{-1} = 2.5 \cdot 10^4 h^2 \Omega \quad \text{for } N_\nu = 3.$$

The first study of the growth of cold-DM fluctuations was the numerical calculations of Peebles [202], who for simplicity ignored neutrinos: $N_\nu = 0$ in the above equation. Subsequent numerical calculations have included the effects of the known neutrino species ($N_\nu = 3$, $m_\nu \approx 0$) both outside and inside

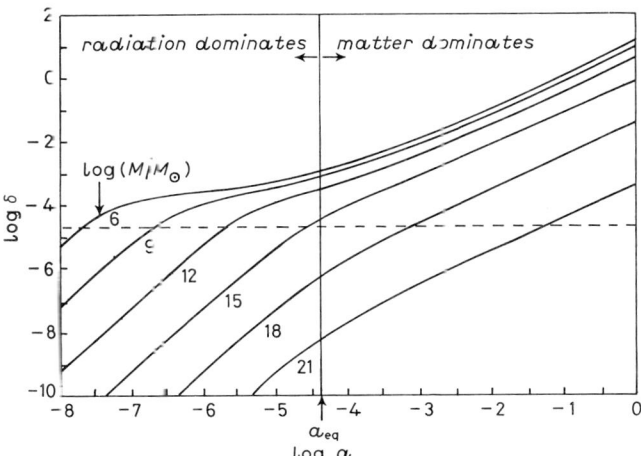

Fig. 4.1. – Numerical results for the growth of $\delta = k^{\frac{3}{2}} \delta_k$ vs. scale factor a for fluctuations of various masses $M = \frac{4}{3}\pi^4 k^{-3} \varrho_c$. The curves are drawn for $n = 1$, $\Omega = h = 1$ and a baryonic to total mass ratio of 0.1. The vertical line represents the value of a when the Universe becomes matter dominated, and the dashed line shows the (constant for $n = 1$) value of δ when each mass scale crosses the horizon. These curves illustrate the stagnation of perturbation growth after small mass scales cross the horizon and show why at late times $\delta(k)$ is nearly flat for large k (small M). (From ref. [203].)

the horizon [135, 174, 175, 203, 204, 222]. Numerically, the largest effect of including neutrinos is the change in z_{eq}.

It is instructive to make the further approximation of setting $\delta_{\gamma+b} = \delta_\nu = 0$ once a fluctuation is inside the horizon. Then one can analytically match the solution for $R > R_{horizon}$

(4.5)
$$\begin{cases} \delta_{DM}(R) = A_1 D_1(R) + A_2 D_2(R), \\ D_1 = 1 + 1.5y, \qquad \text{where } y = R/R_{eq} \\ D_2 = D_1 \ln \frac{(1+y)^{\frac{1}{2}}+1}{(1+y)^{\frac{1}{2}}-1} - 3(1+y)^{\frac{1}{2}}, \end{cases}$$

to the growing mode $\delta_{DM} \sim R^2$ for $R < R_{horizon}$. Matching the derivatives requires $A_2 D_2$ comparable to $A_1 D_1$ but opposite in sign. For $R \gg R_{horizon}$ only the growing solution D_1 survives, which explains the moderate growth in δ_{DM} between horizon crossing and matter dominance. In the limit of large k, one finds [203] $|\delta_k| \propto k^{n/2-2} \ln k$, where as usual $n = 1$ corresponds to a primordial Zel'dovich spectrum. Correspondingly, for $M \ll M_{eq} \approx 10^{16} M_\odot$, the r.m.s. fluctuation in the mass within a random sphere containing average mass M is $\delta M/M \propto |\ln M|^{\frac{1}{2}}$. Some authors have considered only the Meszaros solution D_1 and erroneously inferred that the fluctuation spectrum would be essentially flat for $M < M_{eq}$ for a Zel'dovich primordial spectrum, which would then be inconsistent with observations [223]. One can match the boundary conditions without D_2 only for isothermal primordial fluctuations, for which the amplitude does not grow at very early times since the fluctuating component is subdominant. (This was the case considered by MESZAROS [196].)

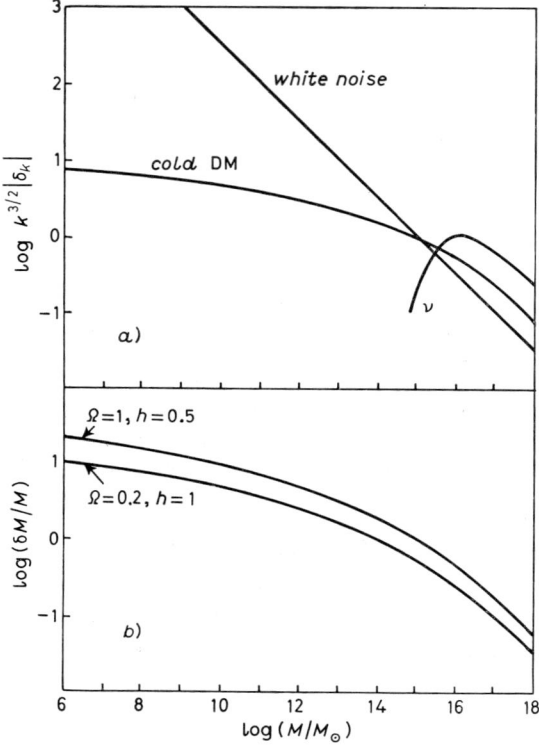

Fig. 4.2. – Density fluctuations as a function of mass. a) $k^{\frac{3}{2}}|\delta_k| = \delta\varrho/\varrho(M)$, where $M = 4\pi^4\varrho_0/3k^3$, for isothermal white noise ($n = 0$), and adiabatic Zel'dovich ($n = 1$) neutrino [198] and cold-dark-matter spectra. b) Root-mean-square mass fluctuation within a randomly placed sphere containing average mass M in a cold-dark-matter universe, for $\Omega = 1$, $h = 0.5$ and $\Omega = 0.2$, $h = 1$. The curves are normalized [202] at 8 Mpc and assume a primordial Zel'dovich ($n = 1$) fluctuation spectrum.

Figure 4.2a) is a sketch of $\delta\varrho/\varrho(M)$ for hot (ν) and cold DM. Figure 4.2b) shows numerical results [204] for $\delta M/M$, again assuming a Zel'dovich ($n = 1$) spectrum and normalized so that $\delta M/M = 1$ at $R = 8h^{-1}$ Mpc [202]. Notice that $\delta M/M$ is relatively flat for $M < 10^9 M_\odot$, and steepens to $\delta M/M = M^{-\frac{2}{3}}$ (that is, $n = 1$, reflecting the primordial spectrum) for $M \gg M_{eq}$ [26, 204].

Galaxy and cluster formation. The key features of galaxy formation in the cold-DM picture are these: after recombination (at $z_{rec} \approx 10^3$) the amplitude of the baryonic fluctuations rapidly grows to match that of the DM fluctuations; smaller-mass fluctuations grow to nonlinearity and virialize, and then are hierarchically clustered within successively larger bound systems; and finally the ordinary matter in bound systems of total mass $\sim 10^{8 \div 12} M_\odot$ cools rapidly enough within their DM halos to form galaxies, while larger-mass fluctuations form clusters.

At any mass scale M, when the fluctuation $\delta M/M$ approaches unity, nonlinear gravitational effects become important. The fluctuation then separates from the Hubble expansion, reaches a maximum radius, and begins to contract. Spherically symmetric fluctuations, for example, contract to about half their maximum radii (see sect. 2). During this contraction, violent relaxation [224] due to the rapidly varying gravitational field converts enough potential energy into kinetic energy for the virial theorem, $\langle PE \rangle = -2\langle KE \rangle$, to be satisfied. After virialization, the mean density within a fluctuation is roughly eight times the density corresponding to the maximum radius of expansion [225].

Since the cold-DM fluctuation spectrum $\delta M/M$ is a decreasing function of M, smaller-mass fluctuations will, on the average, become nonlinear and begin to collapse at earlier times than larger-mass fluctuations. Small-mass bound systems are subsequently clustered within larger-mass systems, which go nonlinear at a later time. This hierarchical clustering of smaller systems into larger and yet larger gravitationally bound systems begins at the baryon Jeans mass ($M_{J,b} \sim 10^5 M_\odot$ at recombination) and continues until the present time.

Although small-mass fluctuations are the first to go nonlinear in the cold-DM picture, pressure inhibits baryons from falling into such fluctuations if $M < M_{J,b}$. More importantly, even for $M > M_{J,b}$, the baryons are not able to contract further unless they can cool by emitting radiation. Without such mass segregation between baryons and DM, the resulting structures will be disrupted by virialization as fluctuations that contain them go nonlinear [226]. Moreover, successively larger fluctuations will collapse relatively soon after one another if they have masses in the flattest part of the $\delta M/M$ spectrum, i.e. (total) mass $\leq 10^9 M_\odot$.

Gas of primordial composition (about 75% atomic hydrogen and 25% helium, by mass) cannot cool significantly unless it is first heated to $\geq 10^4$ K, when it begins to ionize [26]. Assuming a primordial Zel'dovich spectrum normalized so that at the present time $\delta M/M = 1$ at $R = 8h^{-1}$ Mpc, as recommended by PEEBLES [202], the smallest protogalaxies for which the gas is sufficiently heated by virialization to radiate rapidly and contract have total mass $M = 10^9 M_\odot$. One can also deduce an *upper* bound on galaxy masses by requiring that the cooling time be shorter than the dynamical time [227]; this upper bound is $M \leq 10^{12} M_\odot$. These limits are illustrated in fig. 4.3, which shows the density of ordinary (baryonic) matter *vs.* internal kinetic energy (temperature) of typical fluctuations of various sizes, just after virialization.

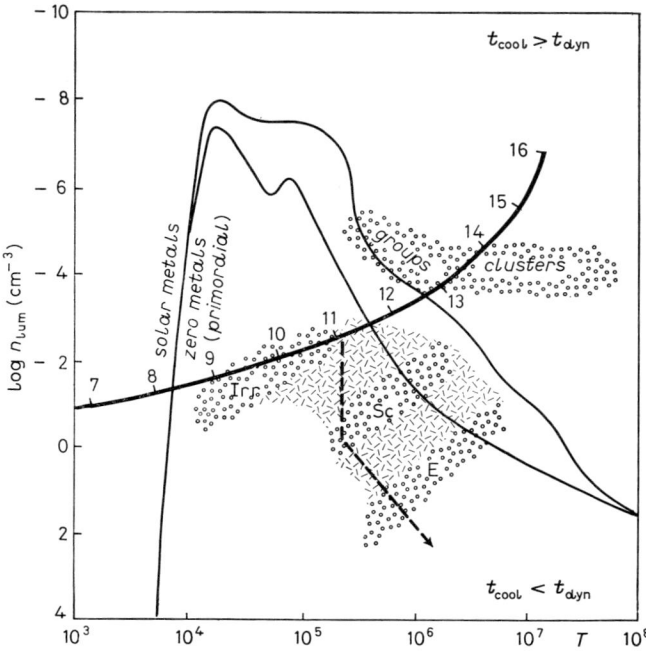

Fig. 4.3. – The baryonic density *vs.* temperature as root-mean-square perturbations having total mass M become nonlinear and virialize. The numbers on the tick marks are the logarithm of M in units of $M_\odot$. This curve assumes $n = 1$, $\Omega = h = 1$ and a baryonic to total mass ratio of 0.07. The region where baryons can cool within a dynamical time lies below the cooling curves. Also shown are the positions of observed galaxies, groups and clusters of galaxies. The dashed line represents a possible evolutionary path for dissipating baryons. (From ref. [32].)

calculated from $\delta M/M$ for $\Omega = h = 1$. This is superimposed upon the Rees-Ostriker cooling curves (for which cooling time equals gravitational free-fall time) and data on galaxies (with kinetic energy determined from rotation velocity for spirals and velocity dispersion for ellipticals) (*).

Only in protogalaxies for which the cooling time is short compared to the dynamical time can the baryons dissipate and contract. This dissipation leads to higher baryonic densities and somewhat higher temperatures.

The collapse of fluctuations having mass $> 10^{13} M_\odot$ leads to clusters of galaxies in this picture. In clusters, only the outer parts of member galaxy halos are stripped off; the inner baryonic cores continue to contract, presumably until star formation halts dissipation.

(*) See ref. [26] for a considerably more elaborate version of fig. 4.3, with virialization curves for several multiples of the r.m.s. fluctuation spectrum for an open as well as an Einstein-de Sitter universe, much more detailed galaxy and cluster data, and discussion of molecular and Compton cooling.

Fluctuations that start with greater amplitude than average will turn around earlier, at higher density, and thus lie below the virialization curve on fig. 4.3. As the baryons in a virialized fluctuation dissipate, their density will initially increase at constant T within the surrounding isothermal halo of dissipationless material (DM), and then T will increase as well when the baryon density exceeds the DM density, as suggested by the dashed line in the figure. The Zel'dovich primordial spectrum is more consistent with the data in fig. 4.3 than an $n = 2$ (or $n = 0$) primordial spectrum, which lies too low (too high) on the figure compared to the galaxies. With the Zel'dovich spectrum, the important conclusion is that one should observe dissipated systems with large halos having total mass $10^8 M_\odot \leqslant M \leqslant 10^{12} M_\odot$. This is essentially the range of observed galaxy masses.

While the n_b-T diagram (fig. 4.3) is useful for comparing data and predictions with the cooling curves, it is also useful to consider total mass M vs. T, as in fig. 4.4. This avoids having to take into account the differing amounts of baryonic dissipation suffered by various galaxies. The heavy solid and dashed curves again correspond to the $n = 1$ cold-DM spectrum, for $\Omega = 1$, $h = 0.5$ and $\Omega = 0.2$, $h = 1$, respectively. It is striking that the galaxies in the M-T diagram lie along lines of roughly the same slope as these curves. This occurs because the effective slope of the $n = 1$ cold-DM fluctuation spectrum in the galaxy mass range is $n_{\text{eff}} \approx -2$, which corresponds to the empirical Tully-Fisher and Faber-Jackson laws: $M \propto v^4$. The light dashed lines in fig. 4.4 are the post-virialization curves for primordial fluctuation spectra with $n = 0$ (white noise) and $n = 2$. Again, the $n = 1$ (Zel'dovich) spectrum is evidently the one that is most consistent with the data.

The points in fig. 4.4 represent essentially all of the clusters identified by GELLER and HUCHRA [228] in the CfA catalogue within 5000 km s^{-1}. The cluster data lie about where they should on the diagram, and even the statistics of the distribution seem roughly to correspond to the expectations represented by the 0.5, 1, 2 and 3σ curves.

Notice that spiral galaxies lie roughly along the 1σ curve, while elliptical galaxies lie along the 2σ curve. Although this displacement is not large compared to the uncertainties, it is consistent with the fact that more than half of all bright galaxies are spirals, while only about 10% are ellipticals. In hierarchical clustering scenarios, it seems likely that the higher-σ fluctuations will develop rather smaller angular momenta, as measured by the dimensionless parameter λ ($\equiv J E^{\frac{1}{2}} G^{-1} M^{-\frac{5}{2}}$). There are two reasons for this: high-overdensity fluctuations collapse earlier than average fluctuations, and are thus typically surrounded by a relatively homogeneous matter distribution [229], also, higher-amplitude fluctuations are typically rounder [230] and consequently have lower quadrupole moments. Both effects result in less torque. This difference appears to exist with either white noise or a flatter spectrum, but to be somewhat larger in the latter case. If high-σ fluctuations have little angular

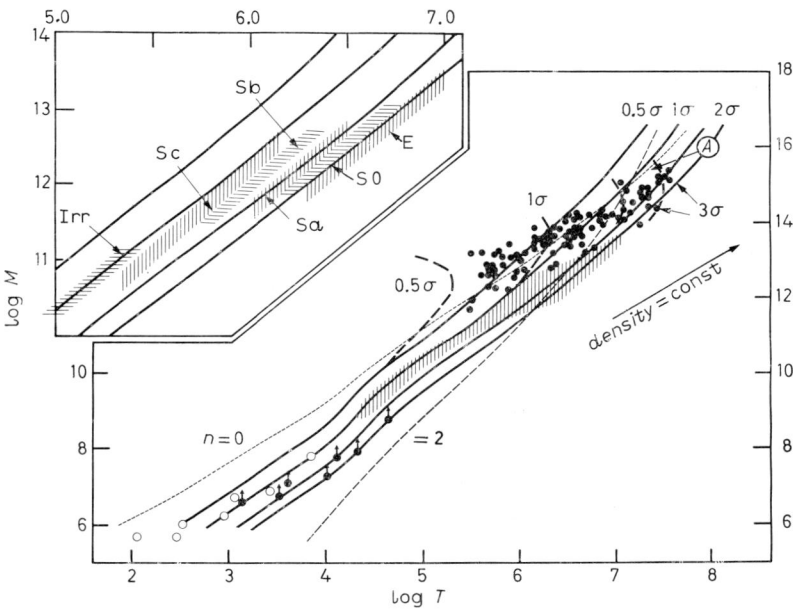

Fig. 4.4. – Total mass M (in units of $M_\odot$) vs. temperature T. The quantity T is $\mu V^2/3k$, where μ is mean molecular weight (≈ 0.6 for ionized, primordial H+He) and k is Boltzmann's constant. M for groups and clusters is total dynamical mass. For galaxies, M is assumed to be $10 M_{\text{lum}}$ (corresponding to fig. 1.11). If dwarf spheroidals actually have $M/L_B^i = 30$, they may have suffered baryon stripping [187], in which case M is a lower limit (arrows). Details of the region occupied by massive galaxies are shown in the inset in upper left. Model curves represent the equilibria of structures that collapse dissipationlessly from the cold-dark-matter initial fluctuation spectra with $n = 1$. The curves labeled 1σ refer to fluctuations with $\delta M/M$ equal to the r.m.s. value. Curves labeled 0.5σ, 2σ and 3σ refer to fluctuations having 0.5, 2 and 3 times the r.m.s. value. Heavy curves: $\Omega = 1$, $h = 0.5$; dashed curves: $\Omega = 0.2$, $h = 1$; these cases were chosen to span the astrophysically interesting range. In addition to the $n = 1$ curves, two 1σ curves for $n = 0$ and $n = 2$ are also shown (light dashes); ——— $n = 1.0$, $\Omega = 1.0$, $h = 0.5$; – – – $n = 1.0$, $\Omega = 0.2$, $h = 1.0$; - - - $n = 0.0$, $\Omega = 1.0$, $h = 0.5$; – - – $n = 2.0$, $\Omega = 1.0$, $h = 0.5$. Major conclusions from the figure: 1) Either set of curves for $n = 1$ (Zel'dovich spectrum) provides a good fit to the observations over 9 orders of magnitude in mass. Curves with $n = 0$ and $n = 2$ do not fit as well. 2) The apparent gap between galaxies and groups and cluster in fig. 4.3 (which stems from baryonic dissipation) vanishes in this figure, and the clustering hierarchy is smooth and unbroken from the smallest structures to the largest ones. 3) The Fisher-Tully and Faber-Jackson laws for galaxies ($M \propto V^4$ or T^2) arise naturally as a consequence of the slope of the cold-DM fluctuation spectrum in the mass region of galaxies. 4) Groups and clusters are distributed around the $n = 1$ loci about as expected. The apparent upward trend among the groups is not physically meaningful but arises from their selection as minimum-density enhancements (see constant-density arrow). 5) The exact locations of galaxies are somewhat uncertain. In particular, the temperatures of E's and S0's may be overestimated owing to the use of nuclear rather than global velocity dispersions. Taken at face value, however, the data suggest that early-type galaxies (E's and S0's) arise from high-$\delta M/M$ fluctuations, whereas late-type galaxies

momentum, their baryons can collapse by a large factor in radius, forming high-density ellipticals and spheroidal bulges, as shown in fig. 4.3. Since, with a flat spectrum, higher-σ fluctuations occur preferentially in denser regions destined to become rich clusters, one expects [231] to find more ellipticals there—as is observed [42, 43, 48]. Note that the rich clusters lie along the same 2 and 3σ curves in fig. 4.4 as do the elliptical galaxies.

Note also that, while the galaxy data lie below the r.m.s. virialization curve, the data on groups and clusters of galaxies lie more or less evenly around it. This suggests that galaxy formation may be an inefficient process, with lower-amplitude fluctuations of galaxy mass not giving rise to visible galaxies [26].

Presumably the collapse of the low-λ protoelliptical galaxies is halted by star formation well before a flattened disk can form, yielding a stellar system of spheroidal shape. The mechanism governing the onset of star formation in these systems is unfortunately not yet understood, but may involve a threshold effect which sets in when the baryon density exceeds the DM halo density by a sufficient factor [181, 226]. Disks (spirals and irregulars) form from average, higher-λ protogalaxies, which, for a given mass, are larger and more diffuse than their protoelliptical counterparts. The collapse of disks thus occurs via relatively slow infall of baryons from $\sim 10^2$ kpc, halted by angular momentum. Infall from such distances is consistent both with the extent of dark halos inferred from observations and with the high angular momenta of present-day disks ($\lambda \sim 0.4$) [232]. The location of the galaxies in fig. 4.3 is consistent with these ideas if the baryons in all galaxies collapsed by roughly the same factor, about an order of magnitude, but somewhat less for late-type spirals and irregulars and somewhat more for early-type E's and spheroidal bulges.

It has been theorized that the Hubble sequence originates in the distribution of either the initial angular momenta [233] or else the initial densities [234] of protogalaxies. However, if overdensity and angular momentum are linked, with the high-σ fluctuations having lower λ, then these two apparently competitive theories become the opposite sides of the same coin.

It is interesting to ask whether the cold-DM picture can account for the wide range of morphologies displayed by clusters of galaxies in X-ray and

---

(Sc's and Irr's) arise from low-$\delta M/M$ fluctuations. 6) Groups and clusters appear to fill a wider band than galaxies. If real, this difference may indicate that very weak, low-$\delta M/M$ fluctuations on the mass scale of galaxies once existed but did not give rise to visible galaxies. This suggests further that galaxy formation, at least in some regions of the Universe, may not have been fully complete and that galaxies are, therefore, not a reliable tracer of total mass. 7) There seems to be a real trend along the Hubble sequence to increasing mass among early-type galaxies. Neither this trend nor the rather sharp demarcation between galaxies and groups and clusters is fully understood. (This figure is from ref. [26].)

optical-band observations, ranging from regular, apparently relaxed configurations to complex, multicomponent structures. (Cf. fig. 1.8 and 1.9.) Preliminary results are encouraging. In particular, simulations show that large central condensations form quickly and can grow by subsequent mergers to form cD galaxies if most of the DM is initially in halos around the baryonic substructures, as expected for cold DM, but not if the DM is initially distributed rather diffusely throughout the whole cluster, as expected for hot DM [47].

Consider finally the difference in fig. 4.4 between the solid and dashed lines. The dashed lines, representing a lower-density universe ($\Omega = 0.2$), curve backward at the largest masses and lie far away from the circle representing the cores of the richest clusters, Abell classes 2 and 3. Since these regions of very high galaxy density contain at least several percent of the mass in the Universe, the circle should lie between the 2 and 3σ lines (assuming Gaussian statistics). It does so for the solid ($\Omega = 1$) lines, but not for the dashed lines. At face value, this is evidence favoring an Einstein-de Sitter universe for cold DM. However, there are at least two reasons why this argument should probably not be taken too seriously. First, the velocity dispersions represented by the Abell cluster circle in fig. 4.4 correspond to the cluster cores. The model curves, on the other hand, refer to the entire virialized cluster, over which the velocity dispersion is considerably lower (as indicated by the arrow attached to the circle in fig. 4.4). Second, the assumption of spherical symmetry used in obtaining both sets of curves in the figure is only an approximation. The initial collapse is probably often quite anisotropic—more like a Zel'dovich pancake than a sphere. It is, therefore, preferable to compare these data with N-body simulations rather than with the simple model represented by the curves in fig. 4.4. This will require N-body simulations of large dynamical range, which can perhaps be achieved by putting many mass points into one cell of the P^{3}M-type simulations. Until this becomes possible, the data in the figure do not allow a clear-cut discrimination between the $\Omega = 0.2$ and $\Omega = 1$ cases, especially if the Hubble parameter h is allowed to vary simultaneously within the observationally allowed range, as has been assumed.

It is important to appreciate that some means of suppressing galaxy formation in regions of lower-than-average density is required in the cold-DM model both for $\Omega = 1$ and for $\Omega \approx 0.2$ [26]. In the former case, this is needed to hide most of the mass. In a low-Ω universe, on the other hand, large regions of much lower density than average cannot form by gravitation alone [235]. The amplitude of linear fluctuations stops growing in an open universe when it goes into free expansion (*i.e.* for $z \lesssim \Omega_0^{-1}$), and large voids would stop expanding in comoving co-ordinates. Thus formation of large galaxies must be inhibited somehow in regions of moderately low density if the number density of galaxies in voids is less than one-quarter of the average, the quoted upper limit for the Boötes void [236]. Since it is possible to imagine various physical phenomena which might contribute to suppression of galaxy formation in underdense regions [26], this is not necessarily a problem for the cold-DM picture. But it does imply that the distribution of luminous galaxies probably cannot be an accurate tracer of the mass distribution on large scales in any version of the cold-DM universe.

4'3. *N-body simulations of large-scale structure.* – Elaborate N-body simulations of the evolution of large-scale structure in a universe dominated by cold DM have recently been carried out by DAVIS, EFSTATHIOU, FRENK and WHITE (DEFW) [201] using the P^{3}M (particle-particle/particle-mesh) scheme [237], in which the equations of motion are integrated directly for nearby particles, while the gravitational effects of more distant masses are calculated by applying fast-Fourier-transform methods to Poisson's equation, with periodic boundary conditions at the edges of the computational volume. A wide range of structures arise in such simulations, including filaments, superclusters and large regions of fairly low density. As fig. 3.6 shows, the distribution of galaxies is qualitatively similar to observations. A more quantitative comparison reveals various problems, however.

Comparison with observations. Ensembles of simulations were run for $\Omega = 1$ and also for $\Omega = 0.2$. In the $\Omega = 1$ simulations, there is no time when the mass autocorrelation function is a good match in both slope and amplitude to that of the galaxies. Moreover, the r.m.s. peculiar velocities of pairs of particles are several times larger than those actually observed for galaxies (discussed in subsect. 2'4 above). For $\Omega = 0.2$, the mass autocorrelation function matches the galaxy autocorrelation function fairly well for $h \approx 1.1$, and the peculiar velocities are in better agreement with observation at separations of $\sim 5h^{-1}$ Mpc although still too large on smaller scales. The parameter Q which relates the galaxy three-point and two-point correlation functions (see eq. (2.26)) is observed to be about unity and weakly (if at all) dependent on the size of the three-point triangle; Q determined from the simulations is too large by almost a factor of 2 on small scales for $\Omega = 1$ and even a little larger for $\Omega = 0.2$, though in better agreement with $Q \approx 1$ on larger scales. An independent set of N-body simulations [238] found that in neutrino models Q varies strongly with scale in a manner completely inconsistent with observations; while the weaker dependence of Q in cold-DM models is more in accord with observations, the residual variation still exceeds any seen in the data.

Thus, while the cold-DM simulations are not in such gross disagreement with observations as the ν simulations, neither the Einstein-de Sitter nor the open-universe cold-DM simulations agree quantitatively with observations of galaxy spatial and velocity distributions. Although the open-universe simulations are perhaps in better agreement, they are in potential conflict with the latest data [144] on small-angle fluctuations in the cosmic background radiation, which imply [174, 175] that $\Omega \geqslant 0.2 h^{-\frac{4}{3}}$ unless there is significant reheating of the intergalactic medium after recombination (cf. eq. (3.8)). There is also a serious problem with the age of the Universe (see fig. 2.4 and 2.5): $t_0 = 8.3 h^{-1}$ Gy for $\Omega = 0.2$ (assuming $\Lambda = 0$), which is in severe conflict with globular-cluster age estimates and even in conflict with nuclear cosmochronometry for $h \geqslant 1$. Finally, $\Omega = 0.2$ is in conflict with the expectations of the inflationary-universe hypothesis.

Flat universe with positive cosmological constant. The age problem can be relaxed and inflationary models allowed if the cosmological constant is nonzero [69, 239]. If, as is usually supposed, there is much more than enough inflation to solve the flatness and other cosmological problems, then the curvature k becomes vanishingly small. Einstein's equation (2.8) evaluated at the present is then

$$(4.6) \qquad 1 = \Omega + \frac{\Lambda}{3H^2}.$$

If $\Lambda = 0$, this, of course, implies $\Omega = 1$, but a positive cosmological constant can compensate for $\Omega < 1$ in an inflationary universe.

In a universe characterized by eq. (4.6), the time dependence of the scale factor in the epoch of nonrelativistic matter domination ($\varrho \propto R^{-3}$) is

$$(4.7) \qquad R(t) \propto \sinh\left[(3\Lambda)^{\frac{1}{2}} t/2\right]^{\frac{2}{3}},$$

and the age of the Universe is

$$(4.8) \qquad t_0 = \frac{2}{3H_0 (1 - \Omega_0)^{\frac{1}{2}}} \sinh^{-1}(\Omega_0^{-1} - 1)^{\frac{1}{2}}$$

for positive Λ. Figure 4.5 shows the corresponding values of cosmological density parameter Ω_0 and Hubble parameter h for various values of t_0. Comparison with the analogous plot for the case $\Lambda = 0$ (fig. 2.5) shows that it is now somewhat easier to bring the predicted age t_0 into consistency with globular-cluster age estimates. Moreover, since fluctuations grow considerably more after $z \leqslant \Omega_0^{-1}$ in the present case compared to the case $\Lambda = 0$ [239], there is now no conflict with the observational constraints [144] on $\Delta T/T$ for $\Omega \approx 0.2$.

DEFW ran one N-body simulation of the case (4.6) with $\Omega_0 = 0.2$, and found that the results closely resembled those found for $\Lambda = 0$ and the same value of Ω_0. Matching the galaxy autocorrelation function again leads to $h \approx 1.1$, and, although this now corresponds to $t_0 = 10$ Gy rather than 7.5 Gy, it still leads to a universe which is embarrassingly young.

« Biased » galaxy formation. Thus far, I have discussed DEFW's attempts to fit their cold-dark-matter N-body simulations to the observed universe

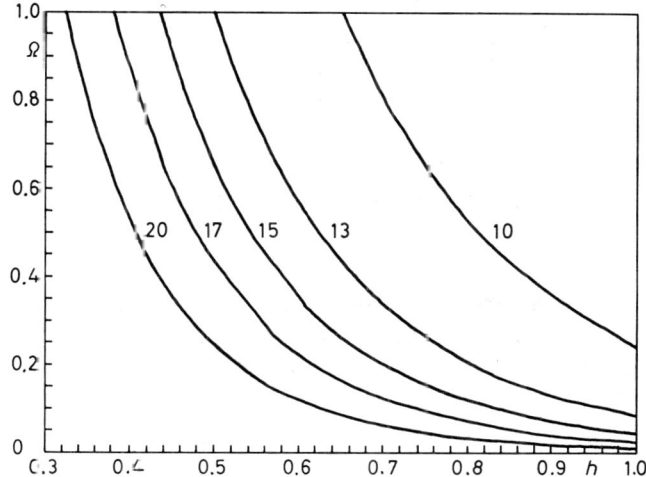

Fig. 4.5. – Cosmological density Ω vs. Hubble parameter for various values of t_0 in a flat universe with cosmological constant given by eq. (4.6).

under the assumption that the distribution of bright galaxies closely approximates the distribution of dark matter. The results are better than for the neutrino simulations, but not really very impressive. DEFW found much better agreement with the data, however, when they tried assuming that bright galaxies form only at relatively high peaks of the linear density distribution. The visible galaxies thus are pictured as being like the unsubmerged peaks of a vast flooded mountain range.

DEFW's biasing procedure depends on two parameters: the threshold ν and the smoothing size r_0. After smoothing, they identified all peaks of amplitude at least ν times the r.m.s. density fluctuation and tagged the nearest particle to each as a «galaxy», whose motion is subsequently followed over the course of the simulation. With a biasing threshold of $\nu = 2.5$ for their Einstein-de Sitter ($\Omega = 1$) simulations, DEFW found that the correlation length of the galaxies exceeds that of the dark matter by a factor of about 2.4, and the observed two- and three-point correlations are fitted quite well for $h = 0.44$ (corresponding to $t_0 = 15$ Gy). The simulations also fit the observed peculiar velocity distribution rather well (about as well as the unbiased $\Omega = 0.2$ simulations did). Since the dark matter is mostly rather smoothly distributed in the biased model, the peculiar velocities generated are much smaller than in the unbiased model with the same density ($\Omega = 1$).

Since both star and galaxy formation are very poorly understood, it is going to be hard to put biased galaxy formation on a solid theoretical footing. But the basic idea that bright galaxies should form only at the highest peaks of the linear density distribution is a rather plausible one, and a number of possible physical mechanisms have been proposed [240]. Moreover, as BFPR already pointed out in connection with fig. 4.4, the greater range of velocity dispersions observed in groups and clusters of a given mass as compared to galaxies is evidence for biasing: Groups of relatively low velocity dispersion, which presumably arose from lower-amplitude fluctuations, are visible because of the galaxies that they contain. The absence of bright galaxies of correspondingly low rotation velocity (for spirals) or velocity dispersion (for ellipticals) suggests that only the fluctuations of larger amplitude form visible galaxies.

Very-large-scale structure. How well can the cold-DM picture account for the largest structures observed in the Universe: superclusters and voids? To answer quantitatively, it is necessary to compare predictions with observations statistically. The challenge is to devise appropriate statistical tests which not only are amenable both to calculation and observation, but also actually discriminate between alternative theories. I will here discuss several statistical tests: cluster autocorrelations, the size distribution of voids, percolation and correlations between the orientation of rich clusters and the directions to nearby ones (Binggeli's statistic).

The autocorrelation function of rich clusters appears to have the same slope as that of galaxies, but with a correlation length r_0 about five times greater [78, 173]. (See fig. 2.8.) As KAISER [241] has emphasized, statistical effects in the formation of rich clusters can lead to just such an effect: $\xi_{\text{clusters}}(r) \approx A \xi_{\text{density}}(r)$, with the amplification factor A increasing with the richness of the clusters, in qualitative agreement with observation. Three-point and higher-order correlations are also enhanced in a calculable way [242], and it may soon be possible to check whether these statistical predictions are in accord with observations. If this is indeed the right explanation for the enhanced clustering of rich clusters, then, if Abell clusters are actually positively correlated at $r \approx 100 h^{-1}$ Mpc, the underlying mass density must also be. DEFW point out that the correlation function corresponding to the cold-DM linear density distribution is negative for $r > 18(\Omega h^2)^{-1}$ Mpc, and thus argue that, for this picture to make sense, $\Omega h < 0.18$. But this is in trouble with $\Delta T/T$ unless $\Lambda > 0$, and with globular-cluster ages. Biased galaxy formation does not help. Of course, it is possible that the strong correlation of rich clusters is at least partly caused by something entirely different from the effect that KAISER discussed, in which case this line of argument is irrelevant. Indeed, N-body simulations of the clustering of clusters [56, 243] suggest that some additional physical effect beyond the Gaussian statistics of rare events will be necessary to account for the large amplitude A that is observed.

The several thousand accurate galaxy red-shifts presently available are not yet enough to tell us very much about the statistics of voids, but DEFW made some preliminary comparisons with their $\Omega = 1$ N-body simulations. They did this by distributing the particles in red-shift space, in which two of the co-ordinates are positions and the third is the corresponding velocity component, and then constructing what amounts to a volume-limited red-shift catalogue. They find that $(8 \div 10)\%$ of the cubes of size $32h^{-1}$ Mpc have density less than half the mean, while only 0.6% of such cubes have density less than 30% of the mean. Large, low-density regions are even less common in openuniverse simulations and in the biased galaxy versions of $\Omega = 1$ simulations. For comparison, an analysis of galaxy counts in randomly placed spheres of radius $20h^{-1}$ Mpc (*i.e.* with the same volume as a $32h^{-1}$ Mpc cube) suggests that 20% of such spheres have density less than half the mean, about twice the number found in the simulations [201]. The $30h^{-1}$ Mpc radius Boötes void appears to have a density of bright galaxies less than 25% of the mean [236]. There is nothing like it in the simulations.

The percolation statistic was advocated by SHANDARIN [244] and his colleagues [179, 245] for comparing models with the observed galaxy distribution. The basic idea is to draw a sphere of radius r around each galaxy, and determine how the length of the longest chain of overlapping spheres depends on r. Unfortunately, percolation turns out not to be a very useful statistic: it is rather sensitive to irrelevant features such as sampling parameters and the depth of the galaxy red-shift survey, but not very sensitive to differences between neutrino and cold-DM simulations [246, 247].

Although there is no statistically significant alignment of ordinary galaxies with larger-scale structures [56], the brightest galaxies in rich clusters are often aligned with their parent clusters, and BINGGELI [53] found that nearby rich clusters tend to point toward each other. More precisely, BINGGELI found a correlation between the position angle of the major axis of a cluster and the position angles of the lines connecting its center to those of neighboring clusters. This correlation was found to be strongest for clusters separated by less than $20h^{-1}$ Mpc. DEKEL and collaborators [56, 248] have looked for similar effects in N-body simulations. They found that there are none in Poisson simulations (*i.e.* simulations that start with randomly distributed mass points), and that in neutrino simulations the correlations are comparable to but a little stronger than those BINGGELI found. Correlations similar to Binggeli's were obtained in Dekel's « AI » simulations (which have Poisson noise superimposed on a neutrino-type power spectrum), and in DEFW's cold-DM simulations only for $\Omega h < 0.2$.

Very recently, STRUBLE and PEEBLES [54] have repeated and extended Binggeli's observations with a sample of 237 clusters, more than five times the number (44) in Binggeli's sample. They find that there is at most marginal alignment, and conclude that BINGGELI most probably hit a statistical fluke. They also remark that in the light of their data, the success of Dekel *et al.* [248] in reproducing the alignments in Binggeli's data sample with neutrino-type simulations is evidence against such models.

Binggeli's statistic does discriminate between different theoretical models, and allows ready comparison with data. Related statistics are being developed by DEKEL, VISHNIAC and others (personal communications). It will be very interesting to see the results of further analysis of models and comparison with data.

4'4. *Summary and prospect.* – I have a hard time keeping all this information straight, so I imagine that you might also. In order to help, I have prepared a « Consumer's Report on Dark Matter », fig. 4.6 (following the notation of the *Consumer Reports* magazine, which is known for its authoritative reviews of American consumer goods). I have tried to be reasonably objective in rating how well each of the three models discussed here—hot, warm and cold DM— accords with each category of data considered.

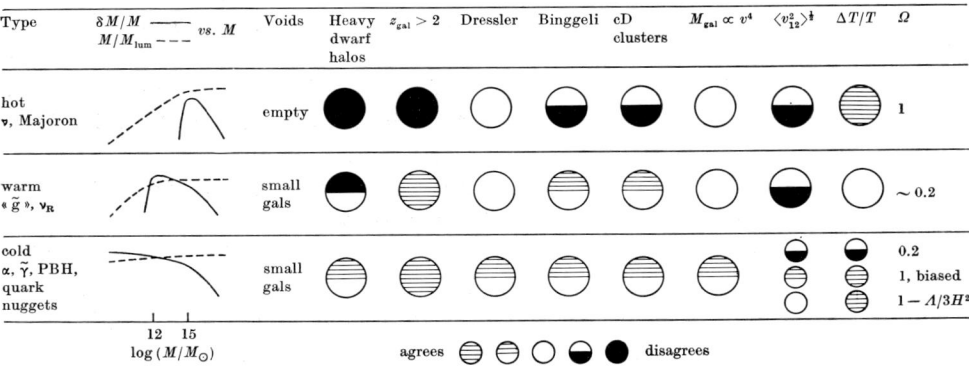

Fig. 4.6. – Consumer's Report on Dark Matter.

The first column of fig. 4.6 lists the three sorts of dark matter, and candidates for each ($\tilde{g}$ = gravitino, $\tilde{\gamma}$ = photino, PBH = primordial black holes). The next two columns summarize theoretical expectations regarding the fluctuation spectrum (solid line), the dependence of the ratio of total to luminous mass on size and the contents of voids. The evidence on M/M_{lum} summarized in fig. 1.11 is evidently most consistent with cold DM. Because free streaming damps out all but supercluster-size structures in the hot-DM model and galaxies can form only after the pancake collapse of these protosuperclusters, the voids (*i.e.* the regions between the pancakes) are expected to contain essentially no galaxies. But since galaxies form before large-scale structure in the warm- and cold-DM models, voids are expected to contain at least some small galaxies. Preliminary, indirect evidence favoring the latter expectation has recently been published [249]: absorption lines corresponding to triply ionized silicon and carbon have been detected at the red-shifts of the Perseus-Pisces and Boötes voids in the ultraviolet spectra of background quasars. In the hot-DM picture, any gas in the voids should be of primordial composition.

According to the « Consumer's Report », the two most severe difficulties of the hot-DM model are the phase-space constraint indicating that the heavy halos that are apparently associated with dwarf spheroidal galaxies cannot be made of light neutrinos, and the mismatch between the recent ($z < 2$) formation of superclusters indicated by density and velocity data and the early formation of galaxies and quasars. I have just discussed the difficulty for hot DM posed by the new data of Struble and Peebles on Binggeli's statistic; and I mentioned earlier in this lecture the indications that it is hard to form cD galaxies by mergers in small clusters unless the protogalaxies are assumed to have individual heavy halos, as expected for cold or warm but not hot DM. Peculiar velocities are too large in all $\Omega = 1$ models unless galaxy formation is « biased »; and although peculiar velocities are smaller in open cold models, their distribution does not agree in detail with observations. Yet another difficulty with hot DM is that the dense clumps that form in this model cannot be correlated with any known population of objects [250]. Even if galaxy formation were for some reason suppressed in the densest regions, they would provide a highly visible population of X-ray sources.

The cold-dark-matter model can be credited with several successes in explaining galaxy and cluster formation [26]. It predicts roughly the observed mass range of galaxies, the dissipational nature of galaxy collapse and dissipationless galactic halos and clusters with the observed Faber-Jackson and Tully-Fisher relations for galaxies. In addition, it may also provide natural explanations for galaxy-environment correlations and for the differences in angular momenta between elliptical and spiral galaxies.

If the cold dark matter is distributed like bright galaxies, then $\Omega \approx 0.2$. However, this version of the cold-DM model is in serious conflict with the observational limit on small-angle $\Delta T/T$ fluctuations, and also with globular-cluster age estimates since $h \approx 1$ for $\Delta T/T$ fitting the galaxy autocorrelation function. The $\Delta T/T$ problem is removed and the age problem is ameliorated, but only a little, in the zero-curvature version of this model, with cosmological constant $\Lambda = 3H^2(1 - \Omega)$. Perhaps the best overall fit to the data is obtained in the $\Omega = 1$ cold-DM model with biased galaxy formation. This version may have more trouble than the open-universe ones in accounting for the distribution of voids, however.

The cold-DM models' possible difficulty in producing voids like that in Boötes and in explaining the enhanced clustering of rich clusters both suggest that there is still an important physical ingredient missing in the cold-DM scheme—maybe a feature in the fluctuation spectrum on a « superpancake » scale of order $(100 \div 150) h^{-1}$ Mpc, as DEKEL has suggested [56, 251]. Perhaps the least ugly way of obtaining this is from the added growth of fluctuations exceeding the photon-baryon Jeans mass in a universe with adiabatic primordial fluctuations and roughly equal amounts of cold dark matter (to preserve the galaxy-size fluctuations, as usual) and baryonic matter (mostly dark now—perhaps « jupiters »): $\Omega_b \approx \Omega_{CDM} \approx 0.1$.

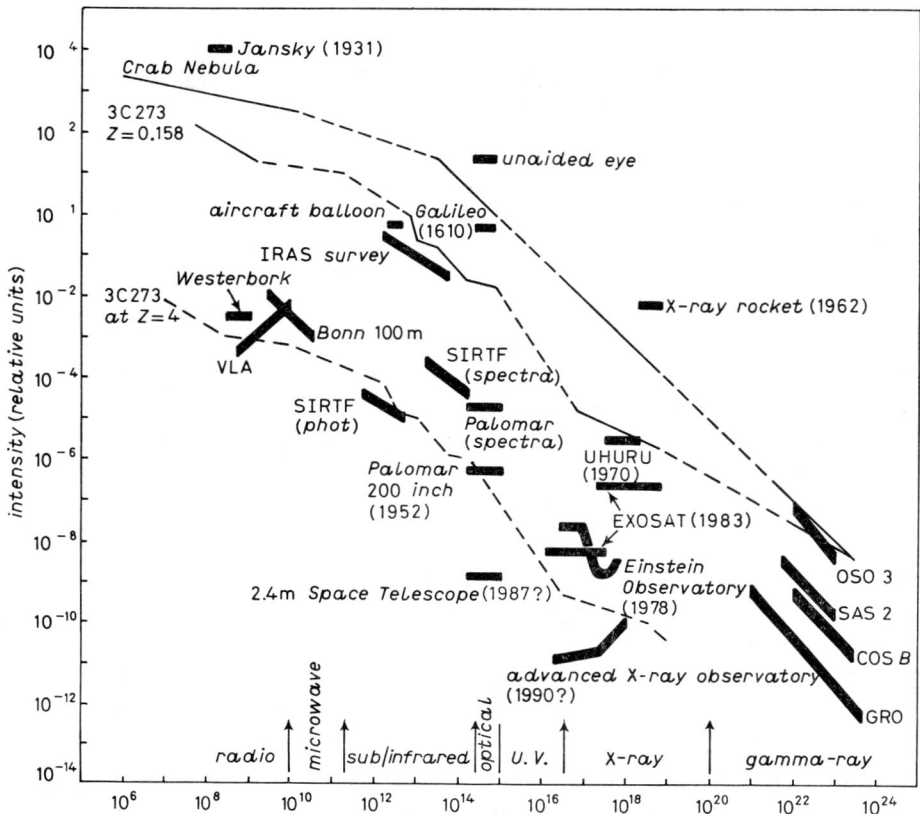

Fig. 4.7. – Relative sensitivities of astronomical instruments. (Figure courtesy of the Space Telescope Science Institute.)

As always, there is work for theorists. But the special excitement of the present era in astrophysics is that there is also plenty of relevant observational data, with the prospect of lots more coming soon. Figure 4.7 summarizes the relative sensitivities of a number of existing and planned astronomical instruments. Just to give one example to illustrate the sorts of new data that will soon be available, consider the observation of ultraviolet Lyman α absorption lines with the Space Telescope. As I explained in connection with fig. 2.9, some of the Ly α absorption lines seen in quasar spectra are apparently caused by clouds of neutral hydrogen distributed along the line of sight from the quasar to us. DEKEL [252] has argued that the absence of autocorrelation in the Ly α absorption line « forest » of individual quasars [253] and of cross-correlation in Ly α forests of two quasars separated by $\sim 1h^{-1}$ Mpc perpendicular to the line of sight [254] indicates that the formation of superclusters did not begin until $z \lesssim 2$. (This conclusion also follows from other lines of argument, as I have discussed.) OORT [255] argues on the contrary that the observation [256] of two pairs of quasars at $z = 2.83$ and 2.85 and $z = 3.14$ and 3.16 suggests that superclustering has existed since $z \approx 3$, and that the lack of cross-correlation in the Ly α forests of the neighboring quasars can be explained by a filamentary distribution of hydrogen clouds in superclusters. If DEKEL is right, then ultraviolet observations of the Ly α forest at red-shifts $z \leqslant 1.5$ will show the onset of superclustering. (It is, of course, also possible that the Ly α clouds are not associated with superclusters, but instead occur only in voids.) Similar observations can also determine the composition of the gas in voids [249]. Such data would obviously be an important constraint on cosmological theories for galaxy formation and large-scale structure.

Remarkably enough, such data may also shed important light on the interactions of elementary particles on very small scales. Figure 4.8 is redrawn from a sketch by Shelley GLASHOW which was reproduced in the *New York Times Magazine* [257]. GLASHOW uses the snake eating its tail—the uroboros, an ancient symbol associated with creation myths [258]—to represent the idea that gravity may determine the structure of the Universe on both the

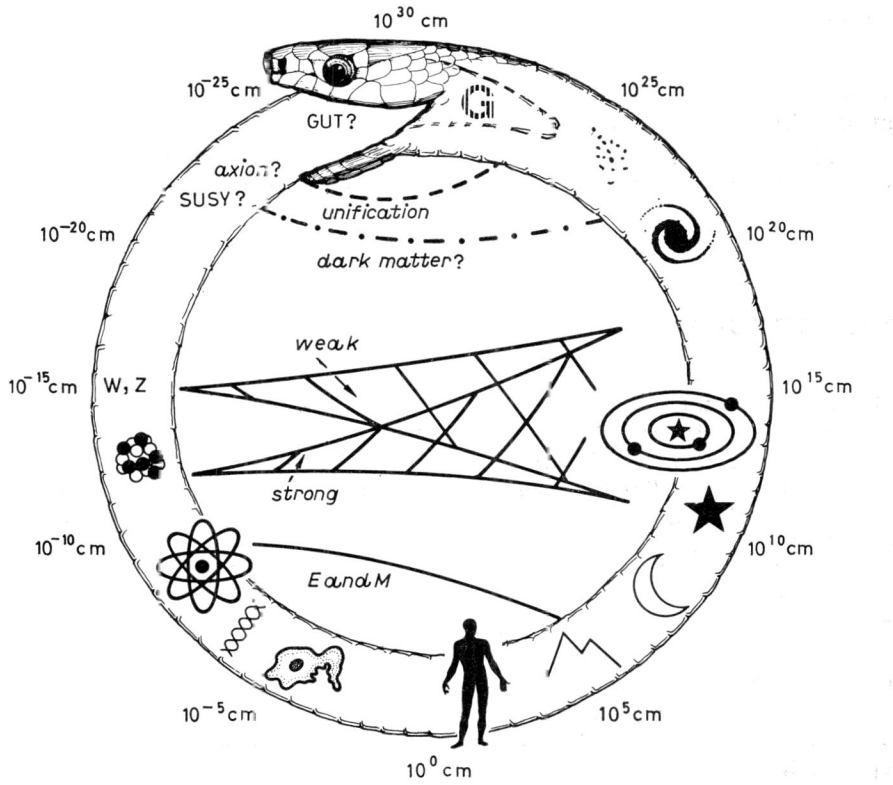

Fig. 4.8. – Cosmological uroboros. (Adapted from ref. [32].)

largest and smallest scales. But there is another fascinating aspect to this figure. There are left-right connections across it: medium small to medium large, even smaller to even larger, and so on. Not only does electromagnetism determine structure from atoms to mountains [259], and the strong and weak interactions control the properties and compositions of stars and solar systems. The dark matter, which may hold the key to understanding the origin of galaxies, clusters and superclusters, may itself reflect fundamental physics below the weak-interaction scale. And if cosmic inflation is to be believed, cosmological structure on scales even larger than the present horizon arose from interactions on the grand-unification scale.

* * *

These lectures are based on research done in collaboration with G. BLUMENTHAL, S. FABER and M. REES—all of whom I would like to thank for their efforts to teach me astrophysics, and also for allowing me to use material from our joint publications, especially ref. [26, 135, 203, 204]. In preparing these lectures, I have also benefitted from G. Blumenthal's unpublished notes for his course on Cosmology at the University of California, Santa Cruz. I thank K. GRIEST, H. HABER and D. HARTMANN for the careful reading of this manuscript and for several helpful suggestions. I am very grateful to S. DRELL for the hospitality of SLAC during 1984, and to N. CABIBBO for inviting me to give these lectures at Varenna—surely one of the loveliest places in the civilized world. Finally, I acknowledge partial support from the National Science Foundation under grant PHY-81-15541.

REFERENCES

[1] J. HEIDMANN: in *Cosmology, History, and Theology*, edited by W. YOURGRAU and A. D. BRECK (Plenum, New York, N. Y., 1977), p. 39.
[2] P. W. HODGE: *Galaxies and Cosmology* (McGraw-Hill, New York, N. Y., 1966).
[3] R. SANCISI: in *The Structure and Evolution of Normal Galaxies*, edited by S. M. FALL and D. LYNDEN-BELL (Cambridge University Press, London, 1981), p. 149.
[4] S. M. FALL: in *The Structure and Evolution of Normal Galaxies*, edited by S. M. FALL and D. LYNDEN-BELL (Cambridge University Press, London, 1981), p. 149.
[5] S. M. FABER and J. S. GALLAGHER: *Annu. Rev. Astron. Astrophys.*, **17**, 135 (1979).
[6] V. C. RUBIN: *Science*, **220**, 1339 (1983).
[7] A. BOSMA: *Astron. J.*, **86**, 1825 (1981).
[8] V. C. RUBIN, W. K. FORD and N. THONNARD: *Astrophys. J.*, **238**, 471 (1980).
[9] F. ZWICKY: *Helv. Phys. Acta*, **6**, 110 (1933).
[10] J. P. OSTRIKER and P. J. E. PEEBLES: *Astrophys. J.*, **186**, 467 (1973); J. P. OSTRIKER, P. J. E. PEEBLES and A. YAHIL: *Astrophys. J. Lett.*, **193**, L1 (1974).

[11] L. Blitz, M. Fich and S. Kulkarni: *Science*, **220**, 1233 (1983), and references therein.
[12] J. Einasto, A. Kaasik and E. Saar: *Nature (London)*, **250**, 309 (1974); D. Lynden-Bell: in *Astrophysical Cosmology: Proceedings of the Study Week on Cosmology and Fundamental Physics*, edited by H. A Brück, G. V. Coyne and M. S. Longair (Vatican, 1982), p. 85.
[13] J. A. Tyson, F. Valdes, J. F. Jarvis and A. P. Mills jr.: *Astrophys. J. Lett.*, **281**, L59 (1984).
[14] C. Frenk and S. D. M. White: *Mon. Not. R. Astron. Soc.*, **193**, 295 (1980).
[15] K. A. Innanen, W. E. Harris and R. F. Webbink: *Astron. J.*, **88**, 338 (1983).
[16] F. D. A. Hartwick and W. L. W. Sargent: *Astrophys. J*, **221**, 512 (1978).
[17] T. Murai and M. Fujimoto: *Publ. Astron. Soc. Jpn.*, **32**, 581 (1980).
[18] D. N. C. Lin and D. Lynden-Bell: *Mon. Not. R. Astron. Soc.*, **198**, 707 (1982).
[19] M. R. S. Hawkins: *Nature (London)*, **303**, 406 (1983).
[20] V. C. Rubin: *Sci. Am.*, **246**(6), 96 (June 1983).
[21] G. Efstathiou and J. Silk: *Fundam. Cosmic Phys.*, **9**, 1 (1983).
[22] J. S. Gallagher and D. A. Hunter: *Annu. Rev. Astron. Astrophys.*, **22**, 37 (1984).
[23] S. M. Faber and R. E. Jackson: *Astrophys. J.*, **204**, 668 (1976).
[24] M. Aaronson, J. Huchra and J. Mould: *Astrophys. J.*, **229**, 1 (1979).
[25] D. Burstein, V. C. Rubin, N. Thonnard and W. K. Ford: *Astrophys. J.*, **253**, 539 (1982).
[26] G. R. Blumenthal, S. M. Faber, J. R. Primack and M. J. Rees: *Nature (London)*, **311**, 517 (1984); (E) **313**, 72 (1985) (BFPR).
[27] S. Faber: in *Astrophysical Cosmology: Proceedings of the Study Week on Cosmology and Fundamental Physics*, edited by H. A. Brück, G. V. Coyne and M. S. Longair (Vatican, 1982), p. 191.
[28] J. E. Gunn: *Philos. Trans. R. Soc. London, Ser. A*, **296**, 313 (1980).
[29] G. Illingworth: in *The Structure and Evolution of Normal Galaxies*, edited by S. M. Fall and D. Lynden-Bell (Cambridge University Press, London, 1981), p. 27.
[30] W. Forman, C. Jones and W. Tucker: *Astrophys. J.*, **293**, 102 (1985).
[31] For additional recent references, see J. R. Primack: in *Second ESO-CERN Symposium on Cosmology, Astronomy and Fundamental Physics*, edited by G. Setti and L. Van Hove (in press).
[32] J. R. Primack and G. R. Blumenthal: in *Fourth Workshop on Grand Unification*, edited by H. A. Weldon, P. Langacker and P. Steinhardt (Birkhäuser, Boston, Mass., 1983), p. 256.
[33] G. O. Abell: *Astrophys. J. Suppl.*, **3**, 211 (1958).
[34] F. Zwicky, E. Herzog, P. Wild, M. Karpowicz and C. T. Kowal: *Catalogue of Galaxies and Clusters of Galaxies*, 6 volumes (Caltech, Pasadena, Cal., 1961-68).
[35] M. J. Geller and J. P. Huchra: *Astrophys. J. Suppl.*, **52**, 61 (1983).
[36] J. P. Huchra, M. Davis, D. W. Latham and J. Tonry: *Astrophys. J. Suppl.*, **52**, 89 (1983).
[37] W. Forman and C. Jones: *Annu. Rev. Astron. Astrophys.*, **20**, 547 (1982).
[38] N. Bahcall: *Annu. Rev. Astron. Astrophys.*, **15**, 505 (1977).
[39] M. J. Geller and T. C. Beers: *Publ. Astron. Soc. Pac.*, **94**, 421 (1982).
[40] J. H. Oort: *Annu. Rev. Astron. Astrophys.*, **21**, 373 (1983).
[41] M. J. Geller: in *Clusters and Groups of Galaxies*, edited by F. Mardirossian, G. Giuricin and M. Mezzetti (Reidel, Dordrecht, 1984), p. 353.
[42] A. Dressler: *Astrophys. J.*, **236**, 351 (1980).
[43] M. Postman and M. J. Geller: *Astrophys. J.*, **281**, 95 (1984).
[44] H. R. Butcher and A. Oemler: *Astrophys. J.*, **226**, 559 (1978).

[45] W. H. PRESS and M. DAVIS: *Astrophys. J.*, **259**, 449 (1982).
[46] S. D. M. WHITE: in *Morphology and Dynamics of Galaxies*, edited by L. MARTINET and M. MAYOR (Geneva Observatory, 1982), p. 289.
[47] T. X. THUAN and W. ROMANISHIN: *Astrophys. J.*, **248**, 439 (1981); A. CAVALIERE, P. SANTANGELO, G. TARQUINI and N. VITTORIO: in *Clusters and Groups of Galaxies*, edited by F. MARDIROSSIAN, G. GIURICIN and M. MEZZETTI (Reidel, Dordrecht, 1984), p. 499.
[48] A. DRESSLER: *Annu. Rev. Astron. Astrophys.*, **22**, 185 (1984).
[49] S. A. GREGORY and L. A. THOMPSON: *Sci. Am.*, **246**(3), 106 (March 1982).
[50] R. B. TULLY: *Astrophys. J.*, **257**, 389 (1982).
[51] S. A. GREGORY, L. A. THOMPSON and W. G. TIFFT: *Astrophys. J.*, **243**, 411 (1981).
[52] R. F. KIRSHNER, A. OEMLER, P. L. SCHECHTER and S. A. SHECTMAN: *Astrophys. J. Lett.*, **248**, L57 (1981); in *Evolution of the Universe and Its Present Structure*, IAU Symposium No. 104, edited by G. O. ABELL and C. GHINCARINI (Reidel, Dordrecht, 1983), p. 197.
[53] B. BINGGELI: *Astron. Astrophys.*, **107**, 338 (1982).
[54] M. F. STRUBLE and P. J. E. PEEBLES: *Astron. J.*, **90**, 582 (1985).
[55] A. DEKEL, M. J. WEST and S. J. AARSETH: *Astrophys. J.*, **279**, 1 (1984).
[56] A. DEKEL: in *Eighth Johns Hopkins Workshop on Current Problems in Particle Theory: Particles and Gravity*, edited by G. DOMOKOS and S. KOVESI-DOMOKOS (World Scientific Publ., Singapore, 1984), p. 191.
[57] M. DAVIS and P. J. E. PEEBLES: *Annu. Rev. Astron. Astrophys.*, **21**, 109 (1983).
[58] L. HART and R. D. DAVIES: *Nature (London)*, **297**, 191 (1982).
[59] YA. B. ZEL'DOVICH, J. EINASTO and S. F. SHANDARIN: *Nature (London)*, **300**, 407 (1982).
[60] S. F. SHANDARIN, A. G. DOROSHKEVICH and YA. B. ZEL'DOVICH: *Sov. Phys. Usp.*, **26**, 46 (1983).
[61] P. J. E. PEEBLES: *Science*, **224**, 1385 (1984).
[62] C. M. WILL: *Theory and Experiment in Gravitational Physics* (Cambridge University Press, London, 1981).
[63] W. RINDLER: *Essential Relativity: Special, General, and Cosmological*, revised second edition (Springer-Verlag, New York, N. Y., 1977).
[64] M. ROWAN-ROBINSON: *Cosmology*, second edition (Oxford, 1981).
[65] S. WEINBERG: *Gravitation and Cosmology: Principles and Applications of the General Theory of Relativity* (Wiley, New York, N. Y., 1972).
[66] C. W. MISNER, K. S. THORNE and J. A. WHEELER: *Gravitation* (Freeman, San Francisco, Cal., 1973).
[67] E. R. HARRISON: *Cosmology: the Science of the Universe* (Cambridge, 1981).
[68] See, *e.g.*, W. RINDLER: *Essential Relativity: Special, General, and Cosmological*, revised second edition (Springer-Verlag, New York, N. Y., 1977), § 9.9.
[69] A. SANDAGE and G. A. TAMMANN: in *First ESO-CERN Symposium: Large-Scale Structure of the Universe, Cosmology and Fundamental Physics*, edited by G. SETTI and L. VAN HOVE (CERN, Geneva, 1984), p. 127.
[70] G. DE VAUCOULEURS: in *Tenth Texas Symposium on Relativistic Astrophysics*, edited by R. RAMATY and F. C. JONES, *Ann. N. Y. Acad. Sci.*, **375**, 90 (1981); in *Clusters and Groups of Galaxies*, edited by F. MARDIROSSIAN, G. GIURICIN and M. MEZZETTI (Reidel, Dordrecht, 1984), p. 29.
[71] W. D. ARNETT, D. BRANCH and J. C. WHEELER: *Nature (London)*, in press.
[72] P. J. E. PEEBLES: *The Large-Scale Structure of the Universe* (Princeton University Press, Princeton, N. J., 1980).
[73] M. DAVIS and P. J. E. PEEBLES: *Astrophys. J.*, **267**, 465 (1983).

[74] A. J. BEAN, G. EFSTATHIOU, R. S. ELLIS, B. A. PETERSON and T. SHANKS: *Mon. Not. R. Astron. Soc.*, **205**, 605 (1983).
[75] P. J. E. PEEBLES and E. J. GROTH: *Astrophys. J.*, **196**, 1 (1975).
[76] M. G. HAUSER and P. J. E. PEEBLES: *Astrophys. J.*, **185**, 757 (1973).
[77] N. A. BAHCALL and R. M. SONEIRA: *Astrophys. J.*, **270**, 20 (1983).
[78] A. A. KLYPIN and A. I. KOPYLOV: *Sov. Astron. Lett.*, **9**, 41 (1984).
[79] A. DEKEL and S. J. AARSETH: *Astrophys. J.*, **283**, 1 (1984).
[80] P. J. E. PEEBLES: *The Large-Scale Structure of the Universe* (Princeton University Press, Princeton, N. J., 1980), eq. (75.10).
[81] P. J. E. PEEBLES: in *Proceedings of the Tenth Texas Symposium on Relativistic Astrophysics*, edited by R. RAMATY and F. C. JONES, *Ann. N. Y. Acad. Sci.*, **375**, 157 (1981).
[82] Cf. also A. DRESSLER: *Astrophys. J.*, **281**, 512 (1984).
[83] G. L. HOFFMAN and E. E. SALPETER: *Astrophys. J.*, **263**, 485 (1982).
[84] R. J. HARMS, H. C. FORD, R. CIARDULLO and F. BARTKO: in *Tenth Texas Symposium on Relativistic Astrophysics*, edited by R. RAMATY and F. C. JONES, *Ann. N. Y. Acad. Sci.*, **375**, 178 (1981).
[85] R. J. HARMS, H. C. FORD and R. CIARDULLO: in *Early Evolution of the Universe and Its Present Structure*, IAU Symposium No. 104, edited by G. O. ABELL and G. CHINCARINI (Reidel, Dordrecht, 1983), p. 285.
[86] J. YANG, M. S. TURNER, G. STEIGMAN, D N. SCHRAMM and K. A. OLIVE: *Astrophys. J.*, **281**, 493 (1984).
[87] R. G. KRON: in *Spectral Evolution of Galaxies*, edited by P. M. GONDHALEKAR (Rutherford Appleton Laboratory, Chilton, Didcot, Oxon, 1984), p. 190.
[88] E. J. WAMPLER, C. M. GASKELL, W. L. BURKE and J. A. BALDWIN: *Astrophys. J.*, **276**, 403 (1984).
[89] H. S. MURDOCH: *Mon. Not. R. Astron. Soc.*, **202**, 987 (1983).
[90] W. H. PRESS and E. T. VISHNIAC: *Astrophys. J.*, **239**, 1 (1980).
[91] J. M. BARDEEN: *Phys. Rev. D*, **22**, 1882 (1980).
[92] P. J. E. PEEBLES: *The Large-Scale Structure of the Universe* (Princeton University Press, Princeton, N. J., 1980), § 11.
[93] G. R. BLUMENTHAL and J. R. PRIMACK: preprint SLAC-PUB-3388.
[94] P. J. E. PEEBLES: *Astrophys. J.*, **75**, 13 (1970).
[95] D. LYNDEN-BELL: *Mon. Not. R. Astron. Soc.*, **136**, 101 (1967).
[96] F. H. SHU: *Astrophys. J.*, **225**, 83 (1978).
[97] S. CHANDRASEKHAR: *Principles of Stellar Dynamics* (Dover, New York, N. Y., 1960).
[98] J. E. GUNN: in *Astrophysical Cosmology: Proceedings of the Study Week on Cosmology and Fundamental Physics*, edited by H. A. BRÜCK, G. V. COYNE and M. S. LONGAIR (Vatican, 1982), p. 233.
[99] See, *e.g.*, P. J. E. PEEBLES: *The Large-Scale Structure of the Universe* (Princeton University Press, Princeton, N. J., 1980), § 20-21 and the references cited there.
[100] YA. B. ZEL'DOVICH: *Astron. Astrophys.*, **5**, 34 (1970).
[101] YA. B. ZEL'DOVICH: in *The Large Scale Structure of the Universe*, IAU Symposium No. 78, edited by M. S. LONGAIR and J. EINASTO (Reidel, Dordrecht, 1978), p. 409.
[102] A. DEKEL: *Astrophys. J.*, **264**, 373 (1983).
[103] S. J. AARSETH and J. BINNEY: *Mon. Not. R. Astron. Soc.*, **185**, 227 (1978).
[104] A. GUTH: *Phys. Rev. D*, **23**, 347 (1981).
[105] S. W. HAWKING and G. F. R. ELLIS: *The Large Scale Structure of Space-Time* (Cambridge University Press, London, 1973).
[106] E. B. GLINER and I. G. DYMNIKOVA: *Sov. Astron. Lett.*, **1**, 93 (1975).

[107] R. BROUT, F. ENGLERT and E. GUNZIG: *Gen. Rel. Grav.*, **10**, 1 (1979); R. BROUT, F. ENGLERT and P. SPINDEL: *Phys. Rev. Lett.*, **43**, 417, (E)890 (1979); R. BROUT, F. ENGLERT, J.-M. FRÈRE, E. GUNZIG, P. NARDONE, C. TRUFFIN and PH. SPINDEL: *Nucl. Phys. B*, **170**, 228 (1980). This work is reviewed by F. ENGLERT: in *Physical Cosmology*, Les Houches Session XXXII (North-Holland Publ. Co., Amsterdam, 1980), and in the 1981 Cargèse Lectures, edited by M. LEVY.
[108] D. KAZANAS: *Astrophys. J. Lett.*, **241**, L59 (1980).
[109] A. H. GUTH and P. J. STEINHARDT: *Sci. Am.*, **250**(5), 116 (May 1984).
[110] A. H. GUTH and E. WEINBERG: *Phys. Rev. D*, **23**, 826 (1981); *Nucl. Phys. B*, **212**, 321 (1983).
[111] A. D. LINDE: *Phys. Lett. B*, **108**, 389 (1982); A. ALBRECHT and P. J. STEINHARDT: *Phys. Rev. Lett.*, **48**, 1220 (1982).
[112] K. A. OLIVE: in *Grand Unification with and without Supersymmetry, and Cosmological Implications*, International School for Advanced Studies Lecture Series No. 2 (World Scientific Publ., Singapore, 1984).
[113] Recent reviews include K. A. OLIVE: in *Grand Unification with and without Supersymmetry, and Cosmological Implications*, International School for Advanced Studies Lecture Series No. 2 (World Scientific Publ., Singapore, 1984), p. 347. M. S. TURNER: in *NATO Advanced Study Institute on Quarks and Leptons*, Max Planck Institute, Munich, September 1983 (Fermilab preprint Conf-84/60-A, to be published); P. J. STEINHARDT: in *Fourth Workshop on Grand Unification*, edited by H. A. WELDON, P. LANGACKER and P. STEINHARDT (Birkhäuser, Boston, Mass., 1983); R. BRANDENBERGER: *Rev. Mod. Phys.*, **57**, 1 (1985).
[114] A. H. GUTH: in *The Very Early Universe*, edited by G. W. GIBBONS, S. HAWKING and S. SIKLOS (Cambridge University Press, London, 1983), p. 171.
[115] L. ABBOTT: *Phys. Lett. B*, **150**, 427 (1985).
[116] A. H. GUTH and S.-Y. PI: *Phys. Rev. Lett.*, **49**, 1110 (1982). Similar calculations were done by S. HAWKING: *Phys. Lett. B*, **115**, 295 (1982); A. STAROBINSKII: *Phys. Lett. B*, **117**, 175 (1982); J. BARDEEN, P. J. STEINHARDT and M. S. TURNER: *Phys. Rev. D*, **28**, 679 (1983).
[117] E. R. HARRISON: *Phys. Rev. D*, **1**, 2726 (1970).
[118] YA. B. ZEL'DOVICH: *Mon. Not. R. Astron. Soc.*, **160**, 1P (1972).
[119] B. J. CARR: *Astrophys. J.*, **201**, 1 (1975); J. D. BARROW and B. J. CARR: *Mon. Not. R. Astron. Soc.*, **182**, 537 (1978).
[120] B. J. CARR: *Comments Astrophys. Space Phys.*, **7**, 161 (1978).
[121] P. J. STEINHARDT and M. S. TURNER: *Phys. Rev. D*, **29**, 2162 (1984).
[122] For example, Q. SHAFI and A. VILENKIN: *Phys. Rev. Lett.*, **52**, 691 (1984).
[123] YA. B. ZEL'DOVICH: *Mon. Not. R. Astron. Soc.*, **192**, 663 (1980); A. VILENKIN: *Phys. Rev. D*, **24**, 2082 (1981). The recent work on cosmic strings has been reviewed by A. VILENKIN: *Phys. Rep.*, **121**, 263 (1985).
[124] L. F. ABBOTT and M. B. WISE: *Astrophys. J. Lett.*, **282**, L47 (1984).
[125] J. TOHLINE: in *Eleventh Texas Conference on Relativistic Astrophysics*, edited by D. E. EVANS, *Ann. N. Y. Acad. Sci.*, **422**, 390 (1984); in *Internal Kinematics and Dynamics of Galaxies*, IAU Symposium No. 100, edited by E. ATHANASSOULA (Reidel, Dordrecht, 1983), p. 205.
[126] M. MILGROM: *Astrophys. J.*, **270**, 365, 371, 384 (1983).
[127] A. DRESSLER and M. LECAR: *Astrophys. J. Lett.* (submitted).
[128] J. BECKENSTEIN and M. MILGROM: *Astrophys. J.*, **286**, 7 (1984); J. E. FELTON: *Astrophys. J.*, **286**, 3 (1984).
[129] S. WEINBERG: *Gravitation and Cosmology: Principles and Applications of the General Theory of Relativity* (Wiley, New York, N. Y., 1972); *Phys. Rev.*, **138**, 988 (1965).

[130] S. WEINBERG: *The First Three Minutes* (Basic Books, New York, N. Y., 1977).
[131] G. STEIGMAN: *Annu. Rev. Astron. Astrophys.*, **14**, 339 (1976).
[132] A. D. DOLGOV and YA. B. ZEL'DOVICH: *Rev. Mod. Phys.*, **53**, 1 (1981).
[133] YA. B. ZEL'DOVICH and I. D. NOVIKOV: *The Structure and Evolution of the Universe* (University of Chicago Press, Chicago, Ill., 1983).
[134] E. WITTEN: *Phys. Rev. D*, **30**, 272 (1984).
[135] J. R. PRIMACK and G. R. BLUMENTHAL: in *Formation and Evolution of Galaxies and Large Structures in the Universe*, edited by J. AUDOUZE and J. TRAN THANH VAN (Reidel, Dordrecht, 1983), p. 163; in *Fourth Workshop on Grand Unification*, edited by H. A. WELDON, P. LANGACKER and P. STEINHARDT (Birkhäuser, Boston, Mass., 1983), p. 256.
[136] D. J. HEGYI and K. A. OLIVE: *Phys. Lett. B*, **126**, 28 (1983).
[137] M. SPITE and F. SPITE: *Astron. Astrophys.*, **115**, 357 (1982).
[138] See, e.g., P. J. E. PEEBLES: *The Large-Scale Structure of the Universe* (Princeton University Press, Princeton, N. J., 1980), § 16.
[139] G. R. BLUMENTHAL, H. PAGELS and J. R. PRIMACK: *Nature (London)*, **299**, 37 (1982).
[140] E. W. KOLB and M. S. TURNER: *Annu. Rev. Nucl. Part. Sci.*, **33**, 645 (1983).
[141] P. J. E. PEEBLES and J. T. YU: *Astrophys. J.*, **162**, 815 (1970); W. H. PRESS and E. VISHNIAC: *Astrophys. J.*, **236**, 323 (1981); M. L. WILSON and J. SILK: *Astrophys. J.*, **243**, 14 (1981); P. J. E. PEEBLES: *Astrophys. J.*, **248**, 885 (1981).
[142] YA. B. ZEL'DOVICH: in *The Large Scale Structure of the Universe*, IAU Symposium No. 78, edited by M. S. LONGAIR and J. EINASTO (Reidel, Dordrecht, 1978), p. 409. For reconsiderations of the adiabatic scheme with neutrinos, see the following reference.
[143] A. G. DOROSHKEVICH, M. YU. KHLOPOV, R. A. SUNYAEV, A. S. SZALAY and YA. B. ZEL'DOVICH: in *Tenth Texas Symposium on Relativistic Astrophysics*, edited by R. RAMATY and F. C. JONES, *Ann. N. Y. Acad. Sci.*, **375**, 32 (1981); H. SATO: in *Tenth Texas Symposium on Relativistic Astrophysics*, edited by R. RAMATY and F. C. JONES, *Ann. N. Y. Acad. Sci.*, **375**, 43 (1981); S. F. SHANDARIN, A. G. DOROSHKEVICH and YA. B. ZEL'DOVICH: *Sov. Phys. Usp.*, **26**, 46 (1983).
[144] R. B. PARTRIDGE: *Astrophys. J.*, **235**, 681 (1980); J. M. USON and D. T. WILKINSON: *Astrophys. J. Lett.*, **277**, L1 (1984); *Astrophys. J.*, **283**, 471 (1984).
[145] J. D. BARROW and M. S. TURNER: *Nature (London)*, **291**, 469 (1981); J. R. BOND, E. W. KOLB and J. SILK: *Astrophys. J.*, **255**, 341 (1982).
[146] M. J. REES: in *The Very Early Universe*, edited by G. GIBBONS, S. HAWKING and S. SIKLOV (Cambridge University Press, London, 1983), p. 29, and references therein.
[147] I thank Dick BOND for suggesting this terminology to me at the 1983 Moriond Conference, where I used it in my talk (ref. [135]). George BLUMENTHAL and I had thought of this classification independently, but not the names.
[148] H. R. PAGELS and J. R. PRIMACK: *Phys. Rev. Lett.*, **48**, 223 (1982).
[149] B. J. CARR: *Comments Astrophys.*, **7**, 161 (1978).
[150] R. CANIZARES: *Astrophys. J.*, **263**, 508 (1982).
[151] C. LACEY: in *Formation and Evolution of Galaxies and Large Structures in the Universe*, edited by J. AUDOUZE and J. TRAN THANH VAN (Reidel, Dordrecht, 1983), p. 351.
[152] D. N. SCHRAMM and G. STEIGMAN: *Astrophys. J.*, **241**, 1 (1981).
[153] G. BLUMENTHAL, A. DEKEL and J. PRIMACK: in preparation.
[154] M. DAVIS, M. LECAR, C. PRYOR and E. WITTEN: *Astrophys. J.*, **250**, 423 (1981).
[155] P. HUT and S. WHITE: *Nature (London)*, **310**, 637 (1984).
[156] S. BLUDMAN and Y. HOFFMAN: *Phys. Rev. Lett.*, **52**, 2087 (1984).

[157] A. G. DOROSHKEVICH and M. YU. KHLOPOV: *Sov. Astron. Lett.*, **11** (in press).
[158] M. TURNER, G. STEIGMAN and L. M. KRAUSS: *Phys. Rev. Lett.*, **52**, 2090 (1984); M. S. TURNER: *Phys. Rev. D*, **31**, 1212 (1985); K. A. OLIVE, D. SECKEL and E. VISHNIAC: *Astrophys. J.*, **292**, 1 (1985).
[159] M. FUKUGITA and T. YANAGIDA: *Phys. Lett. B*, **144**, 386 (1984).
[160] G. GELMINI, D. N. SCHRAMM and J. W. F. VALLE: *Phys. Lett. B*, **146**, 311 (1984).
[161] S. S. GERSHTEIN and YA. B. ZEL'DOVICH: *JETP Lett.*, **4**, 174 (1966); R. COWSIK and J. MCCLELLAND: *Phys. Rev. Lett.*, **29**, 669 (1972); G. MARX and A. S. SZALAY: in *Neutrino '72*, Vol. **1** (Technoinform, Budapest, 1972), p. 123; A. S. SZALAY and G. MARX: *Astron. Astrophys.*, **49**, 437 (1976).
[162] V. A. LYUBIMOV, E. G. NOVIKOV, V. Z. NOZIK, E. F. TRETYAKOV and V. S. KOSIK: *Phys. Lett. B*, **94**, 266 (1980).
[163] J. R. BOND, G. EFSTATHIOU and J. SILK: *Phys. Rev. Lett.*, **45**, 1980 (1980).
[164] F. R. KLINKHAMER and C. A. NORMAN: *Astrophys. J. Lett.*, **243**, L1 (1981).
[165] Y. CHIKASHIGE, R. N. MOHAPATRA and R. D. PECCEI: *Phys. Rev. Lett.*, **45**, 1926 (1980); *Phys. Lett. B*, **98**, 265 (1981).
[166] G. B. GELMINI, S. NUSSINOV and M. RONCADELLI: *Nucl. Phys. B*, **209**, 157 (1982).
[167] V. LUBIMOV, S. BORIS, A. GOLUTVIN, L. LAPTIN, V. NAGOVIZIN, E. NOVIKOV, V. NOZIK, V. SOLOSHENKO, I. TICHOMIROV and E. TRETYAKOV: in *Proceedings of HEP 83*, edited by J. GUY and C. COSTAIN (Rutherford Laboratory, Oxford, 1983), p. 386.
[168] For example, F. BOEHM: in *Fourth Workshop on Grand Unification*, edited by H. A. WELDON, P. LANGACKER and P. STEINHARDT (Birkhäuser, Boston, Mass., 1983), p. 163; cf. also F. BOEHM and P. VOGEL: *Annu. Rev. Nucl. Part. Sci.*, **34**, 125 (1984).
[169] B. W. LEE and S. WEINBERG: *Phys. Rev. Lett.*, **39**, 165 (1977); M. I. VYSOTSKY, A. D. DOLGOV and YA. B. ZEL'DOVICH: *JETP Lett.*, **4**, 120 (1977); P. HUT: *Phys. Lett. B*, **69**, 85 (1977); K. SATO and H. KOBAYASHI: *Prog. Theor. Phys.*, **58**, 1775 (1977); J. E. GUNN, B. W. LEE, I. LERCHE, D. N. SCHRAMM and G. STEIGMAN: *Astrophys. J.*, **223**, 1015 (1978).
[170] P. LANGACKER, G. SEGRÈ and S. SONI: *Phys. Rev. D*, **26**, 3425 (1982); K. FREESE, E. W. KOLB and M. S. TURNER: *Phys. Rev. D*, **27**, 1689 (1983).
[171] G. S. BISNOVATYI-KOGAN and I. D. NOVIKOV: *Sov. Astron.*, **24**, 516 (1980).
[172] See sect. 2, above, and *The Very Early Universe*, edited by G. GIBBONS, S. HAWKING and S. SIKLOV (Cambridge University Press, London, 1983), and references therein.
[173] J. R. BOND and A. S. SZALAY: *Astrophys. J.*, **274**, 443 (1983).
[174] N. VITTORIO and J. SILK: *Astrophys. J. Lett.*, **285**, L39 (1984).
[175] J. R. BOND and G. EFSTATHIOU: *Astrophys. J. Lett.*, **285**, L45 (1984).
[176] A. MELOTT: *Mon. Not. R. Astron. Soc.*, **202**, 595 (1983); J. CENTRELLA and A. MELOTT: *Nature (London)*, **305**, 196 (1983), and references therein.
[177] A. A. KLYPIN and S. F. SHANDARIN: *Mon. Not. R. Astron. Soc.*, **204**, 891 (1983).
[178] C. FRENK, S. M. D. WHITE and M. DAVIS: *Astrophys. J.*, **271**, 417 (1983).
[179] YA. B. ZEL'DOVICH, J. EINASTO and S. F. SHANDARIN: *Nature (London)*, **300**, 407 (1982), and references therein.
[180] N. KAISER: *Astrophys. J. Lett.*, **273**, L17 (1983).
[181] S. FABER: in *First ESO-CERN Symposium: Large-Scale Structure of the Universe, Cosmology and Fundamental Physics*, edited by G. SETTI and L. VAN HOVE (CERN, Geneva, 1984), p. 187.
[182] J. R. BOND, J. CENTRELLA, A. S. SZALAY and J. R. WILSON: in *Formation and Evolution of Galaxies and Large Structures in the Universe*, edited by J. AUDOUZE

and J. Tran Thanh Van (Reidel, Dordrecht, 1983), p. 87; P. R. Shapiro, C. Struck-Marcell and A. L. Melott: *Astrophys. J.*, **275**, 413 (1983).
[183] J. R. Bond, A. S. Szalay and S. D. M. White: *Nature (London)*, **301**, 584 (1983).
[184] S. M. Faber and D. N. C. Lin: *Astrophys. J. Lett.*, **266**, L17 (1983).
[185] M. Aaronson: *Astrophys. J. Lett.*, **266**, L11 (1983); M. Aaronson and K. Cook: *Bull. Am. Astron. Soc.*, **15**, 907 (1983); K. Cook, P. Schechter and M. Aaronson: *Bull. Am. Astron. Soc.*, **15**, 907 (1983).
[186] D. N. C. Lin and S. M. Faber: *Astrophys. J. Lett.*, **266**, L21 (1983).
[187] S. D. M. White: in *Inner Space/Outer Space*, edited by E. W. Kolb et al. (University of Chicago Press, Chicago, Ill., 1986), p. 228.
[188] S. D. Tremain and J. E. Gunn: *Phys. Rev. Lett.*, **42**, 407 (1979).
[189] J. Madsen and R. I. Epstein: *Astrophys. J.*, **282**, 11 (1984).
[190] J. R. Primack: in *Particles and Fields* 2 (Plenum, New York, N. Y., 1983), p. 607.
[191] H. Haber and G. Kane: *Phys. Rep.*, **117**, 75 (1985).
[192] Cf. the preceding reference and S. Dawson, E. Eichten and C. Quigg: *Phys. Rev. D* (in press).
[193] H. Goldberg: *Phys. Rev. Lett.*, **50**, 1419 (1983); J. Ellis, J. S. Hagelin, D. V. Nanopoulos, K. A. Olive and M. Srednicki: *Nucl. Phys. B*, **238**, 453 (1984).
[194] K. A. Olive and M. S. Turner: *Phys. Rev. D*, **25**, 214 (1982).
[195] P. M. Lubin, G. L. Epstein and G. F. Smoot: *Phys. Rev. Lett.*, **50**, 616 (1983); D. J. Fixin, E. S. Cheng and D. T. Wilkinson: *Phys. Rev. Lett.*, **50**, 620 (1983).
[196] M. Guyot and Ya. B. Zel'dovich: *Astron. Astrophys.*, **9**, 227 (1970); P. Meszaros: *Astron. Astrophys.*, **37**, 225 (1974).
[197] A. G. Doroshkevich, Ya. B. Zel'dovich, R. A. Sunyaev and M. Yu. Khlopov: *Sov. Astron. Lett.*, **6**, 252 (1980); A. D. Chernin: *Sov. Astron.*, **25**, 14 (1981).
[198] J. R. Bond, A. S. Szalay and M. S. Turner: *Phys. Rev. Lett.*, **48**, 1636 (1982).
[199] S. D. M. White and M. Rees: *Mon. Not. R. Astron. Soc.*, **183**, 341 (1978).
[200] J. Silk: *Nature (London)*, **301**, 574 (1983).
[201] M. Davis, G. Efstathiou, C. Frenk and S. D. M. White: *Astrophys. J.*, **292**, 371 (1985) (DEFW).
[202] P. J. E. Peebles: *Astrophys. J. Lett.*, **263**, L1 (1982); *Astrophys. J.*, **258**, 415 (1982).
[203] J. R. Primack and G. R. Blumenthal: in *Clusters and Groups of Galaxies*, edited by F. Mardirossian, G. Giuricin and M. Mezzetti (Reidel, Dordrecht, 1984), p. 435.
[204] G. R. Blumenthal and J. R. Primack: SLAC-PUB-3388.
[205] J. Ipser and P. Sikivie: *Phys. Rev. Lett.*, **50**, 925 (1983).
[206] P. Sikivie: *Phys. Rev. Lett.*, **51**, 1415 (1983); (E) **52**, 695 (1984).
[207] B. J. Carr: *Comments Astrophys.*, **7**, 161 (1978); F. W. Stecker and Q. Shafi: *Phys. Rev. Lett.*, **50**, 928 (1983); K. Freese, R. Price and D. N. Schramm: *Astrophys. J.*, **275**, 405 (1983).
[208] R. Peccei and H. Quinn: *Phys. Rev. Lett.*, **38**, 140 (1977).
[209] S. Weinberg: *Phys. Rev. Lett.*, **40**, 223 (1978); F. Wilczek: *Phys. Rev. Lett*, **40**, 279 (1978).
[210] L. Abbott and P. Sikivie: *Phys. Lett. B*, **120**, 133 (1983); M. Dine and W. Fischler: *Phys. Lett. B*, **120**, 137 (1983); J. Preskill, M. Wise and F. Wilczek: *Phys. Lett. B* **120**, 127 (1983); J. Ipser and P. Sikivie: *Phys. Rev. Lett.*, **50**, 925 (1983).
[211] R. Brandenberger: *Phys. Rev. D*, **32**, 501 (1985).
[212] D. Dicus, E. Kolb, V. Teplitz and R. Wagoner: *Phys. Rev. D*, **18**, 1829 (1978); M. Fukugita, S. Watamura and M. Yoshimura: *Phys. Rev. Lett.*, **48**, 1522 (1982).

[213] F. WILCZEK: *Phys. Rev. Lett.*, **49**, 1549 (1982).
[214] G. R. FARRAR and P. FAYET: *Phys. Lett. B*, **76**, 575 (1978); **79**, 442 (1978); G. R. FARRAR and S. WEINBERG: *Phys. Rev. D*, **27**, 2732 (1983).
[215] J. SILK and M. SREDNICKI: *Phys. Rev. Lett.*, **53**, 624 (1984).
[216] J. S. HAGELIN, G. KANE and S. RABY: *Nucl. Phys. B*, **241**, 638 (1984); L. E. IBANEZ: *Phys. Lett. B*, **137**, 160 (1984); S. RABY: *Supersymmetry and Cosmology*, Lectures at the XV GIFT Seminar, June 1984, Los Alamos preprint LA-UR-84-2693.
[217] E. FARHI and R. L. JAFFE: *Phys. Rev. D*, **30**, 2379 (1984).
[218] A. DE RUJULA and S. GLASHOW: *Nature (London)*, **312**, 734 (1984).
[219] C. G. LACEY: in *Formation and Evolution of Galaxies and Large Scale Structures in the Universe*, edited by J. AUDOUZE and J. TRAN THANH VAN (Riedel, Dordrecht 1983), p. 351.
[220] J. IPSER and R. PRICE: *Astrophys. J.*, **216**, 578 (1977).
[221] C. CANIZARES: *Astrophys. J.*, **263**, 508 (1982).
[222] J. BARDEEN: in preparation.
[223] M. S. TURNER, F. WILCZEK and A. ZEE: *Phys. Lett. B*, **125**, 35, (E)519 (1983); T. HARA: *Prog. Theor. Phys.*, **6**, 1556 (1983).
[224] D. LYNDEN-BELL: *Mon. Not. R. Astron. Soc.*, **136**, 101 (1967); F. H. SHU: *Astrophys. J.*, **225**, 83 (1978).
[225] P. J. E. PEEBLES: *The Large Scale Structure of the Universe* (Princeton University Press, Princeton, N. J., 1980); G. EFSTATHIOU and J. SILK: *Fundam. Cosmic Phys.*, **9**, 1 (1983).
[226] S. D. M. WHITE and M. J. REES: *Mon. Not. R. Astron. Soc.*, **183**, 341 (1978).
[227] M. J. REES and J. P. OSTRIKER: *Mon. Not. R. Astron. Soc.*, **179**, 541 (1977).
[228] M. GELLER and J. HUCHRA: *Astrophys. J. Suppl.*, **52**, 61 (1983).
[229] S. M. FABER, G. R. BLUMENTHAL and J. R. PRIMACK: UCSC preprint (1984).
[230] A. G. DOROSHKEVICH: *Astrophysics*, **6**, 320 (1973).
[231] The statistics of such correlations is discussed by G. R. BLUMENTHAL, S. M. FABER and J. R. PRIMACK: in preparation (1986).
[232] G. EFSTATHIOU and B. J. T. JONES: *Mon. Not. R. Astron. Soc.*, **186**, 133 (1979); S. M. FALL and G. EFSTATHIOU: *Mon. Not. R. Astron. Soc.*, **193**, 189 (1980).
[233] A. SANDAGE, K. C. FREEMAN and N. R. STOKES: *Astrophys. J.*, **160**, 831 (1970); G. EFSTATHIOU and J. BARNES: in *Formation and Evolution of Galaxies and Large Scale Structures in the Universe*, edited by J. AUDOUZE and J. TRAN THANH VAN (Riedel, Dordrecht, 1984), p. 361.
[234] J. R. GOTT and T. X. THUAN: *Astrophys. J.*, **204**, 649 (1976).
[235] G. L. HOFFMAN, E. E. SALPETER and I. WASSERMAN: *Astrophys. J.*, **268**, 527 (1983).
[236] R. F. KIRSHNER, A. OEMLER, P. L. SCHECHTER and S. A. SCHECTMAN: *Astrophys. J. Lett.*, **248**, L57 (1981); in *Early Evolution of the Universe and Its Present Structure*, IAU Symposium No. 104, edited by G. O. ABELL and G. CHINCARINI (Riedel, Dordrecht, 1983), p. 197.
[237] G. EFSTATHIOU, M. DAVIS, C. S. FRENK and S. D. M. WHITE: *Astrophys. J. Suppl.*, **57**, 241 (1985).
[238] J. N. FRY and A. L. MELOTT: *Astrophys. J.*, **292**, 395 (1985).
[239] P. J. E. PEEBLES: *Astrophys. J.*, **284**, 439 (1984).
[240] J. BARDEEN: in *Inner Space/Outer Space*, edited by E. W. KOLB *et al.* (University of Chicago Press, Chicago, Ill., 1986), p. 212; N. KAISER: in *Inner Space/Outer Space*, edited by E. W. KOLB *et al.* (University of Chicago Press, Chicago, Ill., 1986), p. 258; J. SILK and R. SCHAEFFER: *Astrophys. J.*, **292**, 319 (1985); J. SILK: *Astrophys. J.*, **297**, 9 (1985).

[241] N. KAISER: *Astrophys. J. Lett.*, **284**, L9 (1984).
[242] H. D. POLITZER and M. B. WISE: Caltech preprint.
[243] J. BARNES, A. DEKEL, G. EFSTATHIOU and C. FRENK: *Astrophys. J.*, **295**, 368 (1985).
[244] S. F. SHANDARIN: *Sov. Astron. Lett.*, **9**, 104 (1983).
[245] A. MELOTT, J. EINASTO, E. SAAR, I. SUISALU, A. A. KLYPIN and S. F. SHANDARIN: *Phys. Rev. Lett.*, **51**, 935 (1983).
[246] J. D. BARROW and S. P. BHAVSAR: *Mon. Not. R. Astron. Soc.*, **205**, 66 (1983).
[247] A. DEKEL and M. J. WEST: *Astrophys. J.* (in press).
[248] A. DEKEL, M. J. WEST and S. J. AARSETH: *Astrophys. J.*, **279**, 1 (1984).
[249] N. BROSCH and P. M. GONDHALEKAR: *Astron. Astrophys.*, **140**, L43 (1984). I thank Adrian MELOTT for bringing this article to my attention.
[250] S. M. D. WHITE, M. DAVIS and C. FRENK: *Mon. Not. R. Astron. Soc.*, **209**, 15P (1984).
[251] A. DEKEL: *Astrophys. J.*, **284**, 445 (1984).
[252] A. DEKEL: *Astrophys. J. Lett.*, **261**, L13 (1982).
[253] W. L. W. SARGENT, P. J. YOUNG, A. BOKSENBERG and D. TYTLER: *Astrophys. J. Suppl.*, **42**, 41 (1980).
[254] W. L. W. SARGENT, P. YOUNG and D. P. SCHNEIDER: *Astrophys. J.*, **256**, 374 (1982).
[255] J. H. OORT: *Astron. Astrophys.*, **139**, 211 (1934).
[256] J. H. OORT, H. ARP and H. DE RUITER: *Astron. Astrophys.*, **95**, 7 (1981).
[257] T. FERRIS: *New York Times Magazine*, September 26, 1982, p. 38.
[258] E. NEUMANN: *Origins and History of Consciousness* (Princeton University Press, Princeton, N. J., 1954).
[259] V. F. WEISSKOPF: *Science*, **187**, 605 (1975); W. H. PRESS and A. P. LIGHTMAN: *Philos. Trans. R. Soc. London, Ser. A*, **310**, 323 (1983).

Supersymmetry and Supergravity.

S. Ferrara

CERN - Geneva, Switzerland
Istituto Nazionale di Fisica Nucleare - Laboratori Nazionali di Frascati, Italia

1. – Introduction.

The present theoretical understanding of particle physics seems to indicate that all fundamental forces of Nature known at present are described by gauge interactions. This is the case for the long-range electromagnetic and gravitational forces as well as for the short-range weak and nuclear forces.

Once such a dynamical framework for the basic interactions is accepted, there are several motivations to think that different low-energy symmetries become unified at higher scales.

This has been recently dramatically confirmed with the discovery at CERN of the intermediate vector bosons $W^{\pm}$, Z^0 which allows us to understand all the electroweak phenomena in terms of a spontaneously broken $SU_2 \times U_1$ gauge theory, the only residual massless vector boson, the photon, being related to the residual unbroken $U_{1,\text{e.m.}}$ electron charge symmetry. On the other hand, the absolute degeneracy of the electric charge of the proton and of the positron strongly indicates that a further symmetry must exist between strongly interacting particles, *i.e.* quarks and leptons, the latter being only subjected to the electroweak force. This fact calls for a unification of the strong gauge group $SU_{3,\text{c}}$ and the electroweak gauge group.

The simplest of these schemes are called GUT's (grand unified theories). These theories are renormalizable field theories in which $SU_{3,\text{c}} \times SU_{2,\text{L}} \times U_1$ becomes unified at some higher scale. Owing to the renormalizable nature of GUT's, such unification, in order to be compatible with the low-energy values of the electroweak and strong-coupling constants, implies that the unification scale should be at least $(10^{14} \div 10^{15})$ GeV, thirteen orders of magnitudes higher that the scale of weak interactions and only three or four orders of magnitude smaller that the Planck scale, *i.e.* the scale at which gravity becomes strong. In spite of their potential success, like the charge quantitatization and the prediction of the electroweak angle θ_{w}, GUT's have many drawbacks, some experimental and some other of pure theoretical nature. For example, the minimal GUT, the Georgi-Glashow SU_5 model, with minimal fermion content and minimal

Higgs structure predicts, due to new gauge interactions which couple quarks to leptons, a proton decay rate ($\sim 10^{30}$ years) by now at least two orders of magnitude faster than the experimental bounds ($\sim 10^{32}$ years). They also predict a monopole production which is not experimentally observed. From a more theoretical sight GUT's do not explain the family repetition and more seriously they have an intrinsic hierarchy problem [1] in the Higgs sector which seems to call for higher symmetries. Last but not least, GUT's predict a unification scale to close to the Planck scale which makes the exclusion of gravity in their formulation doubtful.

On the other hand, the embedding of GUT's in the framework of perturbative renormalizable theories which exclude any link with gravity at the very beginning is strictly related to the fact that in these theories there is an arbitrariness in the choice of the gauge group G which contains $SU_{3,C} \times SU_2 \times U_1$ and this fact in turn implies a lack of uniqueness in the formulation of these theories.

In a quest for a unification of the basic fundamental interactions with gravity one is faced with the problem that the basic force field of gravitational interactions is the metric tensor which should correspond to a new gauge quantum, the spin-2 graviton, strictly massless due to the unbroken Poincaré invariance of the physical laws of Nature. This is in contrast with the other gauge interactions whose gauge quanta have spin one. Therefore, any symmetry which may eventually unify all interactions should contain operators (charges) which change the spin of the particles. At a more fundamental level in a superunified theory there should be no substantial difference between force fields and matter fields. A symmetry making transition between force fields and matter fields should not only change the spin but also the statistics of the particles since we know that usual matter is mostly built out of protons, neutrons and electrons which obey the Fermi-Dirac statistics (fermions). Most surprisingly there is exactly a unique symmetry which can be realized in quantum field theory having all the desired properties, *i.e.* of converting bosons into fermions and *vice versa* and not being in contradiction with Poincaré invariance.

This symmetry, nowadays called supersymmetry [2], is an extension of the Poincaré Lie algebra to a graded Lie algebra (a superalgebra). Such a graded extension of the Poincaré algebra exists in any number of space-time dimensions and has the following general minimal structure. The odd part (fermionic generators) of the GLA contains a set of spinors Q_α^i ($i = 1, ..., N$, $\alpha = 1, ..., 2^{[D/2]}$). These generators form the so-called grading representation of the Lie algebra part of GLA which contains the $D(D+1)/2$ Poincaré generators and possible additional internal symmetries.

The basic commutation relation is

(1) $$\{Q_\alpha^i, \bar{Q}_\beta^j\} = -2\gamma_{\alpha\beta}^\mu P_\mu \delta^{ij} + \text{central charges},$$

which is independent of the dimension of space-time.

Due to the Hermiticity property of the right-hand side of eq. (1) the spinors Q_α^i have to satisfy a reality condition, which generally depends on the space-time dimension D [3]. For dimensions $D = 2$ (mod. 8) and $D = 4$ (mod. 8) spinors can be self-conjugate (Majorana), while for dimensions $D = 2$ (mod. 8) they can be both Majorana and Weil. This means that the supersymmetry algebra can have in this case $2^{D/2-1}$ Hermitian generators.

In supersymmetric theories particle states are classified according to representations of the supersymmetry algebra. In $D = 4$ space-time dimensions a complete classification of supermultiplets has been given. It follows from the algebra given by eq. (1) that for N-extended supersymmetry massless multiplets have helicity content given by $\lambda_{\max}$, $\lambda_{\max} - \frac{1}{2}$ up to $\lambda_{\max} - N/2$, in which $\lambda_{\max} \geqslant N/4$ for even N and $\lambda_{\max} \geqslant (N+1)/4$ for odd N. For massive multiplets the general structure is more complicated and will not be reported here. Let us just mention that in the absence of central charges the smallest massive multiplet of mass M has 2^{2N} states with spin range from $J = 0$ up to $J = N/2$.

Internal symmetries can be incorporated in supersymmetric theories. Let us describe an internal symmetry described by a Lie algebra G.

Two possibilities can occur

(2) $$[G, Q_\alpha^i] = 0,$$

(3) $$[G, Q_\alpha^i] \neq 0.$$

In the first situation all states of a given supermultiplet belong to the same representation of G.

In the second situation states of the same supermultiplet will belong to different representations of G since Q_α^i is not inert under G.

If G is a gauge symmetry, then eq. (3) can only occur if supersymmetry is local, *i.e.* in extended supergravity theories [4]. Generally in this situation the spin-1 gauge fields of G (G connections) belong to the same multiplet of the graviton and of the N spin-$\frac{3}{2}$ gravitinos.

If the gauge group G commutes with the supersymmetry generators, then particle states belonging to the same supermultiplet will transform according to the same representation of the group G.

Since the spin-1 massless gauge bosons of G must belong to the adjoint representation, then it follows that $N \leqslant 4$.

For $N = 1$ the adjoint multiplet contains a massless spin-$\frac{1}{2}$ and a spin-$\frac{1}{2}$ (Majorana) fermion.

For $N = 2$ a massless spin-$\frac{1}{2}$ (Majorana) fermions and two real scalars.

For $N = 4$ (or $N = 3$) a massless spin-1, four (Majorana) fermions and six (real) scalars.

Supermultiplets with spin 0, $\frac{1}{2}$ can only occur for $N \leqslant 2$. They are called « matter » multiplets. For $N = 1$ a Majorana or left-handed fermion is accompanied by a complex scalar. For $N = 2$ a left-handed and a right-handed fermion are accompanied by two complex scalars.

From the above analysis it is, therefore, clear that fermions in complex representations R ($\neq \bar{R}$) of the group G can only occur for $N=1$ matter multiplets. For $N=2$ matter multiplets the right-handed fermion can be identified with the antiparticle of the left-handed fermion only if the representation R of G is pseudoreal.

Supersymmetric models for particle physics use mainly $N=1$ supersymmetry in order to obtain chiral fermions embedded in supersymmetric multiplets without mirror partners.

This is especially important if one wants to describes GUT models with chiral fermion representations in which supersymmetry is unbroken up to the GUT scale.

The present lecture is organized as follows. In sect. 2 the physical motivations why supersymmetry may be relevant in particle physics in the TeV region are given. It is remarkable that a detailed study of supersymmetry breaking leads to the conclusion that phenomenologically viable models can only be constructed in the framework of spontaneously broken $N=1$ supergravity.

In sect. 3 the pure $N=1$ supergravity theory, without matter interactions, will be derived from some general principles of gauge invariance and symmetry. In sect. 4 general techniques to construct arbitrary locally $N=1$ supersymmetric invariant Lagrangians are developed.

In sect. 5 and 6 the most general couplings of $N=1$ matter and Yang-Mills multiplets to $N=1$ supergravity are obtained. Finally in sect. 7 the super-Higgs effect and its general consequences are derived.

2. – The challenge of supersymmetry.

Nongravitational interactions are suitably described at present, in the framework of relativistic quantum field theory, by gauge forces.

In this context one of the main problems is the origin of the several physical energy scales.

In the case of strong interactions, the typical scale, *i.e.* the inverse hadronic size or interaction range, is satisfactorily accounted for by the introduction of the Λ parameter of QCD. This parameter is due to the phenomenon of dimensional transmutation of a scale-invariant Lagrangian, relating this parameter to the dimensionless coupling constant through the renormalization group.

In the case of electroweak interactions the relevant energy scale is related to the Fermi coupling constant G_F

(4) $$G_F^{-\frac{1}{2}} \simeq 293 \text{ GeV}.$$

A possible explanation of a new physical scale, with a naive analogy with QCD, is accounted by a new strong force, based on some non-Abelian gauge theory, with a related Λ parameter of order $G_F^{-\frac{1}{2}}$. This force might give rise to the formation of composite Higgs fields (in analogy with pions in QCD) responsible for the electroweak breaking and perhaps to a subconstituent model for quarks, leptons and gauge vector bosons.

There are many problems with this viewpoint, among them is the fact that one has to rely on strong assumptions on the unknown dynamics of the hypothetical new force and heuristic calculations are based on mere analogies with QCD.

The alternative route in describing the origin of the Fermi scale is through the v.e.v. of a fundamental scalar field (Higgs field) such that

(5) $$\langle \theta \rangle = (\sqrt{2}\, G_F)^{-\frac{1}{2}} = 246 \text{ GeV}.$$

This gives rise to the standard $SU_{2,L} \times U_{1,Y}$ model with a Higgs doublet whose electrically neutral component is θ.

The main advantage of this viewpoint is that it can be realized in the context of a weak-coupling field theory. This in turn makes possible to apply standard perturbative techniques to get physical predictions.

However, in the framework of perturbative field theory, scalar masses usually receive large (quadratically divergent) radiative corrections.

This is not the case for fermion masses which are controlled by chiral symmetries and are for this reason protected from acquiring big radiative corrections. More precisely, the typical expressions for the one-loop contribution to these masses are

(6) $$\delta m_S^2 = A g^2 \Lambda^2, \qquad \delta m_F = B g^2 \, m_F \log \frac{\Lambda}{m_F},$$

where A, B are numbers of order 1, g is a dimensionless gauge or Yukawa coupling, m_F is a tree level fermion mass, and Λ is the momentum cut-off.

Because of eq. (6) the scalar masses are naturally of order $g\Lambda$ and, therefore,

(7) $$\langle \theta \rangle \sim \frac{m_S}{\sqrt{2\lambda}} = O\left(\frac{g\Lambda}{\sqrt{2\lambda}}\right),$$

λ being the dimensionless scalar self-coupling.

Equation (7) together with eq. (5) implies that in a weak-coupling theory ($\lambda < 1$) the cut-off Λ cannot exceed by order of magnitude $G_F^{-\frac{1}{2}}$ unless a reason occurs for which A is extremely small [1]. Actually in renormalizable theories with an additional scale $M \gg G_F^{-\frac{1}{2}}$ the scalar masses get radiative corrections $O(gM)$. In the context of grand unified theories this is called the hierarchy problem [1].

In these theories the problem of light Higgs doublet is even more dramatic since the relation $G_F^{-\frac{1}{2}} \ll M$ (M being the GUT scale) breaks down in perturbation theory.

The essential fact is that the numerical constant A defined by eq. (6) is not positive definite, but it gets contributions with opposite sign from virtual boson and fermion exchanges, as is the case for the vacuum energy of a given matter system of bosons and fermions. This makes the vanishing of A possible, thus avoiding the scale problem. The symmetry which makes the vanishing of A possible must relate in some way bosons and fermions, so that their masses and coupling constants are connected.

The only presently known candidate for such an invariance is an extension of the Fermi-Bose symmetry displayed by the free kinetic Lagrangian of an equal number of bosons and fermions with the same mass

$$(8) \quad \mathscr{L} = -\left(\frac{1}{2}|\partial_\mu z_i|^2 + \bar{\chi}_{Li}\,\partial\!\!\!/\,\chi_{Ri}\right) - \frac{1}{2}m_i^2|z_i|^2 - \frac{m_i}{2}(\bar{\chi}_{Li}\chi_{Li} + \bar{\chi}_{Ri}\chi_{Ri})\,,$$

$$(9) \quad \partial z_i = \bar{\varepsilon}_L \chi_{Li}\,, \quad \partial \chi_L = \tfrac{1}{2}\partial\!\!\!/\,Z_i\varepsilon_R - \tfrac{1}{2}m_i Z_i \varepsilon_L\,,$$

ε_L ($\varepsilon_R = \varepsilon_L^*$) being a constant anticommuting Weyl spinor. Under the variations (9) $\mathscr{L}$ changes by a total derivative. The symmetry given by eq. (9) has an extension to interacting Lagrangians and it gives rise to the algebraic structure of supersymmetry, described in the introduction.

It is obvious that unbroken supersymmetry, by implying relations among boson and fermion couplings and, in particular, degeneracy between bosons and fermions belonging to the same supermultiplet, is not realized by Nature.

However, broken supersymmetry can predict a realistic particle spectrum, as well as saving the hierarchy problem, provided the effective scale of supersymmetry breaking remains in the 1 TeV region.

In fact, while exact supersymmetry would require that the coefficient A in eq. (6) vanish identically, broken supersymmetry may give finite corrections to the scalar masses of order $\delta m_s^2 = O(g^2 \Delta m_i^2)$, where Δm_i^2 is the intramultiplet splitting between boson and fermion particles. Actually the above argument turns out to be true in a renormalizable theory provided supersymmetry is spontaneously broken or at most softly broken in such a way that no new quadratically divergent graphs are produced in perturbation theory [5]. With broken supersymmetry the absence of fine-tuning in δm_s^2 requires that all known particles should find their superpartners in the energy range of the Fermi scale. The detailed prediction of the mass spectrum of all these new particles requires a detailed analysis of supersymmetry breaking.

The most predictive case, and in fact the only possible if one includes gravity, is the case of spontaneous breakdown. In such a case the following low-energy theorem holds:

$$(10) \quad \Delta m_i^2 = g_i M_s^2\,,$$

where g_i is the coupling of the Goldstone fermion (goldstino) ψ_g to the i-th multiplet and M_s is the supersymmetry breaking defined by the linear goldstino term in the spinor current connected to the supersymmetry charge

$$J_\mu = M_s^2 \gamma_\mu \psi_g \,. \tag{11}$$

Two schemes of spontaneous supersymmetry breaking can be considered, depending on the size of the primordial supersymmetry-breaking scale M_s [5].

In the first case

$$M_s^2 = O(\Delta m_i^2) = O(G_F^{-\frac{1}{2}}) \tag{12}$$

and g_i defined by eq. (10) is of the same order of the electroweak gauge or Yukawa couplings.

In the most interesting case

$$M_s^2 = O(M G_F^{-\frac{1}{2}}) \,, \tag{13}$$

where M is a mass scale much higher than $G_F^{-\frac{1}{2}}$. Therefore, g_i has to be very small, i.e. $g_i \simeq O(G_F^{-\frac{1}{2}}/M)$.

In suitable supersymmetric theories the coupling g_i may indeed be given as a ratio of different mass scales, to all orders of perturbation theory (decoupling theorem). As we will see later, the two situations described by eqs. (12) and (13) correspond to two different regimes of the «low-energy» effective theory. In fact, the first case corresponds to an effective spontaneously broken «low-energy» theory, while the latter corresponds to an effective «softly» broken «low-energy» theory.

A general analysis of models for particle interactions based on the idea of «low-energy» supersymmetry, i.e. $\Delta m_i = O(G_F^{-\frac{1}{2}})$, indicates that the situation described by eq. (12) is untenable. In fact, if the theory is «spontaneously broken» at low energy, i.e. $g_i \sim O(1)$, a series of difficulties, both theoretical and phenomenological, arises. The main problem is to obtain a theory with gauge symmetry $SU_{3,L} \times SU_{2,L} \times U_1$ spontaneously broken to $SU_{3,C} \times U_{1,\text{e.m.}}$ and with a realistic mass spectrum for quarks, leptons and their superpartners. If the gauge group is not enlarged, one can show that the mass spectrum is unrealistic because of at least light scalar partners of the quarks are unacceptable.

On the other hand, if the gauge group is enlarged to contain additional U_1 factors as discussed by FAYET [2], it is possible in general to obtain a local minimum of the potential with the desired properties.

However, the new chiral U_1 generally introduces new triangular anomalies whose cancellation needs the introduction of new spin-$\frac{1}{2}$, spin-0 multiplets whose scalar components tend to destabilize the desired $SU_{3,C} \times U_1$ symmetry vacuum to a new vacuum which generally breaks colour or charge.

The main obstruction to a «low-energy» spontaneously broken supersymmetric theory is due to a general tree level mass formula which is true in any renormalizable supersymmetry gauge theory (with semi-simple Yang-Mills group G)

(14) $$\sum_J (-1)^{2J}(2J+1)m_J^2 = 0 \ .$$

The right-hand side of the mass relation can be nonzero only if some U_1 factor with nonvanishing trace over the chiral multiplets is present. If one requires absence of quadratic divergences, then the U_1 must be traceless, i.e.

$$\sum_i q_i = 0 \ .$$

On the other hand, the anomaly cancellation is more complicated since in general it requires more equations. The U_1^3 anomaly cancellation requires

$$\sum_i q_i^3 = 0$$

and the other equations come from the mixed graphs of the new U_1 factor with the $SU_3 \times SU_2 \times U_1$ gauge bosons.

In the alternative point of view $M_S^2 = O(MG_F^{-\frac{1}{2}})$, in which $M \gg G_F^{-\frac{1}{2}}$ can be the grand-unification scale or the Planck mass.

This may happen in theories with an hidden sector which spontaneously breaks supersymmetry at the tree level [5]. « Hidden » means that this sector is not coupled to the « observable » sector at the tree level, or it is coupled only by gravitational interactions.

In the case of a renormalizable theory the « hidden sector » couples to the « observable » sector only through radiative corrections with loop graphs with virtual-particle exchange of masses $O(M)$. This induces an « effective » supersymmetry breaking in the observable sector of a typical scale $\Delta m \simeq$
$\simeq O((g^2)^n M_S^2/M)$.

In supergravity theories the « hidden » sector with renormalizable interactions can be replaced by a « hidden » sector which is only gravitationally coupled to ordinary particle supermultiplets [6].

In this case the effective supersymmetry breaking scale is given by the gravitino mass $m_{\frac{3}{2}} = O(M_S^2/M_P)$ which naturaly splits the boson and fermion masses in each supermultiplet and may also induce the gauge breaking of the $SU_{2,L} \times U_1$ electroweak theory so that $m_{\frac{3}{2}} \simeq O(G_F^{-\frac{1}{2}})$.

The gravitino mass growth is related to the so-called super-Higgs effect, i.e. the spontaneous breakdown of local supersymmetry. In this mechanism the goldstino becomes the spin-$\pm \frac{1}{2}$ component of the massive spin-$\frac{3}{2}$ gravitino, while the scalar superpartner of the goldstino is the « complex » scalar field whose potential energy determines the tree level ground state of the spontaneously broken theory.

From the previous discussion it appears evident that in theories based on supergravity breaking it is quite natural to have some identification (or some relation) between the GUT scale and the Planck mass.

In the next section we will give an elementary derivation of $N=1$ supergravity theory and in sect. 4 we will give the most general interaction of $N=1$ supergravity coupled to an arbitrary matter system.

Before ending this section we would like to point out that, in the presence of supergravity coupling, one can re-express eq. (10) in terms of the gravitino mass scale and the Planck scale as follows:

$$\Delta m_i^2 = \hat{g}_i m_{\frac{3}{2}} M_P . \tag{15}$$

This is due to the fact that $m_{\frac{3}{2}}$ is related to M_S^2 and the Planck scale through the super-Higgs effect.

Equation (15) tells us that, in order that supersymmetry be observable at relatively low energies ($\leqslant 1$ TeV), it must occur that

$$\hat{g}_i = O\left(\frac{M_w^2}{m_{\frac{3}{2}} M_P}\right). \tag{16}$$

Equation (16) implies that the bigger is $m_{\frac{3}{2}}$, the smaller is $\hat{g}_i$. In fact, $m_{\frac{3}{2}}$ could be even of order M_P, in which case $\hat{g}_i = O(M_w^2/M_P^2)$. The natural range of values of $\hat{g}_i$ is, therefore, $g_i \simeq O(1)$ (for $m_{\frac{3}{2}} \sim M_w^2/M_P$) up to $\hat{g}_i \simeq M_w^2/M_P^2$ (for $m_{\frac{3}{2}} \sim M_P$).

The last situation, and more generally $m_{\frac{3}{2}} \gg M_w$, may occur in the so-called no-scale models [7] in which $m_{\frac{3}{2}}$ is essentially a free parameter.

3. – The $N=1$ supergravity theory.

In the present section we will describe, in simple terms, the $N=1$ pure supergravity theory. The later sections will be devoted to its coupling to general matter and Yang-Mills multiplets.

Several motivations can be given for the introduction of supergravity in field theory. Firstly to promote supersymmetry to a local invariance. This requires, because of eq. (1), invariance under general co-ordinate transformations.

Moreover, when the gauge invariance of the free massless Rarita-Schwinger equation is extended to an interacting theory, supergravity emerges as the only solution for a consistent coupling of spin-$\frac{3}{2}$ particles, overcoming the old inconsistencies for coupling higher-spin field equations.

Supergravity provides, for the first time, the principle that fermion fields, which are usually associated with matter, can be associated with gauge fields. This fact gives a subtle interplay between space-time geometry and the quantum-mechanical concept of spin.

Supergravity was also motivated as an attempt to build a meaningful quantum theory of gravity. Indeed the improvement of the ultraviolet behaviour with respect to Einstein theory, due to the short-distance contributions of the supersymmetric partners of the graviton, is spectacular. The ultimate hope is that some version of supergravity (extended supergravity?) or some generalization thereof (superstrings) will lead to a finite theory of quantum gravity unifying all particle interactions.

There have been several reformulations of the basic ($N=1$) supergravity theory. For instance, in first-order form, $N=1$ supergravity can be regarded as the Einstein-Cartan theory for a massless spin-$\frac{3}{2}$ particle.

In spite of significant technical improvements we would like to go through a derivation of the theory which does not require any knowledge of differential geometry or group theory, but is rather based on some very simple physical considerations.

From the representation theory of global supersymmetry it is known that a massless (Majorana) particle with helicity $\lambda = \pm \frac{3}{2}$ can form a multiplet with a partner of helicity $\lambda = \pm 2$, $\lambda = \pm 1$. Indeed in free-field theory both choices are equally possible. The first choice is the supergravity multiplet $(\pm 2, \pm \frac{3}{2})$ with free Langrangian

(17) $$\mathscr{L}^0 = \mathscr{L}^0_{\text{Einstein}}(h) - \tfrac{1}{2} \varepsilon^{\mu\nu\varrho\sigma} \psi_\mu \gamma_5 \gamma_\nu \partial_\varrho \psi_\sigma .$$

$\mathscr{L}^0$ is the linearized Einstein Lagrangian with $g_{\mu\nu} = \eta_{\mu\nu} + kh_{\mu\nu}$; $\mathscr{L}^0$ is invariant under two separate Abelian transformations

(18) $$\delta h_{\mu\nu} = \partial_\mu \xi_\nu + \partial_\nu \xi_\mu , \quad \delta \psi_\mu = \partial_\mu \alpha$$

and global supersymmetry rotations

(19) $$\partial h_{\mu\nu} = \bar{\varepsilon} \gamma_\mu \psi_\nu + \bar{\varepsilon} \gamma_\nu \psi_\mu , \quad \delta \psi_\mu = \partial_\varrho h_{\mu\sigma} \gamma^{\varrho\sigma} \varepsilon .$$

The alternative choice is the multiplet $(\pm \frac{3}{2}, \pm 1)$ with free Lagrangian

(20) $$\mathscr{L}^1 = -\tfrac{1}{4} F_{\mu\nu} F^{\mu\nu} - \tfrac{1}{2} \varepsilon^{\mu\nu\varrho\sigma} \bar{\psi}_\mu \gamma_\varepsilon \gamma_\nu \partial_\varrho \psi_\sigma ,$$

with Abelian gauge invariances

(21) $$\delta A_\mu = \partial_\mu \Lambda , \quad F_\mu = \partial_\mu A_\nu - \partial_\nu A_\mu , \quad \delta \psi_\mu = \partial_\mu \alpha$$

and global supersymmetry rotation

(22) $$\delta A_\mu = \bar{\varepsilon} \psi_\mu , \quad \delta \psi_\mu = \sigma_{\varrho\sigma} \gamma_\mu F^{\varrho\sigma} \varepsilon .$$

The very difference between $\mathscr{L}^0$ and $\mathscr{L}^1$ comes when we try to introduce consistent interactions. If we have $\varepsilon = \varepsilon(x)$ and we perform a local supersymmetry transformation, then a new interaction term is required of the form $K\bar{\psi}_{\mu\alpha} J^{\mu\alpha}$ (K is the gravitational coupling constant) with $\partial^\mu J_{\mu\alpha} = 0$. This is nothing but the Noether coupling. On the other hand, it turns out that, under a supersymmetry variation, the spinor supersymmetry current transforms into the stress tensor of the system $T^{\mu\nu}$. Hence the $K\bar{\psi}_\mu J^\mu$ coupling requires, at the same order, a $Kh_{\mu\nu} T^{\mu\nu}$ term and, therefore, only the ansatz (17) is possible.

This is a general result and is due to the current algebra obeyed by the supercurrent, which is nothing but the nonintegrated form of the basic anticommutation relation given by eq. (1).

The final theory derived by this step-by-step procedure gives the supergravity Lagrangian in the form

$$(23) \quad \mathscr{L}_{\text{SG}} = -\frac{1}{2K^2}\sqrt{-g}\, R(g_{\mu\nu}) - \frac{1}{2}\varepsilon_{\mu\nu\varrho\sigma}\bar{\psi}_\mu \gamma_5 \gamma_\nu D_\varrho \psi_\sigma -$$

$$-\frac{e}{32} K^2 (\bar{\psi}^\mu \gamma^\nu \psi^\varrho)(\bar{\psi}_\mu \gamma_\nu \psi_\varrho + 2\bar{\psi}_\nu \gamma_\mu \psi_\varrho) - 4(\bar{\psi}_\mu \gamma \cdot \psi)^2 \,.$$

$g_{\mu\nu} = e^a_\mu e_{\nu a}$ and $e_{\mu a}$ is the vierbein field. $\mathscr{L}_{\text{SG}}$ is invariant under the following (non-Abelian) gauge transformation with $\varepsilon = \varepsilon(x)$:

$$(24) \quad \delta e_{a\mu} = K \bar{\varepsilon} \gamma_a \psi_\mu \,, \quad \delta \psi_\mu = \frac{2}{K} D_\mu \varepsilon + \frac{1}{4} K \sigma^{ab} \varepsilon (2\bar{\psi}_\mu \gamma_a \psi_b + \bar{\psi}_a \gamma_\mu \psi_b) \,.$$

D_μ being the ordinary gravitational derivative with metric spin connection. For the Lagrangian given by eq. (23) it then happens that at the tree level ($K = 0$) the transformation law (24) degenerates in an Abelian transformation and an independent global supersymmetry rotation as given by eqs. (18) and (19). This is in close analogy with Yang-Mills theories in which the non-Abelian gauge transformation $\delta A_\mu = (1/g) D_\mu \Lambda = (1/g)\partial_\mu + [\Lambda, A_\mu]$ reduces for $g \to 0$ to an Abelian transformation and an independent global rotation.

The only difference is that in supergravity the $\delta \psi_\mu$ transformation is nonlinear in the dynamical field variables (it is actually linear in first order since the $D_\mu \varepsilon$ term in this case contains the Einstein-Cartan spin connection with contorsion terms).

One can obtain an even much simpler derivation of the Lagrangian (23) which does not require any knowledge of the transformation laws of the spin-2, spin-$\frac{3}{2}$ fields but only the fact that the graviton and gravitino fields describe massless particles of appropriate helicities. This derivation is straightforward and only requires the knowledge of the gravitational Born amplitudes for the scattering of two spin-$\frac{3}{2}$ particles. It also emphasizes the interpretation of the four-fermion coupling present in eq. (23) as a seagull term of the same nature of similar terms occurring in ordinary Yang-Mills theories. If we take the Born amplitude for the scattering of two spin-$\frac{3}{2}$ particles through a one-graviton exchange, the polarization tensor $\psi_\mu(p)$ for an external leg of momentum p_μ on the mass shell $(p^2 \psi_\mu(p) = 0)$ satisfies the equations

(25) $$\not{p}\psi(p) = 0, \quad \check{p}\psi_\mu(p) = 0.$$

Equation (25) reduces the number of physical components of $\psi_\mu(p)$ to four. To reduce them further to two the S-matrix elements must vanish when we make the substitution $\psi_\mu(p) = \varepsilon p_\mu$, where ε is an anticommuting (Majorana) spinor.

It is a simple exercise to prove that the Born term does not fulfil this requirement and that the S-matrix vanishes for $\psi_\mu(p) = \varepsilon p_\mu$ only if the appropriate contact term ι (corresponding to the last term in eq. (23)) is introduced.

There is a simple dimensional argument that shows that a four-fermion coupling is all we need to make the theory consistent and that additional contact terms with more spin-$\frac{3}{2}$ fields would never help. Let us consider the scattering, at the tree level, of n spin-$\frac{3}{2}$ particles. The scattering can then proceed through exchanges of gravitons between gravitinos (trilinear coupling). The gravitational constant K in such graphs always appears with power $2(n-1)$. Let us now assume that one replaces one of the external polarization tensors by its momentum, computes the sum of all graphs contributing to the S-matrix and finds that it does not vanish.

This means that one has to introduce an additional term in the Lagrangian of the form $g_{2n} \partial^m (\bar{\psi}\psi)^n$, where m means m derivatives and indices have been omitted. Now a trivial dimensional argument restricts such contact terms very severely: from the kinetic term of the action the dimension of ψ_μ is $(D-1)/2$ in units of mass, D being the space-time dimension. Then the dimension of the coupling g_{2n} is

(26) $$[g_{2n}] = D - n(D-1) - m.$$

However, such contact terms, if required, should compensate possibly non-gauge-invariant terms, and for them $g_{2n} \sim K^{2(n-1)}$. Since $[K] = (2-D)/2$, then we get the consistency equation

$$2(n-1)\frac{2-D}{2} = D - n(D-1) - m, \quad i.e. \ n = 2 - m.$$

The only possible solution is then $m = 0$, $n = 2$ for any D. We conclude that the contact term for supergravity, in any space-time dimension, can only be a four-fermion coupling with no derivatives.

4. – Tensor calculus for $N = 1$ supergravity.

The action and transformation laws of locally supersymmetric systems are constructed by use of the $N = 1$ tensor calculus for supergravity [8].

A major simplification occurs by employing the so-called conformal tensor calculus which uses, for the gravitational sector, the Weyl multiplet, containing the gauge fields of conformal supergravity [9].

The $N = 1$ Weyl multiplet contains the following set of gauge fields

$$(27) \qquad (e_{m\mu}, \psi_{\mu a}, A_\mu, b_\mu)$$

which are, respectively, the connections for general co-ordinate transformations (vierbein), for Q supersymmetry (gravitino), for a chiral U_1 (R-symmetry) and dilatations.

The additional connections $\omega_{\mu m n}$, $\Phi_{\mu a}$, $f_{m\mu}$ for local Lorentz rotations, special supersymmetry and special conformal transformations M_{mn}, S_a, K_m, respectively, can be solved in terms of the gauge fields in eq. (27) by means of the conventional constraints (*)

$$(28) \qquad R^m_{\mu\nu}(P) = 0, \qquad \gamma^\mu R_{\mu\nu}(Q) = 0, \qquad \hat{R}_{\mu\nu mn}(M) e^{\nu n} + \frac{i}{3} \hat{R}_{\mu m}(A) = 0$$

$$\left(\tilde{T}_{mn} = \frac{1}{2} \varepsilon_{mnpq} T_{pq} \right),$$

where $\hat{R}_{\mu\nu mn}$ denotes the fully covariant Lorentz connection (Riemann tensor). The curvatures are covariant, but, since the constraints in eqs. (28) are not Q-invariant, the variations $\delta_Q \omega_{\mu nn}$, $\delta_Q \Phi_{\mu a}$ and $\delta_Q f_{m\mu}$ are not simply given by the group rule. In the conformal tensor calculus we use two basic multiplets other than the one of eq. (27). The scalar (or chiral) multiplet $S = (z, X_L, h)$, with Weyl weight w (w is the Weyl weight of the first component z

---

(*) We use the conventions of ref. [2]. We also use the following conventions: $a = 1, 2$, two-component spinor indices; $m = 1, ..., 4$, flat Lorentz indices; $\mu = 1, ..., 4$, world indices; $\alpha = 1, ..., \dim G$, gauge group indices; $i, j = 1, ..., \dim R$, G-group representation indices, R being a finite unitary representation of G. We also set, in most of our formulae, the gravitational constant $k = 1$; k is related to the Planck mass m_P as follows: $k = \sqrt{8\pi}/m_P$.

of S) transforming as (*)

(29)
$$\begin{cases} \delta z = \bar{\varepsilon}_L \chi_L - \tfrac{1}{3} iw\alpha z\,, \\ \delta \chi_L = \tfrac{1}{2} \check{D} z \varepsilon_R + \tfrac{1}{2} h \varepsilon_L + wz\eta_L + i\alpha(\tfrac{1}{2}-\tfrac{1}{3}w)\chi_L\,, \\ \delta h = \bar{\varepsilon}_R \check{D} \chi_L + 2\eta_L \chi_L (1-w) + i\alpha(1-\tfrac{1}{3}w) h\,, \end{cases}$$

where ε is a space-time-dependent anticommuting Weyl (or Majorana) spinor corresponding to Q transformations, and α is the space-time-dependent parameter of the U_1 chiral group of the superconformal algebra, normalized such that for the U_1 gauge field A_μ

(30)
$$\delta_\alpha A_\mu = \partial_\mu \alpha\,.$$

In eq. (29) η is a space-time-dependent anticommuting Weyl (or Majorana) spinor corresponding to special S_α supersymmetry transformations, and D denotes the superconformal covariant derivative.

The second multiplet is a real vector multiplet V: $V = (C, S_L, H, N_m, \lambda_R, D)$. Its transformation is

(31)
$$\begin{cases} \delta G = \tfrac{1}{2} i \bar{\varepsilon}_L S_L - \tfrac{1}{2} i \bar{\varepsilon}_R S_R\,, \\ \delta S_L = \tfrac{1}{2} i H \varepsilon_L - \tfrac{1}{2} \check{B} \varepsilon_R - \tfrac{1}{2} i \check{D} C \varepsilon_R - iwC\eta_L + \tfrac{1}{2} i\alpha S_L\,, \\ \delta H = -i\bar{\varepsilon}_R \check{D} S_L - i\bar{\varepsilon}_R \lambda_R + i(-2+w)\bar{\eta}_R S_L + i\alpha H\,, \\ \delta B_m = [-\tfrac{1}{2} \bar{\varepsilon}_L D_m S_L - \tfrac{1}{2} \bar{\varepsilon}_L \gamma_m \lambda_R + \tfrac{1}{2}(1+w)\bar{\eta}_L \gamma_m S_R] + \text{h.c.}\,, \\ \delta \lambda_R = \tfrac{1}{2}(\sigma \cdot \hat{F} - iD)\varepsilon_R + \tfrac{1}{2} w(iH\eta_R - \check{B}\eta_L + I\check{D}C\eta_L) + \tfrac{1}{2} i\alpha\lambda_R - w\check{\Lambda}_K S_L\,, \\ \delta D = [\tfrac{1}{2} i\bar{\varepsilon}_L \check{D}\lambda_R - iW\bar{\eta}_L(\lambda_L + \tfrac{1}{2}\check{D}S_R) - w\Lambda_{Km} \hat{D}_m C] + \text{h.c.}\,, \end{cases}$$

with

(31')
$$\hat{F}_{mn} = 2\hat{D}_{[m} B_{n]} + i\varepsilon_{mnpq} \hat{D}_p \hat{D}_q C\,, \qquad [mn] = \tfrac{1}{2}(mn - nm)\,,$$

where the symbol [ ] denotes antisymmetrization, and Λ_{Km} is the space-time-dependent parameter for special conformal transformations.

The scalar multiplet S, for Weyl weight $w = 3$, has its last component h with Weyl weight $w = 4$, and it can be extended to a superconformal density

(32)
$$g^{-1} L_{S(w=3)} = h + \bar{\psi}_R \cdot \gamma \chi_L + \bar{\psi}_{\mu R} \sigma_{\mu\nu} \psi_{\nu R} z\,.$$

(*) For typographical reasons, here and in the following, all the slashed letters have been substituted with the corresponding letter with a caret.

Analogously we can embed a vector multiplet V with $w=2$ into a scalar multiplet S_V with $w=3$, whose components are

$$(33) \qquad (-H^*, \; -i\lambda_L - i\check{D}S_R, \; \hat{\square}D + C + I\hat{D}_m B^m) \,.$$

Then we can apply eq. (32) to the multiplet S_V to obtain a density formula for the vector multiplet with $w=2$:

$$(34) \quad e^{-1}L_{V(w=2)} = D + \hat{\square}C + [\tfrac{1}{2} i\bar{\psi}\cdot\gamma(\lambda_R + \check{D}S_L) - \tfrac{1}{2}\psi_{\mu L}\sigma_{\mu\nu}\psi_{\nu L}H + \text{h.c.}]\,.$$

We can also get a vector embedding of a scalar multiplet S with $w=0$ into a vector multiplet V_S with $w=0$, whose components are

$$(35) \qquad V_S = [i(z-z^*), \; 2\chi_L, \; -2ih, \; -\hat{D}_m(z+z^*), \; 0, \; 0]\,.$$

The multiplication rules of scalar and vector multiplets are given respectively by

$$(36) \quad \begin{cases} S_1 S_2 = (z_1 z_2, \; z_1 \chi_{2L} + z_2 \chi_{1L}, \; z_1 h_2 + z_2 h_1 - 2\chi_{1L}\chi_{2L})\,, \\[4pt] V_1 V_2 = [C_1 C_2, \; C_1 S_{2L} + C_2 S_{1L}, \; C_1 H_2 + C_2 H_1 - \bar{S}_{1L} S_{2L}, \; C_1 B_{m2} + \\ \quad + \tfrac{1}{2} i \bar{S}_{1L}\gamma_m S_{2R} + (1\leftrightarrow 2)\,, \; C_1 \lambda_{2R} + \tfrac{1}{2}(-\check{D}C_1 S_{2L} + H_1 S_{2R} - I\check{B}_1 S_{2L}) + \\ \quad + (1\leftrightarrow 2)\,, \; C_1 D_2 - \bar{S}_{1L}\lambda_{2L} - \bar{S}_{1R}\lambda_{2R} + \tfrac{1}{2}H_1 H_2^* - \tfrac{1}{2} B_1 \cdot B_2 - \\ \quad\quad - \tfrac{1}{2}\hat{D}_m C_1 \hat{D}^m C_2 - \tfrac{1}{2}\bar{S}_{1L}\check{D}S_{2R} - \tfrac{1}{2}\bar{S}_{1R}\check{D}S_{2L} + (1\leftrightarrow 2)]\,,\end{cases}$$

$$(37) \quad \tfrac{1}{2}(S_1 \bar{S}_2 + \bar{S}_1 S_2) = [\tfrac{1}{2} z_1 z_2^*, \; -iz_2^*\chi_{1L}, \; -z_2^* h_1, \; \tfrac{1}{2} i(z_2^* \hat{D}_m z_1 - z_1 \hat{D}_m z_2^*) -$$

$$- i\bar{\chi}_{3R}\gamma_m \chi_{1L}, \; -ih_1 \chi_{2R} + i\hat{D}z_2^* \chi_{1L}, \; h_1 h_2^* - \hat{D}_m z_1 \hat{D}^m z_2^* - \chi_{1L}\overset{\leftrightarrow}{\check{D}}\chi_{1R} + (1\leftrightarrow 2)]\,.$$

The resulting multiplets have Weyl weight $w_1 + w_2$.

The rules we have given so far are sufficient for constructing all superconformal invariant actions. However, if we want to restrict our tensor calculus to Poincaré supersymmetry only, we have to introduce Poincaré rules which will enable us to construct Poincaré supersymmetric local densities for arbitrary Weyl weight of the fields involved. This is most easily achieved by introducing a new compensating chiral multiplet S_0 with Weyl weight $w=1$, and then using it to fix a special superconformal gauge. This choice of compensating multiplet corresponds to the so-called old minimal formulation of $N=1$ supergravity originally proposed in ref. [7]. The connection of this formulation with the new minimal formulation of Sohnius and West [10], which uses a different compensating multiplet to fix the superconformal gauge, has been given in general elsewhere [11].

The chiral compensating multiplet has components

(38) $$S_0 = (a, \xi, \tfrac{1}{3} u),$$

where $u = S - iP$ is the complex auxiliary field of Poincaré supergravity defined by CREMMER et al. [6].

To get Poincaré supergravity from the superconformal tensor calculus we fix a superconformal gauge by fixing the gauges of dilatation, U_1 chiral rotation, special S-supersymmetry transformations and conformal boosts, by the conditions

(39) $$a = 1, \quad \xi = 0, \quad b = 0.$$

These conditions are preserved under a newly defined Poincaré Q local supersymmetry with the parameter

(40) $$\delta^P(\varepsilon) = \delta_Q(\varepsilon) + \delta_S(\eta_L = -\tfrac{1}{6} u \varepsilon_L - \tfrac{1}{6} i \breve{A} \varepsilon_R) + \delta_K(\Lambda_{K\mu} = \tfrac{1}{4} \bar{\varepsilon} \varphi_\mu - \tfrac{1}{4} \bar{\eta} \psi_\mu).$$

Equation (40) provides the rule for computing a Poincaré local supersymmetry transformation in terms of the local superconformal rules.

To compute the density formula for a chiral multiplet S of Weyl weight $w = 0$ we first construct the new chiral multiplet SS_0^3 with Weyl weight $w = 3$. Since in the Poincaré gauge defined by eq. (39), using eq. (36), we have

(41) $$S_0^n = (1, 0, \tfrac{1}{3} nu),$$

we obtain

(42) $$SS_0^n = (z, \chi_L, h + \tfrac{1}{3} nuz).$$

Then the Poincaré density formula for S is given by

(43) $$SS_{0F,SC} = S_{F,\text{Poincaré}} = L_{S(w=0)} = eh + euz + e\bar{\psi}_R \cdot \gamma \chi_L + e\bar{\psi}_{\mu R} \sigma_{\mu\nu} \psi_{\nu R} z.$$

For the vector multiplet V, we take a Weyl weight $w = 0$ and then introduce the compensating vector multiplet $S_0 \bar{S}_0$ with $w = 2$ having components

(44) $$S_0 \bar{S}_0 = [1, 0, -\tfrac{2}{3} u, -\tfrac{2}{3} A_m, 0, \tfrac{2}{9}(uu^* - A_m A^m)].$$

We then get a Poincaré density formula for V by using a superconformal density formula for $VS_0\bar{S}_0$ with $w = 2$, and using the multiplication rule given by eq. (37) in the Poincaré gauge (eq. (39)):

(45) $$[VS_0\bar{S}_0]_{D,SC} = [V]_{D,\text{Poincaré}} = e^{-1} L_V(w=0) =$$
$$= D - \tfrac{1}{2} i\bar{\psi} \cdot \gamma \gamma_5 \lambda - \tfrac{1}{3}(u^* H - u H^*) + \tfrac{2}{3} B_m A^m -$$
$$- \tfrac{1}{3} i \bar{S} \gamma_5 \gamma R^P + \tfrac{1}{4} i \varepsilon^{mnrs} \bar{\psi}_m \gamma_n v_r (B_s - \tfrac{1}{2} \bar{\psi}_s S) - \tfrac{2}{3} C e^{-1} L_{SC},$$

where

(46)
$$\begin{cases} R^{\mathrm{P}}_\mu = e^{-1}\varepsilon_{\mu\nu}{}^{\varrho\sigma}\gamma_5\gamma^\nu[\partial_\varrho + \tfrac{1}{2}\omega_{\varrho mn}(e_1\psi)\sigma^{mn} + \tfrac{1}{2}i\gamma_5 A_\varrho + \tfrac{1}{6}\gamma_\varrho(\omega - i\check{A}\gamma_5)]\,\psi_0\,, \\ L_{sc} = -\tfrac{1}{2}eR[e,\omega(e,\psi)] - \tfrac{1}{2}\varepsilon^{\mu\nu\varrho\sigma}\bar\psi_\mu\gamma_5\gamma_\nu D_\varrho[\omega(e,\psi)]\psi_\sigma - \tfrac{1}{3}euu^* + \tfrac{1}{3}eA_m A^m\,. \end{cases}$$

We remark that the Poincaré density formula as well as the Poincaré local supersymmetry transformations are w-independent. If S has weight w, then $S' = SS_0^{-w}$ has $w = 0$. Equation (41) defines the components of S' with $h' = h - \tfrac{1}{3}wuz$. This defines the w-independent transformation rules of ref. [6]. Analogously for a vector multiplet V with Weyl weight w, we define the new multiplet $V' = V(S_0\bar S_0)^{-w/2}$ with $w = 0$.

The components of V' are easily computed to be

(47)
$$V' = (C, S_{\mathrm L}, H', B'_m, \lambda'_{\mathrm R}, D')\,,$$

with

(48)
$$\begin{cases} H' = H - \tfrac{1}{3}wCu\,, \quad B'_m = B_m + \tfrac{1}{3}wCA_m\,, \\ \lambda'_{\mathrm R} = \lambda_{\mathrm R} + \tfrac{1}{6}w(uS_{\mathrm R} - i\check{A}S_{\mathrm L})\,, \\ D' = D + \tfrac{1}{6}w(Hu^* + H^*u - 2B_m A^m + \tfrac{1}{2}i\bar S\gamma\gamma_5\cdot R^{\mathrm P}) + \\ \qquad\qquad + \tfrac{1}{18}w^2 C(uu^* - A_m A^m)\,. \end{cases}$$

The density formulae for S' and V' as given by eq. (43) (with $h = h'$) and eq. (45) (with $H = H'$, $B_m = B'_m$, $\lambda_{\mathrm R} = \lambda'_{\mathrm R}$, $D = D'$) are local supersymmetric densities with respect to the following Poincaré local supersymmetry transformations:

(49)
$$\begin{cases} \delta^{\mathrm P} z = \bar\varepsilon_{\mathrm L}\chi_{\mathrm L}\,, \\ \delta^{\mathrm P}\chi_{\mathrm L} = \tfrac{1}{2}\check D^{\mathrm P} z\varepsilon_{\mathrm R} + \tfrac{1}{2}h'\varepsilon_{\mathrm L}\,, \\ \delta^{\mathrm P} h' = \bar\varepsilon_{\mathrm R}\check D^{\mathrm P}\chi_{\mathrm L} - \tfrac{1}{3}\bar\varepsilon_{\mathrm L}\chi_{\mathrm L} u - \tfrac{1}{6}i\bar\varepsilon_{\mathrm R}\check A\chi_{\mathrm L}\,, \\ \delta^{\mathrm P} C = \tfrac{1}{2}i\bar\varepsilon_{\mathrm L} S_{\mathrm L} + \text{h.c.}\,, \\ \delta^{\mathrm P} S_{\mathrm L} = \tfrac{1}{2}iH'\varepsilon_{\mathrm L} - \tfrac{1}{2}\check B'\varepsilon_{\mathrm R} - \tfrac{1}{2}i\check D^{\mathrm P} C\varepsilon_{\mathrm R}\,, \\ \delta^{\mathrm P} H' = -i\bar\varepsilon_{\mathrm R}\check D^{\mathrm P} S_{\mathrm L} - i\bar\varepsilon_{\mathrm R}\lambda'_{\mathrm R} + \tfrac{1}{3}i\bar S_{\mathrm L}\varepsilon_{\mathrm L} u + \tfrac{1}{6}\bar S_{\mathrm L}\check A\varepsilon_{\mathrm R}\,, \\ \delta^{\mathrm P} B'_m = [-\tfrac{1}{2}\bar\varepsilon_{\mathrm L}(\check D^{\mathrm P}_m - \tfrac{1}{2}iA_m) S_{\mathrm L} - \tfrac{1}{2}\bar\varepsilon_{\mathrm L}\gamma_m\lambda'_{\mathrm R} + \\ \qquad\qquad + \tfrac{1}{12}\bar S_{\mathrm R}\gamma_m(u\varepsilon_{\mathrm L} + i\check A\varepsilon_{\mathrm R})] + \text{h.c.}\,, \\ \delta^{\mathrm P}\lambda'_{\mathrm R} = \tfrac{1}{2}(\sigma\cdot F^{\mathrm P} - iD')\varepsilon_{\mathrm R}\,, \\ \delta^{\mathrm P} D' = [\tfrac{1}{2}i\bar\varepsilon_{\mathrm L}(\check D^{\mathrm P} - \tfrac{1}{2}i\check A)\lambda'_{\mathrm R}] + \text{h.c.}\,, \end{cases}$$

with

(50) $$\hat{F}^P_{mn} = 2D_{[m} B'_{n]} + \bar{\psi}_{[m} \hat{D}^P_{n]} S + \bar{\psi}_{[m} \gamma_{n]} \lambda' + \bar{S} e^\mu{}_{[m} D_{n]} \psi_\mu \,.$$

We now extend our rules of tensor calculus by introducing a Yang-Mills group [6]. To maintain both supersymmetry and gauge invariance, the space-time–dependent parameters of Yang-Mills transformations must be assigned to an entire chiral multiplet.

The Lie-algebra–valued transformation parameters are

(51) $$\Lambda = \tilde{g}^\alpha \Lambda^\alpha T_i^{\alpha j},$$

where $\tilde{g}^\alpha$ are gauge coupling constants, T^α are representation matrices for the generators of the group G, and Λ^α are scalar multiplets with Weyl weight $w = 0$. For a simple group, $\tilde{g}^\alpha = \tilde{g}$. We will often use the simpler notation $\tilde{g}$ for the set of gauge coupling constants $\tilde{g}^\alpha$ in the case of semi-simple or not semi-simple gauge groups G.

The components of the parameter multiplet Λ^α are denoted by

(52) $$\Lambda^\alpha = (y^\alpha, \varrho^\alpha, v^\alpha)\,.$$

For the Yang-Mills connection (gauge potential) we introduce a Lie-algebra-valued vector multiplet

(53) $$V = \tilde{g} V^\alpha T_i^{\alpha j},$$

which, under a finite Yang-Mills variation, transforms as

(54) $$\exp[2V] \to \exp[-i\bar{\Lambda}] \exp[2V] \exp[i\Lambda]\,.$$

In the infinitesimal, eq. (54) reads

(55) $$\delta V = \frac{i}{2} i(\Lambda - \bar{\Lambda}) + \frac{1}{2} i[V, \Lambda + \bar{\Lambda}] + \frac{1}{6} i[V, [V, \Lambda - \bar{\Lambda}]] + O(V^3)\,.$$

If we are interested in discussing a supersymmetric Yang-Mills theory, we can take gauge choices for all Λ transformations except the real scalar Re y, which actually corresponds to the standard Yang-Mills parameter in ordinary space-time.

The Wess-Zumino gauge is defined by the following gauge choice:

(56) $$C^\alpha = S^\alpha = H^\alpha = 0$$

for the components of V^α.

The constraints in eq. (55) are not invariant under supersymmetry and gauge transformations, since under these combined transformations we get

(57) $$\begin{cases} \delta C^\alpha = \tfrac{1}{2} i(y - y^*), \\ \delta S_L^\alpha = \varrho_L^\alpha - \tfrac{1}{2} \check{B}\varepsilon_R, \\ \delta H^\alpha = -iv^\alpha - \tfrac{1}{2} i\bar{\varepsilon}_R \gamma^\mu \check{B}^\alpha \psi_R - i\bar{\varepsilon}_R \lambda_R^\alpha. \end{cases}$$

Therefore, the Wess-Zumino gauge is only invariant for $y = y^*$, and under a redefined supersymmetry transformation

(58) $$\delta(\varepsilon) = \delta_\varrho(\varepsilon) + \delta(\varrho_L^\alpha = \tfrac{1}{2} \check{B}^\alpha \varepsilon_R) + \delta(v^\alpha = -\tfrac{1}{2} \bar{\varepsilon}_R \gamma^\mu \check{B}^\alpha \psi_{\mu R} - \bar{\varepsilon}_R \lambda_R^\alpha).$$

By applying this rule to the components of V^α in the Wess-Zumino gauge

(59) $$V_{WZ}^\alpha = (0, 0, 0, \lambda_R^\alpha, B^\alpha, D^\alpha),$$

we get

(60) $$\begin{cases} \delta B_\mu = -\tfrac{1}{2} \bar{\varepsilon}_L \gamma_\mu \lambda_R - \tfrac{1}{2} \bar{\varepsilon}_R \gamma_\mu \lambda_L - D_\mu y, \\ \delta \lambda_R = \tfrac{1}{2}(\sigma \cdot \hat{F} - ID)\varepsilon_R + i[\lambda_R, y], \\ \delta D = \tfrac{1}{2} i\bar{\varepsilon}_L \hat{\check{D}} \lambda_R - \tfrac{1}{2} i\bar{\varepsilon}_R \hat{\check{D}} \lambda_L + i[D, y], \end{cases}$$

where $\hat{F}_{\mu\nu}$ and $\hat{D}_\mu$ are now covariant derivatives with respect to local supersymmetry and Yang-Mills transformations.

For chiral multiplets S^i transforming according to some (generally complex) representation of G, we have the finite transformation

(61) $$S \to \exp[i\Lambda] S$$

or, in the infinitesimal,

(62) $$\delta S = -i\Lambda S.$$

Equation (62), when written in components, reads

(63) $$\delta S' = -i(yz, y\chi_L + \varrho_L z, yh + vz - \bar{\varrho}_L \chi_L).$$

Therefore, in the transformation law given by eq. (29) all derivatives are now replaced by Yang-Mills covariant derivatives, and in δh there is an extra term due to the last term in eq. (58):

(64) $$\delta' h = i\bar{\varepsilon}_R \lambda_R z = i\tilde{g}\bar{\varepsilon}_R \lambda_R^\alpha T_i^{\alpha j} z_j.$$

We end this section by giving the vector multiplication of two scalar multiplets which can actually be converted in a chiral multiplication of a chiral multiplet S_1 with the « kinetic multiplet » of S_2, $T(S_2)$, which is the curved generalization of the flat-space superfield $\bar{D}\bar{D}\bar{S}$.

The components of the multiplet $T(S)$ are

(65) $$T(S_i) = (h^{*i} - \tfrac{1}{3}u^*z^{*i},\ \tilde{\psi}_L^i,\ \tilde{H}^i),$$

where

(66) $$\tilde{\psi}_L^i = \check{D}^{\text{P}}\chi_R^i + \tfrac{1}{6}i\check{A}\chi_R^i + \tfrac{1}{6}\gamma \cdot R_R^{\text{P}} z^{*i} - i\tilde{g}z^{*j}T_j^{\alpha i}\lambda_L^\alpha,$$

(67) $$\tilde{H}^i = \hat{\Box}^c z^{*i} - \tfrac{2}{3}u(h^{*i} + \tfrac{1}{3}u^*z^{*i}) + \tilde{g}z^{*j}T_j^{\alpha i}D^\alpha - 2i\tilde{g}\bar{\chi}_R^j T_j^{\alpha i}\lambda_R^\alpha,$$

and $\hat{\Box}^c$ denotes the conformal d'Alembertian with Yang-Mills covariant derivatives.

Using the chiral density formula for the new chiral multiplet

(68) $$C_j^i S_{1i}T(S_{2j}) + C_j^{i*}S_{2j}T(S_{1i}),$$

we get a new local density corresponding to the vector multiplication of chiral multiplets instead of their chiral multiplication as given by eq. (36). The final result for the local density associated with the chiral multiplet eq. (68) is

(69) $e^{-1}LC_j^i S_{1i}T(S_{ij}) + C_j^{i*}S_{2j}T(S_{1i}) =$

$= C_j^i[\tfrac{1}{6}(R + e^{-1}\varepsilon^{\mu\nu\varrho\sigma}\bar{\psi}_\mu\gamma_5\gamma_\nu D_\varrho \psi_\sigma)z_{1i}z_2^{*j} - \hat{D}_\mu z_{1i}\hat{D}^\mu z_2^{*j} -$

$- \bar{\chi}_{1Li}\check{D}\chi_{2R}^j - \bar{\chi}_{2R}^j\check{D}\chi_{1Li} + (h_{1i} + \tfrac{1}{3}uz_{1i})(h_2^{*j} + \tfrac{1}{3}u^*z_2^{*j}) -$

$- \tfrac{1}{9}A_x^2 z_{1i}z_2^{*j} + \tfrac{1}{3}iA^m(-\bar{\chi}_{1Li}\gamma_m\chi_{2R}^j + z_2^{*j}\hat{D}_m z_{1i} - z_{1i}\hat{D}_m z_2^{*j}) +$

$+ \tfrac{1}{8}e^{-1}\varepsilon^{\mu\nu\varrho\sigma}\bar{\psi}_\mu\gamma_\nu\psi_\varrho(z_{1i}\hat{D}_\sigma z_2^{*j} - z_2^{*j}\hat{D}_\sigma z_{1i}) +$

$+ \tfrac{1}{3}\bar{\chi}_{1Li}\sigma^{\nu\mu}D_\mu\psi_\nu z_2^{*j} + \tfrac{1}{3}\bar{\chi}_{2R}^j\sigma^{\nu\mu}D_\mu\psi_\nu z_{1i} +$

$+ \bar{\chi}_{1Li}\sigma^{\nu\mu}\psi_\nu\hat{D}_\mu z_2^{*j} + \bar{\chi}_{2R}^j\sigma^{\nu\mu}\psi_\nu\hat{D}_\mu z_{1i} - D_\mu\bar{\chi}_{1Li}\sigma^{\nu\mu}\psi_\nu z_2^{*j} -$

$- D_\mu\bar{\chi}_{2R}^j\sigma^{\nu\mu}\psi_\nu z_{1i} + 2i\tilde{g}T_i^{\alpha k}\bar{\lambda}_L^\alpha\chi_{1Lk}z_2^{*j} - 2i\tilde{g}\bar{\chi}_{2R}^k T_k^{\alpha j}\lambda_R^\alpha z_{1i} +$

$+ \tilde{g}D^\alpha T_i^{\alpha k}z_{1k}z_2^{*j} - \tfrac{1}{2}i\tilde{g}\bar{\psi}\cdot\gamma\gamma_5\lambda^\alpha T_i^{\alpha k}z_{1k}z_2^{*j}] + \text{h.c.}$,

where $C_j^i z_{1i}z_2^{*j}$ is invariant under G, i.e.

(70) $$C_j^i T_i^{\alpha k}z_{1k}z_2^{*j} - C_j^i T_k^{\alpha j}z_{1j}z_2^{*k} = 0.$$

The other (antisymmetric) combination of the chiral multiplets,

(71) $$C_j^i S_{1i}T(S_{2j}) - C_j^{i*}S_{2j}T(S_{1i}),$$

gives rise to a density formula which is a total derivative.

Equation (69) produces the kinetic terms of the chiral fields and their nonpolynomial generalizations.

5. – Manifestly invariant actions and transformation laws.

In this section we give the matter-coupled supergravity Lagrangian in its superspace form and its component form together with the transformation laws of the component fields under local Poincaré supersymmetry transformation [6].

Let us start with the most general supersymmetric superspace action for a set of n chiral multiplets. This is given in flat superspace by

$$\text{(72)} \qquad \int d^4x\, d^4\theta\, \Phi(S, \bar{S}) + \int d^4x\, \text{Re} \int d^2\theta\, g(S),$$

where $\Phi(S, \bar{S})$ is an arbitrary real function of the chiral multiplets S_i and their complex conjugate $\bar{S}^i$, and $g(S)$ is a chiral function constructed out of the chiral multiplets S^i. The first component of $g(S)$, $g(z_i)$, is often called the superpotential. The first term in eq. (72) gives rise to a D-type density, whilst the second term gives rise to a F-type density. They are, respectively, the nonpolynomial generalizations of the scalar kinetic term $S_i \bar{S}^i$ of global supersymmetry and the self-interaction term (including the mass term) of the chiral multiplets S^i. If the chiral multiplets S_i transform according to some representation of a compact Lie group G, the action is assumed to be G-invariant. This requires the following group properties on the scalar functions $\Phi(z, z^*)$ and $g(z)$:

$$\text{(73)} \qquad \begin{cases} z^{*j} T_j^{\alpha i} \Phi_i = \Phi^i T_j^{\alpha i} z_i, & g^i T_j^{\alpha i} z_i = 0, \\ \left(\Phi^i = \dfrac{\partial \Phi}{\partial z_i}, \quad \Phi_i = \dfrac{\partial \Phi}{\partial z^{*i}}, \quad g^i = \dfrac{\partial g}{\partial z_i} \right). \end{cases}$$

The extension of the action (eq. (72)) to local supersymmetry and to a local Yang-Mills group G needs the following modification to eq. (72) [12]:

$$\text{(74)} \qquad \int d^4x\, d^4\theta E \left\{ \Phi(S, \bar{S} e^{2V}) + \text{Re}\left[\frac{1}{R} g(S)\right] \right\},$$

where V is the gauge vector multiplet of G, E is the superspace determinant, and R is the chiral scalar curvature superfield [12]. The component form of eq. (74) can be computed with the rules of the tensor calculus developed in the previous section.

The remaining part of the Yang-Mills supergravity coupling corresponds to the supergravity extension of the flat Yang-Mills action:

$$\int d^4x \, \text{Re} \int d^2\theta \, W_a^\alpha \varepsilon^{ab} W_b^\alpha \,, \tag{75}$$

where W_a^α is the field strength chiral multiplet.

We can extend eq. (75), including nonpolynomial interaction with the chiral multiplets S_i, as follows [6]:

$$\int d^4x \, \text{Re} \int d^2\theta [f_{\alpha\beta}(S) \, W_a^\alpha \varepsilon^{ab} W_b^\beta] \,, \tag{76}$$

where $f_{\alpha\beta}(S)$ is a chiral superfield transforming as the symmetric product of the adjoint representation of G.

The curved-space generalization of eq. (76) is

$$\int d^4x \, d^4\theta \, E \, \text{Re} \left[\frac{1}{R} f_{\alpha\beta}(S) \, W_a^\alpha \varepsilon^{ab} W_b^\beta \right] \tag{77}$$

and we can compute eq. (77) using the local-density formula for chiral multiplets obtained in sect. 4. In principle, under the requirement that the field Lagrangian contains only first derivatives in fermion fields and second derivatives in scalar fields, we could also have new invariants with higher power of the field strength multiplet W_a^α, such as $f_{\alpha_1 \ldots \alpha_n}(S) W^{\alpha_1} \ldots W^{\alpha_n}$. We will not consider these terms here. All other possible modifications of eqs. (74) and (77) would necessarily contain higher derivatives of boson and fermion fields. These terms cannot appear in the tree level Lagrangian if we require normal propagation of the physical fields, but they may well appear in the effective action as a result of radiative corrections, in exact analogy to the high-curvature terms which appear in the quantum loop expansion of the Einstein theory.

In global supersymmetry, if the gauge group G contains an Abelian factor U_1, there is an extra possible invariant—the so-called Fayet-Iliopoulos term [13]

$$\int d^4x \, d^4\theta \, V \,. \tag{78}$$

The naive extension of eq. (78) to curved space,

$$\int d^4x \, d^4\theta \, EV \,, \tag{79}$$

is not gauge invariant, since under a U_1 transformation $E \to E$ and $V \to V + (\Lambda - \overline{\Lambda})$. Then we have to modify both eq. (79) and the U_1 gauge transformation in order to preserve a U_1 gauge invariance [14].

The Fayet-Iliopoulos term can be introduced in curved superspace as the result of the gauging of an R-symmetry [15], which is defined as follows

(80) $$\begin{cases} V_R \to V_R + i\colon g_R(\Lambda - \bar\Lambda), \\ S_0 \to \exp[i\Lambda]S_0, \quad \bar S_0 \to \exp[-i\bar\Lambda]\bar S_0, \\ S_i \to \exp[-in_i\Lambda]S_i, \quad \bar S^i \to \exp[in_i\bar\Lambda]\bar S_i, \end{cases}$$

and V_R is the gauge multiplet for this U_1 gauge group. The extension of eqs. (74) and (76) to local R-symmetric interactions requires the following condition on the functions $\Phi(S,\bar S)$, $g(S)$ and $f_{\alpha\beta}(S)$ [11]:

(81) $$\begin{cases} \Phi(S_i,\bar S^i) = \Phi\left(\exp[-in_iq]S_i,\exp[in_iq]\bar S_i\right), \\ g(S_i) = \exp[3iq]\exp[-in_iq]S_i, \quad f_{\alpha\beta}(S_i) = f_{\alpha\beta}\exp[-in_iq]S_i, \end{cases}$$

where q is a real constant parameter.

Equations (81) imply that Φ and $f_{\alpha\beta}$ are R-symmetric, but the superpotential $g(S)$ must transform with a given phase under R-symmetry. This is due to the noninvariance of the chiral measure E/R in eq. (74) or, equivalently, to the transformation property of the compensating multiplet S_0 in the density formula given by eq. (43).

Under the restriction (81) the most general Lagrangian density invariant under the Yang-Mills group $C \times U_{R,1}$ is [11]

(82) $$\int d^4x\, d^4\theta\, E\left[\exp[-g_R V_R]\Phi(S_i,\bar S^i \exp[n_i g_R V_R])\exp[2V]\right] + \\ + \mathrm{Re}[\tfrac{1}{2}g(S_i) + f_{\alpha\beta}(S_i)W^\alpha W^\beta + f_R(S_i)W_R^2],$$

or equivalently, using the density formulae of sect. 4,

(83) $$L = -\tfrac{1}{2}\left[S_0 \exp[-g_R V_R]\bar S_0\, \Phi(S_i,\bar S^i \exp[n_i g_R V_R])\exp[2V])\right]_D + \\ + [(S_i)S_0^3]_F - [f_{\alpha\beta}(S_i)W^\alpha W^\beta]_F - [f_R(S_i)W_R^2]_F,$$

where W_R is the field strength multiplet for V_R, $g_R = k^2\xi$, where ξ is a dimensional parameter (dim $\xi = 2$), and g_R is the R-gauge coupling constant. We will non derive the component expression of eqs. (74) and (77), and commment at the end on their generalization to the case of a gauged R-symmetry (eq. (83)).

In order to compute the following density formulae,

(84) $$[S_0 \bar S_0\, \Phi(S,\bar S \exp[2V])]_{\hat D},$$

(85) $$[g(S)S_0^3]_F,$$

(86) $$[f_{\alpha\beta}W^\alpha W^\beta]_F = [fW^2]_F,$$

we have to work out the components of the vector multiplet $\Phi(S, \bar{S} \exp[2V])$ and of the chiral multiplets $g(S)$ and fW^2, respectively.

This can be done using the multiplication rules (36), (37) and (68) and the Wess-Zumino gauge choice for the vector multiplet V^α. In this gauge $V^\alpha V^\beta V^\gamma = 0$, and we get

$$(87) \qquad \Phi(S, \bar{S} e^2) = \Phi(S, \bar{S}) + 2\tilde{g}\Phi_i \bar{S}^j T_j^{\alpha i} V^\alpha + $$
$$+ 2\tilde{g}^2 (\Phi_{ij} \bar{S}^k T_k^{\alpha i} \bar{S}^l T_l^{\beta j} + \Phi_i \bar{S}^k T_k^{\alpha j} T_j^{\beta i}) V^\alpha V^\beta .$$

The components $C(\Phi)$, $S_L(\Phi)$, $H(\Phi)$, $B_m(\Phi)$, $\lambda_R(\Phi)$ and $D(\Phi)$ of the local vector multiplet given in eq. (87) are given by [6]

$$(88) \quad \begin{cases} C(\Phi) = \Phi(z, z^*), \\ S_L(\Phi) = -2i\Phi^i \chi_{Li}, \\ H(\Phi) = -2\Phi^i h_i + 2\Phi^{ij} \bar{\chi}_{Li} \chi_{Lj}, \\ B_m(\Phi) = i\Phi^i \hat{D}_m z_i - i\Phi_i \hat{D}_m z^{*i} - 2i\Phi_i^j \bar{\chi}_R^i \gamma_m \chi_{Lj}, \\ \lambda_R(\Phi) = -2i\Phi_i^j h_j \chi_R^i + 2i\Phi_k^{ij} \chi_R^k \bar{\chi}_{Li} \chi_{Lj} + 2i\Phi_i^j \hat{D} z^{*i} \chi_{Lj} + 2\tilde{g} \lambda_R^\alpha \tilde{D}^\alpha, \\ D(\Phi) = 2\Phi_j^i h_i h^{*j} - 2\Phi_k^{ij} \bar{\chi}_{Li} \chi_{Lj} h^{*k} - 2\Phi_{ij}^k \bar{\chi}_R^i \chi_R^j h_k + \\ \qquad + 2\Phi_{kl}^{ij} \bar{\chi}_{Li} \chi_{Lj} \bar{\chi}_R^k \chi_R^l - 2\Phi_i^j \hat{D}_\mu z^{*i} \hat{D}^\mu z_j - \\ \qquad - 2\Phi_i^j \bar{\chi}_{Lj} \overleftrightarrow{\hat{D}} \chi_L^i - 2\Phi_{jk}^i \hat{D}_\mu z^{*k} - \Phi_j^{ik} \hat{D}_\mu z_k \bar{\chi}_{Li} \gamma^\mu \chi_F^j - \\ \qquad + 2\tilde{g} D^\alpha \tilde{D}^\alpha + 4i\tilde{g} \bar{\lambda}_L^\alpha \tilde{D}^{\alpha i} \lambda_{Li} - 4i\tilde{g} \bar{\lambda}_R^\alpha D_i^\alpha \chi_R^i, \end{cases}$$

where

$$\tilde{D}^\alpha = \Phi_i T_j^{\alpha i} z^{*j} = \Phi^i T_i^{\alpha j} z_j, \quad \tilde{D}^{\alpha i} = \frac{\partial \tilde{D}^\alpha}{\partial z} \quad D_i^\alpha = \frac{\partial \tilde{D}^\alpha}{\partial z^{*i}}$$

(the derivatives $\hat{D}_\mu$ are supersymmetric and Yang-Mills covariant).

Analogously, the components of the two chiral multiplets $g(S)$ and fW^2 are given respectively by [6]

$$(89) \qquad g(S) = [g(z), \chi_L(g) = g^i \chi_{Li}, h(g) = g^i h_i - g^{ij} \bar{\chi}_{Li} \chi_{Lj}],$$

$$(90) \qquad fW^2 = [z(fW^2), \chi_L(fW^2), h(fW^2)],$$

with

(91)
$$\begin{cases} z(fW^2) = -\tfrac{1}{2} f_{\alpha\beta} \bar{\lambda}_L^\alpha \lambda_L^\beta, \\ \chi_L(fW^2) = \tfrac{1}{2} f_{\alpha\beta}(\sigma\cdot \hat{F}^{-\alpha} - iD^\alpha) \lambda_L^\beta - \tfrac{1}{2} f_{\alpha\beta}^i \chi_{Li} \bar{\lambda}_L^\alpha \lambda_L^\beta, \\ h(fW^2) = f_{\alpha\beta}(-\bar{\lambda}_L^\alpha \check{D} \lambda_R^\beta - \tfrac{1}{2} \hat{F}_{\mu\nu}^{-\alpha} F_{\mu\nu}^{-\beta} + \tfrac{1}{2} D^\alpha D^\beta) + \\ \qquad + f_{\alpha\beta}^i \bar{\chi}_{Li}(-\sigma\cdot \hat{F}^{-\alpha} + iD^\alpha) \lambda_L^\beta - \tfrac{1}{2} f_{\alpha\beta}^i h_i \bar{\lambda}_L^\alpha \lambda_L^\beta + \tfrac{1}{2} f_{\alpha\beta}^{ij} \bar{\chi}_{Li} \chi_{Lj} \bar{\lambda}_L^\alpha \lambda_L^\beta. \end{cases}$$

Using the density formulae (43) and (45), we finally get

(92)
$$\begin{aligned} e^{-1} L(\Phi) &= -\tfrac{1}{6} \Phi e^{-1} L_{SO} + \Phi_j^i(-\tfrac{1}{2} D_\mu z_i D^\mu z^{*j} - \bar{\chi}_{Li} \check{D} \chi_R^j + \tfrac{1}{2} h_i h^{*j}) - \\ &\quad - \Phi_k^{ij} \bar{\chi}_{Li} \chi_{Lj} h^{*k} + \Phi_k^{ij} \bar{\chi}_{Li} \check{D} z_j \chi_R^k + \\ &\quad + \tfrac{1}{2} \Phi_k^{ij} \bar{\chi}_{Li} \chi_{Lj} \bar{\chi}_R^k \chi_R^l + \tfrac{1}{3} u^*(\Phi^i h_i - \Phi^{ij} \chi_{Li} \chi_{Lj}) + \\ &\quad + \tfrac{1}{3} i A^\mu [\tfrac{1}{2} \Phi_j^i \bar{\chi}_R^j \gamma_\mu \chi_{Li} + \Phi^i (D_\mu z_i - \bar{\psi}_{\mu L} \chi_{Li})] + \\ &\quad + \Phi_j^i \bar{\psi}_{\mu L} \check{D} z^{*j} \gamma_\mu \chi_{Li} - \tfrac{4}{3} \Phi^i \bar{\chi}_{Li} \sigma^{\mu\nu} D_\mu \psi_{\nu L} - \\ &\quad - \tfrac{1}{8} e^{-1} \varepsilon^{\mu\nu\varrho\sigma} \bar{\psi}_\mu \gamma_\nu \psi_\varrho (\Phi^i D_\sigma z_i + \tfrac{1}{2} \Phi_i^j \bar{\chi}_R^i \gamma_\sigma \chi_{Lj}) - \\ &\quad - \tfrac{1}{2} \Phi_j^i \bar{\psi}_{\mu R} \chi_R^j \bar{\psi}_L^\mu \chi_{Li} + \tfrac{1}{6} \Phi^i \bar{\chi}_{Li} (\bar{\psi}_{\mu L} \bar{\psi}\cdot \gamma \psi^\mu + \\ &\quad + \tfrac{1}{2} \sigma^{\mu\nu} \psi_L^\varrho \bar{\psi}_\nu \gamma_\varrho \psi_\mu + \sigma^{\mu\nu} \psi_{\nu L} \bar{\psi}_\mu \gamma\cdot \psi) + \\ &\quad + \tfrac{1}{2} \tilde{g} \Phi^i T_i^{\alpha j} z_j (D^\alpha + i \bar{\psi}_L\cdot \gamma \lambda_R^\alpha) - 2 i \tilde{g} \Phi_i^j T_j^{\alpha k} z_k \bar{\lambda}_R^\alpha \chi_R^i + \text{h.c.}, \end{aligned}$$

(93)
$$\begin{aligned} e^{-1} L(g) &= -\tfrac{1}{2} g^{ij} \bar{\chi}_{Li} \chi_{Lj} + \tfrac{1}{2} g^i h_i + \tfrac{1}{2} g u + \tfrac{1}{2} \bar{\psi}_R \cdot \gamma \chi_{Li} + \\ &\quad + \tfrac{1}{2} g \bar{\psi}_{\mu R} \sigma^{\mu\nu} \psi_{\nu R} + \text{h.c.}, \end{aligned}$$

(94)
$$\begin{aligned} e^{-1} L(fW^2) &= \tfrac{1}{2} f_{\alpha\beta}[-\tfrac{1}{4} F_{\mu\nu}^\alpha F_{\mu\nu}^\beta - \tfrac{1}{2} \bar{\lambda}^\alpha \check{D} \lambda^\beta + \tfrac{1}{2} D^\alpha D^\beta + \\ &\quad + \tfrac{1}{4} F_{\mu\nu}^\alpha \bar{\psi}_\varrho \sigma^{\mu\nu} \gamma^\varrho \lambda^\beta + \tfrac{1}{4} F_{\mu\nu}^\alpha \tilde{F}_{\mu\nu}^\beta - \tfrac{1}{2} D_\mu(\bar{\lambda}_L^\alpha \gamma^\mu \lambda_R^\beta)] + \\ &\quad + \tfrac{1}{2} f_{\alpha\beta}^i [\bar{\chi}_{Li}(-\sigma\cdot \hat{F}^{-\alpha} + iD^\alpha) \lambda_L^\beta - \tfrac{1}{2} h_i \bar{\lambda}_L^\alpha \lambda_L^\beta - \\ &\quad - \tfrac{1}{2} \bar{\psi}_R \cdot \gamma \chi_{Li} \bar{\lambda}_L^\alpha \lambda_L^\beta] + \tfrac{1}{4} f_{\alpha\beta}^{ij} \bar{\chi}_{Li} \chi_{Lj} \bar{\lambda}_L^\alpha \lambda_L^\beta + \text{h.c.} \end{aligned}$$

Here $\check{D}_\mu \lambda^\alpha$ is a supersymmetric and Yang-Mills covariant derivative but without the $\psi_\mu D^\alpha$ term. The other derivatives are Yang-Mills and general co-ordinate covariant but they have no ψ_μ torsion.

The overall Lagrangian corresponding to the superspace expressions of eqs. (74) and (77) is, therefore, given by [6]

(95) $$L = L(\Phi) + L(g) + L(fW^2).$$

The three terms in eq. (95) are separately invariant under local supersymmetry transformations.

These transformations have the following form, for the several fields involved in eq. (95):

$$(96) \begin{cases} \delta B_\mu^\alpha = -\tfrac{1}{2}\bar{\varepsilon}_L \gamma_\mu \lambda_R^\alpha - \tfrac{1}{2}\bar{\varepsilon}_R \gamma_\mu \lambda_L^\alpha \,, \\ \delta \lambda_R^\alpha = \tfrac{1}{2}\sigma^{\mu\nu}\hat{F}_{\mu\nu}^\alpha \varepsilon_R - \tfrac{1}{2} i D^\alpha \varepsilon_R \,, \\ \delta D^\alpha = \tfrac{1}{2} i \bar{\varepsilon}_L (\check{D}^P - \tfrac{1}{2} i \check{A}) \lambda_R - \tfrac{1}{2} i \bar{\varepsilon}_R (\check{D}^P + \tfrac{1}{2} i \check{A}) \lambda_L^\alpha \,, \\ \delta z_i = \bar{\varepsilon}_L \chi_{Li} \,, \\ \delta \chi_{Li} = \tfrac{1}{2}\check{D} z_i \varepsilon_R + \tfrac{1}{2} h_i \varepsilon_L \,, \\ \delta h_i = \bar{\varepsilon}_R (\check{D}^P - \tfrac{1}{2} i \check{A}) \chi_{Li} - \tfrac{1}{3}\bar{\chi}_{Li}(u\varepsilon_L + i \check{A}\varepsilon_R + i\tilde{g}\bar{\varepsilon}_R \lambda_R^\alpha T_i^{\alpha j} z_j) \,, \\ \delta e_\mu^m = \tfrac{1}{2}\bar{\varepsilon}_L \lambda^m \psi_{\mu R} + \tfrac{1}{2}\bar{\varepsilon}_R \gamma^m \psi_{\mu L} \,, \\ \delta \psi_{\mu L} = [\partial_\mu + \tfrac{1}{2}\omega_{\mu nm}(e,\psi)^{mn} + \tfrac{1}{2} i A_\mu]\varepsilon_L + \tfrac{1}{6}\gamma_\mu(u^* \varepsilon_R + i \check{A}\varepsilon_L) \,, \\ \delta A_\mu = \tfrac{3}{4} i \bar{\varepsilon}_L (R_\mu^P - \tfrac{1}{3}\gamma_\mu \gamma \cdot R^P)_L + \text{h.c.} \,, \\ \delta u = \tfrac{1}{2}\bar{\varepsilon}_R \gamma \cdot R_L^P \,. \end{cases}$$

The overall Lagrangian given by eq. (95) contains auxiliary field components of the multiplets S^i, V^α, and the supergravity multiplet

$$(97) \qquad (e_m, \psi_{\mu a}, u, A_m) \,.$$

In addition, an appropriate Weyl rescaling of the vierbein field, as well as redefinitions of the fermion fields have to be performed in the first term of eq. (95), in its standard form [5]. These redefinitions, as well as the elimination of the auxiliary fields h_i, D^α, u and A_m, will be done in the next section.

These manipulations will allow us to recast the final Lagrangian and transformation rules in a simple form in terms of the particle fields

$$(98) \qquad (B_m^\alpha, \lambda_R^\alpha)\,, \quad (z_i, \chi_{Li})\,, \quad (e_{\mu m}, \psi_{\mu a})\,.$$

We end this section by showing (using simple superspace arguments) that the action given by eq. (95) depends only on a particular combination of the functions $\Phi(S, \bar{S})$ and $g(S)$ [5].

To prove this, we make a super-Weyl rescaling on the supervierbein so that its superdeterminant transforms as

$$(99) \qquad E \to E' \exp[\Sigma + \bar{\Sigma}]\,,$$

where Σ is a chiral parameter superfield. Under the same rescaling the second term in eq. (74) transform as

(100) $$\int d^4x\, d^4\theta\ \mathrm{Re}\, \frac{E'}{R'} \left[(\exp[3\Sigma])\, g(S) \right],$$

whilst the Yang-Mills part (eq. (77)) keeps the same form

(101) $$\int d^4x\, d^4\theta\ \mathrm{Re} \left[\frac{E'}{R'} f_{\alpha\beta}(S)\, W^\alpha W^\beta \right].$$

If we now choose Σ such that

(102) $$\exp[3\Sigma]\, g(S) = 1, \quad i.e.\ \exp[\Sigma] = g(S)^{-\frac{1}{3}},$$

we finally obtain that eq. (74) becomes

(103) $$\int d^4x\, d^4\theta\, E' \exp\left[\frac{1}{3} G(S, \bar{S})\right] \exp[2V] + \int d^4x\, d^4\theta\ \mathrm{Re}\left(\frac{E'}{R'}\right),$$

where the function G is given by

(104) $$G(z, z^*) = 3 \log \Phi(z, z^*) - \log|g(z)|^2,$$

so that

(105) $$\exp\left[\frac{1}{3} G(z, z^*)\right] = \frac{\Phi(z, z^*)}{|g(z)|^{\frac{2}{3}}}.$$

The use of the function $G(S, \bar{S})$ makes the introduction of the Fayet-Iliopoulos term in formula (82) trivial [11, 16]. In fact, in this case we simply get

(106) $$\exp\left[\frac{1}{3} G\big(S, \bar{S} \exp[n_i g_\mathrm{R} V_\mathrm{R}] \exp[2V]\big)\right] =$$
$$= \frac{\Phi\big(S_i, \bar{S} \exp[n_i g_\mathrm{R} V_\mathrm{R}] \exp[2V]\big)}{[g^*(\bar{S} \exp[n_i g_\mathrm{R} V_\mathrm{R}] \exp[2V])\, g(S)]^{\frac{1}{3}}}.$$

This means that in the G superfield we simply have to replace $\bar{S} \exp[2V]$ with $\bar{S} \exp[n_i g_\mathrm{R} V_\mathrm{R}] \exp[2V]$, i.e. we have to covariantize with respect to the full gauge group $G \times U_{\mathrm{R},1}$.

We note that this simple rule does not hold in the case of a vanishing superpotential, $g(S_i) = 0$, since in this case the rescaling given by formula (102) can no longer be performed [11]. At the end of the next section we will give simple substitution rules for the Lagrangian in terms of physical fields which will enable us to go from the general form of $G(S, \bar{S})$ to the limiting case corresponding to $g(S) = 0$.

6. – Final form of the Lagrangian, transformation laws and gauged R-symmetry.

The overall action of the coupled-matter Yang-Mills supergravity multiplets is given in eq. (95).

In order to obtain the final form of the Lagrangian, we have to eliminate, by means of their field equations, the auxilairy fields h_i, D^α, u and A_m of the multiplets involved.

It is important to notice that, after their elimination, only the total Lagrangian (95)—but no the three separate pieces given by eqs. (92) to (94)—will be invariant under local supersymmetry transformations of the physical fields.

The part of the Lagrangian (95) which contains the auxiliary fields is

$$(107) \quad e^{-1} L_{\text{aux}} = \frac{1}{18} \Phi(uu^* - A_m A^m) + \frac{1}{2} \Phi_j^i h_i h^{*j} - \Phi_k^{ij} h^{*k} \bar{\chi}_{Li} \chi_{Lj} + \\
+ \frac{1}{2} g^i h_i - \frac{1}{4} f_{\alpha\beta}^i h_i \bar{\lambda}_L^\alpha \lambda_L^\beta + \frac{1}{3}\left(\Phi^i h_i - \Phi^{ij} \bar{\chi}_{Li} \chi_{Lj} + \frac{3}{2} g^*\right) + \\
+ \frac{1}{3} i A^\mu \left[\frac{1}{2} \Phi_j^i \bar{\chi}_R^j \gamma_\mu \chi_{Li} + \Phi^i(D_\mu z_i - \psi_{\mu L} \chi_{Li}) + \frac{3}{4} \bar{\lambda}_L^\alpha \gamma_\mu \lambda_R^\beta f_{\alpha\beta}\right] + \\
+ \frac{1}{2} \tilde{g} \Phi^i T_i^{\alpha j} z_j D^\alpha + \frac{1}{4} f_{\alpha\beta} D^\alpha D^\beta + \frac{i}{2} f_{\alpha\beta}^i \bar{\chi}_{Li} \lambda_L^\alpha D^\alpha + \text{h.c.}$$

If we define the new expressions

$$(108) \quad J = 3 \log \left[-\frac{\Phi}{3}\right]$$

and

$$(109) \quad \tilde{u} = u + J^i h_i ,$$

then eq. (107) reads

$$(110) \quad e^{-1} L_{\text{aux}} = \frac{1}{18} \Phi(\tilde{u}\tilde{u}^* - A_m A^m) + \frac{1}{6} \Phi J_j^i h_i h^{*j} + \frac{1}{2} g \left[\tilde{u} - h_i\left(J^i - \frac{g^i}{g}\right)\right] - \\
- \frac{1}{g} u^*\left(J^{ij} + \frac{1}{3} J^i J^j\right) \Phi \bar{\chi}_{Li} \chi_{Lj} - \frac{1}{3} \Phi h^{*k} \bar{\chi}_{Li} \chi_{Lj} \left(J_k^{ij} + \frac{2}{3} J_k^i J^j\right) - \frac{1}{4} f_{\alpha\beta}^i h_i \bar{\lambda}_L^\alpha \lambda_L^\beta + \\
+ \frac{1}{3} i A^\mu \left[\frac{1}{2} \Phi_j^i \bar{\chi}_R^j \gamma_\mu \chi_{Li} + \Phi^i(D_\mu z_i - \bar{\psi}_{\mu L} \chi_{Li}) - \frac{3}{4} \bar{\lambda}^\alpha \gamma_\mu \lambda_R^\beta f_{\alpha\beta}\right] + \\
+ \frac{1}{2} \tilde{g} \Phi^i T_i^{\alpha j} z_j D^\alpha + \frac{1}{4} f_{\alpha\beta} D^\alpha D^\beta + \frac{i}{2} f_{\alpha\beta}^i \bar{\chi}_{Li} \lambda_L^\alpha D^\beta + \text{h.c.}$$

The field equations for the auxiliary fields are

(111)
$$\begin{cases}
\tilde{u} = \dfrac{2}{\Phi}\left[-\dfrac{1}{2}g^* + \dfrac{1}{g}\left(J^{ij} + \dfrac{1}{3}J^iJ^j\right)\Phi\bar{\chi}_{Li}\chi_{Lj}\right], \\[4pt]
\dfrac{1}{3}\Phi h_i J^i_k = -\dfrac{1}{2}g^*\left(\dfrac{g^*_k}{g^*} - J_k\right) + \dfrac{1}{3}\Phi\left(J^{ij}_k + \dfrac{2}{3}J^i_k J^j\right)\bar{\chi}_{Li}\chi_{Lj} + \dfrac{1}{4}f^*_{\alpha\beta k}\bar{\lambda}^\alpha_R\lambda^\beta_R, \\[4pt]
\dfrac{2}{3}\Phi A_\mu = \dfrac{1}{2}i\Phi^i_j\bar{\chi}^j_R\gamma_\mu\chi_{Li} + i\Phi^i(D_\mu z_i - \bar{\psi}_{\mu L}\chi_{Li}) + \dfrac{3}{4}if_{\alpha\beta}\bar{\lambda}^\alpha_L\gamma_\mu\lambda^\beta_R + \text{h.c.}, \\[4pt]
-\operatorname{Re}f_{\alpha\beta}D^\beta = \tilde{g}\Phi^i T^{\alpha j}_i z_j + \dfrac{1}{2}if_{\alpha\beta}\bar{\chi}_{Li}\lambda^\beta_L - \dfrac{1}{2}if^*_{\alpha\beta i}\bar{\chi}^i_R\lambda^\beta_R.
\end{cases}$$

The insertion of eqs. (111) into eq. (110) finally gives

(112)
$$e^{-1}L = e^{-1}L_{\text{aux}} + e^{-1}\hat{L},$$

where

(113)
$$\begin{cases}
e^{-1}L_{\text{aux}} = -\dfrac{g}{\Phi}\left|\dfrac{1}{2}g^* - \dfrac{1}{g}\left(J^{ij} + \dfrac{1}{3}J^iJ^j\right)\Phi\bar{\chi}_{Li}\chi_{Lj}\right|^2 - \\[4pt]
\quad - \dfrac{3}{\Phi}(J^{-1})^k_l\left[\dfrac{1}{2}g^*\left(\dfrac{g^*_k}{g^*} - J_k\right) - \dfrac{1}{3}\Phi\left(J^{ij}_k + \dfrac{2}{3}J^i_k J^j\right)\bar{\chi}_{Li}\chi_{Lj} - \dfrac{1}{4}f^*_{\alpha\beta}\bar{\lambda}^\alpha_R\lambda^\beta_R\right]\cdot \\[4pt]
\quad \cdot\left[\dfrac{1}{2}g\left(\dfrac{g^l}{g} - J^l\right) - \dfrac{1}{3}\Phi\left(J^1_{mn} + \dfrac{2}{3}J^1_m J_n\right)\bar{\chi}^m_R\chi^n_R - \dfrac{1}{4}f^1_{\gamma\delta}\bar{\lambda}^\gamma_L\lambda^\delta_L\right] - \\[4pt]
\quad - \dfrac{1}{4\Phi}\bigg[\Phi^i_j\bar{\chi}^j_R\gamma_\mu\chi_{Li} - \Phi^i(D_\mu z_i - \bar{\psi}_{\mu L}\chi_{Li}) - \Phi_i(D_\mu z^{*i} - \bar{\psi}_{\mu R}\chi^i_R) + \\[4pt]
\quad + \dfrac{3}{2}\operatorname{Re}f_{\alpha\beta}\bar{\lambda}^\alpha_L\gamma_\mu\lambda^\beta_R\bigg]^2 - \dfrac{1}{2}(\operatorname{Re}f)^{-1}_{\alpha\beta}\left(\tilde{g}\Phi^i T^{\alpha j}_i z_j + \dfrac{i}{2}f^i_{\alpha\gamma}\bar{\chi}_{Li}\lambda^\gamma_L - \dfrac{1}{2}if^*_{\alpha\gamma i}\bar{\chi}^i_R\lambda^\gamma_R\right)\cdot \\[4pt]
\quad \cdot\left(\tilde{g}\Phi^k T^{\beta 1}_k z_1 + \dfrac{1}{2}if^k_{\beta\delta}\bar{\chi}_{Lk}\lambda^\delta_L - \dfrac{1}{2}if^*_{\beta\delta k}\bar{\chi}^k_R\lambda^\delta_R\right), \\[6pt]
e^{-1}\hat{L} = \dfrac{1}{12}\Phi\bigg\{R[\omega(e)] + \bar{\psi}_\mu\gamma_5\gamma_\nu D_\varrho\psi_\sigma\varepsilon^{\mu\nu\varrho\sigma} + \dfrac{1}{16}(\bar{\psi}_\mu\gamma_\lambda\psi_\varrho + 2\bar{\psi}_\lambda\gamma_\mu\psi_\varrho)\cdot \\[4pt]
\quad \cdot\bar{\psi}^\mu\gamma^\lambda\psi^\varrho - \dfrac{1}{4}(\bar{\psi}_\mu\gamma\cdot\psi)^2 + e^{-1}\partial_\mu(e\psi\cdot\gamma\psi^\mu)\bigg\} + \\[4pt]
\quad + \Phi^i_j\left(-\dfrac{1}{2}D_\mu z_i D^\mu z^{*j} - \bar{\chi}_{Li}\check{D}\chi^j_R\right) + \Phi^{ij}_k\bar{\chi}_{Li}\check{D}z_j\chi^k_R + \\[4pt]
\quad + \Phi^i_j\bar{\psi}_{\mu L}\check{D}z^{*j}\gamma^\mu\chi_{Li} - \dfrac{4}{3}\Phi^i\bar{\chi}_{Li}\sigma^{\mu\nu}D_\mu\psi_{\nu L} - \\[4pt]
\quad - \dfrac{1}{8}e^{-1}\varepsilon^{\mu\nu\varrho\sigma}\bar{\psi}_\mu\gamma_\nu\psi_\varrho\left(\Phi^i D_\sigma z_i + \dfrac{1}{2}\Phi^j_i\bar{\chi}^i_R\gamma_\sigma\chi_{Lj}\right) - \\[4pt]
\quad - \dfrac{1}{2}\Phi^i_j\bar{\psi}_{\mu L}\chi_{Li}\bar{\psi}^\mu_R\chi^j_R + \dfrac{1}{2}\Phi^{ij}_{kl}\bar{\chi}_{Li}\chi_{Lj}\bar{\chi}^k_R\chi^l_R +
\end{cases}$$

$$
\begin{aligned}
(113)\quad &+ \frac{1}{6}\Phi^i \chi_{Li}\left(\bar{\psi}_{\mu L}\,\psi\cdot\gamma\psi^\mu + \frac{1}{2}\sigma^{\mu\nu}\psi_L^\lambda\bar{\psi}_\nu\gamma_\varrho\psi_\mu + \sigma^{\mu\nu}\psi_{\nu L}\bar{\psi}_\mu\gamma\cdot\psi\right) + \\
&+ \frac{1}{2}i\tilde{g}\Phi^i T_i^{\alpha j} z_j\bar{\psi}_L\gamma\lambda_k - 2i\tilde{g}\Phi_k^i T_i^{\alpha j} z_j\bar\lambda_R^\alpha \chi_R^k - \\
&- \frac{1}{2}g^{ij}\bar{\chi}_{Li}\chi_{Lj} - \frac{1}{2}g^i\bar{\psi}_R\cdot\gamma\chi_{Li} + \frac{1}{2}g\bar{\psi}_{\mu R}\sigma^{\mu\nu}\psi_{\nu R} + \\
&+ \frac{1}{2}f_{\alpha\beta}\Bigl[-\frac{1}{4}F^\alpha_{\mu\nu}F^\beta_{\mu\nu} - \frac{1}{2}\bar\lambda^\alpha\check{D}\lambda^\beta + \frac{1}{2}\bar\lambda^\alpha\gamma^\mu\sigma^{\varrho\sigma}\psi_\mu F^\beta_{\varrho\sigma} + \\
&+ \frac{1}{4}\bar\lambda^\alpha\gamma^\mu\sigma^{\varrho\sigma}\psi_\varrho\bar\psi_\sigma\gamma_\sigma\lambda^\alpha - \frac{1}{16}\bar\lambda^\alpha\gamma^\mu\sigma^{mn}\lambda^\beta(2\bar\psi_\mu\gamma_m\psi_n + \bar\psi_m\gamma_\mu\psi_n) + \\
&+ \frac{1}{4}F^\alpha_{\mu\nu}\tilde{F}^\beta_{\mu\nu} - \frac{1}{2}D_\mu(\bar\lambda^\alpha_L\gamma^\mu\lambda^\beta_R)\Bigr] + \frac{1}{2}f^i_{\alpha\beta}\Bigl[-\bar\chi_{Li}\sigma_{\mu\nu}\lambda^\alpha_L\Bigl(F^{-\alpha}_{\mu\nu} - \frac{1}{2}\bar\psi_{\varrho L}\sigma_{\mu\nu}\gamma^\varrho\lambda^\alpha_R + \\
&+ \frac{1}{2}\bar\psi_R\cdot\gamma\sigma_{\mu\nu}\lambda^\alpha_L\Bigr) - \frac{1}{2}\bar\psi_R\cdot\gamma\chi_{Li}\bar\lambda^\alpha_L\lambda^\beta_L\Bigr] + \frac{1}{4}f^{ij}_{\alpha\beta}\bar\chi_{Li}\chi_{Lj}\bar\lambda^\alpha_L\lambda^\beta_L + \text{h.c.}
\end{aligned}
$$

In order to recast the Einstein term and the Rarita-Schwinger action in canonical form, as well as to keep the gaugino and chiral fermion terms in a quasi-canonical form, the following Weyl rescaling on the vierbein and the fermion fields must be performed:

$$
(114)\quad \begin{cases} e_{m\mu}\to e_{m\mu}\exp[\sigma], & e\to\exp[4\sigma]e, \\ \lambda\to\exp[-3\sigma/2]\lambda, & \chi\to\exp[-\sigma/2]\chi, \quad \psi_\mu\to\exp[\sigma/2]\psi_\mu, \end{cases}
$$

with

$$
(115)\quad \exp[2\sigma] = -\frac{3}{\Phi}, \qquad \sigma = -\frac{1}{6}J.
$$

Under the Weyl rescaling of the vierbein we get the standard changes for the Einstein curvature term and of the Lorentz connection:

$$
(116)\quad \begin{cases} \frac{1}{6}e\Phi R \to -\frac{1}{2}eR - \frac{3}{4}e(\partial_\mu\log\Phi)^2 + 4\text{-div}, \\ \omega_{\mu mn} \to \omega_{\mu mn} - 2e^\nu(m^e n)\mu^\delta\nu^\sigma. \end{cases}
$$

After the substitutions (114), the total Lagrangian becomes the sum of a pure bosonic part and a fermionic part:

$$
(117)\quad e^{-1}L = e^{-1}L_B + e^{-1}L_F,
$$

where

(118) $e^{-1} L_B = -\frac{1}{2} R + J^i_j D_\mu z_i Dz^{*j} - \frac{1}{4} \operatorname{Re} f_{\alpha\beta} F^\alpha_{\mu\nu} F^\beta_{\mu\nu} +$
$+ \frac{1}{4} i \operatorname{Im} f_{\alpha\beta} F^\alpha_{\mu\nu} \tilde{F}^\beta_{\mu\nu} + \frac{1}{4} |g|^2 \exp[-J] \left[3 + (J^{-1})^k_1 \left(\frac{g^*_k}{g^*} - J_k \right) \left(\frac{g^1}{g} - J^1 \right) \right] -$
$- \frac{1}{2} \tilde{g}^2 \operatorname{Re} f^{-1}_{\alpha\beta} (J^i T^{\alpha j}_i z_j)(J^k T^{\beta 1}_k z_1)$

and

(119) $e^{-1} L_F = \frac{1}{2} \exp[3\sigma] \left(J^{ij} + \frac{1}{3} J^i J^j \right) g \bar{\chi}_{Li} \chi_{Lj} +$
$+ \frac{1}{2} g \left(\frac{g^1}{g} - J^1 \right) \exp[3\sigma] J^{1k}_1 \left(J^{ij}_k + \frac{2}{3} J^i_k J^j \right) \bar{\chi}_{Li} \chi_{Lj} +$
$+ \frac{1}{2} g \left(\frac{g^1}{g} - J^1 \right) \frac{3}{4\Phi} e J^{-1k}_0 f^*_{\alpha\beta k} \bar{\lambda}^\alpha_R \lambda^\beta_R - \frac{1}{18} \Phi e^2 \cdot$
$\cdot \left(J^{ij} + \frac{1}{3} J^i J^j \right) \bar{\chi}_{Li} \chi_{Lj} \left(J_{k1} + \frac{1}{3} J_k J_1 \right) \bar{\chi}^k_R \chi^1_R -$
$- \frac{1}{6} \Phi e^2 J^{-1k}_1 \left(J^{ij}_k + \frac{2}{3} J^i_k J^j \right) \bar{\chi}_{Li} \chi_{Lj} \left(J^1_{mn} + \frac{2}{3} J^1_m J_n \right) \bar{\chi}^m_R \chi^n_R -$
$- \frac{1}{4} f^1_{\alpha\beta} \bar{\lambda}^\alpha_L \lambda^\beta_L J^{-1k}_1 \left(J^{ij}_k + \frac{2}{3} J^i_k J^j \right) \bar{\chi}_{Li} \chi_{Lj} - \frac{3}{32} \frac{1}{\Phi} \exp[-2\sigma] J^{-1k}_1 f^1_{\alpha\beta} \bar{\lambda}^\alpha_L \lambda^\beta_L f^*_{\gamma\delta k} \bar{\lambda}^\gamma_R \lambda^\delta_R -$
$- \frac{1}{2\Phi} \exp[2\sigma] \Phi^i_j \bar{\chi}^j_R \gamma^\mu \chi_{Li} \Phi^k D_\mu z_k - \frac{1}{2\Phi} \exp[2\sigma] \Phi^i D_\mu z_i (\Phi_j \bar{\psi}_\mu \chi^j_R - \Phi^{\bar{j}} \psi_{\mu L} \chi_{Lj}) -$
$- \frac{3}{4\Phi} \operatorname{Re} f_{\alpha\beta} \Phi^i D_\mu z_i \bar{\lambda}^\alpha \gamma_\mu \lambda^\beta_R - \frac{1}{8\Phi} (\Phi^i_j \bar{\chi}^j_R \gamma_m \chi_{Li})^2 +$
$+ \frac{1}{2\Phi} \Phi^i_j \bar{\chi}^j_R \gamma^\mu \chi_{Li} \Phi^k \bar{\psi}_{\mu L} \chi_{Lk} \exp[2\sigma] - \frac{3}{8\Phi} \Phi^i_j \chi^j_R \gamma^m \chi_{Li} \operatorname{Re} f_{\alpha\beta} \bar{\lambda}^\alpha_L \gamma_m \lambda^\beta_R +$
$+ \frac{3}{4\Phi} \Phi^i \bar{\psi}_{\mu L} \chi_{Li} \operatorname{Re} f_{\alpha\beta} \bar{\lambda}^\alpha_L \gamma^\mu \lambda^\beta_R + \frac{1}{4\Phi} \exp[2\sigma] \Phi^j \bar{\psi}_{\mu L} \chi_{Lj} (\Phi_i \bar{\psi}^i_{\mu R} \chi^i_R - \Phi^i \bar{\psi}_{\mu L} \chi_{Li}) -$
$- \frac{9}{32\Phi} \operatorname{Re} f_{\alpha\beta} \operatorname{Re} f_{\gamma\delta} \bar{\lambda}^\alpha_L \gamma^m \lambda^\beta_R \bar{\lambda}^\gamma_L \gamma_m \lambda^\delta_R \exp[-2\sigma] -$
$- \frac{1}{2} i \operatorname{Re} f^{-1}_{\alpha\beta} \tilde{g} \Phi^i T^{\alpha j}_i z_j f^k_{\beta\gamma} \bar{\chi}_{Lk} \lambda_L \exp[2\sigma] + \frac{1}{8} \operatorname{Re} f^{-1}_{\alpha\beta} f^i_{\alpha\gamma} \chi_{Li} \lambda^\gamma_L (f^j_{\beta\delta} \bar{\chi}_{Lj} \lambda^\delta_L - f^*_{\beta\delta i} \bar{\chi}^j_R \lambda^\delta_R) -$
$- \frac{1}{4} \bar{\psi}_\mu \gamma_5 \gamma_\nu D_\varrho \psi_\sigma \sigma^{\mu\nu\varrho\sigma} e^{-1} - \frac{1}{64} (\bar{\psi}_\mu \gamma_\lambda \psi_\varrho + 2\bar{\psi}_\lambda \lambda_\mu \psi_\varrho) \bar{\psi}^\mu \gamma^\lambda \psi^\varrho +$
$+ \frac{1}{16} (\bar{\psi}_\mu \cdot \gamma \cdot \psi)^2 + 3 \frac{\Phi^i_j}{\Phi} \bar{\chi}_{Li} \check{D} \chi^j_R - 3 \frac{\Phi^{ij}_k}{\Phi} \bar{\chi}_{Li} \check{D} z_j \chi^k_R -$
$- 3 \frac{\Phi^i_j}{\Phi} \bar{\psi}_{\mu L} \check{D} z^{*j} \gamma_\mu \chi_{Li} + \frac{4\Phi^i}{\Phi} \bar{\chi}_{Li} \sigma^{\mu\nu} D_\mu \psi_\nu - \frac{3\Phi^i}{\Phi} \bar{\chi}_{Li} \gamma^\nu \gamma^\mu \psi_{L\nu} D_\mu \sigma +$

$$+ \frac{3}{8} e^{-1} \varepsilon^{\mu\nu\varrho\sigma} \psi_\mu \gamma_\nu \psi_\varrho \left(\frac{\Phi^i}{\Phi} D_\sigma z_i + \frac{1}{2} \frac{\Phi^j_i}{\Phi} \bar{\chi}^i_{\rm R} \gamma_\sigma \chi_{{\rm L}j} \right) + \frac{3}{2} \frac{\Phi^i_j}{\Phi} \psi_{\mu{\rm L}} \bar{\chi}_{{\rm L}i} \bar{\psi}_{\mu{\rm R}} \chi^j_{\rm R} -$$

$$- \frac{3}{2} \frac{\Phi^{ij}_k}{\Phi} \bar{\chi}_{{\rm L}i} \chi_{{\rm L}j} \bar{\chi}^k_{\rm R} \chi^l_{\rm R} - \frac{1}{2} \Phi^i \bar{\chi}_{{\rm L}i} \left(\chi_{\mu{\rm L}} \bar{\psi} \cdot \gamma \psi^\mu + \frac{1}{2} \sigma_{\mu\nu} \psi_{\varrho{\rm L}} \bar{\psi}^\nu \gamma^\varrho \psi^\mu + \sigma^{\mu\nu} \psi_{\nu{\rm L}} \bar{\psi}_\mu \gamma \cdot \psi \right) -$$

$$- \frac{3}{2} i \tilde{g} \frac{\Phi^i}{\Phi} T^{\alpha j}_i z_j \bar{\psi}_{\rm L} \cdot \gamma \lambda^\alpha_{\rm R} + 6 i \tilde{g} \frac{\Phi^j_i}{\Phi} T^{\alpha k}_j z_k \bar{\lambda}^\alpha_{\rm R} \chi^i_{\rm R} - \frac{1}{2} g^{ij} \exp\left[3\sigma\right] \bar{\chi}_{{\rm L}i} \chi_{{\rm L}j} +$$

$$+ \frac{1}{2} g^i \exp\left[3\sigma\right] \psi_{\rm R} \cdot \gamma \chi_{{\rm L}i} + \frac{1}{2} g \exp\left[3\sigma\right] \bar{\psi}_{\mu{\rm R}} \sigma^{\mu\nu} \psi_{\nu{\rm R}} + \frac{1}{2} \mathrm{Re} f_{\alpha\beta} \cdot$$

$$\cdot \left[- \frac{1}{2} \bar{\lambda}^\alpha \check{D} \lambda^\beta + \frac{1}{2} \bar{\lambda}^\alpha \gamma^\mu \sigma^{\varrho\sigma} \psi_\mu \left(F^\beta_{\varrho\sigma} + \frac{1}{2} \bar{\psi}_\varrho \gamma_\sigma \lambda^\beta \right) - \frac{1}{2} \bar{\lambda}^\alpha \gamma_5 \gamma_\nu \lambda^\beta\, e^{-1} \varepsilon^{\mu\nu\varrho\sigma} \psi_\mu \gamma_\varrho \psi_\sigma \right] -$$

$$- \frac{1}{4} i\, \mathrm{Im} f_{\alpha\beta} D^\mu (\bar{\lambda}^\alpha \gamma_\mu \lambda_{\rm R}) + \frac{1}{2} f^i_{\alpha\beta} \left[- \bar{\chi}_{{\rm L}i} \sigma^{\mu\nu} \lambda^\beta_{\rm L} \left(F^{-\alpha}_{\mu\nu} - \frac{1}{2} \bar{\psi}_{\varrho{\rm L}} \sigma_{\mu\nu} \gamma^\varrho \lambda^\alpha_{\rm R} + \frac{1}{2} \bar{\psi}_{\rm R} \cdot \gamma \sigma_{\mu\nu} \lambda^\alpha_{\rm L} \right) -$$

$$- \frac{1}{2} \bar{\psi}_{\rm R} \cdot \gamma \chi_{{\rm L}i} \bar{\lambda}^\alpha_{\rm L} \lambda^\beta_{\rm L} \right] + \frac{1}{4} f^{ij}_{\alpha\beta} \bar{\chi}_{{\rm L}i} \chi_{{\rm L}j} \bar{\lambda}^\alpha_{\rm L} \lambda^\beta_{\rm L} + \mathrm{h.c.}$$

The kinetic part of eq. (119) can be further diagonalized through the redefinition

$$\psi_{\mu{\rm L}} \to \psi_{\mu{\rm L}} - \gamma_\mu \frac{\Phi_i}{\Phi} \chi^i_{\rm R}\,. \tag{120}$$

If, in addition, we perform the following chiral redefinition on the fermion fields,

$$\psi_{\mu{\rm L}} \to \left(\frac{g}{g^*}\right)^{\frac{1}{4}} \psi_{\mu{\rm L}}, \qquad \chi_{{\rm L}i} \to \left(\frac{g}{g^*}\right)^{\frac{1}{4}} \chi_{{\rm L}i}, \qquad \lambda^\alpha_{\rm L} \to \left(\frac{g}{g^*}\right)^{\frac{1}{4}} \lambda^\alpha_{\rm L} \tag{121}$$

and we define the function

$$G = J - \log \tfrac{1}{4} |g|^2 \tag{122}$$

so that

$$G_i = J^i - \frac{g^i}{g}, \qquad G^i_j = J^i_j, \tag{123}$$

the final Lagrangian becomes

$$L = L_{\rm B} + L_{\rm FK} + L_{\rm FM} + L_{\rm (4)F}, \tag{124}$$

where

$$e^{-1} L_{\rm B} = - \tfrac{1}{2} R + G^i_j D_\mu z_i D^\mu z^{*j} - \tfrac{1}{4} \mathrm{Re} f_{\alpha\beta} F^\alpha_{\mu\nu} F^\beta_{\mu\nu} +$$
$$+ \tfrac{1}{4} i\, \mathrm{Im} f_{\alpha\beta} F^\alpha_{\mu\nu} \tilde{F}^\beta_{\mu\nu} + \exp[-G](3 + G_k G^{-1k}_1 G^1) +$$
$$+ \tfrac{1}{2} \tilde{g}^2 \mathrm{Re} f^{-1}_{\alpha\beta} (G^i T^{\alpha j}_i z_j)(G^k T^{\beta l}_k z_l)\,; \tag{125}$$

L_{FK} is the curved generalization of the fermionic terms; L_{FM} is the quadratic part in the fermionic fields without derivative terms; $L_{(4)F}$ is the part of the Lagrangian which contains four-fermion interaction terms (not coming from the spin-$\frac{3}{2}$ torsion contribution already contained in the Lorentz connection $\omega_{\mu mn}(E, \psi)$) and also other interactions of dimension greater than four containing covariant derivatives of boson fields:

$$(126) \quad e^{-1} L_{FK} = -\frac{1}{4} \operatorname{Re} f_{\alpha\beta} \bar{\lambda}^\alpha \check{D} \lambda^\beta - \frac{1}{4} e^{-1} \bar{\psi}_\mu \gamma_5 \gamma_\nu D_\varrho \psi_\sigma \varepsilon^{\mu\nu\varrho\sigma} + $$
$$+ G_j^i \bar{\chi}_{Li} \check{D} \chi_R^j - \frac{i}{8} \operatorname{Im} f_{\alpha\beta} e^{-1} D_\mu (e \bar{\lambda}^\alpha \gamma_5 \gamma_\mu \lambda^\beta) + \text{h.c.},$$

$$(127) \quad e^{-1} L_{FM} = \exp[-G/2] \bar{\psi}_{\mu R} \sigma^{\mu\nu} \psi_{\nu R} - $$
$$- \bar{\psi}_R \cdot \gamma \left(\exp[-G/2] G^i \chi_{Li} - \frac{i}{2} \tilde{g} G^i T_i^{\alpha j} z_j \lambda_L^\alpha \right) + \exp[-G/2] \cdot$$
$$\cdot (G^{ij} - G^i G^j + G^1 G_1^{-1k} G_k^{ij}) \bar{\chi}_{Li} \chi_{Lj} - 2i\tilde{g} G_i^k z^{*j} T_j^{\alpha i} \bar{\lambda}_L^\alpha \chi_{Lk} +$$
$$+ \frac{1}{2} f_{\alpha\beta}^k \left(\frac{1}{2} \exp[-G/2] G_1 G_k^{-11} \bar{\lambda}_L^\alpha + i\tilde{g} \operatorname{Re} f_{\alpha\gamma}^{-1} G^i T_i^{\gamma j} z_j \bar{\chi}_{Lk} \right) \lambda_{Lk}^\beta + \text{h.c.},$$

$$(128) \quad e^{-1} L_{(4)F} = \frac{1}{2} \operatorname{Re} f_{\alpha\beta} \left(\frac{i}{2} \bar{\lambda}^\alpha \gamma^\mu \sigma^{\varrho\sigma} \psi_\mu F_{\varrho\sigma}^\beta - \frac{1}{2} \bar{\lambda}_L^\alpha \gamma_\mu \lambda_R^\beta G^i D_\mu z_i \right) - $$
$$- \frac{1}{2} f_{\alpha\beta}^i \bar{\chi}_{Li} \sigma \cdot F^\alpha \lambda_L^\beta + \frac{1}{8} e^{-1} \varepsilon^{\mu\nu\varrho\sigma} \bar{\psi}_\mu \gamma_\nu \psi_\varrho G^i D_\sigma z_i - $$
$$- G_i^j \bar{\psi}_{\mu L} \check{D} z^{*i} \gamma^\mu \chi_{Lj} - \bar{\chi}_{Li} \check{D} z_j \chi_R^k \left(G_k^{ij} + \frac{1}{2} G_k^i G^j \right) +$$
$$+ \frac{1}{32} G_1^{-1k} f_{\alpha\beta}^1 f_{\gamma\delta k}^* \bar{\lambda}_L^\alpha \lambda_L^\beta \bar{\lambda}_R^\gamma \lambda_R^\delta + \frac{3}{32} (\operatorname{Re} f_{\alpha\beta} \bar{\lambda}_L^\alpha \gamma_m \gamma_R^\beta)^2 +$$
$$+ \frac{1}{8} \operatorname{Re} f_{\alpha\beta} \bar{\lambda}^\alpha \gamma^\mu \sigma^{\varrho\sigma} \psi_\mu \bar{\psi}_\varrho \gamma_\sigma \lambda^\beta + \frac{1}{2} f_{\alpha\beta}^i \left(\bar{\chi}_{Li} \sigma^{\mu\nu} \lambda^\alpha \cdot \psi_{\nu L} \gamma_\mu \lambda_R^\beta + \frac{1}{4} \bar{\psi}_R \cdot \gamma \chi_{Li} \bar{\lambda}_L^\alpha \lambda_L^\beta \right) +$$
$$+ \frac{1}{8} G_i^j \bar{\chi}_k^i \gamma_d \chi_{Lj} (\varepsilon^{abcd} \bar{\psi}_a \gamma_b \psi_c + \bar{\psi}_a \gamma_5 \gamma^d \psi_a) + \frac{1}{16} \bar{\chi}_{Li} \gamma^\mu \chi_R^j \bar{\lambda}_R^\delta \gamma_\mu \lambda_L^\gamma \cdot$$
$$\cdot (-2G_j^i \operatorname{Re} f_{\gamma\delta} + \operatorname{Re} f_{\alpha\beta}^{-1} f_{\alpha\gamma}^i f_{\beta\delta j}^*) - \frac{1}{16} \bar{\chi}_{Li} \sigma_{\mu\nu} \chi_{Lj} \bar{\lambda}_L^\gamma \sigma^{\mu\nu} \lambda_L^\delta \operatorname{Re} f_{\alpha\beta}^{-1} f_{\alpha\gamma}^i f_{\beta\delta}^j +$$
$$+ \frac{1}{16} \bar{\chi}_{Li} \chi_{Lj} \bar{\lambda}_L^\gamma \lambda_L^\delta (-4 G_k^{ij} G_1^{-1k} f_{\gamma\delta}^1 + 4 f_{\gamma\delta}^{ij} - \operatorname{Re} f_{\alpha\beta}^{-1} f_{\alpha\gamma}^i f_{\beta\delta}^j) +$$
$$+ \left(-\frac{1}{2} G_{k1}^{ij} + \frac{1}{2} G_m^{ij} G_n^{-1m} G_{k1}^n - \frac{1}{4} G_k^i G_1^j \right) \bar{\chi}_{Li} \chi_{Lj} \bar{\chi}_R^k \chi_R^1 + \text{h.c.}$$

The final form of the Lagrangian, as given by eq. (124), is invariant under the following local supersymmetry transformations:

$$
(129) \begin{cases}
\delta e_\mu^m = \tfrac{1}{2} \bar{\varepsilon}_L \gamma^m \psi_{\mu R} + \text{h.c.}, \\
\delta \psi_{\mu L} = [\partial_\mu + \tfrac{1}{2} \omega_{\mu mn}(e,\psi)\sigma^{mn}] \varepsilon_L + \tfrac{1}{2} \sigma_{\mu\nu} \varepsilon_L G^j_i \bar{\chi}^i_R \gamma^\nu \chi_{Lj} + \\
\qquad + \tfrac{1}{2} \gamma_\mu \varepsilon_R \exp[-G/2] + \tfrac{1}{4} \psi_{\mu L}(G^i \bar{\varepsilon}_L \chi_{Li} - G_i \bar{\varepsilon}_R \chi_R^i) - \\
\qquad - \tfrac{1}{4} \varepsilon_L(G^i D_\mu z_i - G_i D_\mu z^{*i}) + \tfrac{1}{4}(\sigma_{\nu\mu} + g_{\mu\nu}) \varepsilon_L \bar{\lambda}_L^{*\alpha} \gamma^\nu \lambda_R^\beta \operatorname{Re} f_{\alpha\beta}, \\
\delta B_\mu^\alpha = -\tfrac{1}{2} \bar{\varepsilon}_L \gamma_\mu \lambda_R^\alpha + \text{h.c.}, \\
\delta \lambda_R^\alpha = \tfrac{1}{2} \sigma^{\mu\nu} \hat{F}_{\mu\nu}^\alpha \varepsilon_R + \tfrac{1}{2} i \varepsilon_R \operatorname{Re} f_{\alpha\beta}^{-1}(-\tilde{g} G^i T_i^{\beta j} z_j + \tfrac{1}{2} i f_{\beta\gamma}^i \bar{\chi}_{Li} \lambda_L^\gamma - \\
\qquad - \tfrac{1}{2} i f_{\beta\gamma i}^* \bar{\chi}_R^i \lambda_R^\gamma) - \tfrac{1}{4} \lambda_R^\alpha (G^i \bar{\varepsilon}_L \chi_{Li} - G_i \bar{\varepsilon}_R \chi_R^i), \\
\delta z_i = \bar{\varepsilon}_L \chi_{Li}, \\
\delta \chi_{Li} = \tfrac{1}{2} \check{D} z_i \varepsilon_R - \tfrac{1}{2} \varepsilon_L \exp[-G/2] G_i^{-1j} G_j - \tfrac{1}{8} \varepsilon_L \bar{\lambda}_R^\alpha \lambda_R^\beta G^{-1k} f^*_{\alpha\beta k} + \\
\qquad + \tfrac{1}{2} \varepsilon_L G_i^{-1k} G_k^{j1} \bar{\chi}_{Lj} \chi_{L1} + \tfrac{1}{4} \chi_{Li}(G_j \bar{\varepsilon}_R \chi_R^j - G^j \bar{\varepsilon}_L \chi_{Lj}).
\end{cases}
$$

The transformations (129) are directly obtained from the original transformations (96), with the substitution of eqs. (111) in the fermionic-field variations and with the redefinitions (114), (120) and (121) for the vierbein field and for the fermionic fields. In addition, the local supersymmetry parameter in eqs. (129) has also been redefined, compared with eqs. (96), as follows:

$$
(130) \qquad \varepsilon_L \to \exp\left[-\tfrac{1}{2} J\right] \left(\frac{g}{g^*}\right)^{\frac{1}{4}} \varepsilon_L.
$$

We note that, by using conformal tensor calculus, we could avoid the redefinitions given by eqs. (114), (115), (120), (121) and (130), by suitably choosing a special conformal gauge. This has been shown elsewhere [17].

Finally, we discuss the case in which the gauge group is $G \times U_{R,1}$, corresponding to the superspace action formulae given in eqs. (82) and (83). As a consequence of eq. (106), the presence of a gauged R-symmetry is most easily discussed in the final form of the Lagrangian and transformation laws as given by eqs. (124) and (129).

In fact, the Lagrangian has precisely the same expression as in eq. (124) with suitable gauge covariantization with respect to the $U_{R,1}$ gauge field. The only difference now is the fact that the D-term of the R-multiplet, defined as

$$
(131) \qquad D_R = g_R G^i n_i z_i,
$$

can have a constant term, owing to the noninvariance of the superpotential under R-trasformations:

$$g^i(z) n_i g_R z_i = 3 g_R g(z) \,, \tag{132}$$

where $n_i g_R = q_i$ are the U_1 charges of the scalar fields z_i.

It is obvious that these considerations are no longer valid for a vanishing superpotential function $g(z) = 0$. In this case the final form of the Lagrangian and transformation laws comes directly from eqs. (124) and (131) with the following substitution rules [11, 18]:

i) $\exp[-G]$, $\exp[-G/2] \to 0$, $G^i \nabla_m z_i \Rightarrow J^i \nabla_m z_i + \frac{3}{2} i g_R V^R_m$,

ii) $G_i \nabla_m z^{*i} \Rightarrow J_i \nabla_m z^{*i} - \frac{3}{2} i g_R V^R_m$,

iii) $\tilde{g}_\alpha G^i T^{\alpha j}_i z_j \Rightarrow \tilde{g}_\alpha J^i T^{\alpha j}_i z_j - \frac{3}{2} g_R \delta^\alpha_R$,

iv) for other derivatives of G not in the form given by ii) and iii), replace $G \to J$.

We note, in particular, that the substitution rule ii) produces new minimal gauge coupling of the R-gauge field to the gravitino, gauginos and chiral fermions (but not to the scalars) which are usually eliminated through the chiral redefinitions (121) when $g(z) \neq 0$.

It is important to note that some models discussed in the literature fall within this class [19]. However, these models can only have spontaneously broken supersymmetry with a nonvanishing cosmological constant.

In the more interesting physical situation of spontaneously broken R-symmetry and supersymmetry with a vanishing cosmological constant, the superpotential cannot vanish and the final form of the Lagrangian as given by eq. (124) applies.

7. – The super-Higgs effect in $N=1$ supergravity.

Spontaneous breaking of local supersymmetry implies the super-Higgs effect, *i.e.* the gravitino mass generation and the disappearance of the massless goldstino which is absorbed by the « massive » spin-$\frac{3}{2}$ gravitino.

This phenomenon may or may not occur with a vanishing cosmological constant, *i.e.* in a Minkowski or de Sitter space-time.

The goldstino field is uniquely defined by the spin-$\frac{1}{2}$ fermion which couples to the spin-$\frac{3}{2}$ gravitino in L_{FM} given by eq. (127):

$$-\psi_R \cdot \gamma \eta_L, \quad \eta_L = \exp[-G/2] G^i \chi_{Li} - \frac{i}{2} D^\alpha \lambda^\alpha_L, \tag{133}$$

where

$$D^\alpha = \tilde{g} G^i T^{\alpha j}_i z_j \,. \tag{134}$$

A necessary and sufficient condition for spontaneous supersymmetry breaking, therefore, implies that one of the quantities [6]

$$\exp[-G/2]\,G^i, \quad D^\alpha \tag{135}$$

is different from zero at the minimum of the potential given in eq. (125):

$$V(z, z^*) = -\exp[-G]\,(3 + G^i G^{-1\,j}_i G_j) + \tfrac{1}{2}\operatorname{Re} f^{-1}_{\alpha\beta} D^\alpha D^\beta. \tag{136}$$

In the more interesting situation of a Minkowski space-time, we also demand the vanishing of the cosmological constant

$$V(z_0, z_0^*) = 0 \quad \text{at} \quad \left.\frac{\partial V}{\partial z}\right|_{z=z_0} = 0. \tag{137}$$

Under condition (137), the gravitino mass has a precise meaning, and this is given by the universal formula [5]

$$m_{\frac{3}{2}} = \exp[-G/2]. \tag{138}$$

We now discuss the particularly interesting case of « minimal » coupling of the Yang-Mills system to supergravity. This is defined by the two conditions

$$G^i_j = -\tfrac{1}{2}\delta^i_j, \quad f_{\alpha\beta} = \delta_{\alpha\beta}, \tag{139}$$

in which case all the kinetic terms are canonical and the Lagrangian depends only on the superpotential $g(z)$. In this case we have

$$G(z, z^*) = -\tfrac{1}{2}|z|^2 - \log\tfrac{1}{4}|g(z)|^2. \tag{140}$$

The scalar potential (136) becomes

$$V(z, z^*) = \exp[-G]\,(2G^i G_i - 3) + \tfrac{1}{2}(D^\alpha)^2, \tag{141}$$

with

$$D^\alpha = -\tfrac{1}{2}\tilde{g}_\alpha z^{*i} T^{\alpha j}_i z_j + \xi \delta^\alpha_{\mathrm{E}}.$$

The term L_{FM} given by eq. (127), which determines the fermion mass matrix, is

$$\begin{aligned}e^{-1} L_{\mathrm{FM}} = \exp[-G/2]\,(\bar{\psi}_{\mu\mathrm{R}}\sigma^{\mu\nu}\psi_{\nu\mathrm{R}} - \bar{\psi}_{\mathrm{R}}\cdot\gamma\tilde{\eta}_{\mathrm{L}} - \tfrac{2}{3}\bar{\tilde{\eta}}_{\mathrm{L}}\tilde{\eta}_{\mathrm{L}}) + \\ + \bar{\chi}_{\mathrm{L}i} M^{ij}\chi_{\mathrm{L}j} + 2\bar{\chi}_{\mathrm{L}i} M^{i\alpha}\lambda^\alpha_{\mathrm{L}} + \bar{\lambda}^\alpha_{\mathrm{L}} M^{\alpha\beta}\lambda^\beta_{\mathrm{L}} + \text{h.c.},\end{aligned} \tag{142}$$

where

(143) $$\tilde{\eta}_L = G^i \chi_{Li} - \frac{i}{2} \exp[G/2] D^\alpha \lambda^\alpha_L.$$

The spin-$\frac{1}{2}$ fermion mass matrix has the form

(144) $$\begin{cases} M^{ij} = \exp[-G/2](G^{ij} - \frac{1}{3} G^i G^j), \\ M^{i\alpha} = -\frac{1}{3} i G^i D^\alpha - I D^{\alpha i}, \\ M^{\alpha\beta} = -\frac{1}{6} \exp[G/2] D^\alpha D^\beta. \end{cases}$$

We can now compute the quadratic mass relation which generalizes the result of spontaneously broken, globally supersymmetric Yang-Mills theories. The scalar contribution to the trace of the square mass is

(145) $$\text{Tr } M_0^2 = -(N+4) D^\alpha D^\alpha + 2 G^i G_i D^\alpha D^\alpha + \\ + \exp[-G](4 G_{ij} G^{ij} + N) + 4 D^{\alpha i} D_i^\alpha + 4 D^{\alpha i} D_i^\alpha.$$

The spin-1 contribution is

(146) $$\text{Tr } M_1^2 = 12 D_i^\alpha D^{i\alpha}.$$

The spin-$\frac{1}{2}$ contribution is

(147) $$\text{Tr } M_{\frac{1}{2}}^2 = 8 \exp[-G]\left(G^{ij} - \frac{1}{3} G^i G^j\right)\left(G_{ij} - \frac{1}{3} G_i G_j\right) + \\ + 16\left(D^{\alpha i} + \frac{1}{3} G^i D^\alpha\right)\left(D_i^\alpha + \frac{1}{3} G_i D^\alpha\right) + \frac{2}{g} \exp[G](D^\alpha)^2 (D^\beta)^2,$$

and the gravitino contribution is

(148) $$\text{Tr } M_{\frac{3}{2}}^2 = 4 \exp[-G].$$

Therefore, we finally get the mass formula

(149) $$\text{supertrace } M^2 = \sum_{J=0}^{\frac{3}{2}} (-)^{2J}(2J+1) = \\ = (N-1)(2 m_{\frac{3}{2}}^2 - K^2 D^\alpha D^\alpha) - 2 \tilde{g}_\alpha D^\alpha \text{Tr } T^\alpha.$$

Equation (149) gives back the result of global supersymmetry [20] if we set $m_{\frac{3}{2}} = 0$, $k = 0$. It also reproduces the result of ref. [11] for $N = 1$ and $g = 0$. We also note that the last term in eq. (149) is only possible for Abelian U_1 factors of G with $\text{Tr } T \neq 0$. Equation (149) summarizes the mass relations of spontaneously broken supersymmetry when the simultaneous occurrence of the Higgs and super-Higgs effect takes place.

REFERENCES

[1] E. GILDENER: *Phys. Rev. D*, **14**, 1667 (1976); E. GILDENER and S. WEINBERG: *Phys. Rev. D*, **15**, 3333 (1976); L. MAIANI: in *Proceedings of the Summer School on Weak Interactions, Gif-sur-Yvette* (INP3, Paris, 1980), p. 3; M. VELTMAN: *Acta Phys. Pol. B*, **12**, 437 (1981); E. WITTEN: *Nucl. Phys. B*, **188**, 513 (1981); S. DIMOPOULOS and S. RABY: *Nucl. Phys. B*, **199**, 353 (1981).

[2] For reviews see, for example, P. FAYET and S. FERRARA: *Phys. Rep. C*, **32**, 249 (1979); J. WESS and J. BAGGER: *Supersymmetry and Supergravity* (Princeton University Press, Princeton, N. J., 1983).

[3] J. SCHENK: *Cargése Lectures 1978*, edited by M. LEVY and S. DESER (Plenum, New York, N. Y., 1978), p. 479.

[4] For a review see, for example, P. VAN NIEUWENHUIZEN: *Phys. Rep.*, **68**, 191 (1981).

[5] For review see, for example, R. BARBIERI and S. FERRARA: *Surv. High Energy Phys.*, **4**, 33 (1983); P. H. NILLES: *Phys. Rep.*, **110**, 3 (1984).

[6] E. CREMMER, S. FERRARA, L. GIRARDELLO and A. VAN PROEYEN: *Phys. Lett. B*, **116**, 231 (1982); *Nucl. Phys. B*, **212**, 413 (1983).

[7] E. CREMMER, S. FERRARA, C. KOUNNAS and D. V. NANOPOULOS: *Phys. Lett. B*, **113**, 61 (1983); J. ELLIS, A. B. LAHANAS, D. V. NANOPOULOS and K. KOUNNAS: *Phys. Lett. B*, **134**, 429 (1984); J. ELLIS, K. KOUNNAS and D. V. NANOPOULOS: *Nucl. Phys. B*, **241**, 406 (1984); J. ELLIS, K. ENQVIST and D. V. NANOPOULOS: *Phys. Lett. B*, **147**, 99 (1984).

[8] S. FERRARA and P. VAN NIEUWENHUIZEN: *Phys. Lett. B*, **74**, 333 (1978); **76**, 404 (1978); K. S. STELLE and P. C. WEST: *Phys. Lett. B*, **74**, 330 (1978); **77**, 376 (1978).

[9] M. KAKU, P. K. TOWNSEND and P. VAN NIEUWENHUIZEN: *Phys. Rev. D*, **17**, 3179 (1978); B. DE WIT: in *Supergravity*, edited by S. FERRARA, J. G. TAYLOR and P. VAN NIEUWENHUIZEN (World Scientific, Singapore, 1982), p. 85; T. KUGO and S. UEHARA: *Nucl. Phys. B*, **222**, 125 (1983).

[10] M. F. SOHNIUS and P. C. WEST: *Phys. Lett. B*, **105**, 353 (1981).

[11] S. FERRARA, L. GIRARDELLO, T. KUGO and A. VAN PROEYEN: *Nucl. Phys. B*, **223**, 191 (1983).

[12] See, for example, J. WESS and J. BAGGER: *Supersymmetry and Supergravity* (Princeton University Press, Princeton, N. J., 1983).

[13] P. FAYET and J. ILIOPOULOS: *Phys. Lett. B*, **51**, 461 (1974).

[14] K. S. STELLE and P. C. WEST: *Nucl. Phys. B*, **145**, 175 (1978).

[15] R. BARBIERI, S. FERRARA, D. V. NANOPOULOS and K. S. STELLE: *Phys. Lett. B*, **113**, 219 (1982).

[16] M. T. GRISARU, M. ROČEK and A. KARLHEDE: *Phys. Lett. B*, **120**, 189 (1982).

[17] T. KUGO and S. UEHARA: *Nucl. Phys. B*, **223**, 191 (1983).

[18] J. BAGGER: *Nucl. Phys. B*, **211**, 302 (1983).

[19] D. Z. FREEDMAN: *Phys. Rev. D*, **15**, 1173 (1977); B. DE WIT and P. VAN NIEUWENHUIZEN: *Nucl. Phys. B*, **139**, 531 (1979).

[20] S. FERRARA, L. GIRARDELLO and F. PALUMBO: *Phys. Rev. D*, **20**, 403 (1979).

Ultraviolet Divergences and Supersymmetric Theories (*).

A. SAGNOTTI (**)

Department of Physics and Lawrence Berkeley Laboratory, University of California
Berkeley, CA 94720

1. – Introduction and survey.

Soon after the discovery of supersymmetry [1], it was realized that a symmetry between the bosonic and fermionic degrees of freedom of a field theory would have a beneficial effect on its ultraviolet behavior. For istance, it became clear that supersymmetry, combined with the « minus » sign rule for fermionic loops, would guarantee the absence of corrections to the vacuum energy for an unbroken supersymmetric theory [2]. Moreover, explicit calculations revealed an impressive set of cancellations in specific models. The simplest one of these, known as the Wess-Zumino model [3], is a renormalizable interacting theory of a scalar, a pseudoscalar and a Majorana spinor, all with the same mass and interacting via a single coupling constant. Naively one would expect that, because of supersymmetry, in this case divergences could be absorbed by means of three distinct parameters, one common wave function renormalization for the three fields, one mass renormalization and one coupling constant renormalization. Surprisingly, it was found by explicit calculation that only one wave function renormalization appeared to be needed [4]. Somewhat later it was also shown, again by explicit calculation, that another model, the $N = 4$ Yang-Mills theory [5], possesses no charge renormalization at one and two loops [6, 7], and even at three loops [8]. Though it was natural to conjecture that such a behavior would persist to all orders, there followed a period of impasse, and for about three years no one succeeded in achieving a satisfactory understanding of the results of ref. [4-8], and to draw general conclusions from them.

(*) This work was supported in part by the National Science Foundation under Grant PHY-81-18547, and in part by the Director, Office of Energy Research, Office of High Energy and Nuclear Physics, Division of High Energy Physics of the U.S. Department of Energy under contract DE-AC03-76SF-00098.
(**) Miller Fellow in Theoretical Physics.

It is undeniable that ultraviolet divergences are a rather bizarre phenomenon to get accustomed to. Nonetheless, the successes of quantum electrodynamics first and of Yang-Mills gauge theories later did manage to get physicists accustomed to them, with the result that the improved ultraviolet behavior of supersymmetric theories found in ref. [4-8] appeared rather remarkable and interesting from a theoretical point of view, but did not add much to the rather limited interest that the majority of particle physicists had in these new ideas. A major wave of interest in these models was only aroused by the observation [9] that tying scalars to spinors by supersymmetry would improve the stability of the parameters of low-energy gauge theories upon renormalization. The stability problem for the parameters is often referred to as the gauge hierarchy problem [10].

The situation for gravitational theories, on the other hand, was quite different. In this case it is obvious on dimensional grounds that their perturbation expansion is in danger of being nonrenormalizable, because Newton's constant k^2 has the dimension of a negative power of mass. The possible cancellation of ultraviolet divergences does appear a rather crucial phenomenon here, and leads to hope that minor modifications of Einstein's gravity could lead to cure the problems of its perturbation expansion. Indeed, the early achievements of supergravity theories [11] appeared rather spectacular in this context. It was known for some time, after the work of De Witt [12] and 't Hooft and Veltman [13], that *pure* Einstein quantum gravity has finite one-loop corrections to its S-matrix, but that coupling it to a single scalar field results in nonrenormalizable ultraviolet divergences. The former result requires little calculation to understand. All one needs is to list the possible counterterms of the right dimensionality, which are

(1.1) $\qquad \sqrt{-g}\, R^{\mu\nu\varrho\sigma} R_{\mu\nu\varrho\sigma}, \qquad \sqrt{-g}\, R^{\mu\nu} R_{\mu\nu} \quad \text{and} \quad \sqrt{-g}\, R^2,$

and notice that they all vanish on shell, *i.e.* when the classical field equations are used, on account of the Gauss-Bonnet identity for four-dimensional space-time. In fact, this guarantees that

(1.2) $\qquad \varepsilon^{\mu\nu\varrho\sigma} \varepsilon_{\alpha\beta\gamma\delta} R_{\mu\nu}{}^{\alpha\beta} R_{\varrho\sigma}{}^{\gamma\delta} = -4 R^{\mu\nu\varrho\sigma} R_{\mu\nu\varrho\sigma} + 16 R^{\mu\nu} R_{\mu\nu} - 4R^2$

is a total derivative, and thus vanishes in perturbation theory. The divergence encountered in ref. [13] for the case of a single scalar was shown to persist, again by explicit calculations, for a large number of matter couplings [14], and for some time seemed unavoidable. Actually, this is not the case, as the *pure* supergravity theories do share the one-loop finiteness of Einstein's gravity [15]. However, if one adds extra matter, even of the supersymmetric type, divergences appear again [16]. The former result is a direct consequence of what was said above for Einsten's gravity. In fact, irreducible supersymmetry can be used to relate S-matrix elements to ones corresponding to external gravitons only. These are finite, as their possible divergences correspond to the harmless invariants in eq. (1.1). Remarkably, formal arguments are also available that exclude two-loop divergences for supergravity theories. Two-loop finiteness, however, is somewhat more subtle to establish [17], and stems from the impossibility of turning the only candidate two-loop couterterm for pure gravity into a supersymmetric invariant.

Beyond one loop in pure Einstein gravity the situation is pure mystery, and (almost) the same is true for supergravity beyond two loops. So, even assuming that Einstein gravity does indeed diverge at two loops, we seem very far from having fixed its problems by turning to supergravity. Excluding an infinite number of counterterms of increasing dimensionality on the basis of formal arguments alone seems impossible, and this suggests that trouble is indeed going to show up at the first available opportunity. To be honest, however, the blame is to be put, at present, more on the investigators than on the theories, as our understanding of them is very incomplete. Moreover, there is a subtle point which is easily overlooked. Does it really make sense to write a perturbation series where the expansion parameter is not dimensionless, and actually has the dimensions of a negative power of mass? Just on dimensional grounds, the effective strength of the interaction is bound to increase with momentum, and it is not clear what the small expansion parameter would be in the ultraviolet region. From the mathematical viewpoint, demanding an expansion in integer positive powers of the coupling constant, even of the asymptotic type, is tantamount to demanding analyticity near zero coupling constant of the quantum theory. However, it has been shown that this is actually not the case in toy models. I am referring here to some old work of Parisi [18], where the author infers that the $1/N$ expansion for Φ^4 theory is renormalizable even above four dimensions, the only signature of the renormalizablility being the lack of analyticity of the result in the self-coupling of Φ. It goes without saying that an expansion parameter is needed, as there is no hope of solving the quantum theory exactly for complicated four-dimensional models. Unfortunately, it is rather difficult to envisage what a dimensionless expansion parameter could be for supergravity theories, as they are so tightly constrained in their spectra by supersymmetry.

While keeping this in mind, one must admit that the possibility of a theory with a finite perturbation expansion is so attractive that it deserves attention. Moreover, just as renormalizability served as a very useful guiding principle in the search for theories of strong and electroweak interactions, it is conceivable that finiteness can serve as a guiding principle in the search for a truly unified theory of all interactions. Thus one can proceed and ask the well-defined question of what the perturbation expansion in k looks like for (super)gravitational theories. There are two different approaches to this problem. The first one, obvious in principle but exceedingly difficult in practice, is to proceed to actual calculations, starting from the possibly more accessible case of pure Einstein quantum gravity at two loops. The other consists in trying to gain some insight into the problem by indirect means, and possibly attempting to build up formal arguments.

Going back to renormalizable models with (extended) supersymmetry, I already remarked how until 1981 there was a rather impressive set of « experimental » results with apparently no explanation, the most remarkable of which is the vanishing of the β-function for $N = 4$ Yang-Mills up to three loops. The first half of 1982, on the other hand, saw the occurrence of something new. Fairly rigorous formal arguments were proposed leading to an understanding of the results of ref. [4-8] and to proofs of their persistence to all orders. These arguments all rest on making the supersymmetry more manifest than it is in the usual component formulations, where the balance between the on-shell fermionic and bosonic degrees of freedom is violated by the corresponding off-shell field representations. A crucial observation in this respect came in 1979, and is due to GRISARU, ROČEK and SIEGEL [19]. Motivated by their own attempt to calculate the three-loop β-function for $N = 4$ Yang-Mills using $N = 1$ superfields, they managed to streamline the method of $N = 1$ superspace in dealing with chiral superfields, and it then became obvious why the Wess-Zumino model has only one renormalization constant. The second (and main) step is a paper by GRISARU and SIEGEL [20], where it is shown that combining the familiar properties of $N = 1$ superfields with the background field method and with the *working assumption* that similar manipulations should go through with extended superfields (essentially unknown at the time) leads to a number of rather spectacular conclusions. The $N = 4$ Yang-Mills theory would necessarily be finite to all orders if it were possible to formulate it at least in terms of $N = 2$ extended superfields, and actually all Yang-Mills and matter theories which admitted a formulation in terms of $N = 2$ superfields would be finite beyond one loop. Actually, at the time ref. [20] appeared, there were some problems left before its conclusions could be made effective for $N \geqslant 2$ gauge and matter multiplets. A major difficulty was the need for a suitable formulation of the $N = 2$ Wess-Zumino multiplet (the so-called hypermultiplet), that would allow quantization along the lines of ref. [20]. The long-known off-shell formulation of this model [21] involved off-shell central charges, *i.e.* extra bosonic generators that vanish on shell. The corresponding superspace description contained extra bosonic co-ordinates corresponding to these generators which were not integrated over, with the result that conventional quantization methods ran into difficulties. HOWE, STELLE and TOWNSEND [22] succeeded in arriving at a formulation of the hypermultiplet free from this difficulty, and in fact complete $N = 2$ superfield formulations for gauge and matter multiplets soon became available [23, 24]. There followed, in particular, the existence of a whole class of completely finite *renormalizable* theories with $N = 2$ supersymmetry [25].

There were also several independent attempts along different lines, such as the work of Mandelstam *et al.* [26] that led to the construction of a supersymmetric extension of the light-cone formalism. This provided the first formal argument establishing the finiteness to all orders for $N = 4$ Yang-Mills, without the need of any *ad hoc* assumption. An independent argument [27] strongly suggestive of the all-order finiteness of $N = 4$ Yang-Mills had appeared some time before those mentioned above. The observation is quite elegant. It has to do with the long-known result [28] that, already in $N = 1$ supersymmetric theories, anomalies in internal symmetry currents sit in a supermultiplet together with the trace anomaly. As this is well known to be proportional to the β-function [29] for the gauge coupling, it follows that assuming preservation of the chiral SU_4 in $N = 4$ Yang-Mills implies the vanishing of its (only) β-function, and thus its finiteness. The (minor) weakness has to do with the assumption about the SU_4 symmetry, which is not manifest in an $N = 1$ superfield formulation. This argument is actually equivalent in its conclusion to assuming the preservation of SU_4 and using the nonrenormalization results of ref. [19].

The arguments of ref. [20], even if taken seriously and applied to the most interesting case of maximally extended supergravity, run into the ever-present problem of the coupling constant with negative mass dimension, and manage at most to stretch the first possible onset of divergences to six loops. This, however, is much better than the well-known result of two-loop finiteness, though obviously very far from the complete solution of the problem. The main difficulty with this approach is that its predictions are obtained at the expense of rather strong assumptions on the unknown off-shell structure of models with extended supersymmetry, and the auxiliary fields that serve to close off shell the supersymmetry algebra are not known for most supersymmetric theories. In many cases there are also no-go theorems [30] that, under certain hypotheses, exclude the possibility that any be found. These subjects will be discussed in more detail in sect. **2**, where superspace methods will be described, and the difficulties with off-shell formulations of supersymmetric theories will be reviewed.

Given this rather confused state of affairs, it was found valuable to perform explicit calculations in some controversial case to see how matters would actually turn out be, the idea being that an explicit test of the power counting of ref. [20] would by-pass the difficulties connected with *ad hoc* hypotheses. This work was enterprised by MARCUS and myself [31-33] at Caltech and was carried out there over the last year and a half. The aim was to probe the power counting of ref. [20] by using $N = 4$ Yang-Mills in more than four dimensions. As this theory is nonrenormalizable, local invariants of higher order than those simply ruled out by the $N = 2$ power counting can be probed by evaluating higher-loop corrections to the S-matrix. A novel feature of this two-loop calculation as compared with similar, and simpler, ones in renormalizable models was the need to compute divergent parts of Green functions depending on several (in our case three) external momenta. An interesting by-product of our work is a method [33] to simplify the evaluation of the pole parts of the integrals arising in such calculations. It makes them no more difficult than corresponding ones of two-point functions, and extends to higher orders of the perturbation expansion. The outcome of this work seemed at first very encouraging for supergravity [31], as we found an unexpected cancellation of divergences at two loops in six dimensions. This seemed to indicate that the power counting of ref. [20] was correct beyond the well-established case of $N = 2$ superfields. Clearly, the argument was indirect and at most of suggestive value. However, somewhat later HOWE and STELLE [34] succeeded in giving an alternative, and far more conservative, interpretation of the result of ref. [31] in terms of the available formulation in terms of $N = 2$ superfields [22-24], supplemented by the enforcement of the extra, nonlinearly realized, supersymmetries. This state of affairs undoubtedly made the result of ref. [31] rather empty. However, proceeding with the calculation, after developing the necessary more sophisticated computer techniques [35], we actually obtained a negative, and thus conclusive, result. We found an *explicit violation* of the form of the on-shell effective action assumed in ref. [20]. The conclusion, unfortunately, is rather negative, and we are essentially back to not being able to use the power counting of extended superspace beyond $N = 2$ superfields. As the $N = 2$ power counting carries no useful information for the case of supergravity, it appears that (extended) supergravity theories all diverge starting at three loops. This conclusion is also supported by the counterterm analysis of ref. [36], which excludes further cancellation mechanisms, such as the one noted in ref. [34]. As I have already remarked, it is not completely clear, at the moment, whether a divergent and nonrenormalizable perturbation expansion for gravitational theories should be regarded as a disease of the theories themselves, or rather of our way of calculating them. At any rate, the miracles of supergravity at the first two loop orders [15, 17] are so impressive that they suggest that a very economic solution to the problems of quantum gravity should involve supersymmetry in some way. In this respect, one should keep in mind that multilocal generalizations of supergravity theories, known as superstring theories, have been developed over the past few years by GREEN and SCHWARZ [37].

There are presently *only* three superstring theories in ten dimensions, usually called type I, type IIa and type IIb. Type-II superstring theories contain in their massless sectors the two inequivalent forms of $N = 2$ supergravity [38]. On the other hand, the massless sector of the type-I theory consist of $N = 1$ Yang-Mills coupled to $N = 1$ supergravity [39]. Naively, this theory would sound less attractive than the others, because of the arbitrariness in the choice of the gauge group. It has been known for some time that tree level unitarity already restricts the gauge group to be U_n, SO_n or USp_n [40]. Remarkably, at the time of writing this lecture GREEN and SCHWARZ [41] have just shown that the cancellation of anomalies restricts the gauge group to be SO_{32}, thus making this also a unique theory. Whereas the features of the perturbation expansion of these models are very little understood at the moment, and even their formulation is not at a definitive stage, it has already been shown that at one loop and in $d>4$ they do behave better than their field theory limits [42]. What happens at the next orders is completely unknown at the present time, but this seems a very promising way of departing from the usual description of gravity, perhaps more so than do conformal theories [43], even though these are obviously renormalizable and, in the supersymmetric case, even finite [44].

The conclusion is that, at the moment, we seem to have at our disposal a large class of finite renormalizable theories, with (extended) global supersymmetry. However, the motivation for regarding such models as fundamental theories is far from clear. Whereas demanding finiteness clearly restricts the freedom of the model builder, it would seem that, once a place is to be found for supersymmetry in particle physics, one should go all the way and consider locally supersymmetric theories, as these offer the perspective of unifying all interactions. In this respect, one could conceive of finite renormalizable models as low-energy truncations of (possibly) finite locally supersymmetric theories (*). This could mean superstring theories, as supergravity theories seem candidate to suffering from nonrenormalizable ultraviolet divergences after the first two loop orders.

All the foregoing dicussion has been rather formal, as I have avoided on purpose to emphasize the possible effect of quantization on supersymmetry. In other words, I have not stressed the well-known fact that symmetries which hold at the classical level may be incompatible with the ever-present hidden parameter of the quantum teory, the regulator that disposes of the ultraviolet divergences. This leads to anomalies in classically conserved currents, such as the familiar chiral anomalies [45]. A sufficient (though not necessary) condition in order that a symmetry that holds good at the classical level be free of quantum anomalies is that there exist a regulator compatible with it. Trivially, however, at times a regulator can be lacking only because not discovered yet, as was the case for Yang-Mills gauge theories before the advent of dimensional regularization [46]. Even though, to be precise, one should admit that really rigorous proofs of the renormalizability of Yang-Mills theories have followed the introduction in ref. [46] of a gauge-invariant regulator, one should recall that highly credited work on the subject [47] has actually preceded it. The work of, *e.g.*, ref. [47] was aimed at proving something which, in essence, is equivalent to the existence of a symmetry-respecting regulator.

(*) I thank N. MARCUS for a discussion of this point.

This is the possibility of using an existing symmetry-violating regulator (in that case the Pauli-Villars method [48]), while still achieving the preservation of the gauge symmetry (*i.e.* the fulfillment of the corresponding Ward identities) by means of the addition of suitable extra finite counterterms.

The state of the art for supersymmetric theories can be summarized as follows. There does exist a proposed symmetry-preserving regulator. It was originally designed by SIEGEL [49] in order to maintain the balance between fermionic and bosonic degrees of freedom unaltered upon continuation to a noninteger number of dimensions. This regularization method can thus be regarded as an extension to a noninteger number of dimension of the technique of dimensional reduction, which had so much success in its application to the construction of supersymmetric models (see, most notably, ref. [50]). The prescription is to continue in the number of momenta, while keeping the Lorentz indices on the external fields untouched, which ensures that subtraction of pole parts alone does preserve supersymmetry. Unfortunately, this scheme involves manipulations of different kinds of indices, and some ambiguities do arise. This became clear very early to the author himself [51]. Recently, some work of van Damme and 't Hooft [52] raised the question again. The idea behind it is interesting. It sends us back to the situation for Yang-Mills theories at the beginning of the last decade, and to the work of, *e.g.*, ref. [47]: compute in dimensional regularization, and see whether the results of dimensional reduction can be reached by means of a suitable nonminimal subtraction scheme. This work raised a large interest in the community, especially since it appeared that an incompatibility was arising. Clearly, this would be disastrous for theories like supergravity, for which the supersymmetry current is a gauge current. At the time of this writing the situation has apparently been clarified by JACK and OSBORN [53], and the work of ref. [52] has turned out to contain numerical errors which invalidate its earlier conclusions. Thus there appear to be no difficulties, at the perturbative level, in quantizing supersymmetric theories, and dimensional reduction is equivalent to a nonminimal form of dimensional regularization. Strictly speaking, this has been shown to be case at the two-loop order for renormalizable models. However, it confirms the findings of several groups that have long used dimensional reduction successfully, even with nonsupersymmetric theories [31, 32, 54].

This concludes the historical survey of the problem. The remaining sections are meant to be more technical. They deal in somewhat more detail with some of the points mentioned above. Hopefully, they should make the discussion here more concrete, while at the same time conveying some basic information to the more unexperienced reader. The plan of the remaining sections is as follows. In section **2** superspace methods are rewieved, starting from the very beginning, at least insofar as is needed to discuss the power counting of ref. [20]. This material is mostly well known, and two textbooks are now available, with different levels of completeness (and complexity!) [55, 56].

Thus I will try to be concise. Section **2** also contains brief discussions of the finite models with global supersymmetry [57], and of the difficulties one encounters when attempting to find off-shell formulations for supersymmetric theories. Section **3** addresses in somewhat more detail the issue of the consistency of dimensional reduction and illustrates the result of the work of ref. [31-33]. The discussion summarizes the main points arrived at here. Finally, the appendix contains a brief discussion of two-component formalism, at least insofar as is needed in sect. **2**. Here I have made use of ref. [58], where more details can be found.

An excellent review of these subject matters was written last year by WEST [59]. Consequently, as I have tried to make this discussion self-consistent, some overlapping has been unavoidable.

2. – Superspace, superfields and the superspace power counting.

Superspace [60] is a very useful tool that enables one to make supersymmetry into a manifest symmetry. This is achieved by adjoining to the commuting space-time four-vector co-ordinate x^μ a set of anticommuting spinorial co-ordinates (one of them and its complex conjugate for each supersymmetry). These are actually merely labels, since their square vanishes as a consequence of anticommutativity. Thus (*)

$$(2.1) \qquad x^{\alpha\dot\alpha} \to x^{\alpha\dot\alpha}, \theta^{\alpha i}, \bar\theta^{\dot\alpha}_i.$$

Supersymmetry transformations appear as particular co-ordinate transformation in this extended manifold, such that

$$(2.2a) \qquad \delta x^{\alpha\dot\alpha} = \frac{i}{2}(\bar\varepsilon^{\dot\alpha}_i \theta^{\alpha i} + \varepsilon^{\alpha i} \bar\theta^{\dot\alpha}_i),$$

$$(2.2b) \qquad \delta\theta^{\alpha i} = i\varepsilon^{\alpha i}.$$

Ordinary fields then generalize to superfields, which are functions of all the co-ordinates of the extended manifold. In practice, the θ-dependence is rather trivial as a consequence of anticommutativity, and superfields are just polynomials of finite degree in θ. They provide a convenient way of grouping together the fields of a supersymmetry multiplet. For example,

$$(2.3) \qquad \Phi[x,\theta] = \varphi(x) + \theta^{\alpha i}\psi_{\alpha i}(x) + \dots.$$

(*) For two-component notation see the appendix.

The field components can then be recovered by taking successive derivatives of the superfields at $\theta = 0$. The problem, of course, is to recognize a given supermultiplet inside a superfield and to write down actions in terms of superfields. For simplicity, I will now concentrate on the case of $N = 1$ superfields, for which all is known and writable in a rather accessible form for all models of interest. A superfield without any Lorentz indices, $\Phi[x, \theta]$, transforms as

$$(2.4) \qquad \Phi[x, \theta] \to \Phi[x + \delta x, \theta + \delta \theta]$$

under a supersymmetry transformation. The coefficient of the highest power of θ cannot, so to speak, bear any more θ's. It follows that it must transform as a total divergence under supersymmetry. Inside a space-time integral, this produces a (harmless) surface term. Thus one can write invariant actions as integrals over all superspace of products of superfields

$$(2.5) \qquad \int d^4x \, d^4\theta \, L[\Phi[x, \theta], \partial \Phi[x, \theta]] \, ,$$

where the familiar Berezin integration, which is tantamount to differentiation, picks out the highest θ-component of L.

The problem is to gain control over the formalism, and to write down the proper action for the models of interest. This requires choosing the right superfields, and often achieving a way to truncate their θ-expansion without imposing any condition on the x-dependence of a minimal number of component fields. This truncation is obtained by imposing *constraints* on the superfields, often in the form of differential equations in θ. The constraints are properly distinguished into *on-shell* ones, which do imply the equations of motion for the component fields, and *off-shell* ones, which do not imply the equations of motion for the component fields, and can thus lead to the construction of off-shell Langrangians. I will describe here the superspace formulations of the Wess-Zumino multiplet [61] and of the Yang-Mills multiplet [62]. This material is well known. Therefore, the description that follows is rather sketchy. I hope, however, that it will suffice to make the discussion presented at the end of this section more intelligible, at least insofar as the main ideas are concerned.

First of all, in analogy with what done for usual Poincaré-invariant theories, one introduces representations of the symmetry generators in terms of differential operators on the manifold. Thus, from eqs. (2.2), and specializing to the case of $N = 1$ superspace,

$$(2.6a) \qquad Q_\alpha = i\partial_\alpha + \tfrac{1}{2}\bar\theta^{\dot\alpha} \partial_{\alpha\dot\alpha} \, ,$$

$$(2.6b) \qquad \bar Q_{\dot\alpha} = i\partial_{\dot\alpha} + \tfrac{1}{2}\theta^\alpha \partial_{\alpha\dot\alpha} \, .$$

The supercharges in eqs. (2.6) satisfy

(2.7a) $$\{Q_\alpha, \bar{Q}_{\dot\alpha}\} = i\partial_{\alpha\dot\alpha},$$

(2.7b) $$\{Q_\alpha, Q_\beta\} = 0.$$

For extended supersymmetry eqs. (2.7) would generalize into

(2.8a) $$\{Q_{\alpha i}, \bar{Q}_{\dot\alpha}^j\} = i\delta_i^j \partial_{\alpha\dot\alpha},$$

(2.8b) $$\{Q_{\alpha i}, Q_{\beta j}\} = i\varepsilon_{\alpha\beta} Z_{ij}.$$

The first of eqs. (2.8) is the obvious generalization of the first of eqs. (2.7), but the second of eqs. (2.8) contains new generators, Z_{ij}, which generate the so-called *central charge* transformations [63].

Going back to eqs. (2.6), it is clear that the θ and $\bar\theta$ derivatives do not anticommute with the supersymmetry generators, and are, therefore, not very convenient in the costruction of invariant actions. It is better to work with the covariant spinorial derivatives

(2.9a) $$D_\alpha = \partial_\alpha + \frac{i}{2}\bar\theta^{\dot\alpha}\partial_{\alpha\dot\alpha},$$

(2.9b) $$\bar{D}_{\dot\alpha} = \partial_{\dot\alpha} + \frac{i}{2}\theta^\alpha \partial_{\alpha\dot\alpha},$$

which do anticommute with the supersymmetry charges. They satisfy the algebra

(2.10a) $$\{D_\alpha, \bar{D}_{\dot\alpha}\} = i\partial_{\alpha\dot\alpha},$$

(2.10b) $$\{D_\alpha, D_\beta\} = 0.$$

Then, one can impose differential constraints, for example by applying covariant spinorial derivatives to superfields. For the Wess-Zumino multiplet one needs a complex dimension-one (*i.e.* propagating) scalar, a Weyl spinor and a complex dimension-two (*i.e.* auxiliary) scalar. This set of fields is contained, for example, in a superfield shortened by the chirality condition

(2.11) $$\bar{D}_{\dot\alpha}\Phi = 0.$$

Indeed, expanding Φ in components gives

(2.12) $$\Phi(x) = A(x) + \theta^\alpha \lambda_\alpha(x) + \theta^2 G(x) + \dots,$$

and the remaining components are all space-time derivatives of these. The three fields above are just the propagating scalar, the spinor and the auxiliary scalar of the usual component formulation of the Wess-Zumino model [3].

It should be noted that the constraint in eq. (2.11) does not imply any x-equation for the independent fields inside Φ. The situation would be quite different if, together with eq. (2.11), one had also imposed its complex conjugate. Then, on account of eq. (2.10a), the superfield would contain only fields constant in x-space. The conclusion is that a complex scalar superfield subject to a chirality constraint is what one needs to describe the Wess-Zumino multiplet. The chirality constraint does not imply any component field equations (*i.e.* it is an off-shell constraint), and therefore one can write invariant actions by writing suitable θ-integrals containing Φ and $\bar{\Phi}$. Renormalizable couplings are then selected by the restriction that the corresponding coupling constants be of nonnegative dimension. For example, the Wess-Zumino model with renormalizable couplings would look like

$$(2.13) \qquad \int d^4x \, d^4\theta \, \bar{\Phi}\Phi + \left(m \int d^4x \, d^2\theta \, \Phi^2 + g \int d^4x \, d^2\theta \, \Phi^3 + \text{h.c.} \right),$$

and this expression can be «guessed» using just dimensional analysis. It should be noted that the last two integrals above are only over a subspace of the superspace. They are called chiral integrals, and do produce supersymmetric invariants on account of the constraint (2.11) on Φ. Equivalently, chiral superfields could be regarded as derivatives of unconstrained superfields [64], *i.e.*

$$(2.14) \qquad \Phi = \bar{D}^2 U,$$

where U has the gauge invariance

$$(2.15) \qquad \delta U = \bar{D}^{\dot{\alpha}} \Lambda_{\dot{\alpha}}.$$

As a second, and more complicated, example, consider $N = 1$ Yang-Mills [62]. This four-dimensional model describes the interactions of a multiplet of vectors with one of Weyl spinors, both in the adjoint representation of a gauge group. Thus in components the action is simply

$$(2.16) \qquad S = \int d^4x \left(-\tfrac{1}{4} F_{\mu\nu}^2 + i\bar{\lambda}\nabla\lambda \right).$$

The terms are written in four-component notation, and have the usual definitions

$$(2.17a) \qquad F_{\mu\nu}^a = \partial_\mu A_\nu^a - \partial_\nu A_\mu^a + f^{abc} A_\mu^b A_\nu^c,$$

$$(2.17b) \qquad \nabla_\mu \lambda^a = \partial_\mu \lambda^a + gf^{abc} A_\mu^b \lambda^c.$$

To describe the theory in superspace, one introduces gauge-covariant derivatives ∇_α, $\nabla_{\dot{\alpha}}$ and $\nabla_{\alpha\dot{\alpha}}$, which is tantamount to introducing superfield potentials (much too many!) A_α, $A_{\dot{\alpha}}$ and $A_{\alpha\dot{\alpha}}$. Then one uses the covariantization of the first of eqs. (2.10) to express $A_{\alpha\dot{\alpha}}$ in terms of A_α and $A_{\dot{\alpha}}$, *i.e.* one writes

(2.18) $$\{\nabla_\alpha, \nabla_{\dot\alpha}\} = i\nabla_{\alpha\dot\alpha}.$$

This is actually a constraint, though a trivial one. It amounts to merely resolving the ambiguity in the definition of $\nabla_{\alpha\dot\alpha}$. In fact, eq. (2.18) could at most look like

(2.19) $$\{\nabla_\alpha, \nabla_{\dot\alpha}\} = i\nabla_{\alpha\dot\alpha} + F_{\alpha\dot\alpha},$$

with $F_{\alpha\dot\alpha}$ a field strength. However, adding a covariant object to a covariant derivative produces an equally suitable covariant derivative. On the other hand, the second of eqs. (2.10) and its complex conjugate could generalize to

(2.20a) $$\{\nabla_\alpha, \nabla_\beta\} = F_{\alpha\beta},$$

(2.20b) $$\{\nabla_{\dot\alpha}, \nabla_{\dot\beta}\} = \overline{F}_{\dot\alpha\dot\beta}.$$

The proper way to interpret these equtions is to look at « $\theta = 0$ ». Then $F_{\alpha\beta}$ starts with a self-dual tensor (see the appendix) of dimension one. This clearly cannot be identified with $A_\mu \sim A_{\alpha\dot\alpha}$, nor with the Weyl spinor λ_α. Thus one concludes that the field strengths $F_{\alpha\beta}$ and $\overline{F}_{\dot\alpha\dot\beta}$ must both vanish, which leads to the nontrivial constraint

(2.21) $$\{\nabla_\alpha, \nabla_\beta\} = 0,$$

together with its complex conjugate. These constraints on the covariant derivatives are conditions on the potentials A_α, $A_{\dot\alpha}$ and $A_{\alpha\dot\alpha}$, and allow one to express them in terms of a more basic object, the *prepotential*.

Before heading for the prepotential, let us recall that there exist algebraic relations between the covariant derivatives known as Bianchi identities. These are trivial if no constraints are imposed, but are very useful when working with constrained objects. They allow one to find all the independent field strengths of a theory, both on shell and off shell. These are clearly very useful objects in constructing supersymmetric invariants. One starts by listing all the Bianchi identities in order of increasing dimensionality (∇_α and $\nabla_{\dot\alpha}$ both have dimension $\frac{1}{2}$, whereas $\nabla_{\alpha\dot\alpha}$ has dimension 1):

(2.22a) $[\{\nabla_\alpha, \nabla_\beta\}, \nabla_\gamma] + [\{\nabla_\beta, \nabla_\gamma\}, \nabla_\alpha] + [\{\nabla_\gamma, \nabla_\alpha\}, \nabla_\beta] = 0,$

(2.22b) $[\{\nabla_\alpha, \nabla_{\dot\alpha}\}, \nabla_\beta] + [\{\nabla_{\dot\alpha}, \nabla_\beta\}, \nabla_\alpha] + [\{\nabla_\beta, \nabla_\alpha\}, \nabla_{\dot\alpha}] = 0,$

(2.22c) $[\{\nabla_{\alpha\dot\alpha}, \nabla_\alpha\}, \nabla_\beta] + [\{\nabla_\alpha, \nabla_\beta\}, \nabla_{\alpha\dot\alpha}] + [\{\nabla_\beta, \nabla_{\alpha\dot\alpha}\}, \nabla_\alpha] = 0,$

(2.22d) $[\{\nabla_{\alpha\dot\alpha}, \nabla_\alpha\}, \nabla_{\dot\beta}] + [\{\nabla_\alpha, \nabla_{\dot\beta}\}, \nabla_{\alpha\dot\alpha}] + [\{\nabla_{\dot\beta}, \nabla_{\alpha\dot\alpha}\}, \nabla_\alpha] = 0,$

(2.22e) $[\{\nabla_{\alpha\dot\alpha}, \nabla_{\beta\dot\beta}\}, \nabla_\gamma] + [\{\nabla_{\beta\dot\beta}, \nabla_\gamma\}, \nabla_{\alpha\dot\alpha}] + [\{\nabla_\gamma, \nabla_{\alpha\dot\alpha}\}, \nabla_{\beta\dot\beta}] = 0,$

(2.22f) $[\{\nabla_{\alpha\dot\alpha}, \nabla_{\beta\dot\beta}\}, \nabla_{\gamma\dot\gamma}] + [\{\nabla_{\beta\dot\beta}, \nabla_{\gamma\dot\gamma}\}, \nabla_{\alpha\dot\alpha}] + [\{\nabla_{\gamma\dot\gamma}, \nabla_{\alpha\dot\alpha}\}, \nabla_{\beta\dot\beta}] = 0.$

One then enforces in these otherwise trivial identities the constraints discussed before, and other similar ones involving higher-order field strengths. In the list of eqs. (2.22a)-(2.22f) I have skipped some which are trivially complex conjugates of those listed. Equation (2.22a) is identically satisfied, and eq. (2.22b) implies that

$$[\{\nabla_\alpha, \nabla_{\dot\alpha}\}, \nabla_\beta] = \varepsilon_{\alpha\beta} W_{\dot\alpha}. \tag{2.23}$$

In fact, the l.h.s., being antysimmetric in α and β, can only be proportional to the SU_2 metric $\varepsilon_{\alpha\beta}$ (recall that in this notation $\alpha = 1, 2$ and $\dot\alpha = 1, 2$). This allows one to solve for $\overline{W}_{\dot\alpha}$, and thus for W_α, as

$$W_\alpha = \tfrac{1}{2} [\{\nabla_\alpha, \nabla_{\dot\alpha}\}, \nabla^{\dot\alpha}]. \tag{2.24}$$

At this point one can proceed in two ways. The first is simply to keep on inserting eqs. (2.23) and (2.24) in the remaining Bianchi identities (eqs. (2.22c)-(2.22f)). The other, more handy procedure, consists of taking successive derivatives of W_α. There would seem to be the problem of establishing when to stop. This, however, is clear if one recalls what said above about the constraints. Everything is always interpretable, provided one looks at the $\theta = 0$ part of the superfields. Thus W_α contains the spinor at $\theta = 0$, and differentiating it will say something about the vector field strength. Subsequent differentiations contain information about the field equations, i.e. give a foolproof procedure to establish whether the constraints put the theory on shell, the field equations being easily recognizable using the «$\theta = 0$ trick» mentioned above. The auxiliary fields appear as a whole superfield that, when set to zero, puts the theory on shell. They are in a multiplet with the field equations, and the auxiliary field superfield contains all the field equations at successive orders in θ. To see how all this works let us differentiate eq. (2.24) with respect to $\bar\theta^{\dot\beta}$. This gives

$$\{\nabla_{\dot\beta}, W_\alpha\} = \tfrac{1}{2} \{\nabla_{\dot\beta}, [\{\nabla_\alpha, \nabla_{\dot\alpha}\}, \nabla^{\dot\alpha}]\}, \tag{2.25}$$

which, using the lower Bianchi identities, the constraints and eq. (2.23), can be written

$$\{\nabla_{\dot\beta}, W_\alpha\} = -\tfrac{1}{2} \{\nabla_{\dot\beta}, W_\alpha\}. \tag{2.26}$$

Thus W_α is the gauge-covariant generalization of a chiral superfield. One can also differentiate with respect to θ^β. The result can be trivially decomposed into irreducible pieces as

$$\{\nabla_\beta, W_\alpha\} = i\varepsilon_{\alpha\beta} D + G_{\alpha\beta}, \tag{2.27}$$

where $G_{\alpha\beta}$ is symmetric in its two indices. It is (see the appendix) the self-dual part of the vector field strength, the only object that the component theory of eq. (2.16) contains at dimension 2. Thus the superfield D is the auxiliary field superfield, and its $\theta = 0$ component is the familiar pseudoscalar auxiliary field for the Yang-Mills multiplet [62]. Setting D to zero puts the theory on shell. Indeed, in this case contracting α and β eliminates the r.h.s., and then differentiating with $\nabla_{\dot\alpha}$ and using the chirality of W_α and eq. (2.24) yields

$$(2.28) \qquad \{\nabla_{\alpha\dot\alpha}, W^\alpha\} = 0 \,.$$

This clearly contains the Dirac equation for the spinor at $\theta = 0$. One more differentiation would generate the vector field equation.

We have thus seen how setting the auxiliary field superfield to zero puts the theory on shell. The example I have discussed is rather elementary. The lesson, however, is that deriving on-shell constraints is always conceptually simple, and the component theory serves as a guide all the way through. All one needs to do is set to zero all the field strengths whose $\theta = 0$ parts are not present in the component theory formulated without auxiliary fields. This, however, is by no means a pointless exercise, and provides a very elegant approach for analyzing complicated theories and deriving their complete field equations and supersymmetry transformations (for the application of these techniques to supergravity see, *e.g.*, ref. [65]). For our purposes, the interest in such formulations stems from their allowing one to simply classify on-shell invariants, and thus divergences of the S-matrix as allowed by the theorem of ref. [20].

The really difficult problem is to formulate the theories off shell or, in the corresponding component approach, to find the auxiliary fields that close the supersymmetry algebra off shell. The complete solution to this problem is unknown at present, and many failed efforts have even led to the no-go theorems of ref. [30], which exclude the possibility that auxiliary fields be found for (almost) all multiplets of $N > 2$ supersymmetry. Whereas such theorems always rest on working hypotheses, the lesson they convey is clear. If off-shell formulations exist in the cases excluded by the theorems, they must have a rather unfamiliar look. Thus, also their quantum properties may be very different from what one would expect. I will return to this point at the end of this section.

To complete the discussion of the superspace formulation of $N = 1$ Yang-Mills, I will now show how one goes about finding a prepotential and the superspace form of the action. For this case the solution was originally «guessed» [62]. This was possible because the constraint (2.21) that one must solve is very simple. It is the statement that the covariant derivative D_α is the one suited for the pure gauge case, *i.e.*

$$(2.29) \qquad \nabla_\alpha = \exp[-\Omega] D_\alpha \exp[\Omega] \,,$$

where Ω is a general (complex) scalar superfield. The solution for $\nabla_{\dot\alpha}$ is then obtained by complex conjugation. Here Ω has the gauge invariance

(2.30) $$\delta \exp[\Omega] = \exp[-i\bar\Lambda] \exp[\Omega] \exp[iK],$$

with K a real superfield and $\bar\Lambda$ an antichiral superfield. It should be noted that the $\bar\Lambda$ transformation affects only the prepotential, but not the potential A_α. It is usually called a pre-gauge transformation. Actually, there exists an elegant way of arriving at this result [24] that also works for more complicated cases. All one does is notice that the covariant derivatives are functions of the gauge coupling constant. Differentiating eq. (2.21) with respect to it gives

(2.31) $$\left\{\nabla_{(\alpha}, \frac{d\nabla_{\beta)}}{dg}\right\} = 0,$$

where the parentheses denote symmetrization. On account of eq. (2.21), eq. (2.31) is clearly solved by

(2.32) $$\frac{d\nabla_\alpha}{dg} = [\nabla_\alpha, \Omega],$$

with Ω an arbitrary scalar superfield. Ω is the prepotential. The problem is integrating the matrix differential equation. In this case the solution is very simple, and is eq. (2.29) (once the coupling constant is set back to unity). In general, however, a closed-form solution is not possible, and one must content himself with a power series in g. The superspace action for $N=1$ Yang-Mills is then obtained by noticing that, purely on dimensional grounds, the only gauge-invariant object of the right dimensionality one can write is

(2.33) $$\text{Tr}\int d^4x\, d^2\theta\, (W^\alpha W_\alpha + \text{h.c.}).$$

The reader more familiar with $N=1$ superspace formulations will recognize that this description has some redundancy built in with respect to the conventional one in terms of a *real* scalar superfield V, with gauge invariance

(2.34) $$\delta \exp[V] = \exp[-i\bar\Lambda] \exp[V] \exp[i\Lambda],$$

where Λ is a chiral superfield. The formulation of eq. (2.29), usually called vector representation (see ref. [56] for more details), stands to the usual one as the vierbein formulation of gravity stands to the metric formulation. The conventional formulation can be recovered by means of a partial gauge fixing, whereby the parameter K in eq. (2.30) is used to gauge away the imaginary part of Ω.

I have already remarked that superfields are very convenient objects for deriving power-counting restrictions on the occurrence of particular kinds of counterterms. In this approach there are two steps:

a) an off-shell superfield analysis, which examines the features of the perturbation expansion, and thus sorts out the kinds of structures allowed for the divergences;

b) an on-shell superfield analysis, which shows which of the structures of the type in *a*) survive on shell. This step is, in principle at least, straightforward in all cases.

The really difficult problem is to find a suitable set of off-shell constraints for the covariant derivatives. This is the starting point for quantum calculations. The idea behind the power counting of Grisaru and Siegel [20] is then simple, at least in principle. One gains experience from the rather handy case of $N=1$ superfields by performing calculations « cleverly », *i.e.* by generating the least number of unnecessary terms at intermediate stages (in gauge theories this requires the use of the background field method to enforce gauge covariance at all stages). The result is that some structures are not generated at all, and must thus be absent from the list of all possible divergences. This analysis rests on the work of ref. [19, 20]. The conclusion is the following:

1) All divergences are local *complete* θ-integrals, *i.e.* no integrals over subspaces of the superspace are allowed.

2) The integrands are gauge-invariant functions of connections and field strengths only (*i.e.*, in the example of $N=1$ Yang-Mills discussed above, only A_α, $A_{\dot\alpha}$ and $A_{\alpha\dot\alpha}$ are allowed, not V!).

Since connections and field strengths, as well as θ integration measures, all have positive dimensionality, this puts strong restrictions on the existence of possible counterterms.

In order to have some say about the really interesting case of $N > 1$ supersymmetry, for which the off-shell superfield formulations are mostly unknown (and often believed not to exist [25]), one needs some working hypotheses. In ref. [20] the implicit working hypothesis is that similar manipulations to those possible for $N = 1$ should go through for $N > 1$ formulations, if these exist. This leads to rather surprising results.

For N-extended Yang-Mills at more than one loop in a background field gauge divergences would take the form

$$(2.35) \qquad (g^2)^{L-1} \int d^d x \, d^{4N} \theta \, f(A, W),$$

where A and W denote generically the connections and field strengths of the theory. The lowest-dimensional suitable f is so the form

(2.36) $$AW + \text{higher order in } A,$$

which is gauge invariant modulo a total derivative on account of

(2.37) $$\delta A = D\Lambda + \text{more},$$

and of the typical on-shell Bianchi identity

(2.38) $$DW + \text{more} = 0.$$

Given that f in eq. (2.35) has at least dimension 2, the condition that the effective action be dimensionless yields the restriction

(2.39) $$(4-d)L + 2(N-1) > 0,$$

using the dimensionality of g^2, $4-d$ in d dimensions. Similarly, the lowest-dimensional counterterm for supergravity is the superspace analogue of the cosmological term,

(2.40) $$(k^2)^{L-1}\int d^d x \, d^{4N}\theta s \, \det(E),$$

where E is the superfield generalization of the vierbein, usually called supervierbein (see ref. [55, 56] for more details). Counting powers again and recalling that k^2 has dimensionality $2-d$ in d dimensions then gives

(2.41) $$(2-d)L + 2(N-1) > 0.$$

Equations (2.39) and (2.41) are what I referred to as the power counting of Grisaru and Siegel [20]. It should be stressed that, whereas the derivation above is rather trivial, it rests on a nontrivial analysis of the superspace perturbation theory, which is at the heart of ref. [20] (see also ref. [25]), where it is shown that conditions 1) and 2) above do indeed hold for $N=1$ superfields. Referring to the discussion of $N=1$ Yang-Mills presented above, these statements are proved by showing that one can actually perform calculations in the background field method without ever expressing the background covariant derivatives in terms of the background prepotential.

Sveral remarks are in order here:

I) First of all, a technical one. In ref. [20] it was noticed how a difficulty would arise at one loop. In fact, manifest background covariance would lead in this case to an infinite tower of ghosts coupling only to the background fields, which thus contribute only at one loop. Breaking background covariance would terminate the chain of ghosts, but would also alter the one-loop counterterms by making them noncovariant. Thus, barring more detailed analysis in special cases, the power counting only applies to diagrams with more than one loop. For more details see ref. [25].

II) The second remark is that the power counting stems from a set of sufficient conditions, and can be improved by detailed case-by-case analysis, if the corresponding lowest-dimensional counterterms happen to vanish. As they stand, however, eqs. (2.39) and (2.41) already produce a number of very interesting results:

a) $N \geqslant 2$ Yang-Mills coupled to $N = 2$ matter is finite beyond one loop in four dimensions, if an $N = 2$ superfield formulation respecting the conditions 1) and 2) given above can be found for it. This is a rather impressive result indeed. Since $N = 2$ superfield formulations have been constructed [21], this provides a very simple finiteness proof for $N = 4$ Yang-Mills [20, 25], alternative to the one previously given using light-cone superfields [26]. Moreover, a closer analysis reveals that there is a whole class of globally supersymmetric renormalizable theories with $N = 2$ supersymmetry which are finite to all orders. These were found in the first of ref. [25] by demanding the cancellation of the one-loop contribution to the β-function, which forces one to choose suitable gauge groups and suitable group representations for the matter multiplets. This is sufficient, because finiteness beyond one loop is guaranteed by the power counting of ref. [20] and by the existence of the relevant superspace formulations [22-24]. The condition one finds for one-loop finiteness is

$$(2.42) \qquad \sum_i m_i T(R_i) = C_2(G),$$

where one considers m_i $N = 2$ hypermultiplets in the representations R_i and $\bar{R}_i$, and where $C_2(G)$ denotes the Casimir for the adjoint representation of the gauge group G. Clearly, there are many solutions to this condition. Actually, eq. (2.42) can also be generalized [66]. In fact, if some of the representations for the matter fields are pseudoreal, they do not need to be « doubled », and one obtains the condition

$$(2.43) \qquad \sum_i m_i T(R_i) + \tfrac{1}{2} \sum_j m_j T(R_{pj}) = C_2(G).$$

For more details, see ref. [66], where a list of solutions can be found. Interestingly, one can also add extra terms that violate $N = 2$ supersymmetry, while still preserving the finiteness. This was noticed by PARKES and WEST [57]. The interest in this option is, of course, the greater freedom this allows in the process of model building with finite theories. The solution is not as inelegant as one may think, because supergravity couplings, after spontaneous breaking of local supersymmetry, induce terms which, at low energies, look like explicit breakings [67]. For more details, see ref. [59].

b) For supergravity, the success of a power-counting approach is doomed to be rather limited, because these theories are power-counting nonrenormalizable. Still, eq. (2.41) tells us that, if $N = 8$ supergravity could be formulated in terms of $N = 8$ superfields, the first possible onset of ultraviolet divergences would be *ipso facto* postponed to seven loops, an encouraging improvement with respect to the obvious three-loop barrier. The state of the art, however, is not so encouraging. The theorems of ref. [30] do not allow any conventional formulation of $N = 8$ supergravity in terms of $N = 8$ superfields, leaving at most the possibility of a straightforward formulation in terms of $N = 4$ superfields. This could in principle be done using the known off-shell formulation of $N = 1$ supergravity in ten dimensions and allowing for the existence of a set of four off-shell gravitino multiplets [25], a possibility not excluded by the counting arguments of ref. [30]. At any rate, this would bring the first possible onset of divergences down to three loops again, just as would be the case for a component formulation. The possibility of unconventional formulations is, of course, left open. The very applicability of the power counting beyond $N = 2$ superfields can be examined by direct calculation. I will come back to this point in the next section.

III) The third remark is meant to emphasize what the heart of the problem is. I have already stressed that the power counting rests on a working assumption, namely that the manipulations carried out in the handy (and well-known) case of $N = 1$ superfields do have some bearing on the really interesting case of $N > 1$ superfields. This is a very strong hypothesis, in fact stronger than the hypothesis that the corresponding off-shell superfield formulations exist. It amounts to assuming that $N > 1$ superfield formulations bear a close resemblance to $N = 1$ ones, which appears dubious, in view of the no-go theorems of ref. [30]. The main hypothesis these theorems rest upon is that the spectra of the theories can be read from their quadratic terms, *i.e.* that no cubic or higher-order Lagrange multiplier terms are needed to determine them. Within this assumption, it can be shown that there are no auxiliary fields closing the off-shell algebra for $N = 4$ Yang-Mills and for a number of other models [30]. This is done by showing that two equivalent ways of counting the fermionic auxiliary fields needed for the off-shell closure of the supersymmetry algebra give different, and thus inconsistent, results. One of these uses the known dimension of off-shell representations or superfields, and the other follows from the observation that fermionic auxiliary fields must come in pairs, say

$$\bar{\chi}\psi \, , \tag{2.44}$$

simply because one needs two different fields of half-odd-integer dimension to construct an invariant of dimension four. As a word of caution, one should note that the apparently insignificant restriction on the Lagrange multipliers actually fails for as familiar a case as that of nonlinear σ models, and relaxing it allows off-shell Lorentz covariance [68] for one of the two inequivalent forms of $N = 2$ supergravity in ten dimensions [69], a problem known to be otherwise insolvable [70].

3. – The breakdown of the power counting.

The discussion presented in sect. 2 was meant to emphasize one point. Superspace is a very powerful tool for analizing divergences in supersymmetric theories, with great effectiveness in the case of renormalizable, and thus globally supersymmetric models, where the hypotheses underlying the power counting of ref. [20] have been substantiated by the explicit construction of the corresponding $N = 2$ superfield formulations [22-24]. The reason for this success is twofold. First of all, the models are renormalizable, and thus power-counting arguments have a good chance of being effective. Moreover, one only needs $N = 2$ superfield formulations in order to exploit the power counting of ref. [20] in all its strength. On the other hand, for supergravity theories the nonrenormalizability requires that stronger hypotheses be made on their superspace formulations to arrive at nontrivial, though only partial, results. The same is true for higher-dimensional Yang-Mills theories, again by virtue of their nonrenormalizability. Since $N = 4$ Yang-Mills [5] is one of the models for which no conventional superfield formulation is possible, the study of its ultraviolet behavior in $d > 4$ can in principle lead to useful indications for the more interesting, and far more difficult, case of supergravity. This analysis was carried out in ref. [31-33] by Marcus and the author. There we computed the divergences of the four-spinor S-matrix amplitude at two loops for $N = 4$ Yang-Mills in four, five, six, seven and nine dimensions. We used the Wick rotated action in components written in ten-dimensional notation,

$$(3.1) \qquad S = \int d^{10}x \left(-\frac{1}{4} F_{\mu\nu}^2 - \frac{i}{2} \bar{\lambda} \nabla \lambda \right),$$

where λ is a Majorana-Weyl spinor and the definitions of the terms above are the same as in eqs. (2.17). The corresponding one-loop analysis had already been carried out by Green, Schwarz and Brink [42], by taking the zero-slope limit of the corresponding superstring amplitude. Their result, however, has no direct bearing on the validity of the superspace power counting which, as I have emphasized in the previous section, does not apply to one loop.

The theory was regularized using the dimensional reduction scheme [49], whereby external field indices are kept of fixed number upon regularization, which only affects the momenta. These are continued from d dimensions to $d-\varepsilon$ dimensions, just as in conventional dimensional regularization. Dimensional reduction preserves the equality of Bose and Fermi degrees of freedom upon regularization. Thus it allows for calculations done in the convenient minimal subtraction scheme, whereby only pole parts are subtracted, which at the same time preserve the Ward identities of supersymmetry.

Dimensional reduction is somewhat tricky to use, as it involves an algebra of formal manipulations of objects of two distinct kinds, d-dimensional indices and $(d-\varepsilon)$-dimensional indices. That there are some ambiguities hidden in the corresponding large set of rules has long been noticed by SIEGEL himself [51], shortly after introducing the new regularization scheme. In fact, the ambiguities all have to do with manipulations of ε symbols, objects which are actually not susceptible of a consistent definition in the conventional dimensional regularization scheme [45]. The point is that dimensional reduction can give one the illusion that ε's can be manipulated naively, with potentially disastrous consequences. For example, in d dimensions, the product of two ε symbols can be turned into a product of δ's. Considering for simplicity the case of two-dimensional Euclidean space, one has

$$(3.2) \qquad \varepsilon^{\alpha\beta}\varepsilon_{\gamma\delta} = \delta^{\alpha}_{\gamma}\delta^{\beta}_{\delta} - \delta^{\beta}_{\gamma}\delta^{\alpha}_{\delta}.$$

This, however, yields zero if one specializes α and β to lie in a $(d-\varepsilon)$-dimensional subspace and γ and δ to lie in the orthogonal ε-dimensional one, and leads to different results for the two manifestly identical expressions

$$(3.3) \qquad \varepsilon^{\alpha\beta}\varepsilon_{\gamma\delta}\varepsilon_{\alpha\beta}\varepsilon^{\gamma\delta}$$

and

$$(3.4) \qquad \varepsilon^{\alpha\beta}\varepsilon_{\alpha\beta}\varepsilon^{\gamma\delta}\varepsilon_{\gamma\delta}.$$

On the other hand, dimensional regularization works with $(d-\varepsilon)$-dimensional indices, and does not allow for a consistent definition of ε altogether. In particular, it does not let one write eq. (3.2). Thus it calls for caution whenever such quantities are present. We know that this should be the case, because ε symbols call for axial anomalies. These, in turn, imply an incompatibility between the conservation of two quantities, that as such can be resolved only as a result of a deliberate choice. Whenever anomalies happen to cancel, the ambiguities also resolve because the choice then becomes unique. The conclusion is that ε's have to be dealt with carefully anyway, and the remarks of ref. [51] appear not very significant. This subject matter can actually be investigated in detail, by comparing results obtained by dimensional reduction with corresponding ones obtained by the conventional regularization scheme, where one allows for nonminimal subtractions. This approach was suggested by VAN DAMME and 'T HOOFT [52], and also pursued by others [53]. Whereas ref. [52] contained several numerical errors and correpondingly wrong conclusions, the work of Jack and Osborn [53] is apparently free of errors. Their result is that, at the two-loop level, minimal dimensional reduction is equivalent to a nonminimal form of dimensional regularization. This is tantamount to saying that, to this order, there are no inconsistencies between supersymmetry and quantization, and implies that supersymmetry currents do not suffer from anomalies. This is very relieving, because for supergravity the supersymmetry current is a gauge current, and anomalies in it would destroy such theories completely, in a far more unquestionable way than ultraviolet divergences.

Then, going back to the discussion of ref. [31-33], one recognizes that, when calculating in components, supersymmetry is only broken by the gauge-fixing procedure. The S-matrix, being gauge independent, is guaranteed to be supersymmetric. Calculating a four-spinor S-matrix element thus tells the whole story about all the four-particle amplitudes. The results we found are summarized in table I.

TABLE I. – $N = 4$ *Yang-Mills at one and two loops.*

Loops	Dimensions						
	4	5	6	7	8	9	10
1	F	T	F	T	I_8	T	I_{10}
2	F	F	F	I_7	—	I_9	—

Here T stands for trivially finite, which is the case of all dimensionally regularized amplitudes in odd dimensions at odd numbers of loops, F stands for finite, and I stands for infinite. The subscripts label the form of the divergences encountered. The cases of $d = 8, 10$ at two loops were not considered, because the theory was already known to be one-loop divergent there. A very useful device in analyzing the results was provided by ref. [42]. There it was shown that all one-loop infinities in the four-particle S-matrix of this theory could be recast as products of a common kinematic factor, which can be written

$$(3.5) \quad F_{\mu\nu} F_{\nu\varrho} F_{\varrho\sigma} F_{\sigma\mu} - \tfrac{1}{4} F_{\mu\nu} F_{\mu\nu} F_{\varrho\sigma} F_{\varrho\sigma} + 2i\bar{\lambda}\gamma_\alpha \partial_\beta \lambda F_{\alpha\gamma} F_{\beta\gamma} - $$
$$ - i\bar{\lambda}\gamma_{\alpha\beta\gamma} \lambda F_{\alpha\delta} \partial_\gamma F_{\beta\delta} - \tfrac{1}{3}\bar{\lambda}\gamma_\alpha \partial_\beta \lambda F_{\alpha\gamma} F_{\beta\gamma},$$

and of totally symmetric « group theory » factors, containing at times the Mandelstam-type differential operators $s = 2\, p_1 p_2$, $t = 2p_1 p_3$ and $u = 2p_1 p_4$.

Thus, apart from common kinematic factor of eq. (3.5),

(3.6) $\quad I_8 = \dfrac{g^4}{(4\pi)^4 \varepsilon} \left(\square + \dfrac{1}{6} \left(\rightarrowtail\!\!\!\leftarrowtail + \text{(triangle)} \right) \right)$

and

(3.7) $\quad I_{10} = \dfrac{1}{360} \dfrac{g^4}{(4\pi)^5 \varepsilon} \Bigg(s \left(\rightarrowtail\!\!\!\leftarrowtail + \bowtie \right) +$

$+ t \left(\text{(vertical)} - \bowtie \right) + u \left(-\rightarrowtail\!\!\!\leftarrowtail - \text{(vertical)} \right) \Bigg).$

Here the group theory factors are described graphically using a notation first introduced by CVITANOVIĆ [71], whereby the structure constants f^{abc} are represented by a trilinear vertex:

(3.8) $\qquad\qquad f^{abc} \rightarrow \mathsf{Y}$.

On account of total antisymmetry, the vertex changes sign upon interchange of any two of its three legs. The features of this notation as applied to this case are described in detail in ref. [32]. At two loops the divergences take similar forms with, however, more complicated group theory factors, as allowed by the larger number of structure constants present in two-loop graphs. Thus

(3.9) $\quad I_7 = \dfrac{g^6 \pi}{(4\pi)^7 \varepsilon} \left(\left(\varepsilon \left(\dfrac{4}{9} \right) \text{(pentagon)} \right) + \right.$

$$+ \frac{1}{90}\left(\diagram + \diagram\right)\right) +$$

$$+ t\left(\frac{4}{9}\diagram + \frac{1}{90}\left(\diagram - \diagram\right)\right) +$$

$$+ u\left(\frac{4}{9}\diagram + \frac{1}{90}\left(-\diagram - \diagram\right)\right)$$

and

$$(3.10) \quad I_9 = \frac{g^6 \pi}{(4\pi)^9 \varepsilon}\left(\frac{5}{3024} stu \left(\diagram + \right.\right.$$

$$+ \frac{1}{6}\left(\diagram + \diagram\right)\right) -$$

$$- s^3\left(\frac{13}{4536}\diagram + \frac{5}{133056}\left(\diagram - \diagram\right)\right) -$$

$$-t^3\left(\frac{13}{4536}\ \text{[diagram]}\ +\frac{5}{133056}\left(\text{[diagram]}+\text{[diagram]}\right)\right)-$$

$$-u^3\left(\frac{13}{4536}\ \text{[diagram]}\ +\frac{5}{133056}\left(\text{[diagram]}+\text{[diagram]}\right)\right).$$

Actually, in ref. [32] we have only computed the four-spinor piece of these terms, and we have let supersymmetry determine the other pieces. It should be noted how the two-loop divergences continue the structure of the one-loop ones first found in ref. [42], where this was actually conjectured to occur. Assuming that this pattern continues to all orders (probably a consequence of supersymmetry, but this has not been proved in general), there follow the power-counting rules [42]

$$(3.11) \qquad d < 4 + \frac{2}{L}$$

for $N = 4$ supersymmetric Yang-Mills, and

$$(3.12) \qquad d < 2 + \frac{2}{L}$$

for $N = 8$ supergravity. It should be noted how eq. (3.11) fails to predict the finiteness at two loops in six dimensions. The result (3.12) suggests that $N = 8$ supergravity should diverge at three loops in four dimensions.

The two-loop finiteness in six dimensions would seem to be the most interesting result in the table altogether, because only the far more sophisticated power counting of eq. (2.37) appears to able to explain it, and only for $N = 4$, *i.e.* if a formulation in terms of $N = 4$ superfields is available. This initially led us to an optimistic interpretation as suggesting that the power counting of ref. [20] was correct for all N [31]. This, however, turns out not to be the case. A different, and more conservative, explanation was provided by HOWE and STELLE [34], who managed to explain the result of ref. [31] using the available formulation of six-dimensional $N = 1$ Yang-Mills in terms of six-dimensional $N = 1$ superfields, properly supplemented by the enforcement of the extra, nonlinearly realized, supersymmetries. On the other hand, finiteness in $d = 7$ is not guaranteed by the power counting of ref. [20], even with $N = 4$, and, therefore, naively the corresponding result in the table above sounds rather uninformative. However, matters turn out to be quite different. To understand all this, one should recall the discussion in the previous section, where the main points of the analysis of ref. [20] have been summarized (see remarks 1) and 2)). The point is that, not only do we have a power counting at our disposal, but we also have a very precise statement about the form of the on-shell effective action, namely that its divergent part is a local gauge-invariant functional of connections and field strengths only (no prepotentials explicitly!) and containing a full θ-integral. For the case of $d = 7$, on dimensional grounds, only one term is allowed:

$$\text{(3.13)} \qquad \int d^7 x \, d^{16}\theta \, \{A_A w^A + \text{higher order}\},$$

where (see ref. [32] for more details) A is a ten-dimensional spinor index, and

$$\text{(3.14)} \qquad w^A = (\gamma_\alpha)^{AB} (\gamma^\alpha)^{CD} (D_C D_D A_B - D_B D_C A_D)$$

denotes the linearized form of the dimension-$\frac{3}{2}$ field strength W^A of ten-dimensional Yang-Mills (see the discussion of the four-dimensional case in sect. **2** for comparison). In eq. (3.13) «higher order» stands for

$$\text{(3.15)} \qquad (\gamma_\alpha)^{AB} (\gamma^\alpha)^{CD} \left(-\tfrac{2}{3} A_B [D_C A_D, A_A] + \tfrac{2}{3} A_B [D_C A_A, A_D] - \{A_B, A_C\} \cdot \{A_A, A_D\} \right),$$

derived by requiring invariance under the full nonlinear transformation

$$\text{(3.16)} \qquad \delta A_A = [\nabla_A, \Lambda].$$

This is possible in view of the on-shell Bianchi identity satisfied by W,

$$\text{(3.17)} \qquad \{\nabla_A, W^A\} = 0.$$

The use of the ten-dimensional notation is no real restriction, because the smallest seven-dimensional superspace has the same dimensionality as the smallest ten-dimensional one. The result in eq. (3.9) admits a ten-dimensional notation, in the sense that all restrictions to seven dimensions on the indices come from derivatives, which are necessarily seven-dimensional objects. The same, of course, can be said for all the other divergences found. Thus the contradiction with our result comes up because the term in eqs. (3.13) and

(3.15) has a group theory structure completely fixed by the gauge invariance to be a « tree », in the graphical notation mentioned above. On the other hand, the two-loop divergence in seven dimensions contains a far more intricate group theory factor (see eq. (3.9)), and such a factor can be shown to be independent of the « tree ». The conclusion is that the very essence of the power counting of Grisaru and Siegel [20], the form of the on-shell effective action is *explicitly violated* beyond $N = 2$ superfields. Whereas this is certainly not a proof that supergravity theories diverge at three loops, it is clearly a strong indication in that sense, because the only available reason for some optimism is removed. What is conclusive in this analysis is the statement that the naive extension of the manipulations made with $N = 1$ superfields to $N > 2$ fails. Whether this is due to the nonexistence of the corresponding off-shell superfield formalisms or to their having a different structure from the $N = 1$ case is, of course, not possible to decide upon at this stage. At any rate, the majority of physicists are not so interested in the details of superspace as in its implications for finiteness, and the results I have discussed do tell us that the power counting cannot be trusted beyond $N = 2$. As to the result in eq. (3.9), it appears very plausible for two reasons. First of all, it is of the form suggested by the superstring analysis of ref. [42]. Moreover, one can show that, once it is dimensionally reduced, it can be written in terms of $N = 1$ six-dimensional superfields. This is suggestive of an extension of the available six-dimensional superspace formalism above $d = 6$, along similar lines to what is known to be possible for $N = 1$ four-dimensional superfields [72].

The work of refs. [31, 32] that I have summarized here involved calculations far more difficult than those appeared earlier in the literature. Success in this enterprise rested heavily on the development of a technique for computing divergent parts of Feynman integrals [33] that I would now like to mention, though in a rather sketchy fashion. I actually believe that this technique is possibly more important than the results themselves, as it may open the way to far more complicated direct tests of gravity and supergravity. In the following I will illustrate it, together with some minor, but still useful, points, in a series of remarks. This is meant to convey the general ideas behind the methods. For a more detailed discussion the reader is referred to the original literature [32, 33].

I) Rather than calculating divergences of S-matrix elements, one can calculate the on-shell divergences of the effective action. This implies calculating the parts of the effective action relevant to the given process and enforcing in them the *classical* field equation. Working with the external fields in the co-ordinate representation, no symmetrization is required. Each individual graph has then to be assigned a combinatoric factor equal to the reciprocal of the dimension of the corresponding discrete symmetry group, obtained allowing interchanges of both internal and external legs.

II) Rather than using counterterms determined by lower-order calculations directly, one can more conveniently account for their effect by subtracting subdivergences from Feynman integrals. The minimal subtraction prescription results in a rule for dealing with a pair of mutually contracted Lorentz indices coming from a subtraction. Minkowski metrics generated by any two such indices must be « barred », in the sense that their trace must be taken to be d, rather than $d - \varepsilon$. The contraction with other indices proceeds as though ε were a positive number, *i.e.* with metrics over the « lower »-dimensional space dominant, just as projection operators would be. In this approach finiteness is recognized, at a given loop order, by the vanishing of the corresponding divergent contribution of order $1/\varepsilon$ to the effective action, once the classical field equations are enforced. Terms determined by lower-order subtractions, however, can provide useful checks of the calculations.

III) Once one has settled to work with subtracted Feynman integrals, one can notice that, on general grounds [73], their divergences are *local* in co-ordinate space. There follows a very simple algorithm for extracting pole parts from general loop integrals [33], which can be efficiently implemented on computers, thus allowing for large-scale divergence calculations at higher loops. Basically, there are two steps in this procedure. First of all, since the divergences are guaranteed to be polynomials in momentum space, say of (integer!) degree a, one can use Euler's theorem to lower the degree of divergence of subtracted Feynman integrals by differentiating with respect to the external momenta p_i, *i.e.*

$$(3.18) \qquad I^{(a)} = \frac{1}{a} p_i \frac{\partial}{\partial p_i} I^{(a)}.$$

Then, once the integrals are reduced to logarithmically divergent ones, the divergent parts of these are guaranteed to be *constant*. They are thus independent of the external momenta, which can be arranged at will, with the only care of not running into fake infra-red divergences in the evaluation of the resulting integrals. The conclusion is that computing the pole part of a general n-loop integral is tantamount to computing that of a propagator integral at $n - 1$ loops. For example, consider the (rather simple) integral

$$(3.19) \qquad \int \frac{\mathrm{d}^d k}{(2\pi)^d} \frac{\mathrm{d}^d l}{(2\pi)^d} \frac{1}{k^2 (k+p)^2 (k-l)^2 l^2 (l+p)^2}$$

for the case of six-dimensional space-time, where it is quadratically divergent. Upon differentiation, this can be reduced to

$$(3.20) \qquad \int' \frac{\mathrm{d}^d k}{(2\pi)^d} \frac{\mathrm{d}^d l}{(2\pi)^d} \left\{ \frac{4k \cdot p \, l \cdot p}{k^6 l^6 (k-l)^2} + \frac{8(k \cdot p)^2}{k^8 l^4 (k-l)^2} - \frac{2p^2}{k^6 l^4 (k-l)^2} \right\},$$

where the prime is meant to emphasize that each of the terms above has to be calculated together with the corresponding subtraction. These terms are all logarithmically divergent, and the momentum dependence has effectively « factored out ». I would like to stress that this is possible only because one is looking at subtracted integrals, which are guaranteed to be local. The remaining part of the calculation is very simple, and yields

$$(3.21) \qquad \frac{p^2}{3\varepsilon^2} - \frac{p^2}{9\varepsilon}.$$

Clearly, a far greater simplification is obtained in more complicated cases, where the integrals depend on several external momenta. More details can be found in ref. [33].

4. – Discussion.

I have attempted to outline the present understanding of the ultraviolet behavior of supersymmetric theories. Two things emerge clearly from this discussion. We have at our disposal a class of completely finite renormalizable models with extended global supersymmetry, and we have a number of formal ways of proving their finiteness. However, at present the motivation for looking at such theories is not clear. More precisely, it is not clear why a finite model should be preferred to other infinite, but still renormalizable and predictive, ones. On the other hand, supergravity theories are *a priori* far more interesting, as they offer a perspective for unifying all interactions including gravity. However, their couplings are parametrized by Newton's constant, which is of negative mass dimension. Thus these theories are all potentially nonrenormalizable. At present it does not seem possible to prove that they are finite along the lines of what has been achieved for supersymmetric Yang-Mills theories. All the available arguments fail, in one way or another, due to the presence of a dimensional coupling. Moreover, the indications of the indirect analysis of ref. [32] are rather discouraging, and suggest that divergences should really set in at the « obvious » number of loops, three. Of course, explicit calculations in (super)gravity theories would be most illuminating. Hopefully, the integration technique mentioned in sect. 3, together with the development of a suitable computer software and, at least, the completion of the work of ref. [74], should make this nontrivial task accessible in the near future.

* * *

I am very grateful to N. MARCUS, with whom all the work of ref. [31-33] was done during a long and enjoyable collaboration. We both received very valuable help from M. GOROFF with the most difficult calculations. I learned a large fraction of what I know about superspace formulations from W. SIEGEL. I also long benefited from discussions on these matters with J. KOLLER and B. ZWIEBACH. The appendix is based on the presentation of two-component formalism in Zwiebach's thesis [58]. Finally, I would like to thank Prof. N. CABIBBO for his kind invitation to lecture at this School.

APPENDIX

The two-component notation for four-dimensional space-time can be simply arrived at from the more familiar four-component notation. The basic idea behind the two-component formalism is to build up all representations of the Lorentz group starting from its irreducible Weyl spinors. Thus in some sense it is the most natural formulation. This reflects in the fact that it by-passes the need for explicit γ-matrices altogether. Moreover, it deals with indices running over two values, which can thus be easily symmetrized or antisymmetrized.

Consider an off-diagonal (Weyl) representation of the Dirac algebra (with space-time signature $(+---)$), say

$$\gamma^0 = \begin{pmatrix} 0 & 1 \\ 1 & 0 \end{pmatrix}, \qquad \gamma^i = \begin{pmatrix} 0 & -\sigma^i \\ \sigma^i & 0 \end{pmatrix}. \tag{A.1}$$

These define the matrices σ^μ and $\bar{\sigma}^\mu$ as the internal blocks of the four-dimensional γ-matrices

$$\gamma^\mu = \begin{pmatrix} 0 & (\sigma^\mu)^{\alpha\dot\alpha} \\ (\bar{\sigma}^\mu)_{\dot\alpha\alpha} & 0 \end{pmatrix}, \tag{A.2}$$

and the helicity matrix is simply

$$\gamma_5 = i\gamma^0\gamma^1\gamma^2\gamma^3 = \begin{pmatrix} 1 & 0 \\ 0 & -1 \end{pmatrix}. \tag{A.3}$$

One writes a four-component spinor as

$$\psi = \begin{pmatrix} \psi^\alpha \\ \chi_{\dot\alpha} \end{pmatrix}. \tag{A.4}$$

The Majorana condition, in the representation of eq. (A.1), then implies

$$\begin{cases} \chi^{\dot\alpha} = \bar{\psi}^{\dot\alpha} = (\psi^\alpha)^\dagger, \\ \chi_{\dot\alpha} = \bar{\psi}_{\dot\alpha} = -(\psi_\alpha)^\dagger, \end{cases} \tag{A.5}$$

where the indices are raised and lowered with the metric tensors $\varepsilon^{\alpha\beta}$ and $\varepsilon^{\dot\alpha\dot\beta}$ and their inverses, all proportional to the σ^2 Pauli matrix

$$(A.6) \qquad \varepsilon_{\alpha\beta} = \varepsilon_{\dot\alpha\dot\beta} = -\varepsilon^{\alpha\beta} = -\varepsilon^{\dot\alpha\dot\beta} = \begin{pmatrix} 0 & -i \\ i & 0 \end{pmatrix}.$$

The conversion between four-component notation and two-component notation can then be simply achieved using the definitions in eqs. (A.1)-(A.3). Thus, for example, for two anticommuting Majorana spinors χ and ψ,

$$(A.7) \qquad \bar\chi\psi = ((\chi^c)^* (\bar\chi_{\dot\alpha})^*) \begin{pmatrix} 0 & 1 \\ 1 & 0 \end{pmatrix} \begin{pmatrix} \psi^\alpha \\ \bar\psi_{\dot\alpha} \end{pmatrix} = \chi^\alpha \psi_\alpha + \bar\chi_{\dot\alpha} \bar\psi^{\dot\alpha}.$$

Conventionally, one contracts indices from upper left to lower right. Changing convention for a pair of indices introduces a minus sign. The basic identity of the two-component formalism is that any quantity antisymmetric in two (un)dotted indices is proportional to the appropriate ε tensor. Thus, for example,

$$(A.8) \qquad A_{\alpha\beta} - A_{\beta\alpha} = -\varepsilon_{\alpha\beta} A^\gamma{}_\gamma,$$

where the coefficient has been fixed using

$$(A.9) \qquad \varepsilon^{\alpha\beta} \varepsilon_{\alpha\beta} = 2.$$

All other identities can be derived using the representation in eqs. (A.1) and (A.2). For example, from the γ-matrices, one can construct the Lorentz matrices

$$(A.10) \qquad \sigma^{\mu\nu} = \tfrac{1}{2} \gamma^{[\mu} \gamma^{\nu]}.$$

Their irreducible blocks are

$$(A.11) \qquad \begin{cases} (\sigma^{\mu\nu})^\alpha{}_\beta = \tfrac{1}{2} (\sigma^{[\mu})^{\alpha\dot\alpha} (\bar\sigma^{\nu]})_{\dot\alpha\beta}, \\ (\sigma^{\mu\nu})_{\dot\alpha}{}^{\dot\beta} = \tfrac{1}{2} (\bar\sigma^{[\mu})_{\dot\alpha\beta} (\sigma^{\nu]})^{\beta\dot\beta}. \end{cases}$$

These matrices have definite duality properties. Thus

$$(A.12) \qquad \tfrac{1}{2} \varepsilon^{\mu\nu\varrho\sigma} (\sigma_{\varrho\sigma})^\alpha{}_\beta = i(\sigma^{\mu\nu})^\alpha{}_\beta.$$

Then, writing an antisymmetric tensor $F_{\mu\nu}$ in terms of these projections as

$$(A.13) \qquad F_{\mu\nu} = (\sigma_{\mu\nu})^\alpha{}_\beta F_\alpha{}^\beta + (\sigma_{\mu\nu})_{\dot\alpha}{}^{\dot\beta} F^{\dot\alpha}{}_{\dot\beta},$$

one recognizes the term with undotted indices as corresponding to the self-dual part of $F_{\mu\nu}$, and the term with dotted indices as corresponding to the antiself-dual part. Finally, Hermitian conjugation of superfields is obtained most simply by referring to the same operation on strings of θ's and $\bar{\theta}$'s. Thus, for example,

(A.14) $\qquad (A_{\alpha\dot{\beta}}{}^{\dot{\gamma}})^* \to (\theta_\alpha \bar{\theta}_{\dot{\beta}} \bar{\theta}^{\dot{\gamma}})^* = \theta^\gamma \theta_\beta \bar{\theta}_{\dot{\alpha}} = \theta_\beta \bar{\theta}_{\dot{\alpha}} \theta^\gamma \to A_{\beta\dot{\alpha}}{}^\gamma$.

More details and an extensive list of formulae can be found in ref. [58].

REFERENCES

[1] a) Y. A. GELFAND and E. S. LIKHTMAN: *JETP Lett.*, **13**, 323 (1971); b) D. V. VOLKOV and V. P. AKULOV: *Pis'ma Ž. Èksp. Teor. Fiz.*, **16**, 621 (1972); *Phys. Lett. B*, **46**, 109 (1973); c) J. WESS and B. ZUMINO: *Nucl. Phys. B*, **70**, 139 (1974); for a review of supersymmetry see P. FAYET and S. FERRARA: *Phys. Rep.*, **32**, 249 (1977).
[2] B. ZUMINO: *Nucl. Phys. B*, **89**, 535 (1975).
[3] J. WESS and B. ZUMINO: *Nucl. Phys. B*, **70**, 139 (1974).
[4] J. WESS and B. ZUMINO: *Phys. Lett. B*, **49**, 52 (1974); J. ILIOPOULOS and B. ZUMINO: *Nucl. Phys. B*, **76**, 310 (1974); S. FERRARA, J. ILIOPOULOS and B. ZUMINO: *Nucl. Phys. B*, **77**, 41 (1974).
[5] F. GLIOZZI, J. SCHERK and D. OLIVE: *Nucl. Phys. B*, **122**, 253 (1977); L. BRINK, J. H. SCHWARZ and J. SCHERK: *Nucl. Phys. B*, **121**, 77 (1977).
[6] S. FERRARA and B. ZUMINO: *Nucl. Phys. B*, **79**, 413 (1974).
[7] D. R. T. JONES: *Phys. Lett. B*, **72**, 199 (1977); E. POGGIO and H. PENDELTON: *Phys. Lett. B*, **72**, 200 (1977).
[8] O. TARASOV, A. VLADIMIROV and A. YU: *Phys. Lett. B*, **93**, 429 (1980); M. GRISARU, M. ROČEK and W. SIEGEL: *Phys. Rev. Lett.*, **45**, 1063 (1980); W. E. CASWELL and D. ZANON: *Nucl. Phys. B*, **182**, 125 (1981).
[9] L. MAIANI: in *Proceedings of the Summer School of Gif-sur-Yvette* (IN2P3, Paris, 1980), p. 3; M. VELTMAN: *Acta Phys. Pol. B*, **12**, 437 (1981); E. WITTEN: *Nucl. Phys. B*, **188**, 513 (1981); S. DIMOPOULOS and S. RABI: *Nucl. Phys. B*, **199**, 353 (1981).
[10] E. GILDENER and S. WEINBERG: *Phys. Rev. D*, **13**, 3333 (1976); E. GILDENER: *Phys. Rev. D*, **14**, 1667 (1976).
[11] S. FERRARA, D. Z. FREEDMAN and P. VAN NIEUWENHUIZEN: *Phys. Rev. D*, **13**, 3214 (1976); S. DESER and B. ZUMINO: *Phys. Lett. B*, **62**, 335 (1976). For a review see P. VAN NIEUWENHUIZEN: *Phys. Rep.*, **68**, 192 (1981).
[12] B. S. DE WITT: *Phys. Rev.*, **162**, 1195, 1239 (1967).
[13] G. 'T HOOFT and M. VELTMAN: *Ann. Inst. Henri Poincaré*, **20**, 69 (1974).
[14] S. DESER and P. VAN NIEUWENHUIZEN: *Phys. Rev. D*, **10**, 401, 411 (1974); S. DESER, H. T. TSAO and P. VAN NIEUWENHUIZEN: *Phys. Rev. D*, **10**, 3337 (1974); M. NOURI-MOUGHADAM and J. G. TAYLOR: *Proc. R. Soc. London, Ser. A*, **344**, 87 (1975).
[15] M. T. GRISARU, P. VAN NIEUWENHUIZEN and J. A. M. VERMASEREN: *Phys. Rev. Lett.*, **37**, 1662 (1976).
[16] P. VAN NIEUWENHUIZEN and J. A. M. VERMASEREN: *Phys. Lett. B*, **65**, 263 (1976).
[17] M. GRISARU: *Phys. Lett. B*, **66**, 75 (1977); E. T. TOMBOULIS: *Phys. Lett. B*, **67**, 417 (1977).
[18] G. PARISI: *Nucl. Phys. B*, **100**, 368 (1975).

[19] M. GRISARU, M. ROČEK and W. SIEGEL: *Nucl. Phys. B*, **159**, 42 (1979).
[20] M. GRISARU and W. SIEGEL: *Nucl. Phys. B*, **201**, 292 (1982).
[21] M. SOHNIUS: *Nucl. Phys. B*, **138**, 109 (1978).
[22] P. S. HOWE, K. S. STELLE and P. K. TOWNSEND: *Nucl. Phys. B*, **214**, 519 (1983).
[23] L. MEZINCESCU: JINR Report P2-12572 (1979); P. S. HOWE, G. SIERRA and P. K. TOWNSEND: *Nucl. Phys. B*, **221**, 331 (1983); J. KOLLER: *Nucl. Phys. B*, **222**, 319 (1983).
[24] J. KOLLER: *Phys. Lett. B*, **124**, 324 (1983); K. S. STELLE: Imperial College preprint ICTP/82-83/13.
[25] P. S. HOWE, K. S. STELLE and P. C. WEST: *Phys. Lett. B*, **124**, 55 (1983); P. S. HOWE, K. S. STELLE and P. K. TOWNSEND: *Nucl. Phys. B*, **236**, 125 (1984).
[26] S. MANDELSTAM: *Nucl. Phys. B*, **213**, 149 (1983); L. BRINK, O. LINDGREN and B. E. W. NILSSON: *Phys. Lett. B*, **123**, 323 (1983).
[27] M. SOHNIUS and P. C. WEST: *Phys. Lett. B*, **100**, 45 (1981); S. FERRARA and B. ZUMINO: unpublished.
[28] S. FERRARA and B. ZUMINO: *Nucl. Phys. B*, **87**, 207 (1975).
[29] S. ADLER: *Rev. Mod. Phys.*, **54**, 729 (1982).
[30] M. ROČEK and W. SIEGEL: *Phys. Lett. B*, **105**, 275 (1982); V. O. RIVELLES and J. G. TAYLOR: *Phys. Lett. B*, **121**, 37 (1983).
[31] N. MARCUS and A. SAGNOTTI: *Phys. Lett. B*, **135**, 85 (1984).
[32] N. MARCUS and A. SAGNOTTI: *Nucl. Phys. B*, **256**, 77 (1985).
[33] A. A. VLADIMIROV: preprint JINR-E2-12248 (1979); N. MARCUS and A. SAGNOTTI: *Phys. Lett. B*, **135**, 85 (1984); *Nuovo Cimento*, *A*, **87**, 1 (1985); R. VAN DAMME: *Nucl. Phys. B*, **244**, 105 (1984).
[34] P. S. HOWE and K. S. STELLE: *Phys. Lett. B*, **137**, 175 (1984).
[35] The most difficult calculations in ref. [32] made use of a number of C language programs developed at Caltech by M. GOROFF that implement the integration method of ref. [33].
[36] P. S. HOWE and U. LINDSTRÖM: *Nucl. Phys. B*, **181**, 487 (1981).
[37] The recent paper M. B. GREEN and J. H. SCHWARZ: *Nucl. Phys. B*, **243**, 475 (1984) contains references to other recent works. The two reviews J. H. SCHWARZ: *Phys. Rep.*, **89**, 223 (1982); M. B. GREEN: *Surveys of High Energy Physics*, **3**, 127 (1984) are somewhat out of date, as the field is evolving very rapidly.
[38] One of these theories follows by dimensional reduction from the eleven-dimensional supergravity theory constructed in E. CREMMER, B. JULIA and J. SCHERK: *Phys. Lett. B*, **79**, 231 (1978); the other model was constructed in ref. [69].
[39] E. BERGSHOEFF, M. DE ROO, B. DE WIT and P. VAN NIEUWENHUIZEN: *Nucl. Phys. B*, **195**, 97 (1982); G. F. CHAPLINE and N. S. MANTON: *Phys. Lett. B*, **120**, 105 (1983).
[40] J. H. SCHWARZ: in *Proceedings of the Johns Hopkins Workshop* (1982), p. 233; N. MARCUS and A. SAGNOTTI: *Phys. Lett. B*, **119**, 97 (1982).
[41] M. B. GREEN and J. H. SCHWARZ: *Phys. Lett. B*, **149**, 117 (1984).
[42] M. B. GREEN, J. H. SCHWARZ and L. BRINK: *Nucl. Phys. B*, **198**, 474 (1982).
[43] Conformal supergravity theories are extensions of Weyl's higher-derivative gravity theory. Simple conformal supergravity is described in M. KAKU, P. K. TOWNSEND and P. VAN NIEUWENHUIZEN: *Phys. Rev. D*, **17**, 3179 (1978). See also the review in ref. [11].
[44] M. GRISARU and W. SIEGEL: *Nucl. Phys. B*, **201**, 292 (1982); P. S. HOWE, K. S. STELLE and P. C. WEST: *Phys. Lett. B*, **124**, 55 (1983); P. S. HOWE, K. S. STELLE and P. K. TOWNSEND: *Nucl. Phys. B*, **236**, 125 (1984).
[45] Recent reviews on this vast subject are R. STORA: *Cargèse Lectures* (1983); B. ZUMINO: *Les Houches Lectures* (1983).

[46] G. 'T HOOFT and M. VELTMAN: *Nucl. Phys. B*, **44**, 189 (1972); C. G. BOLLINI and J. J. GIAMBIAGI: *Phys. Lett. B*, **40**, 566 (1972); J. ASHMORE: *Lett. Nuovo Cimento*, **4**, 289 (1972); G. CICUTA and E. MONTALDI: *Lett. Nuovo Cimento*, **4**, 329 (1972).
[47] G. 'T HOOFT: *Nucl. Phys. B*, **33**, 173 (1971); **35**, 167 (1971).
[48] W. PAULI and E. VILLARS: *Rev. Mod. Phys.*, **21**, 434 (1949).
[49] W. SIEGEL: *Phys. Lett. B*, **84**, 193 (1979).
[50] E. CREMMER and B. JULIA: *Nucl. Phys. B*, **159**, 141 (1979).
[51] W. SIEGEL: *Phys. Lett. B*, **94**, 37 (1980). See also L. V. AVDEEV: *Phys. Lett. B*, **117**, 317 (1982).
[52] R. VAN DAMME and G. 'T HOOFT: Utrecht preprint (1984) (unpublished).
[53] I. JACK and H. OSBORN: *Nucl. Phys. B*, **249**, 472 (1985). See also P. S. HOWE, A. PARKES and P. C. WEST: *Phys. Lett. B*, **147**, 409 (1984).
[54] See, for example, P. VAN NIEUWENHUIZEN and P. K. TOWNSEND: *Phys. Rev. D*, **20**, 1832 (1979); D. M. CAPPER, D. R. T. JONES and P. VAN NIEUWENHUIZEN: *Nucl. Phys. B*, **167**, 479 (1980); G. ALTARELLI, M. CURCI, G. MARTINELLI and S. PETRARCA: *Nucl. Phys. B*, **187**, 461 (1981).
[55] J. WESS and J. BAGGER: *Supersymmetry and Supergravity* (Princeton University Press, Princeton, N. J., 1983).
[56] S. J. GATES, M. GRISARU, M. ROČEK and W. SIEGEL: *Superspace* (Benjamin Cummings, New York, N. Y., 1983).
[57] A. PARKES and P. C. WEST: *Phys. Lett. B*, **122**, 365 (1983); **127**, 353 (1983); **138**, 99 (1984); *Nucl. Phys. B*, **222**, 269 (1983).
[58] B. ZWIEBACH: Ph.D. Thesis, Caltech (1983).
[59] P. C. WEST: talk at Shelter Island II (1983).
[60] A. SALAM and J. STRATHDEE: *Nucl. Phys. B*, **76**, 477 (1974).
[61] S. FERRARA, J. WESS and B. ZUMINO: *Phys. Lett. B*, **51**, 239 (1974).
[62] J. WESS and B. ZUMINO: *Nucl. Phys. B*, **78**, 1 (1974); A. SALAM and J. STRATHDEE: *Phys. Lett. B*, **51**, 353, 475 (1974); S. FERRARA and B. ZUMINO: *Nucl. Phys. B*, **79**, 413 (1974).
[63] R. HAAG, J. T. LOPUSZANSKI and M. SOHNIUS: *Nucl. Phys. B*, **88**, 257 (1975),
[64] M. GRISARU, M. ROČEK and W. SIEGEL: *Nucl. Phys. B*, **159**, 42 (1979); P. S. HOWE, K. S. STELLE and P. C. WEST: *Phys. Lett. B*, **124**, 55 (1983); P. S. HOWE, K. S. STELLE and P. K. TOWNSEND: *Nucl. Phys. B*, **236**, 125 (1984); P. P. SRIVASTAVA: *Phys. Lett. B*, **132**, 80 (1983).
[65] S. J. GATES and B. ZWIEBACH: *Nucl. Phys. B*, **238**, 99 (1984); P. S. HOWE: *Nucl. Phys. B*, **199**, 309 (1982).
[66] J. P. DERENDINGER, S. FERRARA and A. MASIERO: *Phys. Lett. B*, **143**, 133 (1984), and references therein.
[67] E. CREMMER, J. SCHERK and J. H. SCHWARZ: *Phys. Lett. B*, **84**, 83 (1979); E. CREMMER, S. FERRARA, L. GIRARDELLO and A. VAN PROYEN: *Nucl. Phys. B*, **212**, 413 (1983).
[68] W. SIEGEL: *Nucl. Phys. B*, **238**, 307 (1984).
[69] M. B. GREEN and J. H. SCHWARZ: *Phys. Lett. B*, **122**, 143 (1983); J. H. SCHWARZ and P. C. WEST: *Phys. Lett. B*, **126**, 301 (1983); J. H. SCHWARZ: *Nucl. Phys. B*, **226**, 269 (1983); P. S. HOWE and P. C. WEST: *Nucl. Phys. B*, **238**, 181 (1984).
[70] N. MARCUS and J. H. SCHWARZ: *Phys. Lett. B*, **119**, 97 (1982).
[71] P. CVITANOVIĆ: *Phys. Rev. D*, **14**, 1536 (1976).
[72] N. MARCUS, A. SAGNOTTI and W. SIEGEL: *Nucl. Phys. B*, **224**, 159 (1983).
[73] See W. E. CASWELL and A. E. KENNEDY: *Phys. Rev. D*, **25**, 392 (1982), and references therein.
[74] S. J. GATES, A. KARLHEDE, U. LINDSTRÖM and M. ROČEK: *Nucl. Phys. B*, **243**, 221 (1984).

From Asymptotic Freedom to Fermion Computer Simulations: Lectures in Lattice Gauge Theory.

J. B. KOGUT (*)

Department of Physics, University of Illinois at Urbana-Champaign
1110 West Green Street, Urbana, IL 61801

Introductory remarks.

Lattice field theory is developing into a productive subfield of high-energy particle physics and statistical mechanics. Both the high-energy and the statistical-physics communities have been enriched by this coming-together. The particle physicists are particularly interested in understanding how nonperturbative dynamics work in gauge theories as well as calculating the mass spectrum and the low-energy properties of the strongly interacting particles. Statistical mechanicians are interested in the physics behind phase transitions, dynamical systems and chaos. In these lectures we shall concentrate on the foundations of the lattice approach to some of these topics. Applications and recent trends will be ignored. However, although we will begin with the formal connections between field theory and statistical mechanics, we end with the latest ideas on fermion computer simulations—both stochastic and moleculardynamics approaches to including the effects of fermion vacuum polarization on gauge field dynamics. On the way we will consider asymptotic freedom, nonperturbative renormalization groups methods, lattice strings, topological excitations and chiral symmetry.

1. – Field theory, statistical mechanics and the transfer matrix.

The lattice approach to field theory exploits several correspondences which exist between field theory and statistical mechanics [1]. These correspondences are useful in practical studies of lattice gauge theory and are important con-

---

(*) Supported in part by the National Science Foundation under Grant NSF PHY82-01948.

ceptually in their own right. We will illustrate them here in the context of simple models and refer to the literature for the general statements. Part of the value of the correspondences lies in their intuitive appeal: it is worthwhile to be bilingual and speak about correlation lengths, thermal fluctuations and spin configuration sums in statistical mechanics and then translate to particle spectra, quantum uncertainties and path integrals in field theory.

Consider a simple quantum-mechanical system: a two-dimensional Hilbert space spin « up » and spin « down » with observables described by Pauli matrices [2]. Let the Hamiltonian be

$$(1.1) \qquad H = -(\sigma_1 - \lambda \sigma_3).$$

This system has quantum fluctuations because $[\sigma_1, \sigma_3] \neq 0$. Consider the operator $\exp[-HT]$, the time evolution operator $\exp[-iHT]$ continued to imaginary time. We want to solve this theory's Schrödinger equation, or, equivalently, to calculate the transition amplitude between states $|a\rangle$ and $|b\rangle$ over a time interval T. To do this follow the ideas of Feynman and divide the time interval into segments of width ΔT, $N\Delta T = T$. We have

$$(1.2) \qquad \langle b|\exp[-HT]|a\rangle = \langle b|\exp[-H(\Delta T + \Delta T + ...)]|a\rangle.$$

Now insert complete sets of states, eigenstates of σ_3, at each time slice,

$$(1.3) \qquad \langle b|\exp[-HT]|a\rangle = \sum_{s_1=\pm 1} \sum_{s_2=\pm 1} ... \sum_{s_{N-1}=\pm 1} \langle b|\exp[-H\Delta T]|s_1\rangle \cdot$$
$$\cdot \langle s_1|\exp[-H\Delta T]|s_2\rangle ... \langle s_{N-1}|\exp[-H\Delta T]|a\rangle.$$

Notice that these sums over intermediate states resemble a configuration sum familiar in statistical mechanical partition functions. To exploit this point define

$$(1.4) \qquad \exp[-V(s, s')] \equiv \langle s|M|s'\rangle \equiv \langle s|\exp[-H\Delta T]|s'\rangle.$$

Equation (1.3) becomes

$$(1.5) \qquad \langle b|\exp[-HT]|a\rangle =$$
$$= \sum_{s_1=\pm 1} \sum_{s_{N-1}=\pm 1} \exp[-(V(b, s_1) + V(s_1, s_2) + ... + V(s_{N-2}, s_{N-1}) + V(s_{N-1}, a))].$$

This resembles the partition function of the 1-d Ising model with nearest-neighbor couplings. To calculate $V(s, s')$ consider the limit ΔT small,

$$(1.6) \qquad \langle s|M|s'\rangle \simeq \langle s|(1 + \Delta T(\sigma_1 - \lambda \sigma_3))|s'\rangle, \qquad M = \begin{pmatrix} 1 - \lambda\Delta T & \Delta T \\ \Delta T & 1 + \lambda\Delta T \end{pmatrix}.$$

Since s and s' can have but two values, it is easy to write down the most general form of V,

$$(1.7) \qquad V(s, s') = \frac{K}{2}(s - s')^2 + \frac{h}{2}(s + s').$$

It is easy to calculate

$$(1.8) \qquad \exp[-2K] = \Delta T, \qquad \exp[h] = 1 + \lambda \Delta T.$$

Equations (1.5) and (1.7) constitute the «path integral» formulation of the original quantum system. The dispersion in σ_3 in the quantum system is reflected in the thermal fluctuations of the s variables in eq. (1.5). When λ is large, the dispersion in σ_3 is small. In the path integral formulation eq. (1.8) implies that this corresponds to the $h \to \infty$ limit in which the s variable becomes locked near $+1$.

Now develop the statistical mechanics-field theory correspondence in greater detail. Consider the quantum-mechanical quantity

$$Z = \operatorname{tr} \exp[-HT] = \sum_{s=\pm 1} \langle s | \exp[-HT] | s \rangle.$$

Inserting intermediate states, we saw above that this is just the partition function of the 1-d Ising model with periodic boundary conditions. The free energy per site of this system in the infinite-volume limit is

$$(1.9) \qquad F = \lim_{N \to \infty} \left(-\frac{1}{N} \ln Z\right) = \lim_{T \to \infty} \left(-\frac{1}{T} \ln \operatorname{tr} \exp[-HT]\right) =$$
$$= \lim_{T \to \infty} \left(-\frac{1}{T} \ln \left(\exp[-E_0 T] + \exp[-E_1 T]\right)\right) =$$
$$= \lim_{T \to \infty} \left(E_0 + \ln\left(1 + \exp[-(E_1 - E_0)T]\right)\right) = E_0, \quad \text{the ground-state energy.}$$

So the statistical mechanical free energy maps F onto the quantum ground-state energy. Be careful to distinguish F from the internal energy of the statistical system—F contains entropy effects and is very nontrivial!

Correlation functions are a standard diagnostic of statistical mechanical systems. The two-spin correlation function is

$$(1.10) \qquad \langle s_0 s_r \rangle = \sum_{s_1, \ldots, s_N = \pm 1} \exp\left[-\sum_i V(s_i, s_{i+1})\right] s_0 s_r / Z.$$

What does this mean in quantum mechanics? Doing the steps which led from eqs. (1.1) to (1.5), we have

$$(1.11) \qquad \langle s_0 s_r \rangle = \operatorname{tr}\left(\sigma_3 \exp[-Hr] \sigma_3 \exp[-H(T-r)]\right) / \operatorname{tr} \exp[-HT].$$

Inserting eigenstates of H, $H(0) = E_0(0)$ and $H(1) = E_1|1\rangle$, we have

(1.12) $\langle s_0 s_r \rangle = \big(|\langle 0|\sigma_3|0\rangle|^2 \exp[-E_0 T] + |\langle 1|\sigma_3|1\rangle|^2 \exp[-E_1 T] +$
$+ |\langle 0|\sigma_3|1\rangle|^2 (\exp[-E_1 r - E_0(T-r)] +$
$+ \exp[-E_0 r - E_1(T-r)])\big) / (\exp[-E_0 T] + \exp[-E_1 T])\,.$

Taking $T \to \infty$ so that only the terms proportional to $\exp[-E_0 T]$ survive, we have

(1.13) $\langle s_0 s_r \rangle = |\langle 0|\sigma_3|0\rangle|^2 + \exp[-(E_1 - E_0)r]\,|\langle 0|\sigma_3|1\rangle|^2\,.$

Notice that this answer is the continuation to imaginary time ($t \to -it$) of $\langle 0|T(\sigma_3(t)\sigma_3(0))|0\rangle$, where T means time-ordered product, $\sigma_3(t)$ is the Heisenberg picture operator $\exp[iHt]\sigma_3 \exp[-iHt]$ and $r = |t|$. Correlation functions of statistical mechanics are, therefore, related to vacuum expectation values of time-ordered products in field theory.

The correlation function (1.13) decays to its asymptotic value exponentially. The decay length, often called the correlation length, is related to the gap between the ground and excited states of H,

(1.14) $$\Delta T \xi = \frac{1}{\Delta}, \qquad \Delta = E_1 - E_0\,.$$

If $\langle 0|\sigma_3|0\rangle \neq 0$ (i.e. $\lambda \neq 0$), then $\lim_{r \to \infty} \langle s_0 s_r \rangle \neq 0$. In this case there is long-range order in the system.

We now generalize from quantum mechanics to quantum field theory. In a quantum field theory we have an infinite number of quantum-mechanical systems, one associated to each point in space. To make things well defined, we will discretize space into a lattice and consider a finite volume, rendering the number of systems finite. We will have more to say later about removing these cut-offs. To illustrate concepts and phenomena common in quantum field theories consider a simple generalization of the example above. A two-level quantum system at each point of one-dimensional space with a Hamiltonian given by [1]

(1.15) $$H = -\sum_n [\sigma_1(n) + \lambda \sigma_3(n)\sigma_3(n+1)]\,.$$

This model is exactly solvable [3], but we can learn about it by studying its properties in perturbation theory. When $\lambda = 0$, H is diagonal in a basis where σ_1 is diagonal. This limit where spatial points are decoupled is often referred to as a strong-coupling or high-temperature limit. If we label states

(1.16) $$\sigma_1|\uparrow\rangle = |\uparrow\rangle, \quad \sigma_1|\downarrow\rangle = -|\downarrow\rangle,$$

the ground state of H looks like

(1.17) $$|0\rangle = |\uparrow\uparrow\ldots\uparrow\rangle, \quad \langle 0|H|0\rangle = -N = E_0.$$

The first excited states are N-fold degenerate (N is the number of sites) and correspond to flipping one spin,

(1.18) $$|r\rangle = |\uparrow\uparrow\ldots\underset{r}{\uparrow\downarrow\uparrow}\ldots\uparrow\rangle, \quad \langle r|H|r\rangle = -N + 2 = E_0 + 2.$$

They are at a finite energy above E_0. These states correspond to a particle localized at site r. The first-order perturbation in λ is diagonalized by the running waves,

(1.19) $$|k\rangle = \sum_r \exp[ikr]|r\rangle, \quad E_k \equiv \langle k|H|k\rangle = E_0 + 2 - 2\lambda \cos k.$$

These are one-particle states with momentum k, reflecting the discrete translation invariance of H

(1.20) $$T\sigma_i(n)T^{-1} = \sigma_i(n+1), \quad [T, H] = 0.$$

Their energy above the ground state is bounded from below by

(1.21) $$\Delta = E_{k=0} - E_0 = 2 - 2\lambda,$$

which is the theory's mass gap. Although derived in perturbation theory, eq. (1.21) turns out to be exact [3]!

Now turn to the limit $\lambda \to \infty$. Here H is diagonal in the basis where σ_3 is diagonal,

(1.22) $$\sigma_3|\uparrow\rangle = |\uparrow\rangle, \quad \sigma_3|\downarrow\rangle = -|\downarrow\rangle.$$

There are two degenerate ground states

(1.23) $$|\uparrow, \uparrow\ldots\uparrow\rangle \quad \text{and} \quad |\downarrow, \downarrow\ldots\downarrow\rangle.$$

These states are carried into each other by the action of the symmetry operator U that commutes with H,

(1.24) $$U = \prod_n \sigma_1(n), \quad [U, H] = 0.$$

Contrast this with the λ-small ground state $|0\rangle$ of eq. (1.17): $U|0\rangle = |0\rangle$. We say that, at λ large, U is a spontaneously broken global symmetry: The ground state is not invariant under the symmetries of H. In a finite-volume system these two ground states would be mixed in perturbation theory and the degeneracy split. This is not true of the infinite-volume theory in which the energy barrier between the two states diverges. The phenomena outlined above, the existence of gaps and the possibility of broken symmetry, are commonly encountered in quantum field theories. They have their analog in the statistical mechanical versions of these theories.

To derive a path integral representation of a system like (1.15) we follow the steps used in the quantum system. Complete sets of states (in the σ_3 diagonal representation) are described by

$$|\{s_x\}\rangle, \qquad \text{where } x = 1, \ldots, N \text{ and } s_i = \pm 1.$$

Inserting such a set at each time interval ΔT, we have

(1.25) $\langle a| \exp[-HT] |b\rangle =$
$$= \sum_{\{s(x,t)=\pm 1\}} \langle a| \exp[-H\Delta T] |\{s(x,1)\}\rangle \langle \{s(x,1)\}| \exp[-H\Delta T] |\rangle \ldots |b\rangle.$$

Define the transfer matrix

(1.26) $\langle \{s(x)\}| \exp[-H\Delta T] |\{s'(x)\}\rangle \equiv$
$$\equiv \langle \{s(x)\}|M|\{s'(x)\}\rangle \equiv \exp[-V(\{s(x)\}, \{s'(x)\})].$$

Evaluating V is a little more difficult than before, but when ΔT is small it is not hard to see that [1]

(1.27) $$V(\{s(x)\}, \{s'(x)\}) = \sum_x \frac{K_t}{2}(s(x) - s'(x))^2 + \frac{K_x}{2}(s(x) - s(x+1))^2$$

with $\exp[-2K_t] \simeq \Delta T$, $K_x = \lambda \Delta T$, in the $\Delta T \to 0$ limit. Note that $\Delta T \to 0$ implies $K_T \gg K_x$. So (1.25) is just the partition function for a two-dimensional Ising model, one with very anisotropic coupling. The existence of a gap (1.21) tells us that the spin-spin correlation functions of this model decay exponentially

$$\langle s(t, x) s(t', x) \rangle \sim \exp[-|t - t'|/\xi], \qquad \Delta T \xi = \frac{1}{\Delta},$$

where $|t - t'|$ is the (dimensionless) number of lattice spacings between the two points. A particularly interesting thing about eq. (1.21) is that Δ vanishes at a specific value of λ, $\lambda = 1$. The vanishing gap implies nonexponential decay. In fact,

$$\langle s(t, x) s(t', x) \rangle \sim \frac{1}{|t - t'|^\eta} \qquad \text{at } \lambda = 1.$$

At large λ the broken-symmetry ground state results in a nonvanishing magnetization of the system

$$\langle s_t s_{t'} \rangle \sim \text{const}, \qquad |t - t'| \to \infty.$$

A final point. A central aspect of many quantum field theories is Lorentz or relativistic invariance. If $c = 1$ and t is analytically continued to $-it$, Lorentz invariance becomes Euclidean rotational invariance in the (x, t)-space. The correlation functions of (1.27) and hence the time-ordered vacuum expectation values of (1.15) do not have this rotational invariance, first because of the lattice and second because of the anisotropic couplings in (1.27). This anisotropy can be removed by redefining the relative length scales in the x and t directions [1]. Alternatively, we can study a modified model with $K_x = K_t = K$. This isotropic model is guaranteed to have at least a discrete rotational symmetry. The Hamiltonian associated with it, defined by $1/\Delta T \log M$, is complicated, however.

The statistical mechanics-field theory correspondences we have illustrated are

$$\begin{array}{rcl}
\text{path integral} & \leftrightarrow & \text{partition function}, \\
\text{ground-state energy} & \leftrightarrow & \text{free energy}, \\
\text{propagator} & \leftrightarrow & \text{correlation function}, \\
\text{mass gap} & \leftrightarrow & \text{correlation length (reciprocal)}.
\end{array}$$

In this discussion the statistical-mechanics models were formulated on a spatial lattice—the classical variables s resided at sites of a regular, Euclidean lattice. However, interesting field theories have two additional important properties—they have nontrivial low-energy mass spectra and are Euclidean (Minkowski) invariant. The lattice spacing a of the statistical-mechanics models acts as an ultraviolet cut-off on the corresponding field theories. Field theories are interesting only in the continuum limit, $a \to 0$. We shall see that it is possible to hold mass gaps constant in physical units as $a \to 0$ and that full Euclidean invariance can occur in the continuum limit. In the language of statistical mechanics, the correlation length ξ is large in these cases—taking $\Delta T \to 0$ in eq. (1.14) and holding Δ fixed implies that ξ diverges! This generic behavior only occurs in statistical-mechanics models if the parameter of the theory's Hamiltonian are chosen at a critical point. We will discuss this in the context of asymptotically free field theories in the next section. It will lead us to a more thorough discussion of renormalization theory, universality, relevant and irrelevant operators, etc. However, it should be clear to the reader at this point that the lattice approach to field theory with its emphasis on the continuum limit leads to one of the hardest problems in statistical mechanics—that of critical phenomena. This approach to field theory is worthwhile, nonetheless, because the statistical mechanics of lattice systems is a highly developed field with techniques for handling difficult dynamics which are beyond the power of traditional approaches to relativistic field theory.

2. – Lattice gauge theory.

Now let us introduce SU_2 lattice gauge theory [4]. It is reasonable to assume that everyone has some familiarity with this subject. The reviews cited in ref. [1] contain more introductory material.

Our discussion will stress several basic properties of the theory. SU_N lattice gauge theory's basic building blocks are SU_N group elements which are assigned to the links of a hypercubic lattice—a discrete form of space-time. In the path integral formulation of the model these variables are freely integrated over. Since the group volumes are finite, such integrals are well defined and have simple invariance properties. This feature of the theory is crucial to its conceptual simplicity. For example, it allows gauge invariance to be stated precisely in the full fluctuating theory. In this sense it is simpler and clearer than the weak-coupling perturbative formulation based on the Lie algebra of SU_N. The theory's formulation in terms of group elements underlies its well-defined strong-coupling, confining features as well.

The idea of local gauge invariance can be stated elegantly on the lattice. It serves several purposes. It dictates the form of the interactions in the theory much as local gauge invariance does in the classical continuum theory. Using local gauge invariance and requiring locality of the interaction, lattice actions can be invented whose classical continuum limits reproduce Yang-Mills theory and whose strong-coupling limits confine quarks. The demonstration of these two points constitutes the core of this section.

Consider a four-dimensional hypercubic Euclidean lattice with spacing a. On each link place an SU_2 matrix, as shown in fig. 1,

(2.1) $$U_\mu(n) = \exp[iB_\mu(n)],$$

where

$$B_\mu(n) = \tfrac{1}{2} ag\tau_i A^i_\mu(n) = \tfrac{1}{2} ag\boldsymbol{\tau}\cdot\boldsymbol{A}_\mu(n).$$

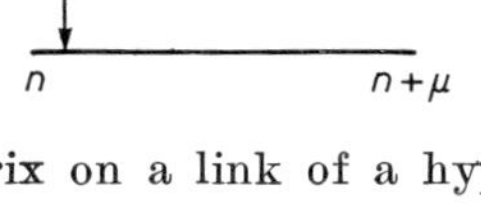

$U_\mu(n) = \exp[iB_\mu(n)]$

$n \qquad n+\mu$

Fig. 1. – An SU_2 rotation matrix on a link of a hypercubic lattice.

Each link carries a direction, (n, μ) with $\mu = 1, 2, 3$ or 4. If we denote a link in the backward direction, we associate with it $U_\mu^{-1}(n)$,

2.2) $$U_{-\mu}(n + \mu) \equiv U_\mu^{-1}(n).$$

The $[U_\mu(n)]_{ij}$ are SU_2 rotation matrices. One can imagine a color frame of reference at each site. Following YANG and MILLS [5] we will require that the orientation of the frames in color SU_2 space be locally arbitrary. A rotation in color space can be done at each site with an SU_2 matrix,

(2.3) $$G(\chi(n)) = \exp[-i\tfrac{1}{2}\boldsymbol{\tau}\cdot\boldsymbol{\chi}(n)].$$

The theory shall have a local gauge invariance and, under a gauge transformation,

(2.4) $$U_\mu(n) \to G(n)\, U_\mu(n)\, G^{-1}(n+\mu),$$

which we recognize as the simplest, local generalization of a global SU_2 invariance, a rotation is applied to the SU_2 axes and SU_2 matrices undergo a similarity transformation. Note the n and $n+\mu$ in eq. (2.4), and the local character of the invariance group. Writing the transformation law out with indices,

(2.5) $$[U_\mu(n)]_{ij} \to \sum_{kl} \left[\exp[-i\tfrac{1}{2}\boldsymbol{\tau}\cdot\boldsymbol{\chi}(n)]\right]_{ik} \left[\exp[i\tfrac{1}{2}\boldsymbol{\tau}\cdot\boldsymbol{\chi}(n+\mu)]\right]_{jl} [U_\mu(n)]_{kl}.$$

Now we need an action which incorporates this local symmetry. It is clear that to incorporate the local symmetry eq. (2.4) S must be constructed out of the traces of products of U-matrices around closed paths. The most local paths on the lattice are plaquettes (fig. 2), so consider

(2.6) $$S = -\frac{1}{2g^2} \sum_{n,\mu,\nu} \operatorname{tr} U_\mu(n)\, U_\nu(n+\mu)\, U_{-\mu}(n+\mu+\nu)\, U_{-\nu}(n+\nu) + \text{h.c.}$$

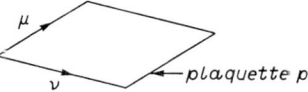

Fig. 2. – A plaquette—a closed path of 4 links.

What is the physics of this model? Let us check that

1) its classical continuum limit ($a \to 0$) is ordinary Yang-Mills theory,

2) its strong-coupling limit confines quarks.

With these results we will have a cut-off, finite, gauge-invariant formulation of gauge fields which can be used to study the theory for all a and all g!

To take the classical continuum limit, we Taylor expand the slowly varying fields $B_\mu(n)$ appropriate to a long-wavelength approximation,

(2.7) $\begin{cases} B_\nu(n + \mu) \simeq B_\nu(n) + a\partial_\mu B_\nu(n) + O(a^2), \\ B_{-\mu}(n + \mu + \nu) = -B_\mu(n + \nu) \simeq -[B_\mu(n) + a\partial_\nu B_\mu(n)] + O(a^2), \\ B_{-\nu}(n + \nu) = -B_\nu(n). \end{cases}$

So,

(2.8) $UUUU \simeq \exp[iB_\mu] \exp[i(B_\nu + a\partial_\mu B_\nu)] \exp[-i(B_\mu + a\partial_\nu B_\mu)] \exp[iB_\nu]$.

Now we use the operator identity

(2.9) $\exp[x]\exp[y] = \exp[x + y + \frac{1}{2}[x, y] + \ldots]$.

So,

(2.10) $UUUU \simeq \exp[i(B_\mu + B_\nu + a\partial_\mu B_\nu) - \frac{1}{2}[B_\mu, B_\nu]] \cdot$
$\cdot \exp[-i(B_\mu + B_\nu + a\partial_\nu B_\mu) - \frac{1}{2}[B_\mu, B_\nu]] \simeq$
$\simeq \exp[ia(\partial_\mu B_\nu - \partial_\nu B_\mu) - [B_\mu, B_\nu]]$.

But

(2.11) $B_\mu(n) = \frac{1}{2}ag\tau_i A_\mu^i(n) \equiv agA_\mu(n), \quad A_\mu(n) \equiv \frac{1}{2}\boldsymbol{\tau}\cdot\boldsymbol{A}(n)$.

So,

(2.12) $ia(\partial_\mu B_\nu - \partial_\nu B_\mu) - [B_\mu, B_\nu] = ia^2 g\{\partial_\mu A_\nu - \partial_\nu A_\mu + ig[A_\mu, A_\nu]\}$.

Identify the conventional Yang-Mills field strength $F_{\mu\nu}$ here. So,

(2.13) $UUUU \approx \exp[ia^2 g F_{\mu\nu}]$,

with corrections in the exponent higher order in a^2. These corrections will not contribute in the classical continuum limit. For smooth classical fields we look at excitations having long wavelengths compared to a, so

(2.14) $a^2 g F_{\mu\nu} \ll 1$

and we can simplify the tr $UUUU$ further,

(2.15) $\operatorname{tr} UUUU \simeq \operatorname{tr} \exp[ia^2 g F_{\mu\nu}] = \operatorname{tr}\{1 + ia^2 g F_{\mu\nu} - \frac{1}{2}a^4 g^2 F_{\mu\nu}^2 + \ldots\} =$
$= \operatorname{tr} 1 - \frac{1}{2}a^4 g^2 \operatorname{tr} F_{\mu\nu}^2 + \ldots,$

where $\operatorname{tr} F_{\mu\nu} = 0$ because tr (generator) $= 0$. Finally,

(2.16) $\operatorname{tr} F_{\mu\nu}^2 = \operatorname{tr}[\frac{1}{2}\tau_i(\partial_\mu A_\nu^i - \partial_\nu A_\mu^i - g\varepsilon_{kli}A_\mu^k A_\nu^l)] \cdot$
$\cdot [\frac{1}{2}\tau_{i'}(\partial_\mu A_\nu^{i'} - \partial_\nu A_\mu^{i'} - g\varepsilon_{k'l'i'}A_\mu^{k'} A_\nu^{l'}]$,

where we used $[\tau_i, \tau_j] = 2i\varepsilon_{ijk}\tau_k$. Next,

$$\text{tr}\,\tau_i\tau_j = 2\delta_{ij}. \tag{2.17}$$

So,

$$\text{tr}\,F_{\mu\nu}^2 = \tfrac{1}{2}(\partial_\mu A_\nu^k - \partial_\nu A_\mu^k - g\varepsilon_{kij}A_\mu^i A_\nu^j)^2,$$

the square of the gauge-covariant field strength tensor. Now, the action becomes

$$S \simeq \frac{1}{2g^2}\int \frac{\mathrm{d}^4 x}{a^4}\, a^4 g^2 \frac{1}{2}(\partial_\mu A_\nu^k - \partial_\nu A_\mu^k - g\varepsilon_{ijk}A_\mu^i A_\nu^j)^2 + O(a^2), \tag{2.18}$$

where we replaced

$$\sum_{n,\mu\nu} \to \sum \frac{\mathrm{d}^4 x}{a^4}\sum_{\mu\nu} \tag{2.19}$$

and a 2 has appeared from the Hermitian conjugate. Finally,

$$S = \frac{1}{4}\int \mathrm{d}^4 x\,(F_{\mu\nu}^i)^2, \tag{2.20}$$

with

$$F_{\mu\nu}^i = \partial_\mu A_\nu^i - \partial_\nu A_\mu^i - g\varepsilon^{ijk}A_\mu^j A_\nu^k, \tag{2.21}$$

which is the usual Euclidean action of classical Yang-Mills.

This result evokes several remarks:

1) The final result is Euclidean O_4 invariant! Where has the difference between the cubic invariance of the original action and the classical continuum limit gone? Into the higher-order a terms! Those operations are classified as «irrelevant»—they do not affect the continuum limit. This underscores the fact that we have much arbitrariness in the construction of lattice actions—one must simply engineer them so that the correct continuum limit occurs.

2) The final result involves $F_{\mu\nu}^i$, the standard gauge-covariant field strength tensor of Yang-Mills. The local invariance built into the lattice action guaranteed this.

What are the interesting physical questions that can be posed in the lattice formulation of the theory? The observables of the pure gauge model must be locally gauge invariant, so they are constructed from the product of U_μ-matrices taken around closed paths,

$$\text{tr}\prod_c U_\mu(n). \tag{2.22}$$

Recognize this object as the lattice form of path-ordered phase factor familiar from continuum formulations,

(2.23) $$\text{tr}\, P \exp\left[ig \oint A_\mu^a \tfrac{1}{2}\tau^a \, dx^\mu\right].$$

These operators have simple physical interpretations which go to the heart of the dynamics of gauge fields. Let us argue that the dependence of the vacuum expectation value of $\text{tr} \prod_c U_\mu(n)$ on the closed contour c determines the heavy-quark potential [4]. Begin with a continuum formulation of the pure gauge theory and imagine the following thought experiment. Adiabatically separate a $Q\overline{Q}$ pair to a relative distance R. Then hold this configuration for a time $T \to \infty$. Finally, bring the quarks back together and let them annihilate. The world-line of the quarks is shown in fig. 3. The Euclidean amplitude for

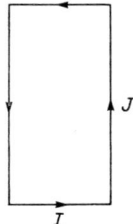

Fig. 3. – World-line of a quark which determines the heavy-quark potential.

the process is the matrix element of the evolution operator $\exp[-HT]$ between the initial final states,

(2.24) $$\langle i| \exp[-HT] |f\rangle.$$

Here $|i\rangle$ and $|f\rangle$ represent a $Q\overline{Q}$ pair a distance $\boldsymbol{R}$ apart and $\boldsymbol{H}$ is the Hamiltonian of the pure gauge theory. Equation (2.23) can be expressed as a path integral, following the ideas of sect. 1,

(2.25) $$\langle i| \exp[-\boldsymbol{H}T] |f\rangle =$$
$$= \int [\mathrm{D}A_\mu^a][\mathrm{D}c_a][\mathrm{D}c_a^*] \exp\left[-S + ig\oint A_\mu^a J_\mu^a \, d^4x\right] \Big/ \int [\mathrm{D}A_\mu^a][\mathrm{D}c_a][\mathrm{D}c_a^*] \exp[-S],$$

where S is the action of the SU_3 gauge theory and J_μ^a ($a = 1, 2, ..., 8$) is an external current describing the world-lines of the heavy quarks and the functional integrals run over the gauge field A_μ^a and Fadeev-Popov ghost fields c_a. For the path shown in fig. 3, eq. (2.24) simplifies to

(2.26) $$\langle i| \exp[-HT] |f\rangle = \int [\mathrm{D}A_\mu^a][\mathrm{D}c_a][\mathrm{D}c_a^*] \cdot$$
$$\cdot \exp\left[-S + ig \oint_c A_\mu^a \tfrac{1}{2}\lambda^a \, dx_\mu\right] \Big/ [\mathrm{D}A_\mu][\mathrm{D}c_a][\mathrm{D}c_a^*] \exp[-S],$$

where the contour c traces out the closed path of the figure. Since $|i\rangle$ and $|j\rangle$ are identical and since the process is static, the left-hand side of eq. (2.26) is simply

$$(2.27) \qquad \exp[-V(R)T]\langle i|f\rangle$$

and $V(R)$ is the heavy-quark potential! So, taking the logarithm of eq. (2.26) we have

$$(2.28) \qquad V(R) = -\lim_{T\to\infty}\frac{1}{T}\ln\left\langle\operatorname{tr} P \exp\left[i\oint A^a_\mu\frac{1}{2}\lambda^a\,\mathrm{d}x_\mu\right]\right\rangle\bigg/\langle\operatorname{tr} 1\rangle,$$

where P (« path ordered ») reminds us that the order of the operators is important and the $\langle\operatorname{tr} 1\rangle$ accounts for the normalization of the initial and final states in eq. (2.25).

Although we have motivated eq. (2.28) in the language of continuum field theory, we will employ it using a lattice gauge theory evaluation of its right-hand side. This leads us to the expectation value of the operator in eq. (2.22)—the lattice version of eq. (2.23). The expectation value in eq. (2.28) will be a summation over U-field configurations and is simpler in the lattice formulation, because of its explicit gauge invariance, than in the continuum where gauges must be chosen and unphysical excitations, the Fadeev-Popov ghosts, appear.

To begin let us illustrate eq. (2.28) and show that the lattice theory has a confining heavy-quark potential for large coupling $g^2\gg 1$ and the mechanism of confinement can be interpreted as the formation of chromoelectric flux into thin tubes. We begin with

$$(2.29) \qquad \left\langle\prod_c U_\mu(N)\right\rangle = \int\prod_{n,\mu}[\mathrm{d}U_\mu(N)]\prod_c U_\mu(N)\exp[-S]\bigg/\int\prod_{n,\mu}[\mathrm{d}U_\mu(N)]\exp[-S].$$

In this formula we are denoting the integral over the SU_2 group on each link generically. In fact, it is the invariant Haar measure which has the properties

$$(2.30) \qquad \int[\mathrm{d}U] = 1, \qquad \int[\mathrm{d}U]f(U) = \int[\mathrm{d}U]f(U_0 U),$$

where U_0 is an arbitrary element of the group and f is an arbitrary but sensible function.

To evaluate $\langle\prod_c U_\mu(n)\rangle$ at strong coupling, $\beta\ll 1$, we can write

$$(2.31) \qquad \exp[-S] = \prod_p \exp[-\beta\operatorname{tr} UUUU],$$

where $\prod_p$ denotes a product over plaquettes and we can expand the exponential. This leads us to integrals of products of U-matrices. To obtain the leading strong-coupling behavior of $\langle \prod_c U_\mu(n) \rangle$ we only need two integrals,

(2.32) $$\int [dU] U_{ij} = 0 \,, \quad \int [dU] U_{ij} U^\dagger_{kl} = \tfrac{1}{2} \delta_{il} \delta_{jk} \,.$$

The first integral is zero since the integrand has no group-invariant piece—U_{ij} transforms as $2 \times \bar{2}$. The second integral is also easy to understand. The product of Kronecker symbols is determined by group invariance and the normalization factor $\tfrac{1}{2}$ follows from the unitary character of each U-matrix.

Now the leading behavior of $\langle \prod_c U_\mu(n) \rangle$ is easy to find. We must expand the factor $\prod_p \exp[-\beta \operatorname{tr} UUUU]$ on each plaquette until the minimal area enclosed in the contour c is « tiled » once! Consider a 2×2 contour (fig. 4) for illustration. The inner plaquettes in the figure indicate the factors $\beta \operatorname{tr} UUUU$

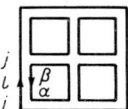

Fig. 4. – A 2×2 contour illustrating the calculation of the loop correlation function.

obtained from the expansion of $\prod_p \exp[-\beta \operatorname{tr} UUUU]$. Consider link No. 1. We have the integral

$$\int U_{ij} U^\dagger_{\beta\alpha} [dU] = \tfrac{1}{2} \delta_{i\alpha} \delta_{j\beta} \,.$$

Doing it, we are left with integrals over the remaining links, as shown in fig. 5. Continuing to the other integrals, we find finally

(2.33) $$\left\langle \prod_c U_\mu(n) \right\rangle = (g^{-2})^{N_c} = \exp[-\ln g^2 \cdot \mathrm{area}] \,,$$

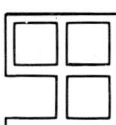

Fig. 5. – The 2×2 contour of fig. 4 after one link has been integrated.

where N_c is the number of plaquettes—its « area » measured in dimensionless units—in the minimal surface determined by the contour c. If we take a rectangular contour of width R and length T and use the relation between the potential and the correlation function discussed above, we find

(2.34) $$V(R) = \sigma |R| \,,$$

and identify the string tension at strong coupling

(2.35) $$\sigma = \ln g^2 + \dots.$$

Other field theories lead to different possible behaviors for the loop integral correlation function. A familiar possibility is the «perimeter law»,

(2.36) $$\left\langle \prod_c U_\mu(n) \right\rangle \sim \exp[-2mT].$$

Then,

(2.37) $$V(R) \sim 2m,$$

for large R. Clearly the underlying physics of this possibility is short-range finite forces between the quarks. The quarks can propagate freely in this case.

The loop correlation function is the order parameter for pure gauge theories. The «area» law labels a disordered phase—clearly the U-matrices are strongly disordered in the strong-coupling calculation illustrated above. By comparison, the «perimeter» law labels an ordered phase.

What about other matrix elements? For example, it is natural to ask whether $\langle U_\mu(n) \rangle$ is an interesting quantity. However, if $\langle U_\mu(n) \rangle \neq 0$ the theory would be spontaneously breaking gauge invariance. Is this possible? Can local symmetries break down spontaneously? Only global symmetries which act on an infinite number of degrees of freedom in a thermodynamic limit can break down spontaneously in systems with local coupling among ordinary degrees of freedom. In fact, there is a theorem by ELITZUR [6] which states this result quite rigorously. Thus gauge-variant operators like $U_\mu(n)$ or $U_\mu(n) U_\nu(n)$ have vanishing matrix elements and are not interesting. The dynamics of the theory lies in the path-ordered products of link variables along closed contours. This makes it a particularly novel statistical system, as we shall see in the next two sections of this lecture series.

3. – Weak-coupling perturbation theory, asymptotic freedom, the renormalization group and continuum limits.

The lattice provides a gauge-invariant regularization scheme for field theories. In asymptotically free theories [7] the continuum limit $a \to 0$ of the lattice model is found in the $g \to 0$ limit. $g^* = 0$ is the infra-red unstable fixed point [8] of the theory. The neighborhood of $g^* = 0$ can be studied using perturbation theory, suitably improved by the renormalization group if necessary, and the approach of the lattice theory to its continuum version can be understood in great detail. The vicinity of $g^* = 0$ where perturbative analysis is adequate is called the «scaling region» of the theory. Asymptotically free scaling laws govern the approach to the continuum limit. We will illustrate these points in a model field theory. But first consider some general results. The simplest such result predicts how the mass gap of the theory depends upon the cut-off and the bare coupling constant. The analysis can be done for any regularization scheme. Require that a physical mass m be a renormalization group invariant,

(3.1) $$\frac{d}{da} m = 0 \,,$$

where a might be a lattice or continuum cut-off. By dimensional analysis,

(3.2) $$m = a^{-1} f(g)$$

in a theory without an intrinsic scale. Substituting into eq. (3.1), we find

(3.3) $$f'(g) = - f(g)/\beta(g) \,,$$

where $\beta(g)$ is the Callan-Symanzik function [9],

(3.4) $$\beta(g) = - a \, \partial g/\partial a \,.$$

In the vicinity of the fixed point $g^* = 0$, $\beta(g)$ can be computed perturbatively,

(3.5) $$a \, \partial g/\partial a = \beta_0 g^3 + \beta_1 g^5 + \dots \,,$$

with

(3.6) $$\beta_0 = \frac{11}{3} \frac{N}{16\pi^2} \,, \qquad \beta_1 = \frac{34}{3} \left(\frac{N}{16\pi^2} \right)^2 ,$$

in SU_N gauge theory. Using eq. (3.5) the formal integral for $f(g)$ following from eq. (3.3)

(3.7) $$f(g) = \exp \left[- \int^g \frac{dg'}{\beta(g')} \right]$$

can be evaluated

(3.8) $$f(g) \propto (\beta_0 g^2)^{-\beta_1/2\beta_0^2} \exp \left[- 1/2\beta_0 g^2 \right].$$

Therefore, the mass gap's dependence on the bare coupling is determined—this is a scaling law. It is analogous to the relationship for the correlation length in a statistical-mechanics setting [8]

(3.9) $$\xi \propto |T - T_c|^{-\nu}, \qquad T \approx T_c \,.$$

The critical index ν is not computable accurately in perturbation theory if the scale-invariant theory at the critical point T_c is interacting. It is a wonderful feature of gauge theories that the scaling laws can be obtained using ordinary perturbation theory!

Note that the mass gap's dependence on g is nonanalytic—mass generation in the theory is a nonperturbative effect. It was necessary to use the renormalization group to obtain this result. The Callan-Symanzik function itself is expandable in powers of g.

One of the aims of lattice gauge theory is the calculation of the mass spectrum of quantum chromodynamics. Equation (3.8) predicts how those masses must depend on g such that finite masses result in the continuum limit. The importance of verifying the scaling laws in nonperturbative lattice calculations is clear.

Consider first the general features of the lattice regularization of a renormalizable field theory. The bare lattice coupling g^2 is related to the renormalized coupling g_R^2,

$$(3.10) \qquad g_R^2 = Z_1^{-1} g^2,$$

where the renormalization constant Z_1 can be expanded perturbatively,

$$(3.11) \qquad Z_1 = 1 + B(a) g^2 + \dots.$$

The renormalizability of the theory indicates that $B(a)$ depend on the lattice spacing logarithmically,

$$(3.12) \qquad B(a) = -\beta_0 \ln \frac{1}{a} + C,$$

where β_0 is a constant which is positive in an asymptotically free theory and C is a finite constant. Contact can be made with the Callan-Symanzik equation by requiring that g_R^2 be a renormalization group invariant,

$$(3.13) \qquad a \frac{\partial}{\partial a} g_R^{-2} = 0 \to a \frac{\partial}{\partial a} (Z_1 g^{-2}) = a \frac{\partial}{\partial a} (g^{-2} + B(a)).$$

So,

$$(3.14) \qquad 0 = \frac{1}{g^4} a \frac{\partial}{\partial a} g^2 - \beta_0 \to a \frac{\partial}{\partial a} g^2 = \beta_0 g^4,$$

which is the one-loop Callan-Symanzik equation. It can be integrated to find the bare coupling's dependence on a which ensures the invariance of g_R^2 to changes in the underlying lattice spacing,

$$(3.15) \qquad g^2(a) \simeq \frac{1}{\beta_0 \ln (1/a\Lambda_L)}, \qquad \ln (1/\Lambda_L) \gg 1.$$

The Λ-parameter here sets the scale of the logarithm. Note that, as $a \to 0$

(continuum limit), the bare coupling vanishes logarithmically—this is asymptotic freedom.

We will now illustrate weak-coupling lattice calculations by deriving asymptotic freedom for a two-dimensional field theory which closely resembles gauge theories in four dimensions. Consider the $SU_N \times SU_N$ spin model in two dimensions [10],

(3.16)
$$\begin{cases} S = -\dfrac{1}{g^2} \sum_{x,\mu} \mathrm{tr}\left[(U(x)\,U^\dagger(x+\mu) - 1) + \mathrm{h.c.}\right], \\ Z = \int \left[\prod_c \mathrm{d}U\right] \exp[-S], \end{cases}$$

where $[\mathrm{d}U]$ is the group-invariant integration measure. The SU_N matrices fluctuate on all length scales from a to ∞. We wish to integrate out the high-frequency modes and obtain the effective action for the low-frequency modes [8]. To make this division we use the background field method. Write

(3.17)
$$U(x) = \exp[ig\varphi(x)]\,U^{\mathrm{cl}}(x),$$

where U^{cl} solves the classical equations of motion and $\varphi(x)$ parametrizes the quantum fluctuations of $U(x)$. Define

(3.18)
$$\varphi(x) = \lambda^\alpha \varphi^\alpha(x), \qquad x = 1, 2, \ldots, N^2 - 1,$$

and

(3.19)
$$[\lambda^\alpha, \lambda^\beta] = if^{\alpha\beta\varrho}\lambda_\varrho, \qquad \mathrm{tr}\,\lambda_\alpha \lambda_\beta = \tfrac{1}{2}\delta_{\alpha\beta}.$$

Substituting into the action, we obtain

(3.20)
$$S = -\frac{1}{g^2} \sum_{x,\mu} \mathrm{tr}\left[(U^{\mathrm{cl}}(x)\,U^{\mathrm{cl}\dagger}(x+\mu) - 1) + \mathrm{h.c.}\right] -$$
$$- \frac{1}{g^2} \sum_{x,\mu} \mathrm{tr}\left[(\exp[-ig\varphi(x+\mu)]\exp[ig\varphi(x)] - 1)\,U^{\mathrm{cl}}(x)\,U^{\mathrm{cl}}(x+\mu) + \mathrm{h.c.}\right].$$

Since we are going to do perturbation theory in g^2, S and $[\mathrm{d}U]$ should be expanded in powers of g^2,

(3.21a) $\exp[-ig\varphi(x+\mu)]\exp[ig\varphi(x)] =$
$$= \exp\left[-ig\nabla_\mu\varphi(x) + \tfrac{1}{2}g^2[\varphi(x+\mu), \varphi(x)] + O(g^3)\right] =$$
$$= 1 - ig\nabla_\mu\varphi(x) + \tfrac{1}{2}g^2[\varphi(x+\mu), \varphi(x)] - \tfrac{1}{2}g^2(\nabla_\mu\varphi(x))^2 + O(g^3),$$

(3.21b)
$$\mathrm{d}U = \prod_{x=1}^{N^2-1} \mathrm{d}\varphi^\alpha[1 + O(g^2\varphi^2)].$$

Equation (3.21b) states that the curvature of the group manifold does not contribute to lowest order.

A convenient parametrization of U^{cl} is

(3.22) $$U^{\text{cl}}(x)\,U^{\text{cl}}(x+\mu) = \exp[iF_\mu(x)], \qquad F_\mu(x) = \lambda^\alpha F_\mu^\alpha(x),$$

where $F_\mu(x)$ is Hermitian. It will prove sufficient to expand this result in powers of $F_\mu(x)$. Collecting everything,

(3.23) $$S = \frac{1}{g^2} S^{\text{cl}} + S_0 + S_{\text{int}} + O(g^2, F^4),$$

where

(3.24) $$\begin{cases} S^{\text{cl}} = -\frac{1}{2}\sum_{x,\mu}(F_\mu^\alpha)^2, \\ S_0 = \sum_{x,\mu}\frac{1}{2}(\nabla_\mu\varphi^\alpha)^2, \\ S_{\text{int}} = \sum_{x,\mu}\{-\frac{1}{2}\operatorname{tr}(\lambda^\alpha\lambda^\beta\lambda^\gamma\lambda^\delta)\nabla_\mu\varphi^\beta\nabla_\mu\varphi^\delta F_\mu^\nu F_\mu^\delta + \frac{1}{2}f^{\alpha\beta\gamma}\varphi^\alpha(x+\mu)\varphi^\beta(x)F_\mu^\gamma\}. \end{cases}$$

So, we have a classical free-field piece, a quantum free-field piece and interactions. It is convenient to modify S_0,

(3.25) $$S_0 \to \sum_{x,\mu}[\tfrac{1}{2}(\nabla_\mu\varphi^\alpha)^2 + \tfrac{1}{2}m^2\varphi^{\alpha 2}],$$

to avoid spurious infra-red problems in the midst of a calculation. The limit $m \to 0$ will be smooth for the gauge-invariant quantities we calculate.

Now we can calculate a few terms in the perturbative expansion of Z. It is

(3.26) $$Z = \int \pi\, d\varphi^\alpha \exp[-g^{-2}S^{\text{cl}} - S_0 - S_{\text{int}}][1 + O(g^2, F^4)] =$$
$$= \exp\left[-\frac{1}{g^2}S^{\text{cl}}\right]\int \pi\, d\varphi^\alpha \exp[-S_0]\{1 - S_{\text{int}} + \frac{1}{2}S_{\text{int}}^2 + \ldots\} =$$
$$= \exp\left[-\frac{1}{g^2}S^{\text{cl}}\right]\{1 + \langle S_{\text{int}}\rangle + \frac{1}{2}\langle S_{\text{int}}^2\rangle + \ldots\}.$$

The first term is

(3.27) $$\langle S_{\text{int}}\rangle = \sum_{x,\mu}[-\tfrac{1}{2}\operatorname{tr}(\lambda^\alpha\lambda^\beta\lambda^\gamma\lambda^\delta)F_\mu^\nu F_\mu^\delta\langle\nabla_\mu\varphi^\alpha\,\nabla_\mu\varphi^\beta\rangle +$$
$$+ \tfrac{1}{2}f^{\alpha\beta\gamma}F_\mu^\gamma\langle\varphi^\alpha(x+\mu)\varphi^\beta(x)\rangle].$$

The second piece in eq. (3.27) is zero because the average of a vector quantity vanishes identically. The first piece does not vanish. Only the terms with

$\alpha = \beta$ and $\gamma = \delta$ contribute. For these indices we have the replacement

$$(3.28) \qquad \operatorname{tr} \lambda^\alpha \lambda^\beta \lambda^\gamma \lambda^\delta = \frac{1}{8} \frac{2}{N} \delta^{\alpha\beta} \delta^{\gamma\delta} + \ldots$$

So,

$$(3.29) \quad \langle S_{\mathrm{int}} \rangle = -\frac{1}{8N} \sum_x \frac{1}{2} (F_\nu^\gamma)^2 \langle (\nabla_\mu \varphi^\alpha)^2 \rangle =$$
$$= -\frac{1}{8N} \sum_x \frac{1}{2} (F_\nu^\gamma)^2 4(N^2 - 1)\big(G(0) - G(1)\big),$$

where

$$(3.30) \qquad G(x) = \langle \varphi^1(x) \varphi^1(0) \rangle = \int_{-\pi}^{\pi} \frac{\mathrm{d}^2 k}{2\pi} \frac{\exp[ikx]}{4 - 2\cos k_1 - 2\cos k_2 + m^2 a^2}$$

is the free scalar propagator on the lattice. From this expression or the differential equation $(\Box + m^2 a^2) G(x) = -\delta(x)$, one easily checks that

$$(3.31) \qquad G(1) - G(0) = -\tfrac{1}{4} + O(m^2 a^2).$$

So,

$$(3.32) \qquad \langle S_{\mathrm{int}} \rangle = \frac{N^2 - 1}{8N} \sum_x \frac{1}{2} (F_\nu^\gamma)^2 = \frac{N^2 - 1}{8N} S^{\mathrm{cl}}.$$

Finally, we have the second-order term in eq. (3.26),

$$(3.33) \quad \frac{1}{2} \langle S_{\mathrm{int}}^2 \rangle = \frac{1}{8} f^{\alpha\beta\gamma} f^{\alpha'\beta'\gamma'} \sum_{yy'\mu\nu} F_\mu^\gamma(y) F_\nu^{\gamma'}(y') \cdot$$
$$\cdot \langle \varphi^\alpha(y+\mu) \varphi^\beta(y) \varphi^{\alpha'}(y'+\nu) \varphi^{\beta'}(y') \rangle + O(F^4) = \frac{1}{8} N \sum_{yy'\mu\nu} F_\mu^\gamma(y) F_\nu^\gamma(y') \cdot$$
$$\cdot [G(y-y'+\mu-\nu) G(y-y') - G(y-y'+\mu) G(y-y'-\nu)],$$

where we used $f^{\alpha\beta\gamma} f^{\alpha\beta\gamma'} = N \delta_{\gamma\gamma'}$. Now we must simplify this collection of propagators. Let $\Delta = 4 - 2\cos k_1 - 2\cos k_2 + m^2 a^2$. In momentum space the propagators in eq. (3.33) become

$$(3.34) \quad \int \frac{\mathrm{d}^2 k}{(2\pi)^2} \int \frac{\mathrm{d}^2 k'}{(2\pi)^2} \cdot$$
$$\cdot \frac{\exp[ik_\mu]\big(\exp[-ik_\nu] - \exp[-ik'_\nu]\big)\big(\exp[ik(y-y')] \exp[ik'(y-y')]\big)}{\Delta(k) \Delta(k')}.$$

The background field is slowly varying, so replace $F_\nu^\gamma(y') \to F_\nu^\gamma(y)$ and the y' integral is trivial,

$$(3.35) \quad \frac{1}{2} \langle S_{\mathrm{int}}^2 \rangle = \frac{1}{8} N \sum_{y\mu\nu} F_\mu^\gamma(y) F_\nu^\gamma(y) \int \frac{\mathrm{d}^2 k}{(2\pi)^2} \frac{\exp[ik_\mu]\big(\exp[-ik_\mu] - \exp[ik_\nu]\big)}{\Delta^2(k)}.$$

The momentum integral vanishes unless $\mu = \nu$, so

(3.36)
$$\tfrac{1}{2}\langle S_{\text{int}}^2\rangle = \tfrac{1}{4}NS^{\text{cl}}G(1),$$

where

(3.37)
$$G(1) = \int \frac{d^2k}{(2\pi)^2} \frac{\exp[ik_1](\exp[-ik_1] - \exp[ik_1])}{\Delta^2(k)},$$

can be shown [10]. Collecting all this, we have the partition function with the high frequencies integrated out,

(3.38)
$$Z = \exp\left[-\frac{Z_{1L}}{g^2} S^{\text{cl}}\right],$$

with

(3.39)
$$Z_{1L} = 1 - g^2\left(\frac{1}{4}NG_L(1) + \frac{N^2-1}{8N}\right) = 1 - g^2\left(\frac{1}{4}NG_L(0) + \frac{N^2-1}{8N} - \frac{N}{16}\right).$$

Since [10]

(3.40)
$$G_L(0) = \frac{1}{4\pi}\ln\frac{32}{m^2 a^2} + O(m^2 a^2),$$

we have

(3.41)
$$Z_{1L} = 1 - g^2\left(\frac{1}{16\pi} N \ln\frac{32}{m^2 a^2} + \frac{N^2-1}{8N} - \frac{N}{16}\right).$$

Equations (3.38) and (3.41) are the final results of this calculation. We see that integrating out the high-frequency modes has led to a renormalization of the theory's coupling contant. Comparing eq. (3.41) with (3.11), we have

(3.42)
$$\beta_0 = N/8\pi$$

in these $SU_N \times SU_N$ models. Since $\beta_0 > 0$, they are, indeed, asymptotically free. We now have learned how to take the continuum limit. As a is scaled to zero, the lattice coupling constant drops to zero logarithmically. This ensures us that the finite-energy features of the theory are independent of the lattice regulator.

Finally, it is instructive to examine eq. (3.26) and see what piece of the operator $1 - S_{\text{int}} + \tfrac{1}{2} S_{\text{int}}^2$ gave rise to the infinite (i.e. logarithmically dependent on the lattice spacing) renormalization. The term in $S_{\text{int}} \sim F_\mu^\gamma F_\mu^\delta \nabla_\mu \varphi^\alpha \nabla_\mu \varphi^\beta$ led only to a finite renormalization of g, as eq. (3.32) shows. The two extra derivatives suppressed this term's effect on the low-energy behavior of the renormalized theory. Such a term is called irrelevant [8] in the renormalization classification of free-field operators and we could have ignored it here. The infinite renormalization came from the term in eq. (3.33) which has the same dimensional behavior as F^2 itself. This term is called marginal [8] in the renormalization group terminology. It was responsible for asymptotic freedom, the logarithmic vanishing of the lattice coupling constant.

This perturbative analysis is only reliable near $g = 0$. In general one needs a renormalization method which applies at all values of g. Then it should be possible to pose questions about the nonperturbative dynamics of the theory (Is there dynamical symmetry breaking, spontaneous mass generation, confinement?) and obtain quantitative answers. In the context of spin models describing second-order phase transitions there are such methods which can be generalized to lattice gauge theory. The crucial physical idea behind these renormalization group methods is the construction of the variables and interactions describing the low-energy (low compared to the cut-off momentum π/a) content of the original theory. Let us review the real-space renormalization procedure [11]. Consider the two-dimensional Ising model as shown in fig. 6

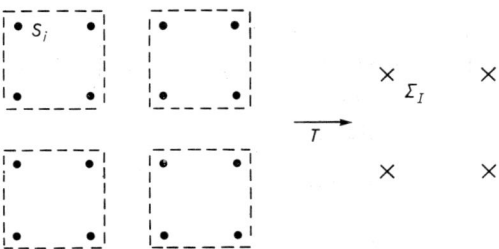

Fig. 6. – Block spin construction on a two-dimensional spin system.

for the purposes of illustration. From the original degrees of freedom s_i we would like to construct spatially averaged variables to describe the lower-energy, longer-wavelength fluctuations of the theory. These are called block spins Σ_i and are constructed from groups of four spins s_i as shown in the figure. The original lattice model is then replaced by one with a lattice spacing twice as large as the original and a new interaction Hamiltonian. However, the construction is required to produce the same physics (*e.g.* correlation functions) as the original model for all momenta up to half the original cut-off. To make this precise consider the original Hamiltonian

$$(3.43) \qquad H = J \sum_{\langle ij \rangle} s_i s_j \,,$$

where $\langle \rangle$ denotes nearest neighbors, and its partition function

$$(3.44) \qquad Z = \sum_{\{s_i\}} \exp\left[-H(\{s_i\})\right].$$

Then the construction of the block spin variables Σ_i is achieved with a mapping $T: \{s_i\} \to \{\Sigma_i\}$ and the interaction between the new variables is described by the Hamiltonian $H'(\{\Sigma_i\})$,

$$(3.45) \qquad \sum_{\{s_i\}} \prod_I T(\Sigma_I, s_i; i\varepsilon \text{ block } I) \exp\left[-H(\{s_i\})\right] = \exp\left[-H'(\{\Sigma_i\})\right].$$

This equation expresses the physical idea of the renormalization group—the variables s_i are summed out (high-energy fluctuations) for given values of the variables Σ_I (low-energy fluctuations) and the dynamics (H') governing the low-energy variables is computed. A crucial condition imposed on T is that the physics of the (Σ, H') system be the same as that of the (s, H), i.e.

$$(3.46a) \qquad \sum_{\{\Sigma\}} \exp\left[-H'(\Sigma)\right] = \sum_{\{s\}} \exp\left[-H(s)\right],$$

which requires

$$(3.46b) \qquad \sum_{\Sigma_I} T(\Sigma_I, s_i) = 1.$$

Of course, the lattice spacing of the new system is twice the old, $a' = 2a$, so the correlation length of the new system is half the old, $\xi' = \frac{1}{2}\xi$. It takes considerable skill to invent practical transformations T, but in addition to the sum rule eq. (3.46b) there are two generic requirements it should fulfill:

1) to respect the symmetries of H,

2) lead to a local effective Hamiltonian H'.

Property 1) restricts the complexity of H' and the need for locality, property 2) is useful in attaining some theoretical control over this procedure.

The crucial point about all this is that it gives us a physical system with half the correlation length of the original. So, if initially $\xi \gg a$ and the system is almost critical and its physics is almost decoupled from the lattice, then T generates a system which is less critical. T itself might be a simple operator because it couples just a few variables s_i into single variables Σ_I. However, it can be iterated many times leading to a sequence of physical systems each with a smaller correlation length,

$$(3.47) \qquad \xi \to \frac{1}{2}\xi \to \frac{1}{4}\xi \to \frac{1}{8}\xi \to \ldots .$$

Near the end of this sequence a physical system with a short correlation length is produced. Presumably, since it only has correlations between almost contiguous block spin variables, it can be analyzed by conventional mathematical-physics methods (cluster expansions, etc.). In this way one obtains information about the original almost critical system which describes the continuum limit of the lattice model.

This is the bare bones structure of block spinning. Now consider it, both conceptually and calculationally in greater detail. Corresponding to each step denoted in eq. (3.47) there will be a new Hamiltonian H'. H' is in general a linear superposition of local operators (requirement 2)) each of which preserves the symmetries of the original system (requirement 1)),

$$(3.48) \qquad H' = \sum J_\alpha O_\alpha(\Sigma),$$

where the set $\{J_\alpha, \alpha = 1, 2, ...\}$ consists of coupling constants one computes using T. Thus T is a mapping in coupling constant space between different sets $\{J_\alpha^{(i)}\}$,

$$(3.49) \qquad \{J_\alpha^{(1)}\} \xrightarrow{T} \{J_\alpha^{(2)}\} \xrightarrow{T} \{J_\alpha^{(3)}\} \xrightarrow{T} \cdots \xrightarrow{T} \{J_\alpha^{(n)}\} \xrightarrow{T} \cdots.$$

Particularly important points in this parameter space are the continuum limits of lattice models, points where $\xi = \infty$. Since T reduces ξ by a factor of 2, they are, in fact, fixed points of the transformation. The behavior of T in the vicinity of its fixed points is crucial to understanding the quantitative features of the theory's continuum limit. Consider a fixed point J^*, in an abbreviated but hopefully clear notation. Then for $J^{(n)}$ near J^* we can linearize the renormalization group trajectories, eq. (3.49),

$$(3.50) \qquad \Delta J_\alpha^{(n+1)} = T_{\alpha\beta} \Delta J_\beta^{(n)},$$

where

$$\Delta J_\alpha^{(n)} = J_\alpha^{(n)} - J_\alpha^*$$

and

$$(3.51) \qquad T_{\alpha\beta} = \frac{\partial J_\alpha^{(n+1)}}{\partial J_\beta^{(n)}} \bigg|_{J^*}.$$

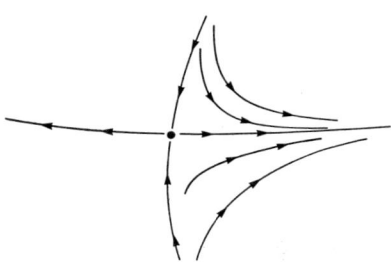

Fig. 7. – Renormalization group flows in a two-parameter space.

A possible visualization of eq. (3.50) is shown in fig. 7. There we have a single fixed point in a two-dimensional parameter space. T should then be diagonalized and its two eigenvalues and eigenoperators (cf. eq. (3.48)) determined. Its eigenvalues are given special names depending on their size compared to unity,

$$\lambda > 1, \quad \text{« relevant » operator (or direction)},$$
$$\lambda = 1, \quad \text{« marginal »},$$
$$\lambda < 1, \quad \text{« irrelevant »}.$$

This terminology reflects the fact read off eq. (3.50) that relevant and marginal operators are amplified by the renormalization group transformation and irrelevant operators are de-amplified. The example in fig. 7 shows one relevant and one irrelevant direction. The relevant direction emerging from the fixed point becomes attached to the «renormalized trajectory» of the theory. Physical systems with (J_1, J_2) values near the fixed point flow under the action of T toward the renormalized trajectory, as shown in the figure.

The classification of operators in asymptotically free theories is particularly simple since the critical point is at zero coupling. The analysis is done reliably with renormalized perturbation theory. In the case of gauge fields $F^\alpha_{\mu\nu} F^{\mu\nu}_\alpha$ is a marginal operator. It is the only one consistent gauge invariance and Lorentz invariance. There are no relevant operators. Irrelevant operators are made from products of gauge-covariant derivatives of $F^\alpha_{\mu\nu}$.

A convenient way to parametrize the action of pure gauge field lattice theories is

(3.52) $$S = -\frac{1}{g^2} \{\theta_1 + c_2 \theta_2 + c_3 \theta_3 + ...\},$$

where θ_1 is the single plaquette term and θ_2 etc. are linear combinations of closed loops which are constructed to be irrelevant. The fixed point lies at $g = 0$, $c_1 = c_1^*$, $c_2 = c_2^*$, etc. Renormalization group transformations which respect gauge invariance, locality and O_4 invariance are not difficult to invent and many schemes are being studied [12, 13]. But how are these calculations used to elucidate these theories' continuum limits? We might begin the lattice calculations with the bare action $g \neq 0$, $c_1 = c_2 = c_3 = ... = 0$. Applying T then drives the theory toward stronger coupling (larger g) and toward the renormalized trajectory, as shown in fig. 8. The same set of calculations can

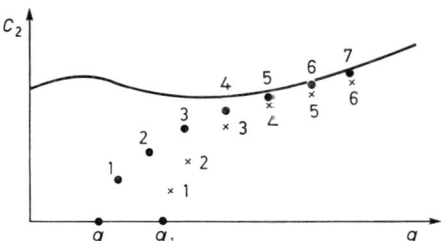

Fig. 8. – Matching procedure for numerical renormalization group studies.

be contemplated, but they might begin at stronger coupling $g = g_1$, $c_1 = c_2 = ... = 0$. After one less renormalization group iteration they might match the previous calculations, as shown in the figure. If g_1 is adjusted to satisfy this matching criterion, then we know that the two initial theories are identical in their actual physics, but they differ in their scales,

(3.53) $$\xi(g^2) = 2\xi(g_1^2).$$

If g and g_1 are sufficiently small, then $\xi(g^2)$ and $\xi(g_1^2)$ satisfy the scaling laws of asymptotic freedom. If g and g_1 are not small, then eq. (3.53) determines the nonperturbative analog of the Callan-Symanzik function. Of course, it is easier to do calculations with the formulations of the theory having small correlation length, *e.g.* point 6 or 7 near the renormalized trajectory at strong coupling in fig. 8. Call this the l-th point in the sequence beginning at g_1. Using this theory, one can calculate the correlation length (or mass gap, say) using strong-coupling methods,

(3.54) $$a(g)\frac{1}{\xi} = c.$$

But $a(g_1^2) = 2^{-l} a(g_l^2)$, so

(3.55) $$a(g_1)\frac{1}{\xi} = 2^{-l} c.$$

And, finally, the dependence of a on g_1 is given by asymptotic freedom,

(3.56) $$\text{const} \cdot (\beta_0 g_1^2)^{-\beta_1/2\beta_0^2} \exp[-1/2\beta_0 g_1^2] = 2^{-l} c.$$

In this way the strong-coupling calculation of c has given us the universal number «const» in eq. (3.56) which is a property of the continuum theorem! For example, adopting this procedure to the calculation of the particle spectrum of the theory yields the universal mass ratios of SU_N pure gauge theories.

The implementation of this program involves numerical methods at this time. It can be done in the context of a Monte Carlo simulation of the gauge theory. This is an active field of research [12, 13] which promises to improve our grasp on the vital properties of the theory including dynamical mass generation and confinement.

4. – Topological excitations in spin and gauge theories.

The importance of topologically nontrivial gauge field configurations has been recognized for the last several years. The list of gauge groups, geometric character of the excitations and physical relevance of each is a long one. Here we will look at just two topics: instantons and vortices in two-dimensional spin models. These illustrations raise several interesting and only partially understood questions of gauge theories and their lattice formulations. First instantons [14]. Since a lattice regulator allows one to study nonperturbative effects systematically, the relevance of instantons to the continuum dynamics should be decided quantitatively. In particular, one wants to know how significant these sexcitation are to the qualitative features of the theory such as confinement and chiral-symmetry breaking. Next vortices [15]. We are particularly interested in monopoles—topologically significant excitations associated with a U_1 subgroup—and their role in determining the character of the gauge theory's ground state.

To begin consider the O_3 spin model in $1+1$ dimensions and the issues of topology and continuity on a lattice [16]. The classical action of the model is

(4.1) $$S_{\rm cl} = \tfrac{1}{2}\int d^2x\, \partial_\mu \mathbf{s}\cdot\partial_\mu \mathbf{s}\,, \qquad |\mathbf{s}(x)|^2 = 1\,,$$

where $\mathbf{s}(x)$ is a unit vector in three-space. Clearly, $\mathbf{s}(x)$ would be a free field were it not for the constraint $|\mathbf{s}(x)|^2 = 1$. This theory has several interesting properties. It is asymptotically free and it has a nontrivial topological charge Q, which is given by the expression [11]

(4.2) $$Q = \tfrac{1}{4}\int d^2x\, \mathbf{s}\cdot(\partial_1 \mathbf{s}\times\partial_2 \mathbf{s})\,,$$

and can assume only integer values $Q = 0$, ± 1, ± 2, ... which label distinct sectors of the theory. The different topological sectors cannot be connected by continuous deformations of the field $\mathbf{s}(x)$. This result can be understood by exposing the geometric meaning of Q. Take a boundary condition on $\mathbf{s}$ so that $\mathbf{s}\to(1,0,0)$ when $x\to\infty$ in any direction. The base space of the model is then essentially a two-sphere, S_2. The field $\mathbf{s}$ becomes a mapping of the base space S_2 to the sphere of $|\mathbf{s}|=1$. Such mappings can cover the unit sphere an integral number of times and we recognize Q as computing this number.

A field configuration with $Q=\pm 1$ can be drawn easily as in fig. 9. It is given by [17]

(4.3) $$\begin{cases} s_1 = 1 - 2\exp[-x^2]\,, \\ s_2 = x_1 f(x^2)\,, \\ s_3 = x_2 f(x^2)\,, \end{cases} \qquad f(x^2) = \frac{4}{x^2}\exp[-x^2]\bigl(1-\exp[-x^2]\bigr)\,.$$

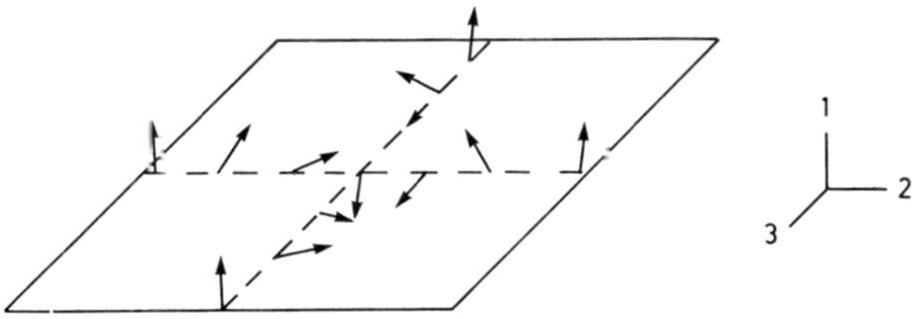

Fig. 9. – An instanton in the O_3 model.

Note that $\mathbf{s}(0) = (-1, 0, 0)$ and that, as $\mathbf{x}$ covers the base space, $\mathbf{s}$ covers S_2 just once.

Field configurations having $Q = \pm 1$ which are local minima of the action are instantons. Their classical action is

(4.4a) $$S_{\text{inst}} = 4\pi, \qquad O_3 \text{ model},$$

and for the generalized CP^{n-1} spin models

(4.4b) $$S_{\text{inst}} = 2\pi n.$$

Since the theory is asymptotically free, any dynamically generated mass in it must depend upon g^2 nonanalytically, as discussed in sect. **3** above.

Consider a Pauli-Villars cut-off M and a coupling $g^2(M)$. Then the correlation length, the reciprocal of the mass gap, depends on M and $g^2(M)$ as

(4.5) $$\xi = \frac{1}{m} \propto \frac{1}{M} g^2 \exp[2\pi/g^2].$$

Now consider the lattice model. Do any of these topological ideas survive? Since continuity concepts are not so clear on a lattice, one naively would think not. However, if we specialize to asymptotically free models where the critical point is at zero coupling, some remnants of the ideas do survive. In the vicinity of the critical point matrix elements of the fields are very smooth in many cases. In particular, consider the lattice action

(4.6) $$S = \sum_{\langle ij \rangle} (1 - \mathbf{s}_i \cdot \mathbf{s}_j), \qquad |\mathbf{s}|^2 = 1.$$

In a theory with the Boltzmann weight $\exp[-\beta S]$, one can compute the internal energy at weak coupling in powers of $g^2 = 1/\beta$,

(4.7) $$\langle 1 - \mathbf{s}_i \cdot \mathbf{s}_j \rangle = \frac{1}{2\beta} + O\left(\frac{1}{\beta^2}\right).$$

This equation implies that as $g^2 \to 0$ the « most likely » fields become arbitrarily smooth and ordinary continuum notions of continuity are useful [16]!

To make the topological properties of the lattice model precise we need a lattice construction of the topological charge Q_L in which [16]:

1) Q_L has only integer values and is composed of a sum of local charge densities.

2) Q_L has the correct continuum limit.

3) The lattice definition of Q_L should be unambiguous except, perhaps, on a set of spin configurations of measure zero and these « exceptional » spin configurations should have an action which is bounded from below by a finite constant.

Property 1) is very important for a useful definition of the topological charge in a lattice model. If Q_L can take on only integral values, we are certain that ordinary perturbative fluctuations will not contribute to it at weak coupling. Since instantons are very dilute (density per site $\sim \exp[-4\pi/g^2]$), $\langle Q_L \rangle$ itself must be very small at weak coupling and, if we used an approximate definition of Q_L which did not exclude perturbative effects exactly, our approximate object would be overwhelmed by trivial perturbative fluctuations. For O_3 the construction of Q_L is particularly transparent. Consider a triangular lattice first. The naive construction which approximates the continuum charge density is the following: At the center of each triangle associate the spherical area enclosed by the three nearest spins as shown in fig. 10. Define $A = \alpha_1 +$

Fig. 10. – Construction of the topological charge in the O_3 model.

$+ \alpha_2 + \alpha_3 + \pi$ in terms of the angles shown in the figure. The sign of the area should be

(4.8a) $$\sigma = \text{sign}\,[\mathbf{s}_1 \cdot (\mathbf{s}_2 \times \mathbf{s}_3)]$$

and the local charge density is

(4.8b) $$q(x^*) = \sigma A/4\pi\,.$$

Then,

(4.9) $$Q_L = \sum_{x^*} q(x^*)$$

is a good lattice definition of the topological charge. Only if $\mathbf{s}_1$, $\mathbf{s}_2$ and $\mathbf{s}_3$ lie in a given plane does the definition become ambiguous if

$$1 + \mathbf{s}_1 \cdot \mathbf{s}_2 + \mathbf{s}_2 \cdot \mathbf{s}_3 + \mathbf{s}_1 \cdot \mathbf{s}_3 \leq 0\,.$$

Such an exceptional configuration has at least one bond which is very unfavorable, *i.e.* the configuration has a large action and is very unlikely at weak coupling.

For a square lattice, one makes triangles and copies this construction. The triangles can be chosen in two distinct ways, so we can construct two local topological charge densities,

(4.10)
$$\begin{cases} q_1(x^*) = \dfrac{1}{4\pi} [\sigma A(s_1 s_2 s_3) + \sigma A(s_2 s_3 s_4)], \\ q_2(x^*) = \dfrac{1}{4\pi} [\sigma A(s_1 s_2 s_3) + \sigma A(s_1 s_3 s_4)]. \end{cases}$$

Similar constructions have been made for CP^{n-1} models in two dimensions and SU_N gauge theories in four dimensions. In each case the exceptional configurations have at least one very-high-energy bond for which a lower bound can be computed.

Now the question of interest is whether in the continuum limit only smooth lattice field configurations survive which have classical partners. The following difficulty can arise. There may be lattice spin configurations having $Q \neq 0$ which have topological charge densities isolated to the neighborhood of only one site and these configurations may survive in the continuum limit. This in fact happens in the O_3 model! The configuration shown in fig. 11 has $Q = 1$ and action $S_d = 6.60 \dots$. Its probability to occur behaves as $\sim \exp[-\beta S_d]$.

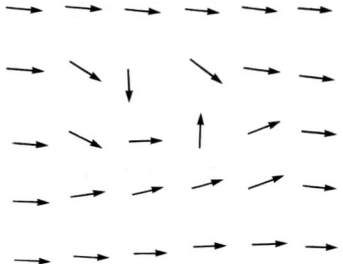

Fig. 11. – A «dislocation» spin configuration in the O_3 model.

In physical units its probability/area

$$\sim \exp[-\beta S_d]/\exp[4\pi\beta] \sim \exp[-\beta(S_d - 4\pi)]$$

(recall from eq. (4.5) that $\xi \sim \exp[2\pi\beta]$). Since $S_d < 4\pi$, these «dislocations» do not become unlikely in the continuum limit. Therefore, the continuum limit of the lattice topological charge is unphysical in this case. Note that this is a violation of universality in the sense that the short-distance lattice artifacts are surviving the continuum limit. This is a serious problem because they will destroy the asymptotic-freedom scaling laws for matrix elements involving the topological charge.

For CP^{n-1} models with $n > 4$, $S > 4\pi$ and the naive continuum limit is obtained. Similarly, for gauge theories in four dimensions these «dislocations» do not contribute in the continuum limit. This is a happy property of the lattice action. If it were not true, a modified lattice action would have had to be invented (if possible) to ensure an interesting continuum limit.

Although the dislocations remain in the O_3 model, they do not contribute significantly to the ordinary magnetic properties such as the magnetic susceptibility. These quantities scale as expected from asymptotic freedom. Why is there such a dramatic difference between ordinary magnetic properties of the model and the model's topological features? We actually answered this question above when we argued for the importance of property 1) of Q_L. Note that the magnetic susceptibility gets contributions from smooth spin configurations when $\beta \to \infty$. However, the topological susceptibility, $\chi_t = \langle Q^2 \rangle/\text{volume}$, does not—it only gets contributions from an exponentially small fraction, $O(\exp[-4\pi\beta])$, of the slowly varying fields. Therefore, field configurations which vary rapidly on the scale of the lattice spacing can contribute significantly to χ_t, while the magnetic suceptibility is, to good approximation, unaffected.

Now consider an even simpler two-dimensional spin model—the U_1 or « planar » model. This is a deceptively simple system which has a subtle phase transition which can be nicely understood in terms of topological excitations —vortices. These are two-dimensional analog of magnetic monopoles, so this model is believed to be a good laboratory for four-dimensional Abelian gauge theory. An interesting and instructive feature of this model is that we can clearly separate its spin configurations into two components: spin waves which are trivial topologically and vortices. Exact expressions can be derived for the system's partition function which shows that the condensation of these vortices gives the system two phases.

To begin, let us discuss the model very qualitatively. We construct the model by placing a planar spin on each lattice site n,

$$(4.11) \qquad \mathbf{s}(n) = \begin{pmatrix} \cos\theta(n) \\ \sin\theta(n) \end{pmatrix}.$$

Let the action have nearest-neighbor coupling,

$$(4.12) \qquad S = -J \sum_{\langle nm \rangle} \mathbf{s}(n) \cdot \mathbf{s}(m) = -J \sum_{\langle nm \rangle} \cos(\theta_n - \theta_m) = -\tfrac{1}{2} J \sum_{\langle nm \rangle} \left(\exp[i(\theta_n - \theta_m)] + \text{h.c.} \right).$$

This can be written more elegantly if we introduce a finite-difference operator in the direction μ,

$$(4.13) \qquad \Delta_\mu \theta(n) = \theta(n+\mu) - \theta(n),$$

so that eq. (4.12) becomes

(4.14) $$S = -J \sum_{n,\mu} \cos\left(\Delta_\mu \theta(n)\right).$$

Now let us examine the spin-spin correlation function at high and low T and argue that the system has two phases. At high T the correlation function,

(4.15) $$\langle \exp[i\theta_0] \exp[-i\theta_n] \rangle = \int \prod_m d\theta_m \exp[i(\theta_0 - \theta_n)] \cdot \exp\left[-\frac{J}{kT} \sum_{\langle nm \rangle} \cos(\theta_n - \theta_m)\right] \Big/ Z,$$

can be estimated using a high-temperature expansion. Since

(4.16) $$\int_0^{2\pi} d\theta_m = 2\pi, \quad \int_0^{2\pi} d\theta_m \exp[i\theta_m] = 0,$$

it is easy to see that the first nonzero term contributing to the high-T expansion of eq. (4.15) is of the order $(J/kT)^{|n|}$. Therefore,

(4.17) $$\langle \exp[i(\theta_0 - \theta_n)] \rangle \simeq (J/kT)^{|n|} = \exp[-|n| \ln(kT/J)],$$

and the correlation function falls off exponentially in the distance between the spins for T sufficiently high.

Next consider the correlation function at low temperature. In this case the absence of large thermal fluctuations suggests that θ_n varies slowly and smoothly throughout the system. Then the cosine in eq. (4.14) can be expanded and only the quadratic term need be accounted for,

(4.18) $$S \approx \tfrac{1}{2} J \sum_{n,i} [\Delta_i \theta(n)]^2.$$

Then the correlation function becomes

(4.19) $$\langle \exp[i(\theta_0 - \theta_n)] \rangle \simeq \int \prod_m d\theta_m \exp[i(\theta_0 - \theta_n)] \exp\left[-\frac{J}{2kT} \sum (\Delta\theta)^2\right] \Big/ Z,$$

which can be evaluated simply because it involves only Gaussian integrals. If $\Delta(n)$ is the lattice propagator for a massless field, then

(4.20) $$\langle \exp[i(\theta_0 - \theta_n)] \rangle \sim \exp[(kT/J)\Delta(n)].$$

For large $|n|$ the lattice propagator is well approximated by the continuum propagator,

(4.21) $$\Delta(n) \approx -(1/2\pi) \ln |n|, \qquad |n| \gg 1,$$

which grows slowly at large distances. Substituting into eq. (4.20), we have

$$(4.22) \qquad \langle \exp[i(\theta_0 - \theta_n)] \rangle \simeq \left(\frac{1}{|n|}\right)^{kT/2\pi J}.$$

This result teaches us several facts. First, it shows that the planar model never magnetizes. This follows because at $|n| \to \infty$ the expectation value of the product of two spins should approach the product of their expectation values,

$$(4.23) \qquad \langle \exp[i(\theta_0 - \theta_n)] \rangle \sim \langle \exp[i\theta_0] \rangle \langle \exp[-i\theta_n] \rangle.$$

But eq. (4.22) falls to zero as $|n| \to \infty$, so

$$(4.24) \qquad \langle \exp[i\theta_0] \rangle = 0,$$

identically. This shows that our approximate analysis is in accord with rigorous theorems which prove that continuous global symmetries cannot break down spontaneously in systems with nearest-neighbor coupling in two dimensions. Second, eq. (4.22) shows that the theory has a line of critical points for T sufficiently small. Recall that at the critical temperature of a system the spin-spin correlation function is expected to be power behaved,

$$(4.25a) \qquad \langle \mathbf{s}(0) \cdot \mathbf{s}(n) \rangle \sim \left(\frac{1}{|n|}\right)^{\eta},$$

where η is a standard critical index. For the planar model, treated in the spin wave approximation,

$$(4.25b) \qquad \eta = kT/2\pi J,$$

and we have a fixed line [3].

Although this analysis of the low-T region of the model is rather informal, it suggests that the planar model has two phases since eqs. (4.17) and (4.25b) are qualitatively different. What then is the nature of the system's phase transition? An intriguing answer was provided by the seminal work of Kosterlitz and Thouless [15]. They suggested that the periodicity of the variable θ_n allowed for singular spin configurations—vortices—to appear in the system at sufficiently high temperatures, and that they disorder the spin-spin correlation function. They considered the Gaussian approximation eq. (4.18) to the model, but they supplemented it with a lattice cut-off and retained the periodicity of its action. That action leads to the equation of motion

$$(4.26) \qquad \nabla^2 \theta = 0 \qquad (0 < \theta < 2\pi).$$

These authors also pointed out that solutions to eq. (4.26) could be labeled by their winding number,

(4.27) $$\oint \nabla\boldsymbol{\theta} \cdot d\boldsymbol{l} = 2\pi q, \qquad q = 0, \pm 1, \pm 2, \ldots.$$

In the spin wave analysis leading to eq. (4.18) we ignored the periodicity problem, so we effectively only dealt with the $q = 0$ sector. A spin configuration having $q = 1$ is shown in fig. 12. It is clear that such a configuration disorders the

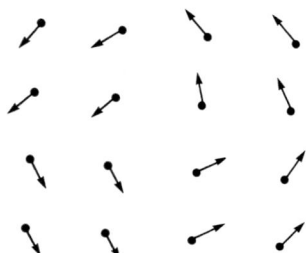

Fig. 12. – A vortex in the planar model.

system significantly, since it stirs the spins over the entire two-dimensional plane. There is a famous plausibility argument suggesting that vortices are irrelevant at low T but drive a phase transition at a moderate T_c [15]. For $T > T_c$ it also suggests that the ground state of the system is a vortex condensate. The argument begins by evaluating the action of a vortex using eq. (4.18),

(4.28) $$S \simeq \pi J \ln (R/a),$$

where R is the linear dimension of the two-dimensional world and a is the lattice spacing. Thus the vortex action diverges as $R \to \infty$. In addition, the fact that S diverges as $a \to 0$ reminds us that a vortex is a singular solution to the continuum version of Laplace's equation. One might think that since S diverges as $R \to \infty$ vortex configurations would be irrelevant to the thermodynamics of the planar model. However, to estimate their importance, consider the free energy of a vortex,

(4.29) $$F = \text{action} - \text{temperature} \times \text{entropy}.$$

We must estimate the entropy of a free vortex. The entropy is just the logarithm of the multiplicity of the configuration. Since a vortex is completely specified by the location of its origin, its multiplicity is just the number of sites of the system,

(4.30) $$\text{entropy} = k \ln (R/a)^2.$$

Therefore,

(4.31) $$F = (\pi J - 2kT) \ln (R/a).$$

The competition between the logarithms of eqs. (4.28) and (4.30) leads us to the interesting conclusion that as T is increased one reaches a point

(4.32) $$T_c = \pi J/2k$$

where F vanishes! Therefore, for all temperatures equal to and greater than T_c, we expect vortices to blend into the ground state—vortex condensation.

We can now summarize the Kosterlitz-Thouless picture of the phases of the model. At low temperature spin waves exhaust the relevant spin configurations of the theory. Spin-spin correlation functions fall off slowly with distance. Free vortices do not exist, but bound states of vortex-antivortex pairs can occur. They do not disorder the system significantly since they affect spins only over small regions. As the temperature is raised, the size of the vortex-antivortex bound states grows until T_c is reached where it diverges. Then free vortices exist and the ground state is a vortex condensate of indefinite global vorticity. The system is disordered and the spin-spin correlation function falls exponentially.

Now let us make the physical picture of the phases of the planar model quantitative. To do this we shall introduce a slightly simpler model, the periodic Gaussian planar model [18], which shares the same symmetries with the action equation (4.12). To motivate the new model, note that the periodic character of the action permits us to write a Fourier series for each bond,

(4.33) $$\exp\left[-\beta[1 - \cos(\theta_n - \theta_m)]\right] = \sum_{l=-\infty}^{\infty} \exp\left[il(\theta_n - \theta_m)\right] I_l(\beta),$$

where $I_l(\beta)$ is the Bessel function of imaginary argument. If β were large (low temperature), $I_l(\beta)$ would be well approximated by a Gaussian and eq. (4.33) would become

(4.34a) $$\exp\left[-\beta[1 - \cos(\theta_n - \theta_m)]\right] \approx$$
$$\approx (1/\sqrt{2\pi\beta}) \sum_{l=-\infty}^{\infty} \exp\left[il(\theta_n - \theta_m)\right] \exp\left[-l^2/2\beta\right].$$

Note that the right-hand side of this expression has the global Abelian symmetry and it preserves the periodicity in the θ_n variables. Therefore, although it is numerically close to the original planar model only for low temperatures, it should have the same phase diagram. It is called the periodic Gaussian model. Of course, the real reason we study it is that it can be analyzed elegantly, while the original cosine model has only been studied numerically.

We shall see that it is possible to separate the spin waves of the model from its vortices without any approximations. We shall also see that the vortex sector of the model is equivalent to the two-dimensional Coulomb gas. This point will shed light on the original spin system, since the phases of the Coulomb gas can be understood physically. To begin, write eq. (4.34a) with more elegant notation,

(4.34b) $$\exp\left[-\beta[1-\cos(\Delta_\mu\theta(r))]\right] \to \frac{1}{\sqrt{2\pi\beta}} \sum_{n_\mu(r)=-\infty}^{\infty} \exp[in_\mu\Delta_\mu\theta]\exp[-n_\mu^2/2\beta],$$

where r labels the two-dimensional lattice and $n_\mu(r)$ is a vector field. Substituting into the partition function, we have

(4.35) $$Z = \int \prod_r d\theta(r') \prod_{r,\mu} \sum_{n_\mu(r)=-\infty}^{\infty} \exp[in_\mu\Delta_\mu\theta]\exp[-n_\mu^2(r)/2\beta],$$

where we have dropped an overall multiplicative constant. The integrals over $\theta(r)$ are now trivial. Some thought shows that each integral generates a constraint,

(4.36) $$\Delta_\mu n_\mu(r) = 0.$$

Now,

(4.37) $$Z = \prod_{r,\mu} \sum_{n_\mu(r)=-\infty}^{\infty} \delta_{\Delta_\mu n_\mu(r),0} \exp[-n_\mu^2(r)/2\beta].$$

It is best to solve the constraint explicitly,

(4.38) $$n_\mu(r) = \varepsilon_{\mu\nu}\Delta_\nu n(r),$$

where $n(r)$ is a scalar, integer-valued field on the lattice. Equation (4.38) expresses the usual fact that a divergence-free vector field is a pure curl. The partition function becomes

(4.39) $$Z = \sum_{n(r)=-\infty}^{\infty} \exp\left[-(1/2\beta)\sum_{r,\mu}[\Delta_\mu n(r)]^2\right],$$

because

(4.40) $$n_\mu(r)n_\mu(r) = [\Delta_\mu n(r)]^2.$$

So, we have transformed our original model into that of a nearest-neighbor coupled integer-valued field subject to the temperature $T^* = T^{-1}$. Equation (4.39) is the « interface roughening » model of crystal growth, but it is not directly useful to us because at low temperatures many terms in the sum over $n(r)$ must be kept in evaluating eq. (4.39). Clearly we would like to replace the integer-valued field $n(r)$ by an ordinary scalar field $\varphi(r)$ and still preserve the periodic character of the original model. The necessary manipulation is provided by the Poisson summation formula,

(4.41) $$\sum_{n=-\infty}^{\infty} g(n) = \sum_{m=-\infty}^{\infty} \int_{-\infty}^{\infty} d\varphi\, g(\varphi) \exp[2\pi i m\varphi],$$

where g is an arbitrary function. The sum over the integer variable m ensures that the periodicity of the original action is preserved. Applying this identity to eq. (4.39) gives

$$(4.42) \quad Z = \int_{-\infty}^{\infty} \prod_r \mathrm{d}\varphi(r) \sum_{m(r)=-\infty}^{\infty} \exp\left[-(1/2\beta)\sum_{r,\mu}(\Delta_\mu \varphi)^2 + 2\pi i \sum_r m(r)\varphi(r)\right].$$

As we analyze this expression further, we shall be able to identify $\varphi(r)$ with the spin waves of the original model and $m(r)$ with its vortices! Equation (4.41) is pleasantly simple: the vortices act as sources for the spin waves, which are ordinary massless fields. The identification of the $m(r)$ with vortex variables is made clearer if we integrate out the spin waves in eq. (4.42). This integration is a standard Gaussian, so we obtain

$$(4.43) \quad Z = Z_{\mathrm{sw}} \sum_{m(r)=-\infty}^{\infty} \exp\left[-2\pi^2\beta \sum_{r,r'} m(r) G(r-r') m(r')\right],$$

where Z_{sw} is the spin wave contribution produced by the Gaussian integrations and $G(r-r')$ is the lattice propagator for a massless field. Since spin waves alone do not drive a phase transition, we shall not discuss Z_{sw}. The lattice propagator $G(r)$ satisfies

$$(4.44) \quad \Delta^2 G(r) = \delta_{r,0},$$

where Δ^2 is a discrete form of the second derivative. For example, we take

$$(4.45a) \quad \Delta^2 = \Delta_x^2 + \Delta_y^2,$$

where

$$(4.45b) \quad \Delta_x^2 f(r) = f(r+x) + f(r-x) - 2f(r).$$

Then one can easily verify that

$$(4.46) \quad G(r) = \int_{-\pi}^{\pi} \frac{\mathrm{d}k_x}{2\pi} \int_{-\pi}^{\pi} \frac{\mathrm{d}k_y}{2\pi} \frac{\exp[ikr]}{4 - 2\cos k_x - 2\cos k_y}.$$

Let us examine some of the properties of $G(r)$. For large r it is well approximated by the continuum massless scalar propagator of two dimensions,

$$(4.47a) \quad G(r) \approx -(1/2\pi)\ln(r/a) - \tfrac{1}{4} \qquad (|r| \gg 1).$$

Therefore, the variables $m(r)$ interact among themselves through a logarithmic potential. One can easily check that vortex spin configurations as sketched in fig. 9 experience the same force law: vortices of opposite (same) vorticity experience an attractive (repulsive) logarithmic potential. Next consider the self-mass of the field $m(r)$. This comes from the $r = r'$ piece of the sum in eq. (4.43). It is easy to see from eq. (4.46) that $G(0)$ is infra-red divergent,

$$(4.47b) \qquad G(0) \approx (1/2\pi) \ln(R/a),$$

where r is the linear dimension of the two-dimensional plane. We recognize this as the action of an isolated vortex. Of course, this infra-red problem means that we should treat eq. (4.46) quite carefully. It is best to decompose $G(r)$ into two pieces, one of which is infra-red finite and the other is not,

$$(4.48a) \qquad G(r) = G'(r) + G(0),$$

where

$$(4.48b) \qquad \begin{cases} G'(r) = \int_{-\pi}^{\pi} \frac{dk_x}{2\pi} \int_{-\pi}^{\pi} \frac{dk_y}{2\pi} \frac{\exp[ikr] - 1}{4 - 2\cos k_x - 2\cos k_y}, \\ G(0) = \int_{-\pi}^{\pi} \frac{dk_x}{2\pi} \int_{-\pi}^{\pi} \frac{dk_y}{2\pi} \frac{1}{4 - 2\cos k_x - 2\cos k_y}. \end{cases}$$

Substituting into the partition function, we have

$$(4.49) \qquad Z = Z_{\text{sw}} \sum_{m(r)} \exp\left[-2\pi^2 \beta G(0) \left[\sum_r m(r)\right]^2\right] \cdot$$
$$\cdot \exp\left[-2\pi^2 \beta \sum_{r,r'} m(r) G'(r - r') m(r')\right].$$

The exponential singles out those configurations which are « neutral »,

$$(4.50) \qquad \sum_r m(r) = 0.$$

Other configurations do not contribute to Z. Therefore, we can write

$$(4.51) \qquad Z = Z_{\text{sw}} \sum_{m(r)}' \exp\left[-2\pi^2 \beta \sum_{r,r'} m(r) G'(r - r') m(r')\right],$$

where the prime on the sum means « neutral configurations of vortices only ». It is enlightening to replace G' by the explicit formula eq. (4.47a). That asymptotic form is quite good even for small $|r| \approx 1$. Then eq. (4.51) becomes

$$(4.52) \qquad Z = Z_{\text{sw}} \sum_{m(r)}' \exp\left[(\pi^2 \beta/2) \sum_{r,r'}' m(r) m(r') + \pi \beta \sum_{r,r'}' m(r) \ln(|r - r'|/a) m(r')\right],$$

where the primes on the sums over sites mean that the $r = r'$ term is omitted. The first term in the exponential can be simplified, since the neutrality condition implies

(4.53) $$0 = \sum_{r,r'} m(r)m(r') = \sum_{r,r'}' m(r)m(r') + \sum_r m^2(r).$$

Now,

(4.54) $$Z = Z_{sw} \sum_{m(r)}' \exp\left[-(\pi^2\beta/2) \sum_r m^2(r) + \pi\beta \sum_{r,r'}' m(r) \ln\left(|r-r'|/a\right) m(r)\right].$$

The first term in the square brackets gives the chemical potential of each vortex and the second gives the logarithmic interactions between different ones.

This Coulomb gas representation suggests the character of the phase diagram discussed in the previous section. If β is very large, then the chemical-potential term in eq. (4.53) suppresses the vortices very effectively, leaving only the spin waves behind. In fact, the strong logarithmic potential between vortices suggests that any vortex-antivortex pairs which might populate the ground state are tightly bound together. However, as the temperature is raised, vortices are not suppressed significantly. Once $\beta = J/kT$ becomes of order unity, one would naively expect vortices to become important configurations of the system. Since the interactions between them are long range, they cannot be thought of simply as free excitations. In fact, they form a plasma, and a screening length is generated dynamically. All this can be made quantitative by a Debye-Hückel picture of the two-dimensional Coulomb gas. A renormalization group can be found to put this physical picture of the phase transition on a firmer footing [19]. A discussion of this would take us too far from our main theme, however. Nonetheless, this example shows very clearly that topological excitations can be significant in a lattice theory's mathematics as well as dynamics. In fact, the phase transition in the planar model is particularly clear when thought of as a molecule-plasma transition.

These model considerations pose the challenge of first identifying the field configurations in gauge theories responsible for quark confinement and chiral-symmetry breaking and then developing a formalism which singles these excitations out in dynamical calculations. Both steps are difficult and have met only partial success to this date.

5. – Hamiltonian lattice gauge theory, flux tube dynamics and continuum string models.

In addition to the partition function approaches to field theory discussed to this point, one might also investigate the space of states and the spectrum of lattice gauge theory. To do this systematically we should consider the transfer matrix and the Hamiltonian, time continuum limit, of the theory [1].

The partition function of lattice gauge theories gave a lattice regularization of

(5.1a) $$L = -\tfrac{1}{4}\int F^\alpha_{\mu\nu} F^\alpha_{\mu\nu}\, \mathrm{d}^3x\,.$$

The formal replacement was

(5.1b) $$F^\alpha_{\mu\nu} F^\alpha_{\mu\nu} \to \frac{1}{4g^2 a^4}\,\mathrm{tr}\, P \exp\left[ig \oint_c A^\alpha_\mu \frac{\tau^\alpha}{2}\, \mathrm{d}x^\mu\right] \to \frac{1}{4g^2 a^4}\,\mathrm{tr}\, UUUU\,.$$

We chose simple plaquettes for the contours c in eq. (5.1b).

To make a Hamiltonian theory, distinguish between plaquettes with a t link and those without. Those without are spatial plaquettes and are left alone in this argument. The temporal links are special—we want to organize things so that the partition function can be written in the form of sect. **1**,

(5.2) $$Z = \int \langle\{U_f\}|\hat{T}|\{U_{N-1}\}\rangle \left[\prod \mathrm{d}U_{N-1}\right] \langle\{U_{N-1}\}|\hat{T}|\{U_{N-2}\}\rangle \ldots \langle\{U_1\}|\hat{T}|\{U_i\}\rangle\,,$$

where $\hat{T}$ is the theory's transfer matrix. Clearly $\hat{T}$ evolves the system one link in the «time» direction (fig. 13). $\hat{T}$ will be simple if we choose temporal links to be trivial, $\exp[iB_0] = 1$. This is called the «temporal gauge» $A_0 = 0$,

Fig. 13. – Transfer matrix decomposition of the partition function.

and is familiar from continuum field theory (fig. 14). Recall that in this case gauge invariance is imposed as a constraint on the Hilbert space of states, i.e. Gauss' law is implemented

(5.3) $$G_\alpha(n)|\mathrm{phys}\rangle = 0\,,$$

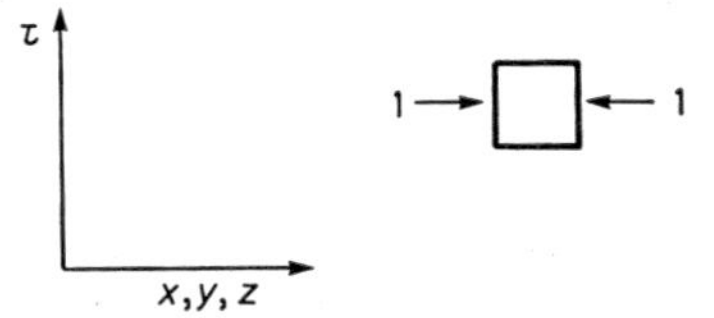

Fig. 14. – Temporal plaquette in the axial gauge $A_0 = 0$.

where $G_\alpha(n)$ = generator of local rotations in color space at the site $n = (x, y, z)$. An explicit construction of G_α will be obtained below.

For temporal plaquettes, eq. (5.2) simplifies,

$$(5.4) \quad \text{tr } U^\dagger(t_{i+1}) U(t_i) - \text{h.c.} = - \text{tr } [U^\dagger(t_{i+1}) - U^\dagger(t_i)][U(t_{i+1}) - U(t_i)] + \text{const},$$

and we can construct the « velocity » term of the Lagrangian. The temporal loops contribute to the Lagrangian density,

$$(5.5) \quad -\frac{1}{4g^2 a^2} \text{tr}\left(\frac{U^\dagger(t_{i+1}) - U^\dagger(t_i)}{a_t}\right)\left(\frac{U(t_{i+1}) - U(t_i)}{a_t}\right),$$

where a_t = temporal lattice spacing. Taking $a_t \to 0$, we obtain a conventional quantum-mechanical picture of the lattice gauge theory,

$$(5.6) \quad L = \sum_{\text{links}} \frac{a}{4g^2} \text{tr } \dot{U}^\dagger \dot{U} + \sum_{\text{sp.plaq.}} \frac{1}{4ag^2} \text{tr } UUUU + \text{h.c.},$$

where the replacement $\int d^3x \to \sum a^3$ has been made. Now we can form the Hamiltonian, passing from « velocity » to « momentum » variables by canonical procedures,

$$(5.7) \quad H = \sum \left(\dot{U}^\dagger_{ij} \frac{\partial L}{\partial \dot{U}^\dagger_{ij}} + \dot{U}_{ij} \frac{\partial L}{\partial \dot{U}_{ij}}\right) - L =$$

$$= \sum_{\text{links}} \frac{a}{4g^2} \text{tr } \dot{U}^\dagger \dot{U} - \sum_{\text{boxes}} \frac{1}{4ag^2} (\text{tr } UUUU + \text{h.c.}),$$

which describes a system of coupled « tops ». To quantize this system we must identify independent degrees of freedom carefully. One might use Euler angles, but this is awkward. Instead we shall eliminate $\dot{U}$ in favor of the generators of local gauge rotations [20]. When an infinitesimal gauge rotation is made at site n,

$$(5.8) \quad U_i(n) \to \left(1 + i\varepsilon \frac{\tau^\alpha}{2}\right) U_i(n).$$

Call the quantum generators of these transformations $E_i^\alpha(n)$. Then

$$(5.9) \quad [E_i^\alpha(n), U_j(m)] = \tfrac{1}{2} \tau^\alpha U_j(m) \delta_{ij} \delta_{nm},$$

in a slightly abbreviated notation. Since the E^α's generate SU_2 rotations,

$$(5.10) \quad [E_i^\alpha(n), E_j^\beta(m)] = i\varepsilon^{\alpha\beta\gamma} E_i^\gamma(n) \delta_{ij} \delta_{nm}.$$

Now we want to eliminate $\dot{U}^\dagger$ and $\dot{U}$ in favor of the E^α variables. Since E^α generates the local gauge rotation which is a symmetry of L, we can obtain it from the Lagrangian using Noether's theorem,

$$(5.11) \quad E^\alpha = \frac{\partial L}{\partial \dot{U}_{ij}}\left(i\frac{\tau^\alpha}{2}U\right)_{ij} + \frac{\partial L}{\partial \dot{U}^\dagger_{ij}}\left(-i\frac{\tau^\alpha}{2}U\right)^\dagger_{ij} = i\frac{a}{4g^2}\left\{\text{tr } \dot{U}^\dagger \frac{1}{2}\tau^\alpha U - \text{h.c.}\right\}.$$

Next consider $E^\alpha E^\alpha$. To compute this, the quadratic Casimir operator, we need two identities,

$$(5.12a) \quad U^\dagger U = 1, \quad \dot{U}^\dagger U + U^\dagger \dot{U} = 0$$

and

$$(5.12b) \quad \tau^\alpha_{ij}\tau^\alpha_{kl} = 2\delta_{il}\delta_{jk} - l\delta_{ij}\delta_{kl},$$

and we compute

$$(5.13) \quad E^\alpha E^\alpha = \frac{a^2}{2g^4}\text{ tr } \dot{U}^\dagger \dot{U}.$$

Collecting everything [21],

$$(5.14) \quad H = \sum_{\text{links}} \frac{g^2}{2a} E^\alpha E^\alpha - \frac{1}{4ag^2}\sum_{\text{boxes}} (\text{tr } UUUU + \text{h.c.}).$$

The basic commutation relations

$$(5.15) \quad \begin{cases} [E^\alpha_i(n), U_j(m)] = \tfrac{1}{2}\tau^\alpha U_j(m)\delta_{ij}\delta_{nm}, \\ [E^\alpha_i(n), E^\beta_j(m)] = i\varepsilon^{\alpha\beta\gamma} E^\gamma_i(n)\delta_{ij}\delta_{nm} \end{cases}$$

specify the model with the definition of physical states

$$(5.16) \quad \sum_{j=-3}^{3} E^\alpha_j(n)|\text{phys}\rangle = 0.$$

Note that in the partition function language time-independent gauge transformations are

$$(5.17) \quad \prod_{n_0=-\infty}^{\infty} G(n_0, n).$$

Since $U = 1$ on temporal links and

$$(5.18) \quad U = 1 \to \exp\left[i\frac{\tau}{2}\chi(n)\right] U \exp\left[-i\frac{\tau}{2}\chi(n)\right] = 1,$$

the axial gauge is preserved appropriately for arbitrary time-independent gauge functions $\chi(n)$. Clearly also

$$(5.19) \quad [E^\alpha_i(n), H] = 0$$

and the H is gauge invariant.

Now we are ready for applications. We want to understand the distribution of electric flux responsible for the linear confining potential. Consider a $Q\overline{Q}$ state. Let the Q and $\overline{Q}$ be heavy and let them be created by the operators $\psi^\dagger(n)$ and $\psi(n)$—they are sources and sinks of color flux of one unit. The generator of a color rotation is now generalized to

$$(5.20) \qquad \sum E_i^\alpha(n) \to \sum E_i^\alpha(n) + \psi^\dagger(n) \tfrac{1}{2}\tau^\alpha \psi(n)$$

at the site n. We need a gauge-invariant operator to describe the heavy $Q\overline{Q}$ state. The operator

$$(5.21) \qquad \psi^\dagger(n)\left[\prod_{\text{path}} U\right]\psi(n+R),$$

where the «path» extends from n to $n + R$ but is otherwise arbitrary, satisfies this requirement. The local contraction of color indices guarantees the local gauge invariance of this operator. Now consider the theory at strong coupling $g^2 \gg 1$. Then the leading term in H is

$$(5.22) \qquad H_0 = \frac{g^2}{2a} \sum_i E_i^2.$$

The vacuum has each link in a color singlet state,

$$(5.23) \qquad E_i^\alpha|0\rangle = 0.$$

So, the $Q\overline{Q}$ state of minimal energy will have $\prod_{\text{path}}$ over the fewest number of links since each link with a U-matrix costs energy,

$$(5.24) \qquad E^2 U|0\rangle = \tfrac{1}{2}\tau^\alpha \tfrac{1}{2}\tau^\alpha U|0\rangle = \tfrac{3}{4} U|0\rangle,$$

as follows from the commutation rule for SU_2,

$$(5.25) \qquad [E^\alpha, U] = \tfrac{1}{2}\tau^\alpha U.$$

So, the energy of the $Q\overline{Q}$ state is

$$(5.26) \qquad V(R) = \frac{g^2}{2a} \frac{3}{4} \frac{R}{a} = \sigma R,$$

which gives us the strong-coupling limit of the string tension in this theory,

$$(5.27) \qquad \sigma = \frac{3}{8} g^2/a^2.$$

Note that requiring σ to be independent of a implies that $g \sim a$—stronger coupling on a coarser lattice. This is the same qualitative behavior as found at weak coupling where asymptotic freedom implied stronger coupling on coarser lattices. This strong-coupling calculation of σ can be carried to high orders in the natural expansion parameters, $1/g^2$, of the Hamiltonian theory [22]. These results complement the Monte Carlo simulations and they also suggest that the continuum limit of the pure gauge theory confines quarks with a finite, fixed, physical string tension. But what about the spatial distribution of the flux responsible for confinement? At strong coupling the flux exists in a thin tube between the two on-axis quarks. Is this a possible state of flux in the continuum limit? To gain some intuition into this question we turn to a simple string model.

Consider a structureless thin string with its ends pinned down at $x = -R/2$ and $x = R/2$. It can be described by a two-component vector field

$$\boldsymbol{\xi}(t, z), \quad -\tfrac{1}{2}R \leqslant z \leqslant \tfrac{1}{2}R, \quad \boldsymbol{\xi}(t, -\tfrac{1}{2}R) = \boldsymbol{\xi}(t, \tfrac{1}{2}R) = 0$$

as in fig. 15. We want an effective action describing the low-frequency modes of this string [23]. Let us make the following assumptions about it:

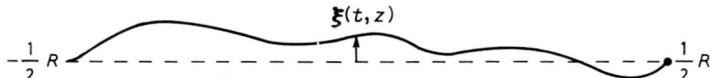

Fig. 15. – A string labeled by the vector field $\boldsymbol{\xi}(t, z)$.

1) S_{eff} must be local, $S_{\text{eff}} = \int dz\, dt\, L$, where L depends on $\boldsymbol{\xi}$ and its derivatives.

2) L should be invariant to the following space-time symmetries:
   a) Poincaré transformations in the (t, z)-plane,
   b) O_2 rotations and translations of ξ.

Now let us write the possible terms in L. Property 2b) precludes mass terms $\sim \xi^2$ for the string. So, L must be made up of derivatives $\partial_\mu \boldsymbol{\xi}$, $\partial_\mu \partial_\nu \boldsymbol{\xi}$, ... ($\mu, \nu = 0, 1$). So,

(5.28) $$S_{\text{eff}} = \int_{-\tfrac{1}{2}R \leqslant z \leqslant \tfrac{1}{2}R} dz\, dt\, \{\partial_\mu \boldsymbol{\xi} \cdot \partial^\mu \boldsymbol{\xi} + b \partial_\mu \partial^\mu \boldsymbol{\xi} \cdot \partial_\nu \partial^\nu \boldsymbol{\xi} + c(\partial_\mu \boldsymbol{\xi} \cdot \partial^\mu \boldsymbol{\xi})^2 + ...\}.$$

The only relevant or marginal operator is $\partial_\mu \boldsymbol{\xi} \cdot \partial^\mu \boldsymbol{\xi}$—the other operators are irrelevant, i.e. the parameters b and c carry dimensions of the underlying spatial cut-off to various powers. Order by order in perturbation theory these irrelevant operators do not contribute to the long-wavelength physics of the string. One can check that when these operators are inserted into Feynman diagrams the infra-red behavior of the graph improves. So, for studying wavelengths $x \gg a$, the intrinsic width of the string, it suffices to only take

$$S_{\text{eff}} = \int dz\, dt\, \partial_\mu \boldsymbol{\xi} \cdot \partial^\mu \boldsymbol{\xi}\,. \tag{5.29}$$

Now, we can ask whether the state with $\langle \boldsymbol{\xi}(x,t) \rangle = 0$ is stable to quantum fluctuations. Calculate the variance in $\boldsymbol{\xi}(z,t)$ for any z between $-R/2$ and $R/2$,

$$\langle \boldsymbol{\xi}^2(z,t) \rangle \sim \int \frac{d^2 k}{k^2} \sim \ln(R/a)\,. \tag{5.30}$$

So, $\langle \boldsymbol{\xi}^2(x,t) \rangle \to \infty$ as $R \to \infty$ and the string is delocalized by fluctuations as shown in fig. 16! This explicit calculation is an example of the more general

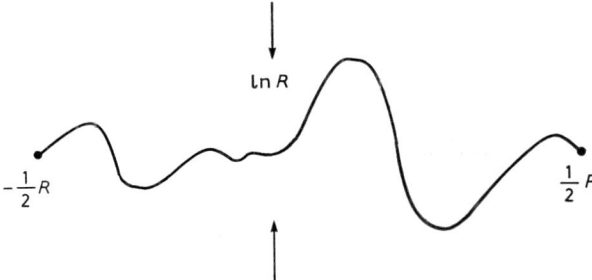

Fig. 16. – A string with a mean distance off axis of $\ln R$.

theorem which states that spontaneous breakdown of a continuous global symmetry in a theory with local couplings in two dimensions is not possible.

A second physical effect of these massless modes is the existence of a universal $1/R$ potential in the string channel of the theory [23],

$$V(R) = \sigma|R| - \frac{(d-2)\pi/24}{R} + \ldots\,. \tag{5.31}$$

The $1/R$ correction term is a one-dimensional Casimir effect which is easily calculated. For thin, structureless strings it is truly universal—it only depends on the space-time dimension d in which the string lives! We begin with a free, massless field in a two-dimensional box of length R,

$$S = \tfrac{1}{2} \int_{0 \leqslant z \leqslant R} dz\, dt\, \partial_\mu \boldsymbol{\xi} \cdot \partial^\mu \boldsymbol{\xi}\,. \tag{5.32}$$

Think of it as a box of massless bosons, one flavor for each transverse direction. Each transverse mode contributes $\tfrac{1}{2}\hbar\omega$ to the ground-state energy,

$$\Delta V(R) = \tfrac{1}{2} \sum_n \varepsilon_n, \qquad \varepsilon_n = \pi n/R,\ n = 0, 1, 2, \ldots, \tag{5.33}$$

when we have counted the massless normal modes for a field with fixed boundary conditions $\xi(0, t) = \xi(R, t) = 0$. The sum over normal modes diverges—the fluctuations shift the ground-state energy density. To separate off this term we first do the sum with a convergence factor $\exp[-\alpha n]$ and let $\alpha \to 0$ at the end. (Actually for a string of thickness a the largest physically sensible value for n is $\sim R/a$ so that $\varepsilon \leqslant \pi/a$. Therefore, the minimal value for α is $\alpha_{\min} \sim a/R$.) Now

$$(5.34) \quad \Delta V(\alpha, R) = \frac{\pi}{2R} \sum_n n \exp[-\alpha n] = -\frac{\pi}{2R} \frac{d}{d\alpha} \sum_n \exp[-\alpha n] =$$

$$= -\frac{\pi}{2R} \frac{d}{d\alpha} \left(\frac{1}{\exp[\alpha] - 1} \right) = -\frac{\pi}{2R} \sum_n B_n \frac{(n-1)\alpha^{n-2}}{n!},$$

$$B_n = \text{Bernoulli numbers}.$$

The $n = 2$ term gives the only finite, nonzero term which survives in the $\alpha \to 0$ limit,

$$(5.35) \quad \Delta V(R) = -\frac{\pi}{2R} \frac{B_2}{2} = -\frac{\pi/24}{R}$$

for each transverse degree of freedom. For $d-2$ transverse directions we have the advertised result

$$(5.36) \quad \Delta V(R) = -\frac{(d-2)\pi/24}{R}.$$

Note that the $n = 0$ term simply renormalizes the string tension additively.

How do these features—the delocalization of the string and the universal $1/R$ potential—occur in the lattice regulated theory? There is only cubic symmetry on the spatial lattice, so straight strings are possible at strong coupling. As $g \to 0$, however, the magnetic effects in the Hamiltonian become important and at some critical coupling, the roughening point, g_R, the string will delocalize [24]. For the relativistic string the lowest-energy transverse excitation costs zero energy in the $R \to \infty$ limit. On the lattice the lowest-energy transverse excitation is shown in fig. 17 and it has an energy

$$\frac{g^2}{2a} \frac{3}{4} (N+1)$$

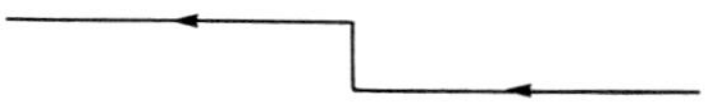

Fig. 17. – Transverse string excitation on the lattice.

in the strong-coupling limit. This is an energy of $\frac{3}{4} g^2/2a$ above that of a straight string of N links. Call the state a «kink» and

$$m_k = \frac{g^2}{2a}\left\{\frac{3}{4} + ...\right\},$$

where the $+ ...$ means that the mass can be calculated in perturbation theory. The kink should be thought of as a particle—its energy is localized and it can carry momentum. The transverse distribution of electric flux is bounded because the transverse wandering of the string is inhibited by a finite-energy barrier, m_k. As g decreases from strong coupling, however, m_k decreases and eventually vanishes at $g = g_R$. Then there is no energy barrier inhibiting the transverse wandering of the string. For all $g \leqslant g_R$ the lattice string is delocalized and resembles more closely the continuum string. Roughening can, therefore, also be seen in a calculation of the mean transverse width of the string,

(5.37) $$\langle r_\perp^2 \rangle = \langle r_\perp^2 E_\parallel^2 \rangle / \langle E_\parallel^2 \rangle,$$

where $E_\parallel^2$ is the electric flux squared on a link parallel to the straight string but displaced in the transverse direction by $r_\perp$ links. One can develop strong-coupling series for

(5.38) $$\langle r_\perp^2 \rangle \sigma,$$

which measures the transverse width of the string in units of the physical length $1/\sqrt{\sigma}$.

One can discuss the roughening transition from the perspective of symmetry restoration as well. For $g \geqslant g_R$ continuous translations perpendicular to the string become good symmetries of the system. This is a corollary to the fact that the transverse excitations of the interface become massless as g passes g_R from below—the long-wavelength surface vibrations have a relativistic massless energy-momentum relation for $g \leqslant g_R$. This fact leads us to consider the restoration of full rotational symmetry in the string sector of the theory. Consider $2 + 1$ dimension SU_3 gauge theory. Let the $Q\bar{Q}$ be off axis as shown in fig. 18. At strong coupling the flux travels by a route of minimal distance.

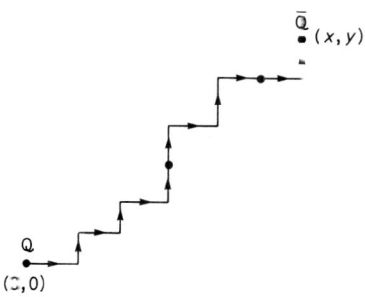

Fig. 18. – A flux tube between off-axis static quarks.

The leading-order heavy-quark potential is

$$(5.39) \qquad V(x, y) = \frac{g^2}{2a} \frac{4}{3} (|x| + |y|),$$

which has equipotentials shown in fig. 19. The equipotentials are not analytic near the axes. It is easy to see that this is because $m_k \neq 0$—the sharp edges

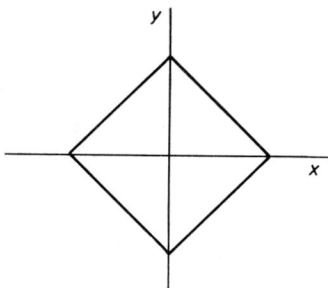

Fig. 19. – An equipotential at strong coupling.

in the equipotentials will disappear at g_R. In general there is bad breaking of rotational symmetry. But rotational symmetry should be restored as $g \to 0$, in the scaling region of the lattice theory. Let us improve the large $g^2 \gg 1$ calculation of V and see this happen in perturbation theory.

We face a problem in degenerate perturbation theory—there are many, $(|x|+|y|)!/|x|!|y|!$, paths of equal length which contribute to the leading-order calculation of V. To calculate the first-order correction to V we must diagonalize the perturbation in this subspace,

$$(5.40) \qquad H = H_0 - x_s H',$$

where

$$(5.41) \qquad H_0 = \frac{g^2}{2a} \sum_l E_l^2, \qquad H' = \sum_{\text{boxes}} (\operatorname{tr} UUUU + \text{h.c.}), \qquad x_s = \frac{2}{g^4}.$$

What is the effect of H' on paths of flux with corners as shown in fig. 18? On the two links in fig. 20 where two U-matrices act one must decompose the product into irreducible representations. Since $3 \times \bar{3} = 1 \times 8$, there is a singlet piece and the flux path shown in fig. 20 results with a weight of $\frac{1}{3}$. These are

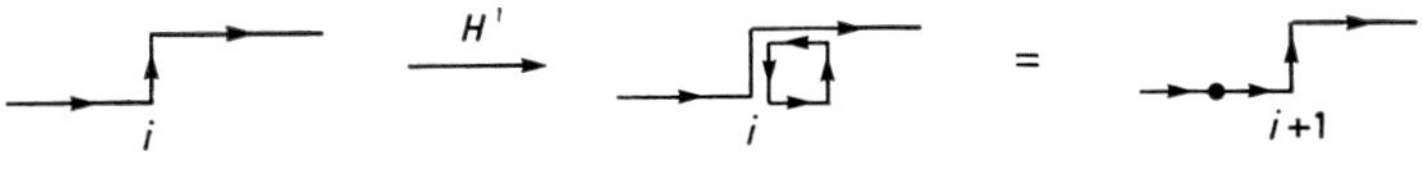

Fig. 20. – The perturbation H' acting on a kink in the string. The kink moves by one link.

the processes of importance—they mix different members of the degenerate subspace. Recognize all this as a familiar problem—particles (fermions) hopping along a chain [25]. To make this observation quantitative, label a path as shown in fig. 21 and make the definitions

y link with flux → fermion at site i,

x link with flux → absence of fermion at site i.

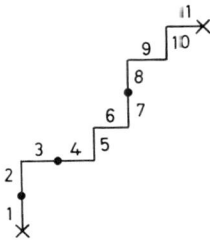

Fig. 21. – An off-axis string with links labeled successively.

So, we imagine a box of length $L = |x| + |y|$ containing $|y|$ fermions. It is clear from the perturbation theory exercises done above that the perturbation allows a « fermion » to hop one site to a nearest neighbor if that site is initially unoccupied. Therefore, the restriction of H' to the degenerate subspace is

(5.42) $$H' = \tfrac{1}{3} \sum_i a_i^\dagger a_{i+1} + \text{h.c.}, \qquad |y| = \sum a_i^\dagger a_i.$$

This standard 1-dimensional hopping Hamiltonian is diagonalized by passing to momentum space,

(5.43) $$\begin{cases} a_i = \sum_n a_n \varphi_n(i), \\ \varphi_n(i) = \dfrac{1}{\sqrt{2(L+1)}} \sin \dfrac{\pi n i}{L+1}, \end{cases} \qquad n = 1, 2, \ldots.$$

Then,

(5.44) $$H' = \frac{2}{3} \sum_n \cos\left(\frac{\pi n}{L+1}\right) a_n^\dagger a_n.$$

For our application, the first $|y|$ fermion levels are filled, so the heavy-quark potential is

(5.45) $$V(x,y) \simeq \frac{g^2}{2a}\left(\frac{4}{3}(|x|+|y|) - x_s \frac{2}{3} \sum_{n=1}^{|y|} \cos \frac{\pi n}{L+1}\right) \simeq$$
$$\simeq \frac{g^2}{2a}\left\{\frac{4}{3}(|x|-|y|) - \frac{4}{3g} \cdot \right.$$
$$\left. \cdot \left(\frac{\sin[\pi(|y|+1)/2(|x|+|y|+1)]\sin[\pi(|x|+1)/2(|x|+|y|+1)]}{\sin[\pi/2(|x|+|y|+1)]} - 1\right)\right\}.$$

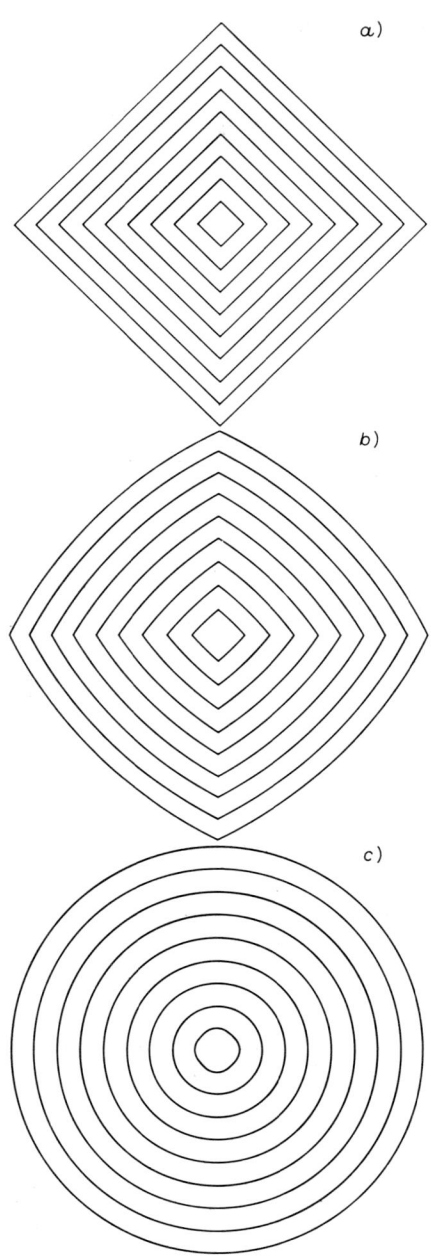

Fig. 22. – The equipotentials of the heavy-quark potential at a) $g = \infty$, b) $g = 1.19$ and c) $g = 1.00$.

It is interesting to plot the equipotentials of V. This is shown in fig. 22. The restoration of rotational symmetry is clear. It occurs at a coupling close to the value where the string tension begins to scale according to asymptotic freedom. This is as expected and is very satisfactory. Higher-order corrections allow one to discuss smaller values of g than can sensibly be handled by our first-order formula. Computer simulations also show restoration of rotational symmetry.

A few final comments. One can expand the formula eq. (5.45) for V in powers of $1/L$ and find power law corrections. The strength of the $1/L$ term is close to that predicted by the thin-string, continuum model discussed. And the off-axis matrix element, therefore, has all the qualitative properties of the continuum string! For all $g \neq \infty$, it is rough. Therefore, the tension of the off-axis string should be an analytic function of g^2, $0 < g^2 < \infty$. This point can be proved in models of fluctuating thin strings [26]. This makes it a more desirable matrix element—but more difficult to work with—than the on-axis string.

6. – Lattice fermions.

Our next task is to understand the Dirac equation, or, more generally, fermions on the lattice. This is a surprisingly tricky subject, as we shall see. It was easy and natural to place bosons, both spin 0 and 1, on the lattice and have sensible continuum limits. However, fermions bring in vital new ingredients: the free Dirac equation is first order in spatial and temporal derivatives and the equation has continuous symmetries such as chirality, $\psi \to \exp[\gamma_5 \theta] \psi$. In addition, if there is a flavor (isospin, say) multiplet, the free equation is also invariant under axial flavor rotations as well. We shall see below that we are not able to formulate conventional lattice theories with all these symmetries, in addition to gauge invariance, when the lattice spacing is different from zero.

To begin, consider $1+1$ dimensions and a free boson field $\varphi(x, t)$. The Klein-Gordon equation is

(6.1) $$\ddot{\varphi} = \nabla^2 \varphi - m^2 \varphi,$$

which implies the energy-momentum relation

(6.2) $$E^2 = \boldsymbol{p}^2 + m^2$$

for plane waves. To place eq. (6.1) on a spatial lattice (time continuum), make the replacement

(6.3) $$a^2 \nabla^2 \varphi \to \varphi(n+1) + \varphi(n-1) - 2\varphi(n) = (d^+ + d^- - 2)\varphi(n),$$

where $d^\pm$ are shift operators, $d^\pm \varphi(n) = \varphi(n \pm 1)$. The lattice Klein-Gordon equation is now

(6.4) $$a^2 \ddot{\varphi} = (d^+ + d^- - 2)\varphi - a^2 m^2 \varphi.$$

Consider a plane-wave solution to eq. (6.4),

(6.5) $$\varphi = \exp[ikna + iEt]$$

on the lattice shown in fig. 23. Construct the Brillouin zone symmetrically about $k = 0$,

(6.6) $$-\pi < ka < \pi.$$

Fig. 23. – A spatial lattice in $1 + 1$ dimensions.

Substituting eq. (6.5) into eq. (6.4) gives

(6.7) $$-E^2 \varphi = \frac{\exp[ika] + \exp[-ika] - 2}{a^2} \varphi - m^2 \varphi.$$

Therefore, the lattice model has an energy-momentum relation

(6.8) $$E^2 = m^2 - 2(\cos ka - 1)/a^2.$$

For small $ka \ll 1$,

(6.9) $$E^2 = m^2 + k^2 + O(k^4 a^2)$$

and the proper energy-momentum relation results in the continuum limit. For finite a, the energy-momentum relation leads to the curve in fig. 24 and

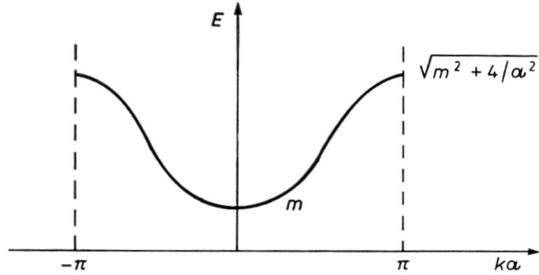

Fig. 24. – Energy-momentum relation for a free boson on the lattice.

its negative. It is clear from the figure that, as $a \to 0$, only the vicinity of $ka \simeq 0$ is relevant and the ordinary energy-momentum relation is obtained and relativity is restored. All is well.

Now consider the Dirac equation ($m_0 = 0$) with the conventions

(6.10) $$\psi = \begin{pmatrix} \psi_1 \\ \psi_2 \end{pmatrix}, \quad \gamma_0 = \begin{pmatrix} 1 & 0 \\ 0 & -1 \end{pmatrix}, \quad \alpha = \gamma_5 = \begin{pmatrix} 0 & 1 \\ 1 & 0 \end{pmatrix}.$$

In the continuum,

(6.11) $$i\dot\psi = -i\alpha\partial_z\psi = -i\gamma_5\partial_z\psi.$$

It is convenient to consider the chiral eigenstates

(6.12) $$\gamma_5\chi_\pm = \pm\chi_\pm.$$

For plane waves we choose

(6.13) $$\psi_\pm = \exp[-ikz + iEt]\chi_\pm,$$

so

(6.14) $$E = \pm k, \qquad -\infty < k < \infty,$$

as shown in fig. 25. We have left- and right-moving fermions and antifermions.

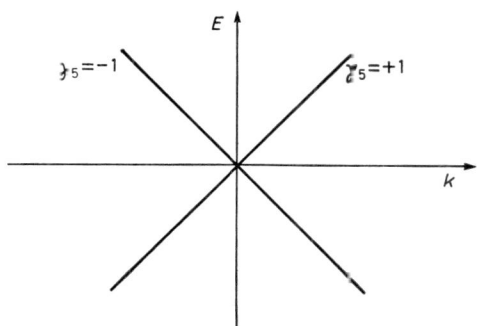

Fig. 25. – Energy-momentum relaation of the continuum Dirac equation in $1+1$ dimensions.

Next consider the simplest lattice form of the Dirac equation. Place

$$\psi = \begin{pmatrix} \psi_1 \\ \psi_2 \end{pmatrix}$$

on each site of a spatial lattice and replace ∂_z by a discrete difference,

(6.15) $$i\dot\psi(n) = \frac{-i}{2a}\gamma_5[\psi(n+1) - \psi(n-1)].$$

Again let

(6.16) $$\psi_\pm = \exp[i(-kna + Et)]\chi_\pm, \quad \gamma_5\chi_\pm = \pm\chi_\pm,$$

so eq. (6.15) implies

(6.17) $$-E = \pm\frac{i}{2a}(\exp[ika] - \exp[-ika]),$$

which gives

(6.18) $$E = \pm\sin(ka)/a.$$

For small $ka \ll 1$, eq. (6.18) becomes

(6.19) $$E \simeq \pm k + O(k^3 a^2).$$

However, for $ka = \pi - k'a$, $k'a \ll 1$, we also have low-energy excitations,

(6.20) $$E \simeq \mp k' + O(k'^3 a^2).$$

The full energy-momentum dispersion relation is shown in fig. 26. Identify

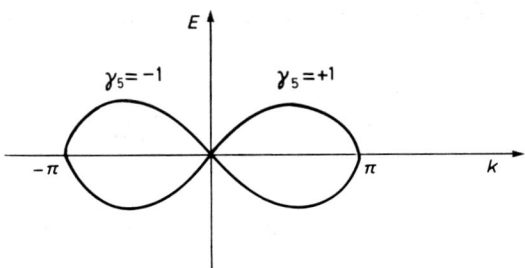

Fig. 26. – Spectrum of the naive lattice Dirac equation.

2(!) two-component Dirac particles in the continuum limit! Note that the total chiral charge of the fermions is zero.

This catastrophe is best seen by considering the field ψ_+ on the lattice. This field is labeled by its chirality—a good quantum number even for $a \neq 0$ in this case. ψ_+ describes the excitations shown in fig. 27. There is a right-mover ($k \simeq 0$) and a left-mover ($k \simeq \pi$). But, since ordinary chirality (helicity for particles) is just velocity in $1+1$ dimensions, the finite energy content of the lattice field χ_+ is a pair of fermions with net chirality zero (« species doubling »)!

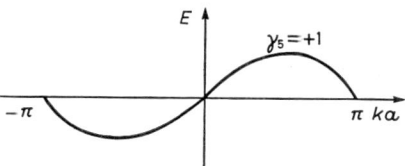

Fig. 27. – Energy-momentum relation of the wave ψ_+.

It is important to ask whether the unexpected properties of the lattice Dirac equation are general in character or particular to the detailed construction of this example. Experience with lattice methods convinced workers in the field long ago that the problem is generic and is closely related to continuous chiral symmetry. The general character of the problem can be appreciated very easily. Compare fig. 27 with fig. 25. The periodicity of $E(ka)$ over the interval $-\pi \leqslant ka \leqslant \pi$ is implied by the regular lattice structure. It causes $E(+\pi) = E(-\pi)$. So, one should think of E defined on a circle. $E(ka)$ must also be a continuous function of ka because the real-space Hamiltonian is local. Note that the three conditions 1) continuity of $E(ka)$, 2) the definite chirality of each branch of $E(ka)$ and 3) the periodicity of $E(ka)$ were the essential ingredients in counting the number of species in the continuum. The detailed shape of $E(ka)$ was irrelevant! A more general E vs. k curve which satisfies conditions 1), 2) and 3) is shown in fig. 28. Note that such a generic

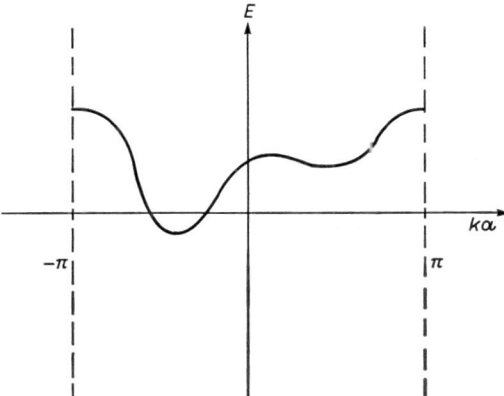

Fig. 28. – A generic energy-momentum relation for ψ_+.

curve must have an even number of zeros since it is closed! The vicinity of one zero describes a left-mover and the other zero a right-mover. Again we have species doubling and a net chirality of zero.

These observations can be organized into theorems stated precisely by NIELSEN and NINOMIYA [27] who presented a number of « no go » results. An interesting example is the following:

Attempt to couple a left-handed fermion ψ_L to the gauge field A_μ,

(6.21) $$S = \int \mathrm{d}^4 x \{ -i\bar{\psi}_\mathrm{L} \dot{\psi}_\mathrm{L} + \bar{\psi}_\mathrm{L} \sigma \cdot (-i\partial - A(x)) \psi_\mathrm{L} \}.$$

Put this theory on a lattice with

1) exact conservation of the chiral charges associated with a compact, continuous group (chiral $SU_3 \times SU_3$, chiral U_1 of QCD, for example);

2) locality in the interactions and hopping terms.

Then it follows using the natural generalizations of the ideas 1)-3) above to $3+1$ dimensions that three are an equal number of left- and right-handed Weyl fermions in the continuum limit of this lattice theory.

Consider two approaches to «beat» the species-doubling problem:

1) thin the degrees of freedom [28],

2) lift the energy at the edges of the Brillouin zone from zero [29].

Approach 1): «Staggered fermions». Place a single-component Fermi field $\varphi(n)$ on each site,

(6.22) $$\{\varphi(n), \varphi(m)\} = 0, \quad \{\varphi^\dagger(n), \varphi(m)\} = \delta_{nm},$$

and describe its dynamics with the «hopping» Hamiltonian,

(6.23) $$H = -\frac{i}{2a} \sum [\varphi^\dagger(n) \varphi(n+1) - \varphi^\dagger(n+1) \varphi(n)].$$

This Hamiltonian gives the equation of motion, $i(\partial/\partial t)\varphi = [H, \varphi]$,

(6.24) $$\dot{\varphi}(n) = -\frac{1}{2a} [\varphi(n+1) - \varphi(n-1)].$$

To identify a single two-component Dirac field decompose the lattice into an even sublattice ($n =$ even integer) and an odd sublattice ($n =$ odd integer),

(6.25) $$\begin{cases} \psi_1(n) = \varphi(n), & n \text{ even}, \\ \psi_2(n) = \varphi(n), & n \text{ odd}, \end{cases} \quad \psi = \begin{pmatrix} \psi_1 \\ \psi_2 \end{pmatrix}.$$

Now the equation of motion becomes

(6.26) $$\begin{cases} \dot{\psi}_1(n) = -\dfrac{1}{2a} [\psi_2(n+1) - \psi_2(n-1)], \\ \dot{\psi}_2(n) = -\dfrac{1}{2a} [\psi_1(n+1) - \psi_1(n-1)], \end{cases}$$

which should be compared to the continuum equation

(6.27)
$$\begin{cases} \dot{\psi} = -\gamma_5 \partial_z \psi = -\begin{pmatrix} 0 & 1 \\ 1 & 0 \end{pmatrix} \partial_z \psi \,, \\ \dot{\psi}_1 = -\partial_z \psi_2 \quad \text{and} \quad \dot{\psi}_2 = -\partial_z \psi_1 \,. \end{cases}$$

The correspondence between eqs. (6.27) and (6.11) is clear.

The staggered-fermion method avoids the species-doubling problem by doubling the size of the unit cube in real space and effectively halving the size of the Brillouin zone. Is this a counter example to the Nielsen-Ninomiya « no go » theorem? No! Since continuous γ_5 rotations mix ψ_1 and ψ_2, chirality is not a good quantum number here!

Let us write fermion bilinears in the new language of $\varphi(n)$ and identify the symmetries of the lattice theory,

(6.28)
$$\begin{cases} \int \psi^\dagger \psi \, dz = \int (\psi_1^\dagger \psi_1 + \psi_2^\dagger \psi_2) \, dz \to \sum_n \varphi^\dagger(n) \varphi(n) \,, \\ \int \bar{\psi} \psi \, dz = \int (\psi_1^\dagger \psi_1 - \psi_2^\dagger \psi_2) \, dz \to \sum_n (-1)^n \varphi^\dagger(n) \varphi(n) \,, \\ \int \psi^\dagger \gamma_5 \psi \, dz = \int (\psi_1^\dagger \psi_2 - \psi_2^\dagger \psi_1) \, dz \to \\ \qquad \to \sum_n (-1)^n [\varphi^\dagger(n) \varphi(n-1) - \varphi^\dagger(n+1) \varphi(n)] \,, \\ \int \psi^\dagger \gamma_5 \psi \, dz = \int (\psi_1^\dagger \psi_2 + \psi_2^\dagger \psi_1) \, dz \to \sum_n \varphi^\dagger(n) \varphi(n+1) + \varphi^\dagger(n+1) \varphi(n) \,. \end{cases}$$

Now we consider the symmetries of

$$H = -\frac{i}{2a} \sum [\varphi^\dagger(n) \varphi(n+1) - \varphi^\dagger(n+1) \varphi(n)] \,.$$

1) Translation of the spatial lattice by an even number of sites. This symmetry should be related to ordinary translations. The generator of ordinary translations in the continuum theory is

(6.29)
$$p_z = -i \int \psi^\dagger \partial_z \psi \, dz = -i \int (\psi_1^\dagger \partial_z \psi_1 + \psi_2^\dagger \partial_z \psi_2) \, dz \,,$$

which does not mix ψ_1 and ψ_2. So, its lattice version should not mix the two sublattices and

(6.30)
$$p_z \to \sum_n \varphi^\dagger(n+2) \varphi(n) + \varphi^\dagger(n) \varphi(n+2)$$

is the corresponding lattice generator. Translations by two lattice spacings corresponds to ordinary continuum translations.

2) Translation by an odd number of sites. Again $H \to H$ by inspection. Note from eq. (6.28) that the generator of translations by one site is $\int \psi^\dagger \gamma_5 \psi \, dz$. So, this lattice fermion method has discrete γ_5 symmetry. Note that

$$\gamma_5 \begin{pmatrix} \psi_1 \\ \psi_2 \end{pmatrix} = \begin{pmatrix} \psi_2 \\ \psi_1 \end{pmatrix}$$

interchanges ψ_1 and ψ_2. These are chiral rotations through $\pi/2$ radians, $\exp[i\gamma_5 \pi/2] = i\gamma_5$.

So, just as the lattice leaves over only discrete translations as symmetries, it admits only discrete chiral rotations. A mass term

(6.31) $$m_0 \sum (-1)^n \varphi^\dagger(n) \varphi(n)$$

is invariant to shifts by an even number of lattice spacings (translations) but not by an odd number of spacings (discrete chiral transformations). Note that the theory with $m_0 = 0$ has discrete γ_5 symmetry, so a bare mass term cannot be generated by interactions, if they also possess the γ_5 symmetry.

So, we have evaded the species-doubling problem at the expense of continuous chiral symmetry. This is consistent with the general Nielsen-Ninomiya constraints.

Approach 2): « Brute force ». Now we will leave two-component fermions on the spatial lattice, but will add terms to H so that E vs. k does not have secondary minima at $ka = \pm \pi$. Let

(6.32) $$H = -\frac{i}{2a} \sum_n \psi^\dagger(n) \alpha [\psi(n+1) - \psi(n-1)] +$$
$$+ \frac{B}{2a} \sum_n \bar{\psi}(n) [2\psi(n) - \psi(n+1) - \psi(n-1)] + \frac{m}{2a} g^2 \sum_n \bar{\psi}(n) \psi(n).$$

Note that the second term is a « boson » kinetic-energy term multiplied by a. In a classical continuum limit,

(6.33) $$\frac{1}{a} \sum_n \bar{\psi}(n)[2\psi(n) - \psi(n+1) - \psi(n-1)] \to \frac{1}{a} \int \frac{dz}{a} a\bar{\psi} a^2 \nabla^2 \psi =$$
$$= a \int \bar{\psi} \nabla^2 \psi \, dz \to 0 \qquad \text{as } a \to 0.$$

If we omit the last term in H for the moment, the equation of motion for ψ is

(6.34) $$i\dot{\psi}(n) = -\frac{i}{2a} \gamma_5 [\psi(n+1) - \psi(n-1)] +$$
$$+ \frac{B}{2a} \gamma_0 [2\psi(n) - \psi(n-1) - \psi(n+1)].$$

For a plane wave,

(6.35) $$\psi = \exp[i(-kna + Et)]\chi,$$

the equation of motion becomes

(6.36) $$\begin{cases} -E\chi = -\dfrac{i}{2a}\gamma_5(\exp[-ika] - \exp[ika])\chi + \\ \qquad\qquad\qquad + \dfrac{B}{2a}\gamma_0(2 - \exp[ika] - \exp[-ika])\chi, \\ -E\chi = -\gamma_5\dfrac{\sin ka}{a}\chi + 2B\gamma_0\dfrac{\sin^2(ka/2)}{a}\chi. \end{cases}$$

So,

(6.37) $$E^2 = \frac{\sin^2 ka}{a^2} + 4B^2\frac{\sin^4(ka/2)}{a^2}.$$

Letting $ka \to 0$, we have $E^2 \simeq k^2 \pm B^2 k^4 a^2 + \ldots$ which is acceptable. Near the edge of the Brillouin zone, $ka \approx \pm\pi$,

(6.38) $$E^2 \simeq 4B^2/a^2$$

and the boson kinetic energy has opened a gap in the spectrum. In this case, the species-doubling problem has been solved with an irrelevant operator. What has the cost been?

1) H explicitly breaks continuous and discrete γ_5 invariance! There is no remnant of one of the most important symmetries of hadron physics.

2) Interactions will induce a bare mass for the fermions—hence the third term is added in eq. (6.32) and a fine tuning must be made so that the continuum limit has finite-mass particles in it!

Now we will devote considerable space to the Euclidean version of « staggered » fermions. The discussion will be made in two dimensions for free fields, but the features of the methods we concentrate on extend to four dimensions in the theory with gauge fields. We are most interested in counting species and exposing the remnants of chiral symmetry in the method. Unlike the Hamiltonian version of the method, there is a continuous remnant of chiral symmetry in addition to discrete γ_5 operations! This is important since it allows lattice studies of the Goldstone mechanism to make sense!

Choose a γ matrix representation in two dimensions,

(6.39) $$\gamma_0 = \begin{pmatrix} 0 & 1 \\ 1 & 0 \end{pmatrix} = \sigma_1, \quad \gamma_1 = \begin{pmatrix} 0 & -i \\ i & 0 \end{pmatrix} = \sigma_2, \quad \gamma_5 = \begin{pmatrix} 1 & 0 \\ 0 & -1 \end{pmatrix} = \sigma_3,$$

in which γ_5 is diagonal. The naive action is

(6.40) $$S = \sum_{i,\mu} \bar{\psi}(i) \gamma_\mu \{\psi(i+\mu) - \psi(i-\mu)\}, \qquad i = (n_0, n_1).$$

To thin the degrees of freedom, «spin diagonalize» [30] this expression. Define $\chi(n_0, n_1)$,

(6.41) $$\psi = \gamma_0^{|n_0|} \gamma_1^{|n_1|} \chi$$

and independently

(6.42) $$\bar{\psi} = \bar{\chi} \gamma_1^{|n_1|} \gamma_0^{|n_0|}.$$

So,

(6.43) $$S = \sum_{i,\mu} \bar{\chi}(i)(-1)^{\varphi(i,\mu)} \{\chi(i+\mu) - \chi(i-\mu)\},$$

where

(6.44) $$\begin{cases} (-1)^{\varphi(i,0)} = \gamma_1^{|n_1|} \gamma_0^{|n_0|} \gamma_0 \gamma_0^{|n_0+1|} \gamma_1^{|n_1|} = +1, \\ (-1)^{\varphi(i,1)} = \gamma_1^{|n_1|} \gamma_0^{|n_0|} \gamma_1 \gamma_0^{|n_0|} \gamma_1^{|n_1+1|} = (-1)^{|n_0|}. \end{cases}$$

Since S is «spin-diagonal», one can thin the degrees of freedom, and one need keep only a single complex field $\chi(i)$ at each site! This trick of spin diagonalization shows the intimate relation between the two fermion methods.

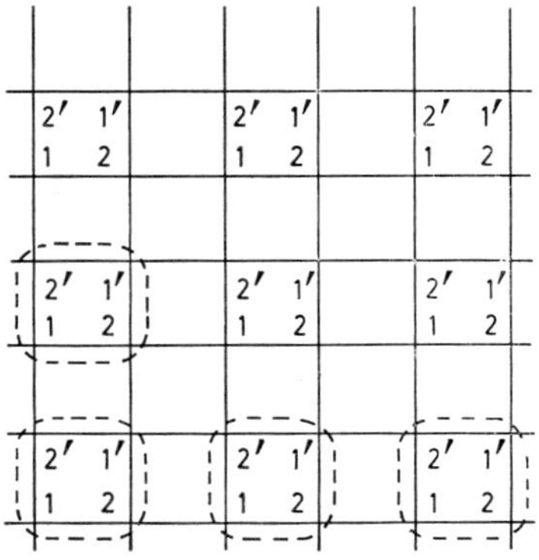

Fig. 29. – Sites, blocks and degrees of freedom for staggered Euclidean fermions.

Label the lattice as two interleaving lattices—one primed and one unprimed as in fig. 29 as an aid to identifying the multiple species hiding in eq. (6.43). The dotted block in the figure repeats itself throughout the lattice by translation. Consider the four sites of the block and label them with

γ-matrices as suggested by eq. (6.41),

(6.45)
$$\begin{cases} \text{site} & & & \\ \text{origin} & \psi_1^a = \chi(1)\begin{pmatrix}1\\0\end{pmatrix}, & & \bar{\psi}_1^a = (1,0)\bar{\chi}(1), \\ \gamma_0 & \psi_2^a = \chi(-)\gamma_0\begin{pmatrix}1\\0\end{pmatrix} = \chi(2)\begin{pmatrix}0\\1\end{pmatrix}, & & \bar{\psi}_2^a = (0,1)\bar{\chi}(2), \\ \gamma_1 & \psi_2^b = \chi(-')\gamma_1\begin{pmatrix}1\\0\end{pmatrix} = i\chi(1')\begin{pmatrix}0\\1\end{pmatrix}, & & \bar{\psi}_2^b = (0,1)[-i\bar{\chi}(2')], \\ \gamma_0\gamma_1 & \psi_1^b = \chi(-')\gamma_0\gamma_1\begin{pmatrix}1\\0\end{pmatrix} = i\chi(1')\begin{pmatrix}1\\0\end{pmatrix}, & & \bar{\psi}_1^b = (1,0)[-i\bar{\chi}(1')], \end{cases}$$

where the labels a and b will help us identify physical degrees of freedom.

The action can be simplified,

(6.46)
$$\begin{aligned} S &= \bar{\chi}(1)\{\nabla_0\chi(2) + \nabla_1\chi(2')\} + \bar{\chi}(2)\{\nabla_0\chi(1) - \nabla_1\chi(1')\} + \\ &+ \bar{\chi}(1')\{\nabla_0\chi(2') - \nabla_1\chi(2)\} + \bar{\chi}(2')\{\nabla_0\chi(2') - \nabla_1\chi(2)\} + \text{other blocks} = \\ &= \bar{\psi}_1^a\{\nabla_0\psi_2^a - i\nabla_1\psi_2^b\} + \bar{\psi}_2^a\{\nabla_0\psi_1^a + i\nabla_1\psi_1^b\} + \bar{\psi}_1^b\{\nabla_0\psi_2^b - i\nabla_1\psi_2^a\} + \\ &+ \bar{\psi}_2^b\{\nabla_0\psi_1^b + i\nabla_1\psi_1^a\} + \text{other blocks}. \end{aligned}$$

where $\nabla_\mu = d_\mu^+ + d_\mu^-$. We can diagonalize the action. Let

(6.47) $$u_i = (\psi_i^a + \psi_i^b)/\sqrt{2}, \quad \tilde{d}_i = (\psi_i^a - \psi_i^b)/\sqrt{2}.$$

So,

(6.48) $$u_i + \tilde{d}_i = \sqrt{2}\psi_i^a, \quad u_2 - \tilde{d}_i = \sqrt{2}\psi_i^b.$$

Substituting into eq. (6.46), we have after some algebra

(6.49) $$S = \bar{u}_1\{\nabla_0 u_2 - i\nabla_1 u_2\} + \bar{u}_2\{\nabla_0 u_1 + i\nabla_1 u_1\} + \bar{\tilde{d}}_1\{\nabla_0\tilde{d}_2 + i\nabla_1\tilde{d}_2\} + \\ + \bar{\tilde{d}}_2\{\nabla_0\tilde{d}_1 - i\nabla_1\tilde{d}_2\} = \bar{u}Du + \bar{\tilde{d}}\tilde{D}\tilde{d},$$

where

(6.50) $$D = \gamma_0\nabla_0 + \gamma_1\nabla_1, \quad \tilde{D} = \gamma_0\nabla_0 - \gamma_1\nabla_1.$$

Finally, we can make a linear transformation on $\tilde{d}$ to obtain a standard Dirac operator

(6.51) $$\tilde{d} = i\gamma_5\gamma_1 d = i\sigma_3\sigma_2 d = \sigma_1 d = \begin{pmatrix}0 & 1\\ 1 & 0\end{pmatrix}d.$$

Then,

(6.52) $$\bar{\tilde{d}}\tilde{D}\tilde{d} = \bar{d}\sigma_1 \tilde{D}\sigma_1 d = \bar{d}\{\gamma_0 \nabla_0 - \gamma_0\gamma_1 \nabla_1 \gamma_0\}d = \bar{d}Dd.$$

Now,

(6.53) $$S = \bar{u}Du + \bar{d}Dd,$$

which suggests that two fermions, an isodoublet, emerge in the continuum limit.

The degrees of freedom relevant to the continuum limit occupy a single square, as shown in fig. 29. If we introduce a «block derivative» [31], the continuum limit physics can be made even clearer than in eq. (6.53). The final result of this exercise will give a flavor-diagonal expression from which we will read off the continuum species and symmetries. Recall

(6.54) $$S = \bar{\psi}_1^a\{\nabla_0\psi_2^a - i\nabla_1\psi_2^b\} + \bar{\psi}_2^a\{\nabla_0\psi_1^a + i\nabla_1\psi_1^b\} + \\ + \bar{\psi}_1^b\{\nabla_0\psi_2^b - i\nabla_1\psi_2^a\} + \bar{\psi}_2^b\{\nabla_0\psi_1^b + i\nabla_1\psi_1^a\}.$$

Let us begin to simplify this expression by examining the discrete differences here in greater detail. Consider three sites $\overset{1}{x}\,\overset{2}{x}\,\overset{3}{x}$ and definitions of a derivative,

(6.55) $$\nabla f = \tfrac{1}{2}[f(3) - f(1)], \qquad \nabla^2 f = f(3) + f(1) - 2f(2).$$

Then,

(6.56) $f(3) - f(2) = \tfrac{1}{2}[f(3) - f(1)] + \tfrac{1}{2}[f(3) + f(1) - 2f(2)] = \nabla f + \tfrac{1}{2}\nabla^2 f,$

(6.57) $f(2) - f(1) = \tfrac{1}{2}[f(3) - f(1)] - \tfrac{1}{2}[f(3) + f(1) - 2f(2)] = \nabla f - \tfrac{1}{2}\nabla^2 f.$

Now introduce a block derivative. In the action, $\bar{\psi}_1^a \nabla_0 \psi_2^a$ occurs. It consists of two terms—one which couples degrees of freedom entirely inside a block and another connecting blocks. Consider the horizontal axis shown in fig. 30.

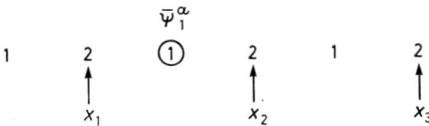

Fig. 30. – Horizontal axis of a block of sites.

The difference $\psi_2^a(x_2) - \psi_2^a(x_1)$ occurs in eq. (6.54). But using the algebra above

(6.58) $\psi_2^a(x_2) - \psi_2^a(x_1) =$
$= \tfrac{1}{2}[\psi_2^a(x_3) - \psi_2^a(x_1)] - \tfrac{1}{2}[\psi_2^a(x_3) + \psi_2^a(x_1) - 2\psi_2^a(x_2)] = \hat{\nabla}\psi_2^a - \tfrac{1}{2}\hat{\nabla}^2\psi_2^a,$

where «hats» indicate intrablock derivatives.

Now the variables ψ_i^a, ψ_i^b ($i = 1, 2$) can be thought to live at the centers of each block. The variables are now naturally organized to account for the multiple flavors which appear in the continuum limit. This formalism also lends itself most naturally to real-space renormalization group methods.

Introducing the block derivatives into S,

(6.59) $\quad S = \bar{\psi}_1^a\{[\hat{\nabla}_0 \psi_2^a - \frac{1}{2}\hat{\nabla}_0^2 \psi_2^a] - i[\hat{\nabla}_1 \psi_2^b - \frac{1}{2}\hat{\nabla}_1^2 \psi_2^b]\} +$
$\quad\quad\quad + \bar{\psi}_2^a\{[\hat{\nabla}_0 \psi_1^a + \frac{1}{2}\hat{\nabla}_0^2 \psi_1^a] + i[\hat{\nabla}_1 \psi_1^b - \frac{1}{2}\hat{\nabla}_1^2 \psi_1^b]\} +$
$\quad\quad\quad + \bar{\psi}_1^b\{[\hat{\nabla}_0 \psi_2^b + \frac{1}{2}\hat{\nabla}_0^2 \psi_2^b] - i[\hat{\nabla}_1 \psi_2^a + \frac{1}{2}\hat{\nabla}_1^2 \psi_2^a]\} +$
$\quad\quad\quad + \bar{\psi}_2^b\{[\hat{\nabla}_0 \psi_1^b - \frac{1}{2}\hat{\nabla}_0^2 \psi_1^b] + i[\hat{\nabla}_1 \psi_1^a + \frac{1}{2}\hat{\nabla}_1^2 \psi_1^a]\}.$

So,

(6.60) $\quad S = \bar{u}\hat{D}u + \bar{d}\hat{D}d + \bar{\psi}_1^a\{-\frac{1}{2}\hat{\nabla}_0^1 \psi_2^a + \frac{1}{2}i\hat{\nabla}_1^2 \psi_2^b\} + \bar{\psi}_2^a\{\frac{1}{2}\hat{\nabla}_0^2 \psi_1^a - \frac{1}{2}i\hat{\nabla}_1^2 \psi_1^b\} +$
$\quad\quad\quad + \bar{\psi}_1^b\{\frac{1}{2}\hat{\nabla}_0^2 \psi_2^b - \frac{1}{2}i\hat{\nabla}_1^2 \psi_2^a\} + \bar{\psi}_2^b\{-\frac{1}{2}\hat{\nabla}_0^2 \psi_1^b + \frac{1}{2}i\hat{\nabla}_1^2 \psi_1^a\}.$

The first two terms in eq. (6.60) can be written as $\bar{u}\hat{D}u + \bar{d}\hat{D}d$ using eqs. (6.51) and (6.52). The terms in braces in eq. (6.60) can also be simplified by transforming the u's and d's. It is convenient to use flavor isospin notation

(6.61) $$f = \begin{pmatrix} u \\ d \end{pmatrix}$$

and flavor matrices

(6.62) $$T_0 = \begin{pmatrix} 0 & -1 \\ 1 & 0 \end{pmatrix}, \quad T_1 = \begin{pmatrix} 0 & -i \\ -i & 0 \end{pmatrix}.$$

Then some algebra yields

(6.63) $$S = \bar{f}\hat{D}f + \bar{f}\gamma_5 T_\mu \tfrac{1}{2} \hat{\nabla}_\mu^2 f.$$

This form of the action has the virtues advertised above. The first term is free of flavor mixing effects and displays the two species in the clearest possible way. The second term is an irrelevant boson kinetic-energy term with an important twist—the matrices $\gamma_5 T_\mu$ occur! The matrix character of the second term has several important consequences. It ensures the existence of a continuous remnant of chiral symmetry and it breaks explicitly the axial flavor-neutral symmetries. These points will be made explicit below in greater detail.

To expose the vector and axial symmetries of the staggered fermions we reconsider the action written in its most primitive form, eq. (6.43).

Consider even and odd sites separately and a $U_1 \times U_1$ global invariance group where the first U_1 acts on even sites, the second U_1 on odd sites,

(6.64a) $$\chi(n) \to \begin{cases} U_e \chi(n), & n \text{ even}, \\ U_o \chi(n), & n \text{ odd}, \end{cases}$$

(6.64b) $$\bar{\chi}(n) \to \begin{cases} \bar{\chi}(n) U_o^\dagger, & n \text{ even}, \\ \bar{\chi}(n) U_e^\dagger, & n \text{ odd}. \end{cases}$$

This is obviously a symmetry of

(6.65) $$S = \bar{\chi}(1)\{\nabla_0 \chi(2) + \nabla_1 \chi(2')\} + \bar{\chi}(2)\{\nabla_0 \chi(1) - \nabla_1 \chi(1')\} + \\ + \bar{\chi}(1')\{\nabla_0 \chi(2') - \nabla_1 \chi(2)\} + \bar{\chi}(2')\{\nabla_0 \chi(1') + \nabla_1 \chi(1)\} + \text{other blocks},$$

since only even (odd) sites are coupled to odd (even) sites. Write this symmetry in terms of conventional, continuum fields,

(6.66) $$\begin{vmatrix} \psi_1 \to U_o \psi_1, \quad \psi_2 \to U_e \psi_2, \\ \begin{pmatrix} u_1 \\ u_2 \end{pmatrix} = \begin{pmatrix} \psi_1^a + \psi_1^b \\ \psi_2^a + \psi_2^b \end{pmatrix} \to \begin{pmatrix} U_o & \\ & U_e \end{pmatrix}\begin{pmatrix} u_1 \\ u_2 \end{pmatrix} = (U_o + U_e)\begin{pmatrix} u_1 \\ u_2 \end{pmatrix} + (U_o - U_e)\begin{pmatrix} u_1 \\ -u_2 \end{pmatrix}. \end{vmatrix}$$

Similarly,

(6.67) $$\begin{pmatrix} \tilde{d}_1 \\ \tilde{d}_2 \end{pmatrix} \to (U_o + U_e)\begin{pmatrix} \tilde{d}_1 \\ \tilde{d}_2 \end{pmatrix} + (U_o - U_e)\gamma_5 \begin{pmatrix} \tilde{d}_1 \\ \tilde{d}_2 \end{pmatrix}.$$

But $\tilde{d} = \sigma_1 d$, and $\gamma_5 = \sigma_3$ in this basis, so

(6.68) $$\begin{pmatrix} \tilde{d}_1' \\ \tilde{d}_2' \end{pmatrix} = \sigma_1 \begin{pmatrix} d_1' \\ d_2' \end{pmatrix} = U_+ \sigma_1 \begin{pmatrix} d_1 \\ d_2 \end{pmatrix} + U_- \gamma_5 \sigma_1 \begin{pmatrix} d_1 \\ d_2 \end{pmatrix},$$

where $U_\pm = U_o \pm U_e$. So,

(6.69) $$d' = \sigma_1 U_+ \sigma_1 d + \sigma_1 U_- \gamma_5 \sigma_1 d.$$

But $\sigma_1 \gamma_5 \sigma_1 = -\gamma_5$, so

(6.70) $$d' = U_+ d - U_- \gamma_5 d, \qquad u' = U_+ u + U_- \gamma_5 u.$$

Let

$$T_3 = \begin{pmatrix} 1 & 0 \\ 0 & -1 \end{pmatrix},$$

and use the isospin notation

(6.71) $$f = \begin{pmatrix} u \\ d \end{pmatrix}.$$

Then,
$$f' = U_+ f + U_- \gamma_5 T_3 f$$

In summary, the action

(6.72) $$S = \bar{f} \hat{D} f + \bar{f} \gamma_5 T_\mu \tfrac{1}{2} \hat{\nabla}_\mu^2 f$$

has the symmetry

(6.73) $$f' = U_+ f + U_- \gamma_5 T_3 f \,.$$

So, the original $U_e \times U_o$ symmetry contains a vector piece $1_o \times 1_F$, fermion number conservation, and axial piece $\gamma_5 \times T_3$, one component of axial isospin. The second piece forbids a mass term in S. When it breaks spontaneously in more than two dimensions, a pion will appear in the usual Goldstone fashion!

Note that the irrelevant operator exposed in the block derivative expression breaks the full axial isospin symmetries to a smaller set of continuous symmetries. It breaks the axial flavor-neutral symmetries explicitly.

In addition to the continuous $U_o \times U_e$ symmetries, there are discrete symmetries of interest in S. If we translate the system by an odd number of sites in either direction and transform phases appropriately, the action will not change. Therefore, these symmetries are square roots of translations and can be identified with λ_5, T_i and their products. These discrete symmetries forbid fermion mass terms, isospin breaking mass terms, etc. One can consult ref. [32] for details.

In summary, this lattice fermion method has the species we want, an isodoublet degenerate in mass, and has the correct symmetries in the continuum limit. On the lattice, however, it only has remnants of the full chiral symmetries. There is one continuous axial flavor symmetry and, if the gauge field dynamics drives spontaneous symmetry breaking, a Goldstone pion should occur even in the cut-off model. In the continuum limit, full chiral symmetry should appear and should be spontaneously broken as well. Then additional Goldstone pions should evolve as $a \to 0$. The lattice fermion method has enough symmetry to guarantee this quite naturally.

The generalization of this lattice fermion method to $3 + 1$ dimensions leads to four Dirac particles in the continuum limit. This formulation of the theory is under intense study both conceptually and numerically at this time.

7. – Fermion computational methods.

So far we have exploited the analogy between field theory and statistical mechanics for bosons. The four-dimensional Euclidean field theory version of quantized boson field operators was simply commuting C number fields. This correspondence was essential for the usefulness of the analogy. We meet a roadblock, however, when we attempt a generalization to fermions. What is a classical, Euclidean fermion? They are anticommuting numbers, Grassmann fields. These are very abstract formal objects which do not have an immediate statistical-mechanics interpretation. Our boson fields could be thought of as stochastic variables and the path integral could be interpreted as a partition function with a probabilistic weight given by a Boltzmann factor, $\exp[-S]$. All of this is lost in the case of fermions! The formulation is very formal and probabilistic interpretations are unclear.

In the lattice theory we have the generic action

(7.1) $$S = \sum_{ij} \bar{\psi}_i (\check{D}(U) + m)_{ij} \psi_j + S_0(U),$$

where $\check{D}(U) + m$ denotes the hopping matrix of staggered fermions, for example, and $S_0(U)$ is the pure gauge field piece of the action we have already studied in detail. The hopping matrix $\check{D}(U) + m$, which we studied in the last section for the free-field case, now depends upon the gauge field U. This ensures that S is locally gauge invariant. Let us check this. In eq. (7.1) the fermion hopping term is nonzero only if sites i and j are identical or are nearest neighbors. The hopping term $\bar{\psi}_i \check{D}(U) \psi_j$ has pieces of the form

(7.2) $$\text{phase} \cdot \bar{\psi}_i U_\mu(i) \psi_{i+\mu},$$

where a contraction over color indices, $\sum_{\alpha\beta} \bar{\psi}^\alpha U^{\alpha\beta} \psi^\beta$, is implied. The « phase » in eq. (7.2) is that of the staggered fermions. Equation (7.2) is constructed to be invariant under local gauge transformations,

(7.3) $$\begin{cases} \psi_{i+\mu} \to G(i+\mu)\psi_{i+\mu}, \\ U_\mu(i) \to G(i) U_\mu(i) G^{-1}(i+\mu), \\ \bar{\psi}_i \to \bar{\psi}_i G^{-1}(i). \end{cases}$$

Equation (7.1) is the natural lattice generalization of the gauge-invariant kinetic energy of the continuum formulation of quantum chromodynamics.

The path integral of lattice QCD now reads

(7.4) $$Z = \int [d\psi][d\bar{\psi}][dU] \exp[-S].$$

The functional integral over the Grassmann variables is familiar formalism from continuum field theory. This mathematics was developed and used extensively by SCHWINGER in his pioneering work on quantum electrodynamics. Since S is a quadratic form, we can integrate out the Grassmann variables. The crucial identity is the Grassmann equivalent of the ordinary Gaussian integral. The general result is the following. Let A be a linear operator and consider the action $S_\mathrm{f} = -\int d^4 x \bar\psi A \psi$. Then,

(7.5a) $$\int [d\psi][d\bar\psi] \exp[-S_\mathrm{f}] = \det A \,.$$

Compare this result to the boson problem $S_\mathrm{b} = -\int d^4 x \varphi^* A \varphi$ and

(7.5b) $$\int [d\varphi^*][d\varphi] \exp[-S_\mathrm{b}] = (\det A)^{-1},$$

where φ is a complex field. Recall from your experience with perturbation theory that these determinants contain the vacuum polarization effects of the matter fields. The fact that $\det A$ appears in eq. (7.5a) rather than $(\det A)^{-1}$ is a reflection of Fermi statistics. When one develops the perturbative expansion for the fermion (boson) path integral, there is a minus (plus) sign for each closed fermion (boson) loop. Since the logarithms of the path integrals eqs. (7.5a) and (7.5b) give the connected graphs and since the connected graphs consist of one fermion (boson) loop interacting with the fixed gauge field in all possible ways, the power of $(\det A)^{\pm 1}$ is obviously correct.

If we apply eq. (7.5a) to eq. (7.1) and (7.4), we find

(7.6) $$\int [d\psi][d\bar\psi] \exp[-S] = \det[\check{D}(U) + m] \exp[-S_0(U)] =$$
$$= \exp[-S_0 - \operatorname{tr} \ln[\check{D}(U) + m]],$$

so we can identify an effective gauge field action with the effect of the fermion vacuum polarization included,

(7.7) $$S_\mathrm{eff} = -S_0(U) - \operatorname{tr} \ln [\check{D}(U) + M].$$

Although the fermions do not appear in eq. (7.7), a dear price has been paid—S_eff is nonlocal because of the logarithm and is very difficult to deal with. However, a brute-force numerical method has been proposed for calculating with eq. (7.7) and we will discuss it briefly.

Suppose that we simply want to calculate the fermion propagator in a given gauge field configuration $\{U\}$ without the inclusion of fermion vacuum polarization. Since this is a one-body problem (one fermion), the subtleties of Fermi statistics can be evaded. The fermion propagator is

(7.8) $$G(i,j;\,U) = (\check{D}(U) + m)^{-1}_{ij}.$$

It can be calculated using only boson fields by considering the action [33]

(7.9) $$S(\varphi) = \sum_{ij} \varphi_i^* (\check{D}(U) + m)_{ij} \varphi_j$$

and calculating the φ-φ correlation function

(7.10) $G(x, y; U) = \langle \varphi(x) \varphi^*(y) \rangle =$
$$= \frac{1}{Z} \int [d\varphi^*][d\varphi] \varphi(x) \varphi^*(y) \exp\left[-\sum_{ij} \varphi_i^*(\check{D}(U) + m)_{ij} \varphi_j\right].$$

The calculation of eq. (7.10) can be done using standard Monte Carlo methods. The only technicality one must face concerns the numerical convergence of the integral. Since $\check{D}(U) + m$ is not generally positive, one should replace it by

(7.11) $$(-\check{D} + m)(\check{D} + m) = -\check{D}^2 + m$$

in eq. (7.10) and thereby compute $(-\check{D}^2 + m^2)^{-1}$. Finally, $-\check{D} + m$ can be applied to the result and the first-order fermion propagator $(\check{D} + m)^{-1}$ can be obtained.

These methods—the pseudofermion technique—are particularly effective in studies of chiral-symmetry breaking and the hadron mass spectrum in the approximation that vacuum polarization is dropped from the QCD dynamics. This is called the quenched approximation [34] and it is believed, on the basis of considerable quark model phenomenology, to be quite faithful to many (but not all!) aspects of the real theory. Calculations of the hadron spectrum which obtain the composite propagators of mesons and baryons from appropriate products of fermion propagators by Monte Carlo methods are quite realistic and interesting. However, our interest here is in the theory with vacuum polarization. Even here the pseudofermion method can be applied [33]. Consider a stochastic, Monte Carlo calculation in which only *small* changes $\{U\} \to \{\tilde{U}\}$ are made. Return to the expression

(7.12) $$S_{\text{eff}} = S_0(U) - \text{tr} \ln[\check{D}(U) + m].$$

If $\{\tilde{U}\}$ and $\{U\}$ are almost identical, then we can expand

(7.13) $S_{\text{eff}}(\tilde{U}) - S_{\text{eff}}(U) =$
$$= S_0(\tilde{U}) - S_0(U) - \sum_{i,j} G_{ij}(U) \frac{\delta \check{D}(U)_{ij}}{\delta U} [(\tilde{U} - U) + O[(\tilde{U} - U)^2]].$$

This result suggests a modified Metropolis algorithm. Consider a gauge field configuration $\{U\}$. Calculate $G_{ij}(U)$ by the pseudofermion method. In other words, update the pseudofermion field φ_i with the action

(7.14) $$S(\varphi) = \sum_{i,j} \varphi_i^*[\check{D}(U) + m]_{ij} \varphi_j$$

n times and collect $\overline{\varphi_i^* \varphi_j}$ in the configuration $\{U\}$. Next update the gauge field with the action

(7.15) $$S(U) = S_0(U) + \sum_{i,j} \overline{\varphi_i^* \varphi_j}[\check{D}(U) + m]_{ij}$$

and obtain a new field configuration $\{\tilde{U}\}$. Repeat the procedure many times. Neglecting errors $O[(\tilde{U} - U)^2]$, the correct results are obtained when $n \to \infty$.

Unfortunately the limit $n \to \infty$ is replaced by $n = 10 \div 50$ in practice. One can check that $\overline{\varphi_i^* \varphi_j}$ is only obtained with huge variances (uncertainties by $(100 \div 500)\%$ are typical!) for small, but typical, values of m. The method then relies on cancellations of errors in lattice averages and these cancellations are not understood at this time. Although recent applications of this method to interesting physics problems in QCD look quite reasonable, no error estimates can be provided.

There is another approach to fermion dynamics which avoids stochastic methods entirely. It is based on molecular-dynamics approaches to microcanonical formulations of statistical mechanics. To begin consider a pure Bose field problem described by a lattice action $S(\varphi)$ [35]. We know how to calculate the expectation value of any operator $\theta(\varphi)$ in this theory in the partition function approach. Alternatively we can think of $S(\varphi)$ as a potential energy $S(\varphi) \equiv \beta V(\varphi)$ acting between the Bose degrees of freedom. A classical statistical mechanics based on the invariant phase space $\prod_n \mathrm{d}p_n \mathrm{d}\varphi_n$ and the Boltzmann factor $\exp[-\beta H]$ with

(7.16) $$H = T + V = \sum_n \tfrac{1}{2} p_n^2 + V$$

yields the same expectation values as the original problem. This is the traditional classical canonical ensemble if we introduce a time τ with

(7.17) $$p_n = \mathrm{d}\varphi_n/\mathrm{d}\tau$$

taken as a definition.

It proves convenient to replace this canonical formulation with a microcanonical one. One considers a system at fixed energy $H = E$ rather than at fixed temperature. Expectation values are computed with the weight $\prod_n \mathrm{d}p_n \mathrm{d}\varphi_n \delta(H - E)$,

(7.18a) $$\langle \theta \rangle = \frac{1}{Z} \int \prod_n \mathrm{d}p_n \mathrm{d}\varphi_n \theta(p, \varphi) \delta(H - E)$$

with

(7.18b) $$Z = \int \prod_n \mathrm{d}p_n \mathrm{d}\varphi_n \delta(E - H).$$

Standard arguments of statistical mechanics guarantee that these microcanonical expectation values are identical to their canonical expectation values in the thermodynamic limit in which the number of degrees of freedom grows large (the difference between the two expectation values is $O(N^{-1})$, $N =$ number of independent degrees of freedom). In practice the microcanonical expectation values are computed by the methods of molecular dynamics. One assumes the ergodic hypothesis—that the system's trajectories will cover the $E = H$ surface of phase space uniformly in the $\tau \to \infty$ limit—so that expectation values can be replaced by τ-averages,

$$(7.19) \qquad \langle \theta \rangle = \lim_{T \to \infty} \frac{1}{T} \int_0^T \theta\big(p(\tau), \varphi(\tau)\big) \, d\tau \,.$$

The last ingredient in the microcanonical-molecular dynamics approach is the determination of the system's temperature. The equipartition theorem can be used to find

$$(7.20) \qquad \tfrac{1}{2} \beta^{-1} N_{\text{indep}} = \langle T \rangle$$

which completes the mapping between the ensembles. Here N_{indep} is the number of independent degrees of freedom in the ensemble—N_{indep} may be different from N because of constraints which can occur in mechanical systems, gauge field theories, etc. This is an important detail since β is the coupling appearing in the canonical ensemble—it is determined by the microcanonical dynamics through eq. (7.20).

The generalization of this formalism to pure gauge fields is not difficult [35]. Consider the gauge groups SU_2 and the simplest action

$$(7.21a) \qquad S = \beta V, \qquad\qquad \beta = \frac{1}{g_0^2},$$

with

$$(7.21b) \qquad V \sum_{\text{plq}} (1 - \tfrac{1}{2} \operatorname{tr} UUUU) \,.$$

The form of the differential equations of the molecular dynamics will depend upon the parametrization chosen for the U matrices. A natural choice uses the Pauli matrices

$$(7.22a) \qquad U_\mu(n) = a_0 \mathbf{1} + i \boldsymbol{a} \cdot \boldsymbol{\sigma}$$

subject to the unitarity constraint

$$(7.22b) \qquad a_0^2 + \boldsymbol{a}^2 = 1 \,.$$

The Hamiltonian equations for motion are then written for the constrained variables $(a_0, \boldsymbol{a})$

(7.23) $$v_i = da_i/d\tau, \quad dp_i/d\tau = -\partial V/\partial a_i.$$

These molecular-dynamics equations can then be integrated numerically and expectation values of field-theoretic interest computed. The inclusion of the constraint $a_0^2 + \boldsymbol{a}^2 = 1$ can be done in several ways. Following standard classical-mechanics techniques, the constraint can be incorporated into the dynamics with a Lagrange multiplier, *i.e.* add to the potential a new term

(7.24) $$V \to V + \tfrac{1}{2} \sum_{\boldsymbol{a}} \lambda(\boldsymbol{a})(1 - \boldsymbol{a}^2)$$

and rederive the equations of motion. The resulting set of coupled first-order equations is still manageable by numerical methods and produces an algorithm which yields expectation values with computer requirements comparable to the stochastic, canonical ensemble technique. Finally, to make contact with the original field theory problem we must calculate the coupling β from the equipartition theorem, eq. (7.24). For an SU_2 gauge field, N_{indep} is calculated by observing that 1) each SU_2 matrix is labeled by three independent variables, and 2) local gauge invariance constrains one link variable emanating from each site. Therefore,

(7.25) $$N_{\text{indep}} = 3(d-1)L^d.$$

The next task is to include fermions into the microcanonical-molecular dynamics scheme. Grassmann variables must be avoided, but the det A must be generated by the equations of motion. An exceedingly naive, but instructive stochastic method to do this considers the pseudofermion action

(7.26) $$\tilde{S} = S_0(U) + \sum_{ij} \varphi_i^* (A^\dagger(U) A(U))_{ij}^{-1} \varphi_j.$$

Because of the explicit appearance of the *inverse* of the fermion hopping matrix $A(U) \equiv \check{D}(U) + m$, the functional average over the boson field φ generates the fermion determinant. Equation (7.26) can also be studied numerically by Monte Carlo methods if only local and *small* changes are made in both variables U and φ in the updating procedure. However, the method (whose details will not be discussed here) is very inaccurate unless very, very small changes in U and φ are made. Then it is, in fact, too slow to be useful! The nonlocal form of eq. (7.26) is also unpleasant. However, eq. (7.26) suggests an interesting and useful microcanonical formulation [36]. As discussed above for the pure Bose theory, we can form a classical canonical ensemble with the

Hamiltonian [33]

(7.27) $$H = \tfrac{1}{2}\sum_i p_i^2 + P(A^\dagger A)^{-1}P^\dagger + S_0(U) + \omega^2 \varphi^\dagger \varphi,$$

if we define the canonical momenta

(7.28) $$p_i = \mathrm{d}U/\mathrm{d}\tau \equiv \dot U, \qquad P = \dot\varphi^\dagger A^\dagger A,$$

so the Hamilton equations of motion are

(7.29) $$\begin{cases} \dfrac{\mathrm{d}}{\mathrm{d}\tau} P^\dagger = \dfrac{\mathrm{d}}{\mathrm{d}\tau}\left(A^\dagger(U)A(U)\dot\varphi\right) = -\omega^2 \varphi, \\ \dfrac{\mathrm{d}}{\mathrm{d}\tau} p_\mathrm{L} = \ddot U_i = -\dfrac{\partial}{\partial U_i} S_0(U) + \varphi^\dagger \dfrac{\partial}{\partial U_i}\left(A^\dagger(U)A(U)\right)\dot\varphi. \end{cases}$$

This classical system has the canonical partition function

(7.30) $$Z = \int [\mathrm{D}U][\mathrm{D}p][\mathrm{D}\varphi][\mathrm{D}\varphi^\dagger]\mathrm{D}[P]\mathrm{D}[P^\dagger] \exp\left[-\frac{1}{T} H\right].$$

All the integrals are trivial except for that over the pure gauge field U,

(7.31) $$Z = \mathrm{const}\cdot \int [\mathrm{D}U]\, \det\left(A^\dagger(U)A[U]\right)\exp\left[-\frac{1}{T} S_0(U)\right].$$

For staggered fermions $\det A$ is real, so eq. (7.31) is

(7.32) $$Z = \mathrm{const}\cdot \int [\mathrm{D}U]\, \det{}^2 A(U)\exp\left[-\frac{1}{T} S_0(U)\right].$$

The square in eq. (7.32) was forced upon us by the necessity of having the Hamiltonian eq. (7.27) positive. Since staggered Euclidean fermions lead to 4 massless Dirac particles in the continuum limit, eq. (7.32) doubles this and leads to eight flavors, $N_f = 8$. However, we can make a model with $N_f = 4$ by setting $\varphi_i = 0$ on all even (or odd) lattice sites in the staggered-fermion method [34]. This is permissible because $A^\dagger A$ couples only even (odd) sites to even (odd) sites. With this understood in eq. (7.27) $N_f = 4$ results in eq. (7.32).

The final ingredient is the generalization of the equipartition theorem, eq. (7.20),

(7.33) $$\tfrac{1}{2}\beta^{-1} N_\mathrm{indep} = \langle T \rangle = \left\langle \dot\varphi^\dagger A^\dagger A \dot\varphi + \tfrac{1}{2}\sum_i \dot U_i^2 \right\rangle.$$

In practice one integrates the molecular-dynamics equations eq. (7.29) numerically. By the ergodic theorem this procedure is equivalent to the fixed-energy, microcanonical ensemble. One finds that the method is very practical. The equations of motion are local and are well treated by ordinary numerical methods. The fermions do slow down the algorithm by an order of magnitude over the pure gauge field case. The reason is that $A^\dagger A$ must be inverted in each iteration of eq. (7.29). Since $A^\dagger A$ is a sparse matrix, this can be done effectively by iterative methods. Note that it is the repeated inversion of $A^\dagger A$ in the molecular-dynamics approach which generates the fermion determinant! The accuracy of the inversion of $A^\dagger A$ and integration of eq. (7.29) can be controlled and the method is subject to ordinary error analysis. In this sense it is superior to the pseudofermion stochastic method discussed earlier.

The microcanonical fermion method is being applied to a wealth of phyies problems now [37]—the spectrum of color-singlet states in QCD, the topological susceptibility, the finite-temperature deconfinement/chiral-symmetry restoring transitions, etc. Similar projects are also being done using the pseudofermion technique [38]. Workers in the field expect considerable progress on this difficult problem in the coming year!

REFERENCES

[1] Recent more detailed reviews include *Rev. Mod. Phys.*, **51**, 659 (1979); **55**, 775 (1983).
[2] S. SHENKER: Les Houches Summer School, 1982.
[3] P. PFEUTY: *Ann. Phys. (N. Y.)*, **57**, 79 (1970).
[4] K. G. WILSON: *Phys. Rev. D*, **14**, 2455 (1974); A. M. POLYAKOV: *Phys. Lett. B*, **59**, 82 (1975).
[5] C. N. YANG and R. L. MILLS: *Phys. Rev.*, **96**, 1305 (1954).
[6] S. ELITZUR: *Phys. Rev. D*, **12**, 3978 (1975).
[7] H. D. POLITZER: *Phys. Rev. Lett.*, **30**, 1346 (1973); D. J. GROSS and F. WILCZEK: *Phys. Rev. Lett.*, **30**, 1343 (1973).
[8] For a review of the renormalization group, fixed points, scaling laws, see K. G. WILSON and J. KOGUT: *Phys. Rep. C*, **12**, 75 (1974).
[9] K. SYMANZIK: *Commun. Math. Phys.*, **49**, 424 (1970); C. CALLAN: *Phys. Rev. D*, **3**, 1541 (1970).
[10] J. SHIGEMITSU and J. KOGUT: *Nucl. Phys. B*, **190**, 365 (1981).
[11] S.-K. MA: *Modern Theory of Critical Phenomena* (W. A. Benjamin Inc., Reading, Mass., 1976). This monograph contains references to the original literature on the subject, as well as more complete discussions of this and related renormalization group methods.
[12] R. H. SWENDSEN: *Phys. Rev. Lett.*, **47**, 1775 (1981).
[13] A. PATEL, R. CORDERY, R. GUPTA and M. A. NOVOTNY: *Phys. Rev. Lett.*, **53**, 527 (1984).
[14] A. M. POLYAKOV: *Phys. Lett. B*, **59**, 79 (1975).
[15] J. M. KOSTERLITZ and D. J. THOULESS: *J. Phys. C*, **6**, 118 (1973).
[16] B. BERG and M. LÜSCHER: *Nucl. Phys. B*, **190**[FS3], 412 (1981).
[17] L. SUSSKIND: reprinted in *Les Houches Summer School Proceedings*, edited by R. BALIAN and C. H. LLEWELLYN SMITH (North-Holland, Amsterdam, 1978).

[18] J. VILLAIN: *J. Phys. C*, **36**, 581 (1975).
[19] J. M. KOSTERLITZ: *J. Phys. C*, **7**, 1046 (1974).
[20] S. COLEMAN: unpublished.
[21] J. KOGUT and L. SUSSKIND: *Phys. Rev. D*, **11**, 395 (1975).
[22] J. KOGUT, R. PEARSON and J. SHIGEMITSU: *Phys. Lett. B*, **98**, 63 (1980).
[23] M. LÜSCHER: *Nucl. Phys. B*, **180**, 317 (1981).
[24] A. HASENFRATZ, E. HASENFRATZ and P. HASENFRATZ: *Nucl. Phys. B*, **180**, 353 (1981).
[25] J. KOGUT, R. PEARSON, J. RICHARDSON, J. SHIGEMITSU and D. K. SINCLAIR: *Phys. Rev. D*, **23**, 2945 (1981).
[26] J. KOGUT and D. K. SINCLAIR: *Phys. Rev. D*, **24**, 1610 (1981).
[27] H. B. NIELSEN and M. NINOMIYA: *Nucl. Phys.*, **185**, 10 (1981).
[28] T. BANKS, J. KOGUT and L. SUSSKIND: *Phys. Rev. D*, **13**, 1043 (1976).
[29] K. G. WILSON: *Proceedings of the Erice Summer School*, edited by A. ZICHICHI (Plenum Press, New York, N. Y., 1975).
[30] N. KAWAMOTO and J. SMIT: *Nucl. Phys. B*, **192**, 100 (1981).
[31] A. DUNCAN, R. ROSKIES and H. VAIDYA: University of Pittsburgh preprint PITT-82-6 (1982).
[32] J. KOGUT, M. STONE, H. W. WYLD, S. H. SHENKER, J. SHIGEMITSU and D. K. SINCLAIR: *Nucl. Phys. B*, **225**[FS9], 326 (1983); H. KLUBERG-STERN, A. MOREL, O. NAPOLY and B. PETERSON: *Nucl. Phys. B*, **190**[FS3], 504 (1981). The Saclay group is particularly active elucidating symmetries of lattice fermions.
[33] R. FUCITO, E. MARINARI, G. PARISI and C. REBBI: *Nucl. Phys. B*, **180**[FS2], 369 (1981).
[34] H. HAMBER and G. PARISI: *Phys. Rev. Lett.*, **47**, 1792 (1981); D. WEINGARTEN: *Phys. Lett. B*, **109**, 57 (1982).
[35] D. CALLAWAY and A. RAHMAN: *Phys. Rev. Lett.*, **49**, 613 (1982).
[36] J. POLONYI and H. W. WYLD: *Phys. Rev. Lett.*, **51**, 2257 (1983).
[37] J. POLONYI, H. W. WYLD, J. KOGUT, J. SHIGEMITSU and D. K. SINCLAIR: *Nucl. Phys. B*, **251** [FS13], 311 (1985).
[38] F. FUCITO and S. SOLOMON: *Phys. Rev. Lett.*, **55**, 2641 (1985).

The Deconfinement Transition in Finite-Temperature Lattice Gauge Theory.

F. KARSCH (*)

CERN - Geneva, Switzerland

1. – Introduction.

It is generally expected that at high temperatures and/or densities strongly interacting matter undergoes a fundamental change in behaviour—hadronic matter turns over into a quark-gluon plasma where quarks and gluons are free to move around in a colour screened environment.

During the recent years lattice gauge theory has contributed a lot to our understanding of the behaviour of QCD at high temperatures. With the lattice approach to quantum field theories [1] methods well known from statistical mechanics became available for the analysis of the phase structure of QCD. This allowed us to make contact between the behaviour of SU_N gauge theories and well-studied spin systems [2] and for the first time led to a rigorous existence proof of a deconfining transition in pure SU_N gauge theories [3] for arbitrary bare gauge couplings g.

The nature of the SU_N phase transition has been explored in detail by Monte Carlo simulations for $N=2$ [4-6], 3 [7-12] and recently also for $N=4$ [13-15]. Complete agreement about the order of the phase transition has been established between these MC results and theoretical considerations based on universality arguments [16] as well as mean-field studies [17-20]. In addition the influence of lattice artifacts on the transition parameters has been analysed [21] and for SU_3 the scaling behaviour of the deconfinement temperature at intermediate couplings [22, 23] has been shown to be in agreement with the nonperturbative SU_3 β-function determined by Monte Carlo renormalization group methods [24].

Compared to this rather complete picture gained for the finite-temperature behaviour of pure SU_N theories the situation in the presence of virtual quarks is still in an exploratory stage. First results [25-30] on the phase structure of SU_3 with fermions differ in their conclusions on the existence and/or order

---

(*) Address after October 1, 1984: Department of Physics, University of Illinois at Urbana-Champaign, 1110 West Green Street, Urbana, IL 61801.

of a phase transition in the presence of quarks of finite mass. We will discuss the present status of these calculations in sect. **4**. After a brief introduction in the formalism of finite-temperature lattice gauge theories in sect. **2**, we will review in sect. **3** the status of the deconfinement transition in pure SU_N gauge theories concentrating on a discussion of the order of the deconfinement transition and the scaling behaviour of thermodynamic quantities. Section **5** contains our conclusions.

2. – Thermodynamics of SU_N gauge theories.

A formulation of the thermodynamics of quantum field theories suitable for the discussion of the lattice approach utilizes the path integral representation of the partition function $Z = \text{Tr} \exp[-\beta H]$ [31], where $T = \beta^{-1}$ is the temperature and H is the Hamiltonian describing the quantum system. In the Euclidean path integral formulation the partition function is given by

$$(2.1) \qquad Z(\beta, V) = \int \mathcal{D}A \exp[-S(\beta, V)]$$

with the action

$$(2.2) \qquad S(\beta, V) = -\int_0^\beta dx_0 \int_V d^3x \, \mathcal{L}(A(x_0, \boldsymbol{x})).$$

The bosonic (fermionic) fields A obey periodic (antiperiodic) boundary conditions at the limits of the temporal integral in eq. (2.2), i.e. $A(\beta, x) = +(-)A(0, x)$.

For the moment we will only consider bosonic fields A. The Lagrangian $\mathcal{L}$ of a SU_N gauge theory is then given by

$$(2.3) \qquad \mathcal{L}(A) = -\tfrac{1}{4} F^a_{\mu\nu} F^a_{\mu\nu}$$

with

$$(2.4) \qquad F^a_{\mu\nu} = \partial_\mu A^a_\nu - \partial_\nu A^a_\mu - g f^{abc} A^b_\mu A^c_\nu, \qquad \mu, \nu = 1, ..., 4, \quad a, b, c = 1, ..., N.$$

Here f^{abc} are the structure functions of the SU_N group. From the partition function we obtain all thermodynamic quantities as suitable derivatives with respect to the temperature or volume. For instance, the energy density is obtained as

$$(2.5) \qquad \varepsilon = -\frac{1}{V} \frac{\partial}{\partial \beta} \ln Z(\beta, V).$$

At high temperatures the SU_N gauge theories are asymptotically free and can be analysed by ordinary perturbation theory. At lowest order this yields the thermodynamics of an ideal Bose gas with $2(N^2-1)$ degrees of freedom. The energy density is then given by the Stefan-Boltzmann form

$$(2.6) \qquad \varepsilon_{\rm SB} = \frac{N^2-1}{15}\pi^2 T^4.$$

However, at finite but still large T there are corrections due to gluon interactions. To $O(g^2)$ one finds for the partition function [32, 33]

$$(2.7) \qquad \frac{1}{V}\ln Z = \frac{1}{45}\pi^2 T^3(N^2-1)\left[1 - g^2\frac{25}{64\pi^2}\right],$$

which leads to deviations from the ideal-gas limit. In order to study the thermodynamics of the gluonic gas at lower temperatures, perturbation theory is not suitable. A nonperturbative analysis of this regime is, however, possible in the lattice approach proposed by WILSON [1]. Introducing a lattice of size $N_\sigma^3 \times N_\tau$ and lattice cut-offs a (a_t) in space (time) direction such that temperature and volume are given by $\beta = N_t a_t$ and $V = (N_\sigma a_\sigma)^3$ the lattice regularized partition function can be written as

$$(2.8) \qquad Z(\beta, V) = \int \prod_{x,\mu} {\rm d}U_{x,\mu} \exp\left[-S(U)\right]$$

with

$$(2.9) \qquad S(U) = 2N\left[\xi g_\sigma^{-2}\sum_{\{P_\sigma\}}\left(1 - \frac{1}{N}{\rm Re\,Tr\,}U_{x,\mu}U_{x+\mu,\nu}U^+_{x+\nu,\mu}U^+_{x,\nu}\right) + \right.$$
$$\left. + \xi^{-1}g_\tau^{-2}\sum_{\{P_\tau\}}\left(1 - \frac{1}{N}{\rm Re\,Tr\,}U_{x,0}U_{x+0,\mu}U^+_{x+\mu,0}U^+_{x,\mu}\right)\right].$$

In eq. (2.8) $U_{x,\mu} \in SU_N$, ${\rm d}U_{x,\mu}$ denotes the Haar measure of the group, the summation in the action runs over all spacelike $(\{P_\sigma\})$ and timelike $(\{P_\tau\})$ plaquettes of a hypercubic lattice and $\xi = a/a_\tau$ is the ratio of lattice spacings in space and time directions. Notice that the temperature dependence is now hidden in the asymmetry parameter ξ as well as in the temporal extent N_τ of the lattice. In addition the couplings $g_{\sigma(\tau)}$ depend both on the asymmetry ξ and on the lattice spacing a. For $\xi = 1$ they are identical, $g_\sigma(a, 1) = g_\tau(a, 1) = g(a)$.

The energy density is then given by

$$(2.10) \qquad \varepsilon = \frac{\xi}{N_\sigma^3 N_\tau a^4}\frac{\partial}{\partial \xi}\ln Z|_{\xi=1} = 6N\{g^{-2}(\langle P_\sigma\rangle - \langle P_\tau\rangle) - c'_\sigma\langle P_\sigma\rangle - c'_\tau\langle P_\tau\rangle\},$$

where $\langle P_{\sigma(\tau)}\rangle$ denotes the average values for space (time)-like plaquettes and the coefficients $c'_{\sigma(\tau)}$ are the derivatives of the space (time)-like couplings with respect to ξ. They can be evaluated in weak-coupling perturbation theory and for $\xi = 1$ they are given by [34]

$$(2.11) \quad \begin{cases} c'_\sigma = 4N \left\{ \dfrac{N^2-1}{32N^2} 0.586\,844 + 0.000\,499 \right\}, \\ c'_\tau = 4N \left\{ -\dfrac{N^2-1}{32N^2} 0.586\,844 + 0.005\,306 \right\}. \end{cases}$$

The energy density given by eq. (2.10) should be normalized by subtracting the zero-temperature contribution [6], which in practice is taken to be the result of eq. (2.10) for $N_\tau = N_\sigma$. A perturbative calculation of the energy density on a finite lattice [35, 36] shows that it is generally strongly affected by finite-size effects, especially due to the discretization of the temperature direction. This has to be taken into account when comparing Monte Carlo data with continuum results.

Besides the local gauge invariance the gluonic action eq. (2.9) has also a global Z_N symmetry which allows us to transform, in a given hyperplane of fixed time, all timelike link variables by multiplying them with an element of the centre of the group

$$(2.12) \quad U_{x,0} \to z U_{x,0}, \quad z \in Z_N, \quad \forall x \equiv (x_0, \boldsymbol{x}) \text{ with } x_0 \text{ fixed}.$$

While the action is invariant under this global Z_N transformation, expectation values of Wilson loops with a net winding number n around the temporal direction (which is closed due to periodic boundary conditions) transform nontrivially under this transformation if $\text{mod}_N(n) \neq 0$. The simplest observable of this type is the Polyakov line

$$(2.13) \quad L(\boldsymbol{x}) = \prod_{x_0=0}^{N_\tau} U_{(x_0,\boldsymbol{x}),0}.$$

It measures in addition the excess free energy of a static colour source in the gluonic heat bath, $\langle \text{Tr}\, L(\boldsymbol{x}) \rangle \sim \exp[-\beta F]$. Thus there is a close relation between the deconfinement transition and the breaking of the global Z_N centre symmetry:

$$\langle L \rangle \begin{cases} = 0, \text{ confinement phase} \leftrightarrow \text{unbroken } Z_N \text{ symmetry}, \\ \neq 0, \text{ deconfined phase} \leftrightarrow \text{broken } Z_N \text{ symmetry}. \end{cases}$$

The expectation value of the Polyakov line $\langle L \rangle = \langle N^{-1} \sum \text{Tr}\, L(\boldsymbol{x}) \rangle / N_\sigma^3$ is thus an order parameter for the deconfinement transition in pure SU_N gauge model. In the following section we will discuss in detail the nature of this transition for different N.

3. – The SU_N deconfinement transition.

The existence of a deconfinement transition in all SU_N lattice gauge theories has been proved rigorously by BORGS and SEILER [3]. On a given lattice of size $N_\sigma \times N_\tau$ and asymmetry ξ they establish a bound on the bare coupling g^2 below which $\langle L \rangle$ is nonzero. The SU_N theory is thus in the deconfined phase for

$$(3.1) \qquad g^2 \leqslant g_c^2 = \frac{1}{\xi N}\left[\left(1 + \frac{\alpha}{1.5164}\right)^{1/N_\tau} - 1\right]^{-1}, \qquad \alpha = \begin{cases} 2, & SU_2, \\ 1, & N > 2. \end{cases}$$

However, the above bound is quite weak and does not rule out the possibility that the physical transition temperature moves to infinity in the continuum limit. Thus one has to verify that the critical temperature stays finite when the lattice cut-off is removed, i.e. we have to verify that physical quantities scale according to the SU_N β-function when the cut-off is changed. This means that T_c should behave like

$$(3.2) \qquad T_c^{-1} = N_\tau a(g^2)$$

with $a(g^2)$ given by

$$(3.3) \qquad a(g^2)\Lambda_L = \left(\frac{11 N g^2}{48\pi^2}\right)^{-51/121} \exp\left[-\frac{24\pi^2}{11 N g^2}\right].$$

In addition we are interested in the details of the critical behaviour at the phase transition like the order of the transition and critical exponents.

3'1. The order of the deconfinement transition.

As has been pointed out by SVETITSKY and YAFFE [16], the critical behaviour of the SU_N gauge models in $d + 1$ dimensions is closely related to an effective d-dimensional Z_N spin theory. The basic idea is to integrate out in the partition function, eq. (2.8), those degrees of freedom which do not have long-range fluctuations at the critical coupling. The resulting effective theory will be a spin model in terms of the Polyakov line L with short-range interactions only which lies in the same universality class as the original theory. The integration over irrelevant degrees of freedom can be performed explicitly in the strong-coupling limit. Integrating out the spacelike plaquettes leads in lowest-order strong coupling to the partition function [16-20, 37]

$$(3.4) \qquad Z_{\text{eff}} = \int \prod_x dL(x) \exp\left[\beta' \sum_x \sum_{e=1}^3 \operatorname{Tr} L(x)\operatorname{Tr} L(x + e) + \text{c.c.}\right]$$

with

$$(3.5) \qquad \beta' = \beta^{N_\tau} + \dots$$

This is a Z_N spin model with nearest-neighbour interaction only. However, due to the Haar measure $dL(x)$ the spins $\text{Tr } L(x)$ feel a complicated potential.

In the following we will discuss theoretical expectations for the order of the deconfinement transition based on universality arguments [16] and a mean-field analysis of the above effective spin model [17-20] and compare with MC results obtained for different SU_N groups.

SU_2. – As the effective spin model has a global Z_2 symmetry, it lies in the same universality class as the 3-dimensional Ising model. One thus expects a second-order phase transition for the SU_2 gauge theory.

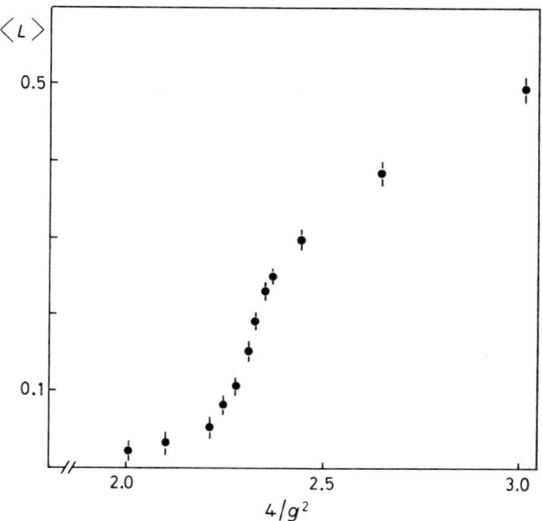

Fig. 1. – The order parameter of the SU_2 gauge model vs. $4/g^2$ on a $10^3 \times 4$ lattice. The data in the cross-over region are taken from ref. [38].

In fig. 1 we show the MC results for the Polyakow line for the SU_2 gauge model on a $10^3 \times 4$ lattice. There is a sudden increase of $\langle L \rangle$ around $\beta \approx 2.3$ which signals the onset of the deconfined phase. Using the asymptotic scaling relation eq. (3.3) leads to a critical temperature $T_c = 43 \Lambda_L$. The second-order nature of the transition is supported by a detailed analysis of the order parameter near the critical point [38]. Here $\langle L \rangle$ is expected to vary like

$$(3.6) \qquad \langle L \rangle \sim |T - T_c|^\beta, \qquad T \to T_c^+,$$

with a critical exponent $\beta = 0.33$ like in the 3-d Ising model. The MC data [10, 38] are consistent with this scaling law.

In fig. 2 we show the energy density for the SU_2 model on a $10^3 \times 3$ lattice (normalized to the free-gas limit ε_{SB} calculated on a lattice of the same

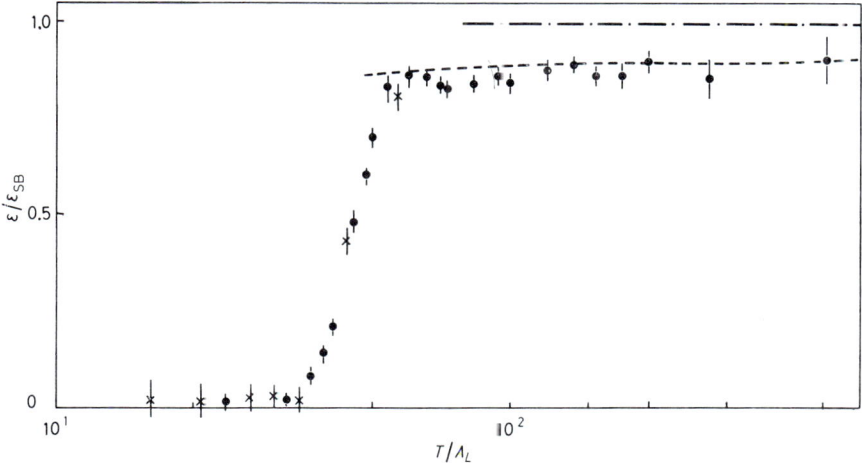

Fig. 2. – The energy density of the SU_2 gluon gas on a $10^3 \times 3$ lattice normalized to the free gas on a lattice of same size. Also shown are the $O(g^2)$ corrections ($---$) to the ideal-gas behaviour.

size [35]). Again there is a sudden rise around $T_c \approx 43 \Lambda_L$. Above the phase transition, the energy density rapidly approaches the free-gas limit ε_{SB} given by eq. (2.6). The approach to this limit is in agreement with the $O(g^2)$ weak-coupling corrections [36] shown also in fig. 2.

SU_3. As no fixed point is known in the class of globally Z_3 symmetric theories, universality arguments suggest a first-order phase transition for SU_3 [16]. Indeed this is what has been found in MC simulations [9, 10]. The first-order nature of the transition has been confirmed by the observation of

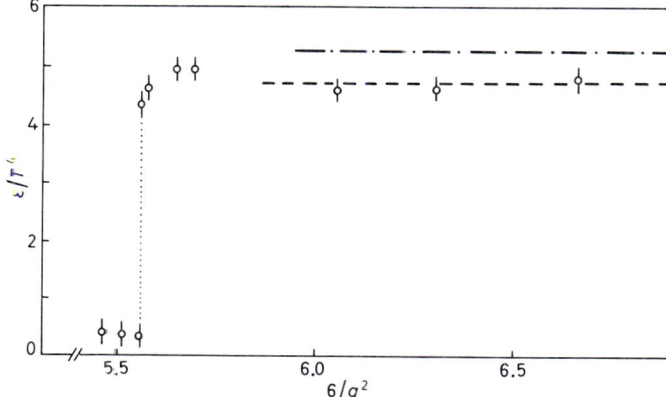

Fig. 3. – The energy density of the SU_3 gluon gas on a $8^3 \times 3$ lattice. Also shown are the $O(g^2)$ corrections ($---$) to the ideal-gas behaviour (dash-dotted line).

coexisting states [9] and recently also metastable states [39] in microcanonical MC simulation. Thermodynamic quantities like the order parameter or the energy density are discontinuous at the transition point. In fig. 3 we show the energy density of the SU_3 gluon gas on a $8^3 \times 3$ lattice. There is a clear jump at $6/g^2 = 5.55$ with a latent heat $\Delta \varepsilon / T_c^4 \simeq 4$. Above the transition the energy density is again in agreement with perturbative weak-coupling results [36] and rapidly approaches the ideal-gas limit. The application of the asymptotic scaling relation eq. (3.3) leads to a critical temperature $T_c = 86 \Lambda_L$ for $N_\tau = 3$. In the next subsection we will discuss in detail the scaling behaviour of T_c when N_τ is varied.

SU_N, $N \geqslant 4$. For $N > 3$ universality arguments are not applicable as the order of the transition is not unique in the class of globally Z_N symmetric models for $N \geqslant 4$ [16]. The mean-field and MC analysis of the effective spin model, however, favours a first-order phase transition [17-20, 40] for all $N \geqslant 3$. In fig. 4 we show the result of a MC study of the order parameter for the SU_4 effective spin model [40]. The broad hysteresis clearly indicates a strong first-order phase transition.

The MC analysis of the SU_N gauge models for $N \geqslant 4$ is complicated due to the fact that there is an additional first-order bulk phase transition for the

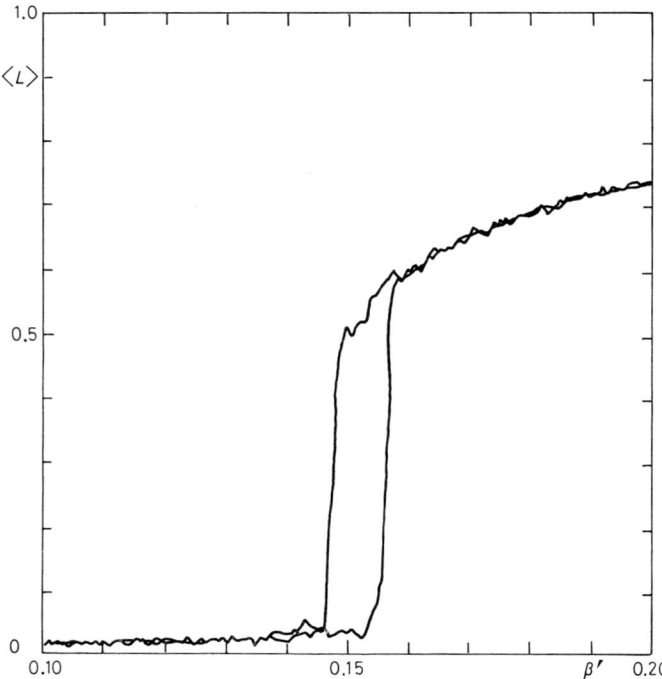

Fig. 4. – The order parameter for the SU_4 effective spin model on a 7^3 lattice vs. β'. Shown is a thermal cycle where β' was increased from 0.1 to 0.2, and then decreased back to 0.1, in steps of $\Delta\beta' = 0.001$, with 50 iterations per step.

standard Wilson action. One thus has to go to rather large lattices to separate the deconfinement transition from the bulk transition, use a mixed action [14] or work on a lattice with large asymmetry ξ [41]. Recent MC simulations for the SU_4 gauge model [13-15] and reduced models [41, 42] support the first-order nature of the transition.

3'2. *Scaling properties of thermodynamic quantities.* – For SU_3 the critical temperature has by now been measured on lattices with various extent N_τ in the time direction (= various values of g^2). For $N_\tau = 2, 3, 4$ T_c could be determined by a direct observation of coexisting states at the critical coupling β_c [9]. However, this becomes more and more difficult for increasing N_τ as the discontinuity in the observables between the ordered and disordered phase becomes smaller (*). For $N_\tau = 5, 6$ only bounds on β_c have been established this way. However, a recent measurement of T_c based on the observation of phase flips among the different Z_3 sectors [23] in combination with a finite-size analysis determines β_c for lattices up to $N_\tau = 10$. These results lead for $N_\tau = 6$ to a smaller β_c than determined in ref. [22] which partly may be due to the application of the finite-size scaling for nearly symmetric lattices. However, also the large relaxation time on the large lattices used in ref. [22] might lead to a misidentification of the phases. The existing data are summarized in table I.

TABLE I. – *Summary of critical couplings of the SU_3 deconfinement transition on lattices of size $N_\sigma^3 \times N_\tau$. The rows with $N_\sigma = \infty$ refer to the extrapolated data of ref. [23]. The other data are taken from ref. [21, 22]. Also given are the critical temperatures T_c/Λ_L obtained by assuming the validity of eq. (3.3) and the ratio T_c over square root of the string tension. The string tension data are taken from ref. [43, 44].*

N_τ	N_σ	$6/g^2$	T_c/Λ_L	$T_c/\sqrt{\sigma}$
2	12	5.11 ±0.01	78 ±1	
	∞	5.097±0.001		
3	10	5.55 ±0.01	86 ±1	
4	10	5.70 ±0.01	76 ±1	0.56±0.03
	∞	5.696±0.004		
5	12	5.79 ÷5.82	68.5±1	0.60±0.02
6	16	5.92 ÷5.94	65.5±1	0.61±0.02
	∞	5.877±0.006	62 ±0.3	0.56±0.01
8	∞	6.00 ±0.02	53 ±1.5	0.50±0.02
10	∞	6.22 ±0.07	55 ±4	0.54±0.05

(*) Of course, the gap stays constant in physical units. But the observables are measured in lattice units a. Thus, for instance, the latent-heat gap in the energy density εa^4 drops like N_τ^{-4}. Also the gap in the order parameter is decreasing.

The fourth column of table I gives T_c in units of Λ_L assuming the validity of the asymptotic scaling relation, eq. (3.9). Clearly T_c/Λ_L is not constant in the coupling regime $\beta = 5.5 \div 6.0$ but drops significantly. This, however, is not specific for thermodynamic quantities. The same is true for other observables, for instance the string tension [43, 44]. The fact that these observables are not independent of g^2 shows that the asymptotic scaling relation is not valid in the coupling regime covered by these measurements. However, these observables show similar deviations and in fact their ratio stays more or less constant in the range $\beta = 5.7 \div 6.0$. The ratio $T_c/\sqrt{\sigma}$ is given in the fifth column of table I. From the spread of these values

$$(3.7) \qquad T_c/\sqrt{\sigma} = 0.5 \div 0.6$$

we find for the critical temperature $T_c = (220 \pm 20)$ MeV (with $\sqrt{\sigma} = 400$ MeV). Similarly the dimensionless ratio $\Delta\varepsilon/T_c^4$ stays practically constant when β is varied [21] and gives

$$(3.8) \qquad \Delta\varepsilon/T_c^4 \simeq 4$$

or $\Delta\varepsilon = (1.2 \pm 0.4)$ GeV.

The g^2 independence of dimensionless ratios of physical observables suggests that lattice artifacts are strongly suppressed in this intermediate-coupling regime and continuum physics is already visible. However, physical quantities do not scale according to the asymptotic scaling relation but according to a more general nonperturbative β-function valid at intermediate couplings. This is supported by a recent measurement of the SU_3 β-function using MCRG techniques [24, 45].

4. – Deconfinement in the presence of virtual quarks.

While in many spectroscopical problems the effect of virtual quarks is expected to be small and only quantitative, the introduction of quarks into thermodynamic problems leads to qualitative changes: As fermionic fields in the fundamental representation (quarks) destroy the global Z_N symmetry present in the pure gauge sector, the close relation between symmetry breaking and deconfinement is lost. The phase structure of QCD thus cannot be predicted only on the basis of universality arguments. However, we can analyse the influence of quarks in limiting cases like the heavy-mass approximation.

In the presence of quarks the partition function becomes

$$(4.1) \qquad Z = \int \prod_{x,\mu} dU_{x,\mu} \prod_x d\psi_x d\bar{\psi}_x \exp\left[-S(U, \bar{\psi}, \psi)\right],$$

where the action is

(4.2) $$S(U, \bar{\psi}, \psi) = S_G(U) + S_F(U, \bar{\psi}, \psi).$$

The gluonic contribution S_G to the action is given by eq. (2.9), while the fermionic part S_F is

(4.3) $$S_F(U, \bar{\psi}, \psi) = \sum_{x,y} \bar{\psi}_x Q_{xy} \psi_y$$

with

(4.4a) $$Q \equiv 1 - K \sum_{\mu=0}^{3} M_\mu,$$

(4.4b) $$M_\mu = (r - \gamma_\mu) U_{x,\mu} \delta_{x,y-\mu} + (r + \gamma_\mu) U^+_{x-\mu,\mu} \delta_{x,y+\mu}.$$

Here $r = 1$ for Wilson fermions and $r = 0$ for naive fermions, which can be easily related to the Kogut-Susskind formulation using a spin diagonalizing transformation. The parameter K is given by $K = 1/(8r + 2ma)$ in the free theory.

As the action is quadratic in the fermion variables $\bar{\psi}, \psi$, the corresponding Grassmann integrals in eq. (4.1) can be done leaving «only» a determinant in terms of the gluonic fields $U_{x,\mu}$. The partition function then reads

(4.5) $$Z = \int \prod_{x,\mu} dU_{x,\mu} \exp\left[-S_G(U) - n_f \operatorname{Tr} \ln\left(1 - K \sum_\mu M_\mu\right)\right].$$

For heavy quarks the logarithm can be expanded in terms of the hopping parameter K. The lowest-order contribution (for $N_\tau < 4$) is given in terms of the Polyakov line and we find

(4.6) $$S_{\text{eff}}(U) = S_G(U) - H \sum_x (\operatorname{Tr} L(x) + \text{c.c.})$$

with $H = 2n_f(2K)^{N_\tau}$ for n_f flavours.

Thus in the heavy-quark limit the fermions contribute a Z_N symmetry-breaking term to the action, very similar to an external magnetic field of strength H in spin models. Indeed, in the strong-coupling limit one obtains an effective Z_N spin model in an external field. This led to the speculation that the deconfinement transition present in the pure gauge sector would not survive in the presence of quarks when it was second order in the pure gauge sector, like in the case of SU_2, and would disappear below a critical mass (above a critical field strength) when it was first order [46].

A MC analysis of the heavy-quark model defined by eq. (4.6) showed that for SU_3 indeed the critical mass below which the first-order deconfining transition disappears seems to be large, of the order of ~ 1 GeV [25]. However, one should keep in mind that, besides the deconfining transition, QCD is expected to undergo also a chiral-symmetry-restoring transition at $m = 0$ (*), which at least in the quenched approximation is first order [10]. In the presence of light quarks the chiral transition may still influence the thermodynamics significantly. Thus a generic phase diagram in the (T, m)-plane might look like the one shown in fig. 5. Whether the lines of phase transitions coming

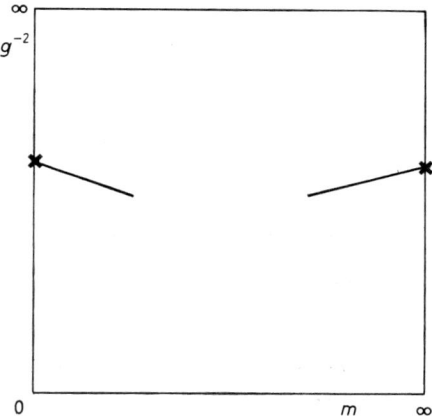

Fig. 5. – Possible phase diagram for the SU_3 gauge theory with fermions. The crosses indicate the deconfinement transition in the pure gauge sector ($m = \infty$) and the chiral transition at $m = 0$.

from the deconfinement transition at $m = \infty$ (pure gauge sector) and the chiral transition at $m = 0$ really enters the phase diagram, how far they reach and whether they are connected in the interior are open questions. Future MC studies will certainly try to explore this phase diagram in more detail. At present there exist results for light quarks by several groups [26-30] which, however, do not allow to draw definite conclusions on the existence or order of a phase transition for light quarks. However, there is no doubt that the change in thermodynamic quantities is still very rapid in a very narrow temperature range. In fig. 6 we show the Polyakov line obtained by GAVAI *et al.* [28] on a $6^3 \times 2$ lattice for two different masses. As can be seen, the change in both quantities is still very dramatic, although there is no sign for a discontinuous transition like in the pure gauge sector [10], which is also shown in

---

(*) For SU_2 lattice gauge theories with Kogut-Susskind fermions, TOMBOULIS and YAFFE have shown that there exists a chiral phase transition on asymmetric lattices [47].

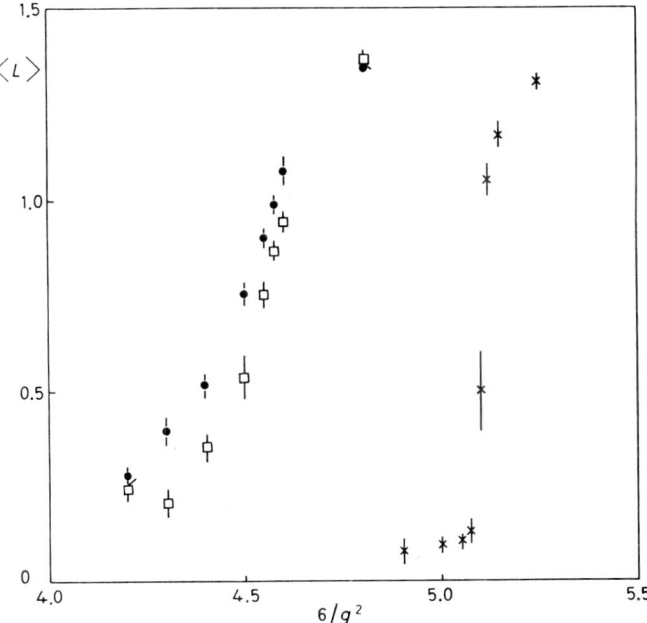

Fig. 6. – The expectation value of the Polyakov line (normalized to 3) in the presence of fermions of mass $ma = 0.1$ (dots) and $ma = 0.2$ (squares) on a $6^3 \times 2$ lattice. The data are taken from ref. [28]. Also shown is $\langle L \rangle$ for the pure gauge theory (crosses) [10].

this figure. It might, however, well be that deeper in the continuum limit (larger N_τ) the transition turns out to be still first order as has been stated in ref. [29].

5. – Conclusions.

The nature of the deconfinement transition in pure SU_N gauge theories is by now well understood. The transition turns out to be second order for SU_2 with critical exponents of the 3-dimensional Ising model. For all $N \geqslant 3$ the transition is first order. The quantitative analysis of the SU_3 transition led to an understanding of the observed scaling violations in terms of the nonperturbative β-function at intermediate couplings and allowed a determination of the critical temperature with an accuracy on the 10% level.

The phase diagram of QCD in the presence of virtual quarks is presently under investigation by many groups. Whether a phase transition as a true singularity of the partition function persists over the entire mass range is uncertain. However, thermodynamic quantities still seem to vary rapidly in a small temperature interval. This indicates that the change from hadronic matter to an asymptotically free quark-gluon plasma will be very rapid and might well lead to observable signals in future heavy-ion experiments.

* * *

I would like to thank J. ENGELS, R. V. GAVAI, P. HASENFRATZ, U. HELLER and H. SATZ for many helpful discussions.

REFERENCES

[1] K. G. WILSON: *Phys. Rev. D*, **10**, 2445 (1974).
[2] A. M. POLYAKOV: *Phys. Lett. B*, **72**, 477 (1978); L. SUSSKIND: *Phys. Rev. D*, **20**, 2610 (1979).
[3] C. BORGS and E. SEILER: *Commun. Math. Phys.*, **91**, 329 (1983).
[4] L. MCLERRAN and B. SVETITSKY: *Phys. Lett. B*, **91**, 195 (1981).
[5] J. KUTI, J. POLONYI and K. SZLACHANYI: *Phys. Lett. B*, **98**, 199 (1981).
[6] J. ENGELS, F. KARSCH, I. MONTVAY and H. SATZ: *Phys. Lett. B*, **101**, 83 (1981); *Nucl. Phys. B*, **205** [FS5], 545 (1982).
[7] K. KAJANTIE, C. MONTONEN and E. PIETARINEN: *Z. Phys. C*, **9**, 148 (1982).
[8] I. MONTVAY and E. PIETARINEN: *Phys. Lett. B*, **110**, 148 (1982).
[9] T. CELIK, J. ENGELS and H. SATZ: *Phys. Lett. B*, **125**, 411 (1983); **129**, 323 (1983).
[10] J. KOGUT, M. STONE, H. W. WYLD, W. R. GIBBS, J. SHIGEMITSU, S. H. SHENKER and D. K. SINCLAIR: *Phys. Rev. Lett.*, **50**, 393 (1983).
[11] J. KOGUT, H. MATSUOKA, M. STONE, H. W. WYLD, S. H. SHENKER, J. SHIGEMITSU and D. K. SINCLAIR: *Phys. Rev. Lett.*, **51**, 869 (1983).
[12] B. SVETITSKY and F. FUCITO: *Phys. Lett. B*, **131**, 165 (1983).
[13] A. GOCKSCH and M. OKAWA: *Phys. Rev. Lett.*, **52**, 1751 (1984).
[14] G. G. BATROUNI and B. SVETITSKY: *Phys. Rev. Lett.*, **52**, 1751 (1984).
[15] J. WHEATER and M. GROSS: *Nucl. Phys. B*, **240**, 253 (1984).
[16] B. SVETITSKY and L. G. YAFFE: *Nucl. Phys. B*, **210** [FS6], 423 (1982).
[17] J. BARTHOLOMEW, D. HOCHBERG, P. H. DAMGAARD and M. GROSS: *Phys. Lett. B*, **133**, 218 (1983).
[18] F. GREEN and F. KARSCH: *Nucl. Phys. B*, **238**, 297 (1984).
[19] J. M. DROUFFE, J. JURKIEWICZ and A. KRZYWICKI: *Phys. Rev. D*, **29**, 2982 (1984).
[20] M. OGILVIE: *Phys. Rev. Lett.*, **52**, 1369 (1984).
[21] T. CELIK, J. ENGELS and H. SATZ: *Z. Phys. C*, **22**, 301 (1984).
[22] F. KARSCH and R. PETRONZIO: *Phys. Lett. B*, **141**, 105 (1984).
[23] A. D. KENNEDY, J. KUTI, S. MEYER and B. J. PENDLETON: *Phys. Rev. Lett.*, **54**, 87 (1985).
[24] A. HASENFRATZ, P. HASENFRATZ, U. HELLER and F. KARSCH: *Phys. Lett. B*, **143**, 193 (1984).
[25] P. HASENFRATZ, F. KARSCH and I. O. STAMATESCU: *Phys. Lett. B*, **133**, 221 (1983).
[26] T. CELIK, J. ENGELS and H. SATZ: *Phys. Lett. B*, **133**, 427 (1984).
[27] M. FISCHLER and R. ROSKIES: *Phys. Lett. B*, **145**, 99 (1984).
[28] R. V. GAVAI, M. LEV and B. PETERSSON: *Phys. Lett. B*, **140**, 397 (1984); **149**, 492 (1984).

[29] F. FUCITO, C. REBBI and S. SOLOMON: *Nucl. Phys. B*, **248**, 615 (1984); *Phys. Rev. D*, **31**, 1461 (1985).
[30] J. POLONYI, H. W. WYLD, J. B. KOGUT, J. SHIGEMITSU and D. K. SINCLAIR: *Phys. Rev. Lett.*, **53**, 644 (1984).
[31] C. BERNARD: *Phys. Rev. D*, **9**, 3312 (1974).
[32] J. I. KAPUSTA: *Nucl. Phys. B*, **148**, 461 (1979).
[33] O. K. KALASHNIKOV and V. V. KLIMOV: *Phys. Lett. B*, **88**, 328 (1979).
[34] F. KARSCH: *Nucl. Phys. B*, **205**[FS5], 285 (1982).
[35] J. ENGELS, F. KARSCH and H. SATZ: *Nucl. Phys. B*, **205**[FS5], 239 (1982).
[36] U. HELLER and F. KARSCH: *Nucl. Phys. B*, **251**[FS13], 254 (1985).
[37] J. POLONYI and K. SZLACHANYI: *Phys. Lett. B*, **110**, 395 (1982).
[38] R. V. GAVAI and H. SATZ: *Phys. Lett. B*, **145**, 248 (1984).
[39] J. KOGUT, J. POLONYI, H. W. WYLD, J. SHIGEMITSU and D. K. SINCLAIR: *Nucl. Phys. B*, **251**, 311 (1985).
[40] F. GREEN and F. KARSCH: *Phys. Rev. D*, **29**, 2986 (1984).
[41] S. R. DAS and J. KOGUT: *Phys. Lett. B*, **144**, 105 (1984); **145**, 375 (1984).
[42] K. FABRICIUS, O. HAAN and F. R. KLINKHAMER: *Phys. Rev. D*, **30**, 2227 (1984).
[43] F. GUTBROD, P. HASENFRATZ, Z. KUNSZT and I. MONTVAY: *Phys. Lett. B*, **128**, 415 (1984).
[44] D. BARKAI, K. J. M. MORIARTY and C. REBBI: *Phys. Rev. D*, **30**, 1293 (1984).
[45] R. GUPTA, G. GUERALNICK, A. PATEL, T. WARNOCK and C. ZEMACH: *Phys. Rev. Lett.*, **53**, 1721 (1984).
[46] T. BANKS and A. UKAWA: *Nucl. Phys. B*, **225** [FS9], 145 (1983).
[47] E. T. TOMBOULIS and L. G. YAFFE: *Phys. Rev. Lett.*, **52**, 2115 (1984).

Spin Glasses.

G. Parisi

Dipartimento di Fisica della II Università « Tor Vergata » - Roma, Italia
Istituto Nazionale di Fisica Nucleare - Laboratori Nazionali di Frascati, Italia

1. – Introduction.

Why spin glasses? Different justifications are possible: new concepts and new tools have been introduced in the study of spin glasses [1-4]; while a serious mathematical treatment is not yet available (present-day mathematics is not enough advanced), the ideas used in spin glasses may be useful for the study of other amorphous materials like glasses, rubber and (why not?) a random lattice space-time or a foamlike space-time. If we exclude these last speculations [5], which depend on an unfortunately lacking synthesis of quantum theory and general relativity, the domain of applications is very large; moreover, if we accept the perverse logic of broken replica symmetry, the basic ideas are relatively easy to grasp also from a particle physicist, who will likely be amused (or confused) by the unusual group theory.

In sect. **2** we will introduce spin glass, and the concept of frustration.

In sect. **3** we will describe the Sherrington-Kirkpatrick model for spin glass and present its solution in the high-temperature phase.

In sect. **4** we will see how to solve the model using the formalism of broken replica symmetry. We will describe the group-theoretical implications of the formalism and we will see how the solution of the model can be reformulated in term of a stochastic differential equation.

Finally, in sect. **5** we will present the physical interpretation of the formalism of the previous section.

2. – Spin glasses.

Spin glasses belong to a large category of models for which we need to compute the free energy without knowing the Hamiltonian, but only the probability distribution of the Hamiltonian. More precisely we write the Hamiltonian of our system (which for definiteness we suppose to be an Ising model, where the spins σ can take only the ± 1 values) as a function of some control parameters J's; the J's are not known, but their probability distribution $P[J]$ is known. For each choice of the J's we can compute the partition function Z_J and the free-energy density F_J defined as usual

$$(1) \qquad Z_J = \sum_{\{\sigma\}} \exp[-\beta H_J[\sigma]], \qquad F_J = -\frac{1}{\beta V} \ln Z_J,$$

V being the total number of spins in the system. A typical form of $H_J[\sigma]$ is

$$(2) \qquad H_J[\sigma] = \sum_{i,k} J_{ik} \sigma_i \sigma_k - \sum_i h\sigma_i,$$

h being the external magnetic field.

Our goal is to compute the average of the free-energy density, i.e.

$$(3) \qquad F = \overline{F_J} = \int dP[J] F_J.$$

In the infinite-volume limit, if J_{ik} decrease faster than $(i-k)^{-\frac{3}{2}}$ when $i-k$ goes to infinity, it can be proved that

$$(4) \qquad \lim_{V\to\infty} \overline{F_J^2} = \left(\lim_{V\to\infty} \overline{F_J}\right)^2.$$

In other words, the free-energy density does not fluctuate from sample to sample (i.e. does not depend on the choice of the J's) with probability one for a given form of $P[J]$: as often happens in thermodynamics, the most likely values coincide with the average values in the infinite-volume limit.

Different models of spin glasses have different form of the J's: in the simplest Sherrington-Kirkpatrick (S-K) model [6] the interaction is infinite range; all $J_{i,k}$, for any pair i, k, have the same probability:

$$(5) \qquad dP[J] = \prod_{(i,k)} dJ_{ik} \exp\left[-\frac{V}{2} \sum_{(i,k)} J_{ik}^2\right].$$

In other words, the J's have a Gaussian distribution with zero average and covariance

$$(6) \qquad \overline{J_{ik} J_{i'k'}} = \frac{1}{V} \delta_{ii'} \delta_{kk'}.$$

In the Edwards-Anderson model [7], which is not very far from being realized in nature, we have the same relations as eqs. (5), (6), where now the sum in eq. (5) runs over the nearest-neighbour pairs (i, k) and all $J_{i,k}$ which are not nearest neighbour are zero; moreover, we have $1/Z$ at the place of $1/V$,

Z being the co-ordination number of the lattice ($2D$ for an hypercubic lattice in D dimensions).

In a very realistic model [8] the position of the spins in real space is random: *i.e.* we associate to each spin a vector $\boldsymbol{x}_i$, which denotes its position; the J's are given by

$$\tag{7} J_{ik} = \frac{\sin(r_{ik} k_{\mathrm{F}})}{r_{ik}^3}, \qquad r_{ik} = |\boldsymbol{x}_i - \boldsymbol{x}_k|.$$

This model represents what is going on in alloys like $\mathrm{Fe_5 Au_{95}}$ where the spins of iron interact only by deforming the Fermi surface of gold: k_{F} in (7) is indeed the Fermi momentum of the host material (in this case gold).

The expert reader has certainly realized that at $h = 0$ (if the probability distribution of the J's is symmetric) the theory is invariant under the local Z_2 gauge transformations: at fixed i we can change

$$\tag{8} J_{ik} \to -J_{ik} \quad \forall k, \; \sigma_i \to -\sigma_i,$$

without any effects on the free energy. The average of quantities which are not invariant under the transformation (8) are obviously zero, *e.g.* $\langle \sigma_i \rangle = 0$. Examples of gauge-invariant quantities are (apart from obvious quantities like internal energy, magnetic susceptibility and so on)

$$\tag{9} q_{\mathrm{EA}} = \overline{\langle \sigma_i \rangle^2}, \qquad \chi_R = \sum_k |\langle \sigma_i \sigma_k \rangle - \langle \sigma_i \rangle \langle \sigma_k \rangle|^2.$$

It is reasonable to suppose that for each choice of the J's at low temperature (at least in high dimensions as we shall see) the system will have a spontaneous local magnetization and $\langle \sigma_i \rangle = m_i$ will be different from zero; however, the sign of m_i will change from point to point and/or from realization to realization of the J's, so that it averages to zero; at low temperature $q_{\mathrm{EA}} = \overline{m_i^2}$ will be different from zero, while it will be zero at higher temperatures. The reader must note that, although the definition of q_{EA} seems clear, we shall see later that it is not as unambiguous as it looks.

One of the typical features of spin glass is that, for a given choice of the J's, it is very difficult to find the ground state, *i.e.* the configuration of spins σ_1 which minimizes the Hamiltonian. Indeed, as a consequence of the randomness of the $J_{i,k}$, it is not possible to fix the σ_i's in such a way that

$$\tag{10} J_{i,k} \sigma_i \sigma_k > 0, \qquad \forall i, k.$$

If eq. (10) could be satisfied, the product of all J_{ik} along a closed loop (the Toulose-Wilson loop [9]) should be positive, but that is impossible, the J's having zero mean. (If the product of the J's along a loop is negative, the loop is said to be frustrated.)

We must decide which bond must be frustrated (*i.e.* is $\sigma_i \sigma_k J_{i,k} < 0$). Different arrangements of the frustrated bonds may differ very little in energy but correspond to very different spin configurations. Briefly, in more than two dimensions all known algorithms for finding the ground state of an N-site spin glass take a time proportional to $\exp[aN]$. This multiplicity of ground states, which are nearly equivalent from the energetic point of view, is responsible for many of the strange properties of spin glasses.

Different models can be analysed in terms of the behaviour of the Toulouse-Wilson loop, which are the only gauge-invariant characterizations of the J's. If for simplicity we consider only loops in which no J appears twice, we have both in the S-K model and in the E-A model

$$\overline{W(C)} = 0 \, , \tag{11}$$

$W(C)$ being the Toulouse-Wilson loop associated to the contour C. We can consider a more general model where the J's are nearest neighbours and they can written as

$$J_{ik} = \frac{1}{\sqrt{M}} \sum_{l=1}^{M} r_i^{(l)} r_k^{(l)} \, , \tag{12}$$

where the r's are random numbers taking only the values ± 1. The $M = 1$ model is gauge equivalent to the Ising model indeed

$$W(C) = 1 \qquad \forall C. \tag{13}$$

For generic M we have that

$$W(C) = \frac{1}{(M)^{L/2}} \, , \tag{14}$$

where L is the length of the loop: for M going to infinity we recover the original E-A model. It is important to note that in more realistic models like those corresponding to eq. (7) the expectation values of the Wilson loops are nonzero, so that they have an intermediate status between the E-A model and the simplified model (12).

The main approach to spin glasses is based on the introduction of replicas: we consider the system with Hamiltonian

$$H^{(n)} = \sum_{i,k} \sum_{a=1}^{n} \sigma_i^a \sigma_k^a J_{ik} \, , \tag{15}$$

where the spins σ's have an additional index a going from 1 to n. We define

$$
(16) \quad \begin{cases} Z^{(n)} = \int \mathrm{d}P[J] \sum_{\{\sigma\}} \exp\left[-\beta H^{(n)}\right] = \int \mathrm{d}P[J]\, [Z_J]^n\,, \\ F^{(n)} = -\dfrac{1}{n\beta V} \ln Z^{(n)}\,. \end{cases}
$$

Equation (16) allows us to continue analytically over n; it is also evident that $F^{(n)}$ in the limit $n \to 0$ reduces to the F of eq. (3); indeed

$$
(17) \quad -\frac{1}{n\beta V} \ln \overline{[Z_J]^n} \simeq -\frac{1}{n\beta V} \ln \overline{[1 + n \ln Z_J]} \simeq -\frac{1}{\beta V} \overline{\ln Z_J}\,.
$$

The strategy consists in formally using the representations (15), (16) also for noninteger n by analytically continuing all relevant formulae from integer n to n equal to zero.

Now eq. (16) describes the normal partition function of a system of n replicas of the same spins coupled to the J's. The integration over the J's can often be done; we generate a direct coupling among the replicas; for example, in the case of the E-A model one finds

$$
(18) \quad \begin{cases} Z^{(n)}_{\text{EA}} = \sum_{\{\sigma\}} \exp\left[-\tilde{H}^{(n)}_{\text{EA}}\right], \\ \tilde{H}^{(n)}_{\text{EA}} = -\dfrac{1}{4D} \beta^2 \sum_{(i,j)} \sum_{a=1}^{n} \sum_{b=1}^{n} \sigma_i^a \sigma_i^b \sigma_j^a \sigma_j^b\,, \end{cases}
$$

and the sum over (i, j) runs over all nearest-neighbour pairs. The reader must recall that the integration over the J's was one of the major difficulties in the usual formalism, which is now by-passed by the introduction of the replicas; we have just put the dirt under the carpet: it will come out soon.

We can now proceed as in the Ising model to derive the mean-field approximation. Using the relation $\sigma^2 = 1$, we can write

$$
(19) \quad Z^{(n)}_{\text{EA}} = \sum_{\{\sigma\}} \int \mathrm{d}[Q] \exp\left[\beta^2 \sum_i \sum_{a=1}^{n} \sum_{b=1}^{n} Q^i_{ab} \sigma_i^a \sigma_i^b - \frac{\beta^2}{D} \sum_{i,k} \sum_{a=1}^{n} \sum_{b=1}^{n} Q^i_{ab} Q^k_{ab} G(i-k)\right],
$$

where $G(i)$ satisfies the relation

$$
(20) \quad \sum_j G(i-j) = \delta_{i,0}
$$

and the sum over j in eq. (20) runs over the nearest points of i; in other words,

$$
(21) \quad G(k) = \frac{1}{(2\pi)^D} \int_B \mathrm{d}^D p \exp\left[ipk\right] \tilde{G}(p)\,, \qquad \tilde{G}(p) = \sum_{\nu=1}^{D} 2\cos p_\nu\,.
$$

The correctness of eq. (19) can be checked by integrating back over the Q's.

As usual, the perturbative expansion can be constructed by starting from the mean-field theory: *i.e.* we set

$$Q_{ab}^i = Q_{ab} + \tilde{Q}_{ab}^i \,, \tag{22}$$

Q_{ab} being a constant (*i.e.* it does not depend on i).

If $\tilde{Q}_{ab}^i$ is neglected, the value of Q_{ab} is found by maximizing the argument of the exponent in eq. (19); we easily find that it can be written as $-VA(Q)$, where $A(Q)$ is given by

$$A[Q] = \frac{\beta^2}{4} \sum_{a,b} Q_{ab}^2 + \ln\left[\sum_{\{\sigma\}} \exp\left[-\sum_{a,b} \beta^2 Q_{ab} \sigma^a \sigma^b\right]\right] \,, \tag{23}$$

where the sum over the σ's is done over the 2^n configuration of the n replicas of a single spin. The condition of minimum of $A[Q]$ implies that

$$\frac{\partial A}{\partial Q_{cb}} = 0 = \beta^2[Q_{ab} - \langle \sigma^a \sigma^b \rangle \beta^2 Q_{ab}] \,, \tag{24}$$

which can be written as

$$Q_{ab} = \sum_{\{\sigma\}} \sigma_a \sigma_b \exp\left[-\sum_{c,d} \beta^2 Q_{cd} \sigma^c \sigma^d\right] \Big/ \sum_{\{\sigma\}} \exp\left[-\sum_{c,d} \beta^2 Q_{cd} \sigma^c \sigma^d\right] \,. \tag{25}$$

After the maximum is found, the correction to this approximation can be computed by taking care of $\tilde{Q}$. As usual, these corrections disappear when D goes to infinity. The standard loop expansion can be constructed provided that the Hessian matrix

$$M_{ab,cd} = \frac{\partial^2 A}{\partial Q_{ab} \partial Q_{cd}} \tag{26}$$

has no negative eigenvalues as should be at the point of minimum.

We shall see that the necessity of the analytic continuation in n makes the evaluation of the point of minimum and of the eigenvalues of the Hessian particularly difficult.

3. – The Sherrington-Kirkpatrick model.

The advantage of the S-K model is that the mean-field approximation is exact. Indeed with the same notation as in the previous section we find after the integration over the $J_{i,k}$ [6]

$$
(27) \begin{cases} Z_n = \sum_{\{\sigma_i^a\}} \exp\left[\beta^2 \sum_{i=1}^{N} \sum_{k=1}^{N} \sum_{a=1}^{n} \sum_{b=1}^{n} \sigma_i^a \sigma_k^b + h\beta \sum_{i=1}^{N} \sum_{a=1}^{n} \sigma_i^a\right] = \int_{-\infty}^{+\infty} dQ_{ab} \exp\left[-NA(Q)\right], \\ A[Q] = -\frac{\beta^2}{4} + \frac{1}{4} \sum_{a=1}^{n} \sum_{b=1}^{n} \beta^2 Q_{ab}^2 + \ln\left[\text{tr}\left[\exp\left[\sum_{a=1}^{n} \sum_{b=1}^{n} Q_{ab} S_a S_b + \beta h \sum_{a=1}^{n} S_a\right]\right]\right], \end{cases}
$$

where the indices a and b label the different replicas of the spin system; Q_{ab} is an $n \times n$ matrix, zero on the diagonal, and tr stands for the sum over the 2^n possible values of the n Ising spin variables S_a.

Up to now, we have done legal operations. When N goes to infinity we would like to use the saddle point method and write

$$
(28) \qquad \tilde{F} = -\frac{1}{\beta} \lim_{n \to 0} \left[\min_Q A(Q)\right],
$$

as discussed in the previous section. This means that, if pathological results are found, this cannot be due to the inadequacy of the mean-field approximation.

Let us start and evaluate eq. (28). We have been not very careful in defining what we mean when we say that we should solve the problem at integer n and analytically continue the solution in n up to $n = 0$. We will make strong use of the fact that $A[Q]$ is invariant under the action of the symmetry group P_n, the permutations of n objects: indeed it is evident that, if we permute different rows and columns of Q, $A[Q]$ does not change; in particular, we will pay attention only to the definition of P_n symmetric functionals of Q. Very often the P_n group is called the replica symmetry group because it exchanges different replicas.

The meaning of eq. (28) is no clearer than that of a Delphic oracle: should we find the minimum at $n \neq 0$ and analitically continue to $n = 0$? No. The analytic continuation of a minimum may be a maximum. The minimum should be found directly at $n = 0$. However, the number of variables corresponding to the Q_{ab} is $n(n-1)$ which is negative for $0 < n < 1$: the concept of a minimum of a function depending on a negative number of variables is rather subtle! Everybody would say that the minimum of $n^{-1} \text{Tr} Q^2$ is at $Q = 0$; however, if we set

$$
(29) \qquad Q_{ab} = q \quad \forall a, b \qquad (Q_{a,a} = 0),
$$

we find

$$
(30) \qquad n^{-1} \text{Tr} Q^2 = (n-1) q^2 .
$$

The point $q = 0$ is (for $n < 1$) a maximum as a function of q. The solution to this apparent paradox is quite simple: the condition that the Hessian matrix (26) has positive eigenvalues does not imply that $\langle x|H|x \rangle$ is positive, if $|x\rangle$ belongs to a negative dimensional space (e.g., the trace of the identity is equal to the dimension of the space). A moment of reflection is needed to understand that the necessary condition for the use of the saddle point method is that all the eigenvalues of the Hessian must be nonnegative. This also guarantees that all the susceptibilities, which are positive definite, are positive indeed.

The final interpretation of eq. (28) is the following: we must consider all possible analytic families of matrices $Q_{ab}^{(n)}$ which may depend on real parameters q_i or integer parameters m_i. An analytic family is an infinite set of matrices (one for each n multiple of n_0) such that the P_n invariant quantities, e.g.

$$(31) \qquad \text{Tr}\, Q^k \quad \text{or} \quad \sum_{a=1}^{n} \sum_{b=1}^{n} (Q_{ab})^k ,$$

are analytic functions of n. For each analytic family we should compute the analytic continuation in n up to $n = 0$ of the function $F(Q)$ and of the eigenvalues of the Hessian. The final result will be given by that analytic family (hopefully unique) whose elements are stationary points of $A[Q]$, for all n multiples of n_0, and the eigenvalues of the Hessian analytically continued at $n = 0$ are nonnegative. As far as one can construct analytic families of matrices which depend analytically on the integer parameters m_i, one is allowed also to consider noninteger values of the m_i's.

While it is not clear if this interpretation gives the correct result, it does make the problem well defined from the mathematical point of view but very hard to control from the practical point of view: the space of all analytical families is very large. Up to now the only approach has been to construct ansätze. Let us see how this works.

The first case we study is

$$(32) \qquad Q_{ab} = q, \qquad\qquad a \neq b.$$

After some simple algebraic manipulations [6] we find that

$$(33) \qquad A(q) = -\frac{\beta^2}{4}(1+q^2) - \frac{1}{(2\pi)^{\frac{1}{2}}} \int_{-\infty}^{+\infty} dz \, [\exp[-z^2/2] \ln[2\cosh(\beta q^{\frac{1}{2}} z + \beta h)]] .$$

The Hessian will have one eigenvector in the one-dimensional space. One can immediately check that for $n < 1$ the corresponding eigenvalue is positive if $A(q)$ is a maximum (not a minimum!) as a function of q.

One finally finds at $h = 0$ (after a rescaling of q)

$$(34) \begin{cases} q = 0, & T > T_c = 1, \\ q_{EA} = q = \int_{-\infty}^{+\infty} \frac{dz}{(2\pi)^{\frac{1}{2}}} [\exp[-z^2/2] \operatorname{tgh}^2(\beta q^{\frac{1}{2}} z)] \neq 0, & T < 1, \\ U(T) \to -\frac{1}{2}\sqrt{\pi} = -0.798, \quad q(T) \sim 1 - \sqrt{\pi T} & (T \sim 0), \\ C(T) \sim T & (T \sim 0), \\ \chi(0) \sim \sqrt{\pi}/2, \quad S(0) = -1/2\pi, \end{cases}$$

where U, C, χ and S are the internal energy, the specific heat, the susceptibility and the entropy. Now the Monte Carlo results tell us that $U(0) \simeq \simeq -(0.76 \div 0.77)$ in small but definite disagreement with eq. (34), and the specific heat is quadratic in T; on the other hand, the dependence of $q(T)$ is qualitatively correct (apart from the fact that Monte Carlo data suggest $q(T) \sim 1 - aT^2$). Unfortunately the entropy becomes negative at low T and a negative Ising system entropy cannot be tolerated.

The theoretical explanation of this failure has been found by DE ALMEIDA and THOULESS [10]; they noticed that when $T < \tilde{T}(h)$, where

$$(35) \begin{cases} \tilde{T}(h) = T_c - |h|^{\frac{2}{3}} & \text{for } |h| \ll 1, \\ \tilde{T}(h) \sim \ln |h| & \text{for } |h| \to \infty, \end{cases}$$

one of the eigenvalues of the Hessian becomes negative: this is certainly not acceptable, also because χ_R is proportional to the inverse of this eigenvalue. The previous computation is wrong when $T < \tilde{T}(h)$ and eq. (32) is not the correct choice. Unfortunately eq. (32) is the only possible ansatz which is P_n symmetric, we need, therefore, to break spontaneously the replica symmetry.

4. – The solution.

It must be clear that at the present moment the only way to solve eq. (28) is by trying different ansätze and check if the positivity condition on the eigenvalues of the Hessian is satisfied.

The only known construction [1, 11] (up to equivalences) is the following: as a preliminary step we divide the n replicas in n/m groups of m replicas (of course, n must be multiple of m). We set $Q_{ab} = q_0$ if a and b belong to the same group, $Q_{ab} = q_1$ if a and b belong to different groups (Q_{aa} is always zero). In other words,

$$(36) \begin{cases} Q_{ab} = q_0 & \text{if } I(a/m) = I(b/m), \\ Q_{ab} = q_1 & \text{if } I(a/m) \neq I(b/m), \end{cases}$$

where $I(x)$ is an integer-valued function: its value is the smallest integer greater than or equal to x. Equation (36) provides us with an example of an analytic family of matrices, depending on the parameters q_0, q_1 and m.

It is evident that

$$\lim_{n \to 0} \left(\frac{1}{n} \operatorname{Tr} Q^2 \right) = (1-m) q_0^2 - m q_1^2 \tag{37}$$

is not negative definite if $m > 1$; we must maximize it with respect to q_0 and minimize it with respect to q_1: this automatically leads to a free energy worse than the one obtained in the previous section. However, if $0 < m < 1$, it is always negative (obviously m is no more an integer, but we are allowed to do this). After some tedious algebra we get at $h = 0$

$$\begin{cases} A(q_0, q_1, n) = -\dfrac{\beta^2}{4} [1 + m q_1^2 + (1-m) q_0^2 - 2 q_0^2] - \\ \qquad\qquad - \int dp(z) \dfrac{1}{m} \ln \left[\int dp(y) \cosh^m [\beta (q_1^{\frac{1}{2}} z - (q_0 - q_1)^{\frac{1}{2}} y)] \right], \\ dp(z) \equiv \exp[-z^2/2] dz/(2\pi)^{\frac{1}{2}}. \end{cases} \tag{38}$$

Maximizing A with respect to q_0 and q_1 and m (restricted to the interval 0-1), we obtain the following surprising results: the curves for V, C and q_{EA} (assuming $q_{\text{EA}} = \max Q_{ab} = q_0$) are in very good agreement with the Monte Carlo data (e.g., $V(0) = -0.7652$); the free energy is obviously higher than that obtained in the previous section. The entropy at zero temperature has collapsed from $S(0) \simeq -0.16$ to $S(0) \simeq -0.01$.

We are clearly on the right track! In order to generalize eq. (36), let us do some unusual group theory. Equation (36) corresponds to breaking the P_n group in the following way

$$P_n \to (P_m)^{\otimes n/m} \otimes P_{n/m'}. \tag{39}$$

Indeed we can permute both the replicas inside each group (and this leads to the product of n/m times P_m) and the groups among themselves (this leads to $P_{n/m}$). In the limit $n \to 0$ we have the following pattern of symmetry breaking

$$P_0 \to (P_m)^{\otimes 0} \otimes P_0. \tag{40}$$

In other words, P_0 contains itself as a subgroup! It is clear now that we can go on and repeat the same operation many times; we introduce a set of integer numbers m_i ($i = 0, ..., k-1$) such that $m_0 = 0$ and $m_{k+1} = n$ and m_i/m_{i-1} is an integer (for $i = 1, ..., k+1$). We can divide the n replicas in n/m_k groups of m_k replicas, each group of m_k replicas is divided in m_k/m_{k-1} groups of m_{k-1}

replicas and so on. The matrix Q will be given by

(41) $\begin{cases} Q_{ab} = q_i & \text{if } I\left(\dfrac{a}{m_i}\right) \neq I\left(\dfrac{b}{m_i}\right) \\ \text{and} \\ I\left(\dfrac{a}{m_{i+1}}\right) = I\left(\dfrac{b}{m_{i+1}}\right), & i = 0, ..., k, \end{cases}$

and the q_i's are a set of $k+1$ real parameters. For $k = 1$ recover the previous example and for $k = 0$ we recover the unbroken symmetry theory.

An easy computation shows that

(42) $$-\lim_{n \to 0}\left(\frac{1}{n}\operatorname{Tr} Q^2\right) = \sum_{i=1}^{k}(m_i - m_{i+1})q_i^2.$$

Condition $\operatorname{Tr} Q^2 < 0$ is satisfied only if

(43) $$0 \leqslant m_{i+1} \leqslant m_i \leqslant 1.$$

From now on let us assume that this condition is valid.

For each value of k, one can compute the free energy by maximizing it with respect to the q's and the m's. An explicit computation shows that, near T_c, $F^{(k)}$ contains a term proportional to

(44) $$(T - T_c)^5/(2k+1)^4;$$

we are naturally led to consider the case $k \to \infty$. In order to keep track of the parameters q_i and m_i, it is convenient to consider the function

(45) $$q(x) = q_i, \qquad m_{i+1} < x < m_i.$$

There is a one-to-one correspondence between the piecewise constant functions with k discontinuities and the parameters q_i and m_i. In the limit $k \to \infty$, $q(x)$ becomes a generic L^2-function over the interval 0-1.

Let us now compute the free energy: after some calculations one arrives to the surprising result [11]

(46) $\begin{cases} -\beta F = \max\limits_{q(x)} A[q], \\ A[q] = -\dfrac{1}{4}\beta^2\left[1 + \displaystyle\int_0^1 q^2(x)\,dx + 2q(1)\right] - a[q], \\ a[q] = f(0, h), \end{cases}$

where the function $f(x, y)$ satisfies the following differential equation:

$$\text{(47)} \qquad \frac{\partial f}{\partial x} = -\frac{1}{2}\frac{dq}{dx}\left[-\frac{\partial^2 f}{\partial y^2} + x\left(\frac{\partial f}{\partial y}\right)^2\right]$$

with the boundary condition

$$\text{(48)} \qquad f(1, y) = \ln[2\cosh(\beta y)].$$

Equation (46) is correct only if $q(0) = 0$, otherwise

$$\text{(49)} \qquad a[q] = \int_{-\infty}^{+\infty} dp(z) f(0, h + \sqrt{q(0)}\, z).$$

A long argument shows that $q(0) \sim |h|^{\frac{2}{3}}$ and

$$\chi = 1 - O(h^{\frac{2}{3}}).$$

When we cross the AT line $x_\mathrm{m} \to x_\mathrm{M} \neq 0$, the following semi-empirical rules are exact, or well satisfied:

$$\text{(50)} \qquad \begin{cases} q_\mathrm{m}(\beta, h) = q_\mathrm{m}(h), \quad q_\mathrm{M}(\beta, h) = q_\mathrm{M}(\beta), \quad q(x) = q_\mathrm{m}, & x < x_\mathrm{m}, \\ q(x, \beta, h) = q(x\beta), & x_\mathrm{m} < x < x_\mathrm{M}, \\ q(x) = q_\mathrm{M}, & x > x_\mathrm{M}. \end{cases}$$

Numerical investigations support the hypothesis that $S(0) = 0$ in this scheme; the ground-state energy estimated is $U(0) \simeq -0.7633 \pm 10^{-4}$. Apart from the region of very small temperature, the results are very similar to those obtained for $k = 1$.

Before discussing in the next section the physical implications of the replica symmetry breaking, we can associate to eq. (46) a stochastic differential equation.

We first define the function $x(q)$ as the inverse of the function $q(x)$; we assume for simplicity that $q(0) = q_\mathrm{m} = 0$: the function $x(q)$ will be defined in the range 0-q_M. We consider a function $w(q)$ which satisfies the following stochastic differential equation [4, 12-14]:

$$\text{(51)} \qquad \dot{w}(q) = \eta(q) - \beta x(q)\, m(w, q),$$

where $\eta(q)$ is a white noise,

$$\text{(52)} \qquad \overline{\eta(q_1)\eta(q_2)} = \delta(q_1 - q_2),$$

and $w(0) = 0$.

The function $m(h, q)$ is fixed by the self-consistency condition

$$(53) \qquad m(h, q) = \overline{\operatorname{tgh}\left(\beta w(q_\mathrm{M})\right)}\Big|^{q_\mathrm{M}}_{q}\Big|_{w(q)=h},$$

where the strange notation indicates that $m(h, q)$ is the average of the $\beta w(q)$ over all the trajectories such that $w(q) = h$, the average being done over the noise η in the interval q-q_M. This self-consistency condition can be solved for any choice of $x(q)$; the function $x(q)$ is fixed by the extra self-consistency relation

$$(54) \qquad q = \overline{m^2(w(q), q)} = \int \left(\overline{\operatorname{tgh}\,\beta w(q_\mathrm{M})}\Big|^{q_\mathrm{M}}_{q}\Big|_{w(q)=h}\right)^2 \mathrm{d}h\,\Big|^{q}_{0}.$$

It is a relative simple exercise in the theory of stochastic differential equations to show the equivalence of eqs. (46)-(48) and eqs. (51)-(54).

As was stressed in ref. [13], the following remarkable relations hold

$$(55) \qquad \langle\sigma_i\rangle^k = \overline{m^k(w(q_\mathrm{M}), q_\mathrm{M})}\Big|^{q_\mathrm{M}}_{0} = \overline{\operatorname{tgh}^k(w(q_\mathrm{M}))}\Big|^{q_\mathrm{M}}.$$

Although these last equations are very suggestive of a simple physical interpretation and are related to the time evolution of the system, their precise meaning is not yet fully understood.

The most important result is that an explicit evaluation shows that the Hessian has negative eigenvalue and that the scheme here described is free of the inconsistencies of the previous replica symmetric approach [15, 16].

5. – The physical interpretation.

In this section we show how to interpret in a physical way the strange form of the matrix Q. Although the information contained in this section should be enough to derive eqs. (46)-(48) in a simple straightforward way, I never succeeded in finding such a derivation: I sincerely hope that some of the readers will solve this problem.

The first thing to do is to be more careful on the precise definition of $q_\mathrm{EA} = \langle\sigma_i\rangle^2$. Indeed we must define the magnetization $m_i \equiv \langle\sigma_i\rangle$ for a given choice of the J's.

If we consider a real system (or a computer simulation), the magnetization m_i is defined [16] by

$$(56) \qquad m_i = \frac{1}{t}\int_0^1 \mathrm{d}\tau\,\sigma_i(\tau),$$

where $\sigma_i(\tau)$ is the value that σ_i takes at the time τ and t is a large (macroscopic) but not too large observation time. For example, in a d-dimensional ferromagnetic system of size L, t must satisfy the conditions

$$(57) \qquad t_{\mathrm{m}} \ll t \ll t_{\mathrm{M}} \approx t_{\mathrm{m}} \exp[L^{d-1}],$$

where t_{m} is the microscopic relaxation time, e.g. one Monte Carlo step. When we change the initial conditions (e.g. $\sigma_i(\tau)$ at $\tau = 0$), we may obtain different results: in the Ising ferromagnetic case below T_{c} there are essentially two possibilities ($m_i > 0$ or $m_i < 0$); it is important to note that, if the initial state is disordered, the approach to equilibrium for quantities invariant under the global Z_2 is slow (there are corrections proportional to powers of t), while one needs a time t at least proportional to the volume ($t > L^d$) in order to establish a translationally invariant state.

What do we expect for a spin glass? There will be many minima of the thermodynamic potential which are separated by very high walls. At relative short times the system will remain near one minimum, later on a at very large (macroscopic?) time the system will start jumping from one minimum to the other one, by thermodynamic tunnel effects [17].

Let us denote by $m_i^{(\alpha)}$ the expectation value of σ_i in the state labelled by α. We have approximately

$$(58) \qquad q(t) \equiv \frac{1}{V} \sum_{i \in V} \left[\frac{1}{t} \int_0^t \sigma_i(t') \, dt' \right]^2 = \frac{1}{V} \sum_{i \in V} \left[\sum_{\alpha=1}^{M(t/t_{\mathrm{M}})} (m_i^{(\alpha)})^2 / M(t/t_{\mathrm{M}}) \right],$$

where t_{M} is a macroscopic time $M[1] = 1$ and when $t \to \infty$ the sum runs over all possible states of the system. In this language $q(t_{\mathrm{M}}) = q_{\mathrm{EA}}$, while $q(\infty)$ is obviously much smaller ($q(\infty)$ is zero at $h = 0$).

The concept of average around a minimum of the thermodynamic potential can be sharpened in more general terms [2]: if we denote by $\langle\ \rangle_{\mathrm{G}}$ the expectation values in the Gibbs state obtained by considering the infinite-volume limit of the usual canonical distribution, it is possible that the Gibbs state is not clustering, in other words the connected correlation functions do not go to zero at large distances; for example, in the usual Ising model at $H = 0$, $T = T_{\mathrm{c}}$ we have

$$(59) \qquad \langle \sigma_i \rangle = 0, \quad \langle \sigma_i \sigma_j \rangle \xrightarrow[i-j \to 0]{} m_{\mathrm{s}}^2,$$

where m_{s} is the spontaneous magnetization.

Under general hypotheses the Gibbs state can be decomposed as the sum of pure clustering equilibrium state

$$(60) \qquad \langle\ \rangle_{\mathrm{G}} = \sum_{\alpha} w_{\alpha} \langle\ \rangle_{\alpha},$$

where the connected correlation functions vanish at large distances and the DLR equations [18] are satisfied for each state α. In the case of the Ising model we have

(61) $$\langle\;\rangle_G = \tfrac{1}{2}\langle\;\rangle_+ + \tfrac{1}{2}\langle\;\rangle_-, \qquad \langle\sigma_i\rangle_+ = -\langle\sigma_i\rangle_- = m_s.$$

Let us assume, as a working hypothesis, that the decomposition is nontrivial for spin glasses and let us try to see the implications of this assumption. In each state α we can define an E-A order parameter

(62) $$q_\alpha = \lim_{V\to\infty} \frac{1}{V} \sum_{i=1}^{V} (m_i^\alpha)^2, \qquad m_i^\alpha = \langle\sigma_i\rangle_\alpha,$$

which is likely to be independent of α.

Similarly we would like to define the overlap $q_{\alpha\beta}$ (and the distance $\delta_{\alpha\beta}$) of two states:

(63) $$\begin{cases} q_{\alpha\beta} = \lim_{V\to\infty} \dfrac{1}{V} \sum_{i=1}^{V} m_i^\alpha m_i^\beta, \\ \delta_{\alpha\beta} = \lim_{V\to\infty} \dfrac{1}{V} \sum_{i=1}^{V} (m_i^\alpha - m_i^\beta)^2 = q_\alpha + q_\beta - 2q_{\alpha\beta} = 2(q_{\text{EA}} - q_{\alpha\beta}). \end{cases}$$

Our aim is to compute the probability $P_J(q)$ for two states having overlap q, i.e.

(64) $$P_J(q) = \sum_{\alpha\beta} w_\alpha w_\beta\, \delta(q - q_{\alpha\beta}).$$

We could also define the average over the J's of $P_J(q)$

(65) $$P(q) = \overline{P_J(q)}.$$

In principle (and also in practice) the function $P_J(q)$ may change when we change the realization of the J's.

It is convenient to consider the quantities

(66) $$Q_J^{(k)} = \int \mathrm{d}q\, P_J(q)\, q^k.$$

An easy computation shows that

(67) $$Q_J^{(k)} = \frac{1}{V^k} \sum_{i_1} \cdots \sum_{i_k} |\langle \sigma_{i_1} \ldots \sigma_{i_k}\rangle_G|^2 =$$
$$= \frac{1}{V^k} \sum_{i_1} \cdots \sum_{i_k} \sum_\alpha \sum_\beta w_\alpha w_\beta \langle\sigma_{i_1}\ldots\sigma_{i_k}\rangle_\alpha \langle\sigma_{i_1}\ldots\sigma_{i_k}\rangle_\beta,$$

where we have used the clustering properties of the pure state to do the approximation

$$\langle \sigma_{i_1} \ldots \sigma_{i_k}\rangle_\alpha = m_{i_1}^\alpha \ldots m_{i_k}^\alpha, \tag{68}$$

which becomes exact when the points $i_1,\ldots, i_k$ are widely separated.

In the replica formalism we could naively write

$$\overline{Q_J^{(k)}} = \frac{1}{V^k} \sum_{i_1} \cdots \sum_{i_k} \langle \sigma_{i_1}^1 \sigma_{i_1}^2 \ldots \sigma_{i_k}^1 \sigma_{i_k}^2 \rangle = Q_{1,2}. \tag{69}$$

However, when the replica symmetries are broken, the off-diagonal elements of Q depend on the indices and the choice of the first and second replicas is arbitrary. The correct formula is obtained by summing over all possible choices of the two replicas [19]

$$\overline{Q_J^{(k)}} = \lim_{n\to 0} \frac{1}{n(n-1)} \sum_{ab=1}^n Q_{ab}^k = \int dx\, q^k(x), \tag{70}$$

where the last equation follows from the assumed form of the matrix Q [20] (cf. eq. (42)).

We finally find that

$$P(q) = \frac{dx}{dq}, \tag{71}$$

where $x(q)$ is is the inverse of $q(x)$.

When the replica symmetry is exact, $q(x)$ is a contant and $P(q)$ is a delta-function (only one pure state); otherwise $P(q)$ is not a delta-function and the state structure is much more complex.

Much information can be obtained on the structure of the states in a similar way; for example [3].

$$P^2(q_1, q_2) \equiv \overline{P_J(q_1)P_J(q_2)} = \tfrac{2}{3} P(q_1)P(q_2) + \tfrac{1}{3} P(q_1)\delta(q_1 - q_2). \tag{72}$$

We can also compute the probability for the states α, β, γ of having overlap q_{12}, q_{23}, q_{31}, i.e.

$$P(q_{12}, q_{23}, q_{31}) = \sum_{\alpha,\beta,\gamma} w_\alpha w_\beta w_\gamma \delta(q_{\alpha\beta} - q_{12})\delta(q_{\beta\gamma} - q_{23})\delta(q_{\gamma\alpha} - q_{31}) = \tag{73}$$

$$= \lim_{n\to 0} \frac{\sum_{a,b,c=1}^n \delta(q_{12} - Q_{ab})\delta(q_{23} - Q_{bc})\delta(q_{31} - Q_{ca})}{n(n-1)(n-2)} =$$

$$= \tfrac{1}{2} P(q_{12}) x(q_{12}) \delta(q_{12} - q_{23})\delta(q_{12} - q_{13}) +$$
$$+ \tfrac{1}{2} [P(q_{12}) P(q_{23}) \delta(q_{12} - q_{23})\delta(q_{23} - q_{13}) + 2 \text{ permutations}].$$

If we go from the overlap to distances, this last equation implies that

$$P(\delta_{12}, \delta_{23}, \delta_{13}) = 0 \quad \text{if } \delta_{13} > \max(\delta_{12}, \delta_{23}). \tag{74}$$

In other words, the states form an ultrametric space. Briefly the pure states have very interesting properties when the replica symmetry is broken and these properties can be computed using the same techniques as in eqs. (72), (73) [3].

The reader will certainly realize that the whole machinery is too heavy and lot of work is needed to simplify the approach.

It is evident that we have in our hands the information needed to compute the corrections to the mean-field theory. The first results show that in momentum space [15]

$$（75) \qquad \sum_{ab} \tilde{Q}_{ab}(k)\tilde{Q}_{ab}(-k) \sim \frac{1}{k^3} \qquad (k \approx 0),$$

where, for simplicity, we have neglected logarithmic factors. Apparently the $1/k^3$ behaviour is the effect of the «Goldstone boson» associated to the spontaneous breaking of the replica symmetry.

Equation (75) suggests that the lower critical dimension is 3 and that we can associate to the three-dimensional spin glass an asymptotically free theory. A careful analysis is needed to prove or to disprove these suggestions.

A very interesting and open field is the generalization of this approach to other systems where the Hamiltonian is fixed. In these cases it is possible that the system has many equilibrium states, the number of these equilibrium states may depend on the volume also in the infinite-volume limit. In this case the substitute of eqs. (65), (72) should be likely

$$(76) \qquad P(q) = \lim_{V \to \infty} \frac{1}{V} \int^V dV' P_{V'}(q),$$

$$(77) \qquad P^2(q_1, q_2) = \lim_{V \to \infty} \frac{1}{V} \int^V dV' P_{V'}(q_1) P_{V'}(q_2).$$

It is likely that eqs. (75), (76) and eqs. (65), (72) are equivalent in the spin glass case.

Apparently we have in our hands a tool which may be used to describe and to predict the behaviour of many amorphous systems. Further investigations are needed to find its precise range of applications.

REFERENCES

[1] G. Parisi: *Phys. Rev. Lett.*, **43**, 1754 (1979).
[2] G. Parisi: *Phys. Rev. Lett.*, **50**, 1946 (1983).
[3] M. Mèzard, G. Parisi, N. Sourlas, G. Toulouse and M. Virasoro: *Phys. Rev. Lett.*, **52**, 725 (1984); *J. Phys. (Paris)*, **45**, 843 (1984).

[4] For a review on spin glasses see the various contributions to the *Heidelberg Colloquium on Spin Glasses, Heidelberg, June 1983*, edited by L. VAN HEMMEN and I. MORGENSTEIN (Springer-Verlag, Berlin, 1984); see also G. PARISI: in *Recent Advances in Field Theory and Statistical Mechanics*, edited by J. B. ZUBER and R. STORA (North-Holland, Amsterdam, 1984), which we follow rather closely in many points (we warn the reader that eq. (6.22) of that paper is partially wrong).
[5] G. PARISI: *Phys. Lett. A*, **73**, 531 (1979).
[6] D. SHERRINGTON and S KIRKPATRICK: *Phys. Rev. Lett.*, **35**, 1792 (1975); *Phys. Rev. B*, **17**, 4384 (1978).
[7] S. F. EDWARDS and P. W. ANDERSON: *J. Phys. F*, **5**, 965 (1975); **6**, 1927 (1976).
[8] For a realistic review of spin glasses see R. RAMMAL and J. SOULETIE: *Spin Glasses* (Grenoble, 1981).
[9] G. TOULOUSE: *Commun. Phys.*, **2**, 115 (1977).
[10] J. R. J. DE ALMEIDA and D. J. THOULESS: *J. Phys. A*, **11**, 983 (1978).
[11] G. PARISI: *J. Phys. A*, **13**, L115, 1101, 1887 (1980).
[12] G. PARISI: *Phys. Rep.*, **67**, 25 (1980).
[13] J. R. L. DE ALMEIDA and E. J. S. LAGE: *J. Phys. C*, **16**, 939 (1982).
[14] H. J. SOMMERS and W. DUPONT: Julich preprint (1984).
[15] C. DE DOMINICIS and I. KONDOR: *J. Phys. Lett.*, **45**, L210 (1984), and references therein; A. V. GOLTSEV: *J. Phys. A*, **14**, 237 (1984).
[16] This discussion is taken from G. PARISI: in *Disordered Systems and Localization*, edited by C. CASTELLANI, C. DI CASTRO and L. PELITI (Springer-Verlag, Berlin, 1981), p. 107.
[17] A. P. YOUNG: *Phys. Rev. Lett.*, **49**, 685 (1982).
[18] R. L. DOBRUSHIN: *Theory Probab. Its Appl. (Eng. Transl.)*, **13**, 194 (1969); O. E. LANFORD III and D. RUELLE: *Commun. Math. Phys.*, **13**, 194 (1969).
[19] C. DE DOMINICIS and A. P. YOUNG: *J. Phys. A*, **46**, 2063 (1983).
[20] Equation (70) has been firstly derived by C. DE DOMINICIS and C. ITZYCKSON: private communication (1980).

Computers and Theoretical Physics.

A. E. TERRANO

Physics Department, Columbia University - New York, N.Y. 10027

The subject of these lectures is Computers and Theoretical Physics. In them, I will describe two innovative uses being made of computers in scientific research. These lectures are introductory; they assume only modest programming experience, and no prior knowledge of the design and operation of computers. They include an overview of the organization of computers and an introduction to a number of programming techniques which are not widely known among physicists, and proceed to a discussion of two advanced topics. The first topic is the design and construction of machines with unconventional architectures. These machines are being used to investigate problems which cannot be addressed with commercially available machines. The second topic is the design and implementation of programs for manipulating algebraic expressions. With these programs, it is possible to obtain the exact solution of problems which are too massive to be carried out by hand. These two topics have been chosen because I have detailed experience with each of them, and because discussing them leads to a broad view of computation in theoretical physics; they are not meant to be an exhaustive list of the innovative uses of computers being made by physicists.

The outline of the lectures is as follows:

1) The architecture of conventional computers.

2) The design of special-purpose machines.

3) Elements of modern programming.

4) Algebraic and interactive programs.

1. – The architecture of conventional computers.

The organization of any computer system, of both the hardware and the software, is highly stratified. At each stratum, a program is able to control the operation of the computer by calling subroutines which make use of the lower levels of the system. The use of a computer at any given level does not require an understanding of the lower levels of the system, but only of these system functions at the present level. In this lecture, we will start at the topmost level of a conventional or von Neumann computer and descend quite a distance into its innards.

One's usual interaction with a computer is through a programming language. The elements of the language are, typically, numbers, arrays of numbers and arithmetic statements. The digital circuitry in a computer, on the other hand, is able only to perform logical operations on individual bits of data. Thus none of what appears in a program is directly intelligible to a computer. Instead, it serves only as the input to another program, the compiler, which transforms the (nearly) mathematical expressions in the program into the numerical codes, or machine instructions, which can then be decoded by the circuitry in the computer, finally resulting in the desired sequence of Boolean operations being carried out. Before taking a closer look at what these machine instructions are and how they work, we need to have an overall picture of the elements of a computer.

The « intelligent » element of a computer is the central processing unit, called the CPU for short, which performs the two tasks of interpreting the machine instructions in a program and transforming the data as required. The data are organized into words containing a number of bits: the precise number varies from one computer to another; common values are 16 bits in microcomputers and 32 bits in minicomputers. The memory in a CPU is typically quite limited: in a general-purpose computer, the CPU will contain on the order of 10 general-purpose registers, *i.e.* it will be able to remember only 10 words at a given moment. The main storage of the computer is provided in two forms: a small amount of fast semiconductor memory and a large amount of slower magnetic memory on disks and tapes. The memory is typically organized as a linear series of words (or sometimes 8-bit chunks called bytes). Each location has an address, which is just the number of words between the beginning of memory and its location in the series: when the contents of a memory location is needed, the CPU sends the desired address to the memory, and, in return, the memory sends the data stored at that address to the CPU. In a general-purpose computer the interconnection between these elements is provided by a single common data path, called a bus, which allows any one pair of the elements—the CPU and the memory, or the disk and the memory, or the CPU and a terminal, etc.—to exchange data at any given moment. Such an organization, while being completely general purpose, can result in inefficient use of the CPU, and, indeed, one of the principal tasks in designing a computer is deciding on a bus structure and distribution of memory which allows the processors to work continuously.

While most compilers translate statements in a programming language directly into machine instructions, it is useful to consider an intermediate level of computer control called assembly language, or, simply, assembler. Assembler differs from normal programming languages in several respects. In the first place, high-level system subroutines like read or write are not available. Typically, the programmer has to do all of the data formatting and may even have to know hardware details to be able to transfer data between the various components of the computer. However, full access to the hardware is available. In the second place, all of the machine instructions in the instruction set of the CPU are directly accessible. High-level languages may in fact use the entire set of machine instructions, but the user may be unable to control when they are employed. For example, in most processors, a word stored in a register can be used in two ways: as numerical data, or as the memory address of another word. When a variable is used to store the location of another variable, it is called a pointer, and is said to point to the variable whose location it contains. In FORTRAN, although the compiler will use a register to hold a pointer when making array references, there is no way for the programmer to explicitly use a number as an address value or pointer. The final difference between assembler and a high-level language is the absence of compound instructions, like if(...) then(...) else(...) or do i = 1, 10, or even automatic array indexing. These high-level constructions are accomplished by a series of machine instructions.

As a concrete example, let us consider the following simple FORTRAN program:

```
              j=0
              do 100 i=1, 10
   100        j=j+i
```

In PDP-11 assembler this becomes (comments follow; 's)

```
    i:      word                ;allocate memory location i
    j:      word                ;allocate memory location j
            mov     0, j        ;move the value 0 into location j
            mov     0, i        ;move the value 0 into location i
    loop:                       ;put a label at the beginning of the repeated code
            add     i, j        ;add the value at i to that at j
            add     1, i        ;add one to the value at i
            cmp     i, 10       ;compare the value at i to 10
            jlt     loop        ;if i is less than 10, go to loop
```

Each executable line of assembler corresponds to a single machine intruction, and the sequence of machine instructions is stored in order in memory. The CPU has a dedicated register, called the program counter, or PC for short, which contains the address of the next instruction to be executed. When the execution of this program begins, the PC contains the address of the first instruction: mov 0, j. The data stored at the location pointed to by the PC are fetched from memory and decoded, to determine which operation is to be performed. In this case the constant 0 is part of the instruction, as is the address of the memory location j. The number 0 is fetched from the program area and then stored in the data area corresponding to j. The final step in the execution of the instruction is to update the PC so that it now points to the address of the next instruction. This is done by simply adding the number of bytes in the instruction which has just been executed to the PC. The next instruction is executed similarly. The label «loop:» simply associates a symbolic name with the location in the code, just as the declarations for i and j associated names with locations in memory. The next instruction, add i, j, causes the value stored at location i to be added to that stored in location j, with the result being stored at j. Not all processors are able to carry out this operation in a single machine instruction. Instead, i first would have to be moved into a register with one instruction, and then the contents of that register added to j with the next.

The final two instructions show how conditional branching is achieved. The CPU maintains several single-bit flags—a typical set would be one each for carry, overflow, zero and negative—which are set after each instruction has been executed, reflecting the outcome. Thus, if the result of adding two numbers was zero, the zero flag would be set, and so forth for each of the flags. The cmp a, b (compare) instruction subtracts the value of b from that of a, and sets the flags accordingly, but without saving the actual result. The conditional jump instruction jlt, or jump if less than, looks at the appropriate combination of flags, in this case the negative flag, and, if the condition is met, replaces the contents of the PC with the address of its argument, namely that of «loop»; otherwise PC is updated in the usual manner. In this way, the instructions inside the do-loop are executed the required number of times.

The flow control involved in subroutines is more complex. The additional complications are due to the fact that a subroutine can be called from many different locations in a program, and must return control to the correct location in each case. This difficulty is solved through the introduction of a second dedicated register, called the stack pointer or SP for short. Most programs maintain a region in memory, called the stack, which serves as a scratch pad. The stack is organized as a last-in, first-out data buffer, and the stack pointer contains the address of the next free location on the stack. When a word needs to be stored temporarily, it is pushed onto the stack, *i.e.* written into the location pointed to by the SP, and the SP is then updated to point at the next free location. When a word which has been stored on the stack is needed, it is popped off of the stack, *i.e.* moved into a register and the SP updated to point at the location where the word was previously stored, which is no longer in use and has become the next free location on the stack. The procedure for calling a subroutine consists of first pushing the present value of the PC onto the stack and then loading the PC with the location of the first instruction of the subroutine being called. The final instruction in a subroutine is a return, which pops the old value of the PC off the stack into the PC and updates it to point at the next instruction after the original subroutine call. The stack is also commonly used for passing arguments to a subroutine as well: they are pushed onto the stack in order before the subroutine call; the subroutine can then access them by popping them off the stack.

It should be clear that, far from being an elementary operation, the execution of a single machine instruction involves a number of distinct actions:

1) Fetch the instruction from memory.

2) Decode the instruction.

3) Get data, if necessary.

4) Operate on the data, if necessary.

5) Store data, if necessary.

6) Update the program counter.

Indeed, each machine instruction corresponds to a program, called a microcode program, which is stored in the CPU and re-executed each time the machine instruction is decoded. Each bit of the output of the microcode memory controls the operation of part of the circuitry in the CPU. As an elementary example, we could add to a CPU the operation of inverting the bits in a word by inserting an exclusive-or gate in the data path of each of the bits in a word and connecting the remaining input of the XORs to a microcode bit. Then, the possible states of the circuitry are:

microcode	data	output
low	low	low
low	high	high
high	low	high
high	high	low

Whenever the microcode bit is low, the output of each XOR is the same as the data input, while, whenever the microcode bit is high, the outputs are the opposite of the inputs. The control of a complete CPU is much more complex, but each of the possible states of the circuitry is associated with a unique combination of microcode bits, and, by stepping through a sequence of such states, a complete machine instruction can be performed.

To conclude this section, I want to discuss two architectural concepts which arise in attempts to improve the performance of serial or von Neuman machines. While these techniques involve performing operations in parallel, computers which employ them still execute the machine intructions in a program serially. I will discuss some of the issues involved in true parallel processing in the next section.

The first technique is that of pipelining. If we look back at the sequence of actions involved in the execution of a machine instruction, we can see that it would be possible for several instructions to be processed at once. While one instruction is operating on the data, the next one can be being decoded, and a still later one can be being fetched. If the CPU is designed with separate processors for each task, all of the stages in the execution of a machine instruction can be being carried out simultaneously, but on different instructions. For the sake of simplicity, let us suppose that each of the stages takes one clock cycle to complete. Then, in a nonpipelined CPU, it will take six cycles to complete each machine instruction. In a pipelined CPU, since six instructions can be worked on in each cycle, one instruction will be completed in each clock cycle. It will still take six cycles from when a given instruction is fetched until it is completed, but the effective speed of the CPU will be six times greater than that of the unpipelined CPU. The actual benefit is somewhat less in practice, since, when a jump or a subroutine call is encountered, all of the pre-fetched and decoded, but unexecuted instructions will not be used, and also because the amount of time required by the data transformation step can vary greatly between instructions.

The technique of pipelining can be applied to any compound operation. Another common instance is in floating-point addition, which involves several sequential steps:

1) unpack the numbers—floating-point numbers are often stored in an encoded format with hidden bits;

2) compare exponents;

3) shift the significand of the number with the smaller exponent in order to line up the decimal points;

4) add the significands;

5) normalize and repack the result.

In order to achieve very high floating-point performance, the pipeline can be broken into a large number of very fast stages. If the same operations are to be performed on a lot of data, such a pipeline can be filled, and the effective floating-point rate becomes one operation per clock cycle. Any calculation which can be reformulated as a linear algebra problem, involving the multiplication of matrices and vectors, can be performed efficiently using such a machine, hence the name of vector or array processors. For a given vector machine, there will be a smallest lower limit on the size of the matrices and vectors which can be manipulated efficiently. However, the speed-up can be substantial. For example, a Cray-I supercomputer can carry out floating-point arithmetic at a maximum rate of 160 million operations per second in a fully vectorized calculations, while, for an unvectorized program, the rate can be reduced by a factor of twenty.

2. – Special-purpose machine design.

In this section, we will study the fruits of a fairly recent activity in theoretical physics: the design of special-purpose, high-performance computers. Such machines have been built for studying the properties of the Ising model, of random spin systems and of lattice gauge theories, and for carrying out high-precision integrations of Feynman diagrams. In the first two cases, the computation involves bit operations only, and the CPU can be simple. The number of possible input states is small, and the corresponding output state can simply be tabulated. In the pure Ising case, the «tabulation» is hardwired into the CPU: the output is a simple logical operation on the input [1]. The random spin or link case is more complex, and requires the use of a lookup table. This table can also be hardwired in a sense: the logic table for a given problem can be stored in a programmable chip, and changing problems then simply requires replacing the programmable logic array defining the calculation which the computer is performing [2]. While some parallelism is achieved by packing several bits into a single computer word and operating on the entire word at once, the basic issue in designing these machines is in organizing the accesses to memory so that fetching all of the data needed for an operation takes no longer than the operation itself requires.

The latter two problems are dominated by floating-point arithmetic. With the availability of inexpensive floating-point processors, the goal is to bring many such arithmetic units to bear on the same problem: these machines are true parallel processors. The basic issue which faces one in designing such a computer is the bus structure, or, alternatively, the organization of memory.

In the case of few-dimensional numerical integrations, of the sort which arise in evaluating Feynman diagrams, the parallelism in the problem can consist of evaluating the integrand at several points simultaneously. A simple strategy would be to use several independent computers and statistically combine the results. A more ingenious approach has been devised by LEVINE at Carnegie Mellon University [3], in which a single general-purpose CPU is used to control four floating-point processors simultaneously. With this approach, the bus is busy on every cycle.

The machines used for numerical lattice gauge theory (see [4] for a review) calculations need to provide for sharing data between the processing elements. Two interconnection schemes have been used so far. In a machine which has been designed and built at Caltech [5], $N = 2^k$ processors are connected as a k-dimensional hypercube with two processors along each axis. Such an arrangement guarantees that the time required to move data between any two processors grows only as the logarithm of the number of processors. However, since the number of interconnections also grows logarithmically, the complexity and cost of the individual processors will as well, at least when N is large. It is, therefore, desirable to use a simple internode bus. The Caltech machine uses a bit-serial interconnection, resulting in a fairly low rate of data transfer. This in turn limits the rate at which arithmetic can be performed when one of the operands must be transferred from a neighboring node. However, since the amount of data which needs to be shared in a lattice gauge theory calculation is fairly small, and since each number is used in many operations, the low internode bandwidth is reasonably well matched to the relatively slow processors used in this machine. However, this system will require substantial redesign in order to be able to make full use of the faster processors which will be available in the future.

A rectangular mesh interconnection is being used in a machine which is under construction at Columbia University [6]. Although the time to transfer data between any two nodes in such an arrangement grows as the square root of the number of nodes, the cost of a node is fixed, and a more expensive but much faster internode bus can be afforded. In addition, for lattice gauge calculations, as indeed for many physics calculations, most of the operations are local, involving data associated with physically nearby points. Thus, if the processors are organized in a planar array, and the plane of the processors corresponds to one of the physical planes of the problem, an interconnection scheme which allows nearest-neighbor pairs of processors to exchange data directly will be sufficient. In the remainder of this section, I want to describe this machine in some detail.

Each element of the array consists of a processor and memory. It is useful, however, to think of the system as an array of memories joined to one another by the processing elements, as shown in fig. 2.1. The memory nodes are mapped uniformly onto the (x, y)-plane of the problem: each node is associated with a specific rectangular region in the (x, y)-plane. The regions cover the entire plane and must have the same size, shape and orientation. The data associated with the lattice point (x, y, z, t) are stored in the memory associated with the region in the (x, y)-plane containing the point (x, y).

In lattice gauge theories, the only possible gauge-invariant operations are those which only involve quantities associated with nearest-neighbor sites. Therefore, the interconnections between the nodes need to provide direct communication only between nearest-neighbor pairs of memories. With the interconnections shown in fig. 2.1 there is a unique processor with direct access to any such pair of operands. Although each memory is accessible to more than one processor, no memory access contention will arise if all processors execute the same program completely synchronously. Since lattice gauge calculations are spacially homogeneous, it is possible for each node in the array of processors to be executing identical code.

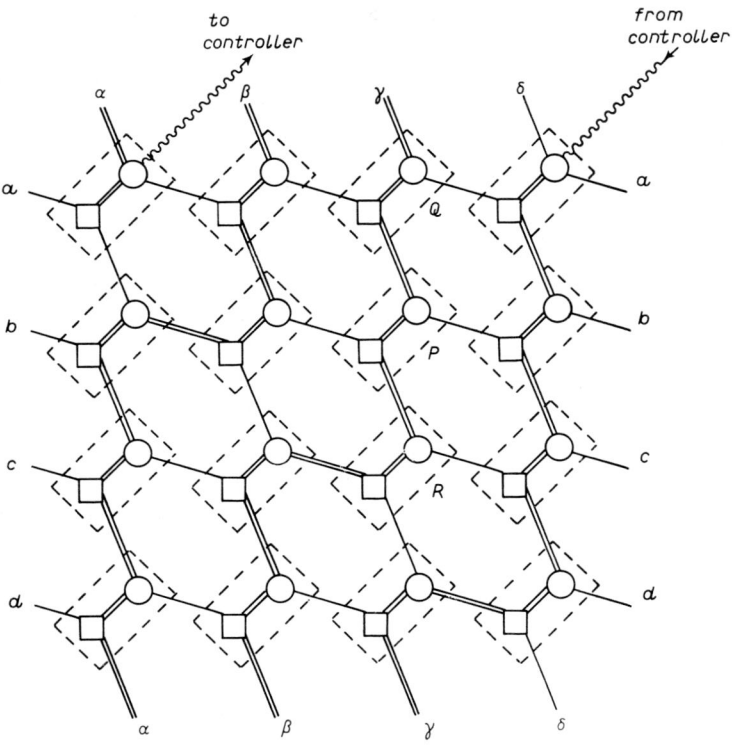

Fig. 2.1.

The requirement of synchronicity makes the use of conditional branching impossible, since different nodes might take different branches. To allow the use of data-dependent branching or node-dependent subroutines, an asynchronous mode is provided, in which only local activity is permitted. As each processor completes a period of asynchronous operation, it issues a request for re-synchronization. When all of the processors have finished, they can then be restarted synchronously.

Often in the problem of interest one imposes « periodic » boundary conditions, effectively joining with links those planes of sites lying on opposite boundaries. These conditions can be realized by our array if the connections indicated in fig. 2.1 between top and bottom and between left and right are provided. Thus the connections in the array form the surface of a two-dimensional torus.

Let us now consider the implementation of this architectural design. Before beginning to build a computer, a decision must be made regarding the level at which the engineering is to be performed. The choices range from networking commercially manufactured computers, to designing circuit boards using off-the-shelf VLSI arithmetic units and CPUs, to building processors from LSI and MSI components, all the way to designing one's own custom integrated circuits. In this machine, we took the approach of using the most powerful VLSI components available, and designing a board which would allow them to be used to their full capacity. The choice for the CPU is one of the microprocessors which are generically referred to as a « VAX on a chip ». These computers can typically execute 1 million instructions per second and cost on the order of $ 100. The arithmetic unit is composed of VLSI floating-point arithmetic chips. The first such unit sold was a 22 + 22 bit adder with a cycle time of 125 ns, for a maximum rate of 8 million floating-point operations per second, or 8 megaflops. These chips use a special format consisting of a 16-bit significand and a 6-bit exponent. A comparable multiplier can be assembled from a 16×16 bit integer multiplier coupled with an integer adder for the exponents. An arithmetic unit built around these two units could possibly perform 16 million floating-point operations each second, at a price in the neighborhood of $ 500. More recently, a chip set which performs full 32 bit arithmetic and runs at the same speed has become available for about $ 1000. With these elements, it is possible to build inexpensive units which will operate in the 10 Mflop range for realistic programs.

The processing element at each node of the array consists of an Intel 80286/287 microprocessor with a $16 K \times 16$ bit program memory and a microprogrammable floating-point vector processor. Each node also contains 128 KB of data memory divided into two independent, simultaneously addressable $32 K \times 16$ bit banks, which can be accessed by both the microprocessor and the vector processor, and a switch to allow the desired pair of (local and/or remote) memories to be addressed. All of the data paths are 16 bits wide. In addition, each node is provided with a standard bus connection to allow inclusion of additional local memory, accessible by the microprocessor. The operation of the array and the transfer of data and code to and from the host computer is directed by a simple central controller which contains a 16 KB data buffer, and provides the control signals and common clock for the array. Mechanically, each node occupies a separate board. The separate nodes are connected to each other and to the central controller by ribbon cable. The architecture of a single node is shown in fig. 2 2.

The microprocessor supervises all the activity on the node. It can read from and write to all the memory on the node as well as the data memories on two of the adjacent nodes. In addition to controlling the vector processor, the CPU must perform all of the scalar processing required to complete the calculation. The CPU is provided with its own independent program memory so that it can operate concurrently with the vector processor: after starting a microcode program, the CPU will continue running until it is ready to start the next program, when it will stop until the vector processor completes the first program.

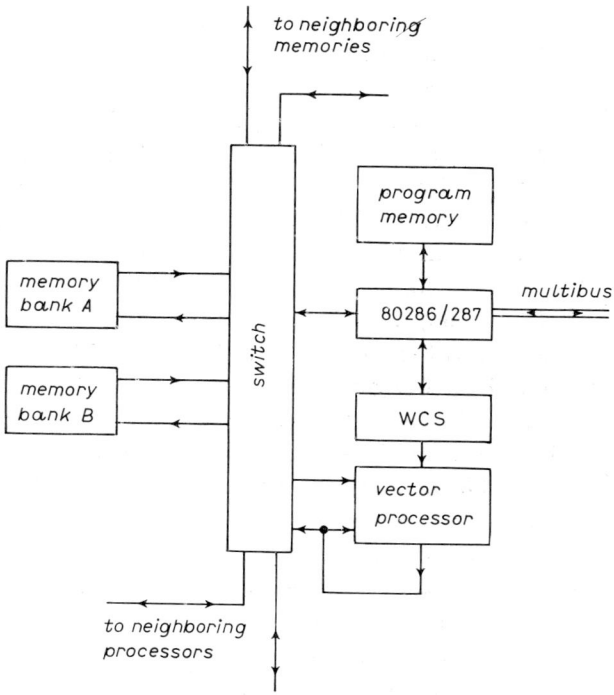

Fig. 2.2.

The floating-point vector processor is pipelined, with 125 ns stages, and is microprogrammed to perform the needed arithmetic operations. It has two independent inputs, each of which can accept a new operand every 125 ns. Each input to the exponent adder has enough temporary memory to store four exponents. The inputs of the multiplier are also separately controlled: by suitably ordering the complex multiplication no additional temporary storage is needed for the incoming significands. The outputs of the multiplier and exponent adder attach directly to the input of the floating-point adder. Since matrix operations involve the accumulation of a series of products, the adder can be run in its accumulate mode and requires only one new operand on every clock pulse. Further, since the adder requires two cycles to complete an addition, it can accumulate both the real and imaginary parts of a sum simultaneously without any external storage. With this arrangement the vector processor can sum a string of products of complex numbers at the maximum rate allowed by the 8 MHz clock: 16 million floating-point additions and multiplications each second.

The operation of the vector processor is controlled by 56-bit microcode. In addition to providing the signals necessary to control the arithmetic unit and switch, it determines the addresses of the source and destination operands. The CPU initially loads up to four base addresses for the operands. In each cycle, for each of the A and B memories, the microcode chooses one of these 16-bit base addresses and supplies an 8-bit offset to be added to it to determine the effective address of the operand. The microcode sequencer is simply a counter driven by the system clock. The CPU initializes the counter to point at the first instruction of the microcode subroutine to be executed. The subroutine must be a simple sequence of operations; no branching is possible. The vector processor is stopped by a microcode bit which disables the counter and releases the CPU from its stopped state, if necessary.

In order to run the vector processor at full speed, two numbers must be obtained from memory every 125 ns. To accomplish this, the memory at each node is split into two independent banks. Since each bank can be accessed every cycle, there is no need for further high-speed registers to feed the vector processor. Either memory bank provides one input to the vector processor on the same node. The other bank feeds the second input of the vector processor on that or on one of the two adjacent nodes. Switching between banks is accomplished at the same time as switching between nodes, and the delay due to all switching is buried in the pipeline. There is no delay for data coming from a neighboring node. Thus the vector processor is fed by internode data transfer at a rate of 16 million bytes per second.

Each node is provided with an independent, general-purpose bus, allowing the possibility of adding additional memory. The conventional memory is not fast enough to feed the vector processor, and, since dynamic memory requires refreshing at random intervals, it is unsuitable for use during synchronous operations. However, it can be used for code which runs asynchronously, in particular for the operating system, and, with some sort of swapping scheme, it can be used for bulk data storage. Similarly, the bus also allows the direct connection of magnetic-disk storage to some or all of the nodes in the array. Data transfer between the host computer and the array is performed in bucket brigade fashion via the bus, with two of the nodes providing the external input and output ports.

The final element is the central controller, which provides some simple global communication and commands: i) It broadcasts a synchronous clock signal to all nodes of the array. ii) It initiates synchronous program execution. iii) It resynchronizes the processors after they have executed differing subroutines. iv) Finally it receives any error signals, and a finished signal from each node and transmits the information to the host.

We now turn to the question of programming this machine. One of the significant benefits of using a commercial microprocessor for the CPU is that the software already exists for compiling high-level languages. This software runs on the host computer, where all of the program development is performed. Writing the microcode programs can be quite arduous, however, and it is advisable to write an assembler for the vector processor which will convert arithmetic operations, such as multiply two complex numbers and add the result to a third, into the proper sequence of microcode instructions. Once this has been done, the programming stategy is fairly straightforward. A conventional program must be organized so that the bulk of the arithmetic is carried out by subroutines. Since these subroutines will be converted into microcode programs using the assembler, they cannot have any branching and must involve only accumulations of products of pairs of numbers. The subroutine calls must be converted into code which starts the corresponding microcode program. The remainder of the program can simply be compiled and used « as is ».

At the present time, benchmarks have been performed using only a single node. For real programs, the nominal maximum speed of 16 Mflops will be reached only momentarily. A more practical measure of its performance is given by running lattice gauge theory Monte Carlo programs. The product of two SU_3 matrices can be taken in 20 µs, corresponding to a speed of 10 Mflops; a 10-hit Metropolis update takes less than 2.0 ms at present. For comparison, a Cray-I has a nominal speed of 160 Mflops, multiplies two SU_3 matrices at a rate of 100 Mflops, and can perform a 10-hit Metropolis update in 75 µs. The cost of producing a single board is less than $ 2500; the cost for the planned system of 256 boards, including the necessary disks and other equipment, will be less than $ 800 K. This system will provide approximately 25 times the performance of a Cray at roughly a tenth of the cost of one.

Lattice gauge theory calculations, in common with a large number of interesting problems in physics, have a number of features which make them well suited to solution by a special-purpose computer. i) The calculation is dominated by floating-point arithmetic, which can be efficiently carried out by a pipelined multiplier-adder. ii) All of the operations are local: only matrices and vectors associated with contiguous links and sites are to be combined. iii) The problem is homogeneous: the same products of the variables associated with the links and sites are to be carried out for all the points in the lattice. The machine which I have described exploits each of these properties; it can be programmed to employ any of the popular lattice gauge theory Monte Carlo algorithms. Nowhere in the design is it restricted to performing SU_3 lattice gauge theory calculations. It can be efficiently used to solve any problem which requires massive amounts of floating-point arithmetic, and which can be formulated so that a given sequence of operations can be performed on several streams of data at once.

3. – Elements of modern programming.

In this section, I want to introduce some modern programming techniques. One thing which all of these techniques have in common is that they are straightforward to employ in modern structured languages and nearly impossible to implement in FORTRAN. You will hear it said that any program which can be written in a modern language can also be written in FORTRAN. This statement is true. It is also true that writing it will take significantly longer, and it will be much harder to get it right. The examples in this section will be presented using the programming language C [7]. While other modern languages such as PASCAL or PL/I are also powerful enough to use, C has become sufficiently popular that compilers are available for essentially all computers from PCs through VAXes and IBMs to Crays. And, being conceptually clearer than FORTRAN, it is easier to learn. I strongly encourage anyone who must undertake a serious computer programming effort to learn it.

In order to provide a context for the techniques under discussion, I want to discuss the design of a special-purpose symbolic-algebra program. In this section, we will design the internal workings of the program; in the next section, we will solve the problem of interacting with the program. The algebraic problem which I want to address is that of the evaluation of the products of gamma-matrices which arise in Feynman diagram calculations. Computers have been used to address this particular problem for over twenty years, and have become indispensable in performing high-precision calculations in quantum electrodynamics. This algebraic system is fairly simple: it can be completely specified in terms of two definitions and two rules. For purposes of controlling the ultraviolet divergences which arise in four space-time dimensions, let us work in N space-time dimensions. We consider a set of N objects, γ_μ, $\mu = 0$, $N-1$, forming an N-vector. We denote the Minkowski-space dot product of γ_μ with an N-vector p_μ by $\hat{p}$. We have the following two axioms:

i) $\hat{p} \cdot \hat{q} - \hat{q} \cdot \hat{p} = 2 p \cdot q$,

ii) $\gamma_\mu \gamma^\mu = N$.

The gammas have a realization as $2^{N/2} \times 2^{N/2}$ matrices; the explicit form of the gammas is not usually needed.

Recursion. The rules i) and ii) above are not particularly suitable as an algorithm for evaluating the trace of products of $\hat{p}$. What we need is a reformulation of the rules which is recursive, *i.e.* which can be applied iteratively until the expression has been completely simplified. It is straightforward to verify that the following relation holds:

iii) $\operatorname{tr}\left(\hat{a} \prod_{i=1}^{n} \hat{b}_i\right) = \sum_{i=1}^{n} (-1)^i (a \cdot b_i) \operatorname{tr} \prod_{j \neq i} \hat{b}_j$.

Applying this relation to an expression will transform it into a sum of similar but simpler expressions: each of the products in the traces on the right-hand side of iii) has two fewer terms than that on the left. If we add the rule that tr (I) = const, then we can recursively apply this rule to an expression until the right-hand side no longer contains any traces.

The implementation of a recursive algorithm is most directly accomplished in terms of a subroutine which calls itself. As a trivial example, consider a function to numerically evaluate the gamma-function. We can use the recursion relation $\Gamma(x+1) = x\Gamma(x)$ to shift the argument of the Γ-function into the interval $[1, 2)$, where an efficient power series expansion exists. A C program implementing this procedure is shown in fig. 3.1.

```
/* declare gamma to be a real-valued function:  */
float gamma(x)
/* declare x to be a real number:  */
float x;
{
/* if x < 1, use the recursion relation to increase
the argument, and call gamma again:  */
if ( x < 1.0 )   return gamma(x+1.0)/(x+1.0);

/* if x > 2.0, use the recursion relation to decrease
the argument, and call gamma again:  */
else if ( x > 2.0 )   return (x-1.0)*gamma(x-1.0);

/* otherwise use the power series to evaluate
gamma(x):  */
else {
    x = x-1.0;
    x = x*(-0.57486+x*(0.95123+x*(-0.69985+
        x*(0.42455-0.10106*x))));
    return 1.0+x;
    }
}
```

Fig. 3.1. – Recursive evaluation of $\Gamma(x)$. Comment statement are enclosed between /* and */ pairs.

This program could not be written in FORTRAN, since a FORTRAN function or subroutine may not call itself. However, a program could have been written to perform the same calculation nonrecursivery by using a do-loop. The replacement of a recursive function call with a do-loop is usually possible; the cost, however, is an increase in complexity and a loss in clarity.

The complete implementation of a recursive program requires the use of a recursive data format. In this example, where the data are a single floating-point number, it could not be simpler. In the symbolic case, where the data are substantially more complicated than in the numerical case, the choice of data type is critical. Before we can discuss the data formats for our gamma-matrix algebra program however, we need to discuss direct address manipulation and structures.

Pointers. In the first section, we saw that a given number, when moved into a register in the CPU, could be regarded as being the address of a piece of data, as well as simply a piece of data. In most modern languages, it is possible to explicitly define a variable to be a pointer to a location. In C, for instance, two unary operators are introduced to allow combining numerical and pointer data types. The address of the number x is given by &x; the value pointed to by p is refered to as *p.

In fig. 3.2, note that the declaration for y allocates space only for an address: until y is made to point to an available location, it is not possible to write to *y.

```
/* x is an integer, y is a pointer to an integer: */
int x, *y;
/* x is given the value 3: */
x = 3;
/* make y point to the location of x.  Then *y equals 3:
*/
y = &x;
/* set x equal to 4: */
*y = 4;
```

Fig. 3.2.

Structures. The data types provided in most programming languages are typically of two sorts: numbers—integers and real numbers, and sometimes complex numbers—of varying precision, and magnitudes, again of varying sizes—bytes or characters, and words or unsigned integers. Most modern languages allow the program to define additional data types, or structures. As a simple example, let us consider adding a complex number type to C, which, unlike FORTRAN, does not contain complex numbers. A complex number is, of course, simply a pair of real numbers. The syntax in C is shown in fig. 3.3. Such an implementation of complex numbers is not particularly elegant. The real power of structures is seen in problems where several different elementary data types need to be co-ordinated, as we will see below.

Stacks, linked lists and trees. One useful data structure which we have already discussed in the first section is the stack. A stack is simply an array, where the data are stored, and a pointer, giving the location of the last element added to the stack. To use the stack, two operations are needed: push(numb), which increments the stack pointer and copies the value of numb into the location which it points to, and pop(&numb), which copies the contents of the location pointed to by the stack pointer into the location &numb, and

```
/* define a data type called cmplx to represent a
complex number consisting of two floating point numbers
called re and im:  */
typedef struct {
    float re;
    float im;
} cmplx;

/* declare x, y and z to be complex numbers:  */
cmplx x, y, z;

/* initialize x and y:  */
x. re = 2.1;        x. im = 0.7;
y. re = 13.9;       y. im = − 1.2;
/* set z equal to the complex product of x and y: */
z. re = x. re * y. re − x. im * y.im;
z. im = x. re * y. im + x. im * y. re;
```

Fig. 3.3.

then decrements the stack pointer. As in the case of subroutine calls, when the data in a problem are used in a last-in, first-out order, it can be stored on a stack.

When the order of data access is more complicated, a linked list may be useful. An element in a linked list consists of the data and a pointer to the next element in the list. By pointing to the next element directly, rather than referring to it implicitly by order, it becomes a simple matter to add and delete elements in the list. To insert an element k between elements i and j, one simply copies the pointer in i into k, and resets the pointer in i to point to k. Similarly, to delete the element following i, one copies the pointer in the element being removed into the pointer in i.

Even if the size and order of the list does not change, using a linked list can be more efficient than using a simple one. For example, consider a Fibonacci-type random number generator. The algorithm requires maintaining a list of the last n random numbers generated. The next number is defined to be the sum of the n-th and $(n-k)$-th previous numbers, modulo l. The precise choice of n, k and l is made to produce the highest-quality random sequence. A good choice is $n = 55$, $k = 24$, $l =$ word length of the computer. (For further discussions of this and other random number generators, see [8].) Our implementation of this algorithm involves organizing the table as a circular linked list, with the pointer at element i pointing to element $i + 1$ for i not equal to n, and the pointer at element n pointing to the first element. We maintain two pointers, namely the pointers to the next two terms to be added. The addition is performed modulo 2^{16} (10000 in hexadecimal notation). The complete program is shown in fig. 3.4. The initialization routine is used to fill the table with numbers and set up the pointers which implement the linked list.

```
/* recursive definition of the structure rand, an
element of which is a pointer to another rand: */
typedef struct{
      int numb;
      struct rand *ptr;
      } rand;
/* define a table of 55 rands, and initialize pt1 and
pt2 to point two rands separated by 21 = 55 − 24: */
rand table[55], *pt1 = table, *pt2 = table+21;

int ran( )
{
/* a→b means (*a).b where a points to a structure
with element b. *
pt1 = pt1→ptr;
pt2 = pt2→ptr;
          /* "/" is the mod operator */
return (pt1→numb = (pt1→numb+pt2→numb) "/" 0×10000);
}

/* initialize the table: make the table into a linked
list and put in an
initial set of numbers (they don't have to be random):
*/
init( )
{
int i = −1;
while(i++ < 55){
      table[i].numb = i*i*i+17;
            /* nearly any initial number will do */
      if(i==54) table[i].ptr = table;
      else table[i].ptr = table+i+1;
      }
return;
}
```

Fig. 3.4. – Random number generator using a linked list.

The choice of the initial numbers is not too critical; ran can be called to completely randomize the table.

Let us consider the general expression which we will encounter in our gamma algebra program. The program will generate arithmetic expressions involving sums and products of numbers and dot products of vectors. In addition, there must be a representation of an unsimplified trace, which serves as the input to the next iteration of the algorithm. While we could use a linked list to represent the expressions, it is much more convenient to use a data structure which has more pointers. We will use a binary tree. The fundamental element, or node, of a binary tree is a structure with three elements: a head, a left branch and a right branch. The branches are pointers to other nodes, while the head is a label indicating what sort of junction the node represents. For arithmetic expressions, the head represents the operator whose arguments are represented by the two branches. A simpler format is adequate for the unsimplified traces: a structure containing the number of arguments and a list of pointers to them is all that is needed.

Dynamic memory allocation. We are now ready to outline the program which evalutes the traces. First, formula iii) must be applied to the unsimplified trace. This is done recursively with a subroutine that returns a pointer to the linked list which forms the answer.

Schematically, we have the following algorithm:

>   let ng be the number of factors in the unsimplified
>   trace
>   if ng is 0 then return a pointer to the constant 4.
>   let p be the final factor in the trace
>   loop over the remaining ng—1 factors
>   let q be the ith factor
>   make a node representing the dot product of p
>   and q; call it pq
>   make a trace with ng—2 terms, omitting p and q.
>   let res be the result of applying this routine
>   to te new trace.
>   make a node representing the product of pq and
>   res; call it prod.
>   if this is the first q, let ans be res
>   otherwise make a node representing the sum of
>   ans and prod.
>   continue
>   return ans

In writing down the algorithm, we see that we need to be able to make new nodes several times. There are two approaches which can be taken to provide dynamic memory allocation. On some computers, notably those running **UNIX** or those providing a reasonable degree of **UNIX** emulation in their C-compiler, there are system functions which will allocate and reclaim memory. The function calloc(N) returns a pointer to N contiguous bytes of memory which has been set to zero. The inverse function, which returns N bytes of memory to the system, making it available for subsequent calls to calloc, is cfree(buffer, bytes). A home-made approach which is suitable for our application is to declare a large global array and use the next free element in it when we need more memory. It is still good practice to use a function to allocate memory even if using the global-array technique.

```
typedef struct{
    int op;
    struct node *left;
    struct node *right;
    } node;

/* maximum number of gammas in a trace: */
#define MAXTRACE 40
typedef struct {
    int ng;
    node *gp[MAXTRACE];
    } trace;
```

Fig. 3.5.

```
#define MAXSYMS 1000
#define MAXLEN 10
int next_sym = -1;
char symbol_table[MAXSYMS][MAXLEN];

symbol *getsym()
{
if( ++next_sym == MAXSYMS ) yyerror("symbol table full");
return( symbol_table+next_sym );
}

#define MAXNODES 1000
int next_node = -1;
node node_table[MAXNODES];

node *getbr()
{
if( ++next_node == MAXNODES ) yyerror("no free nodes");
return( node_table+next_node );
}

node *mknod(o, l, r)
node *l, *r;
char o;
{
node *node;
node = getbr();
node->op = o;
node->left = l;
node->right = r;
return node;
}
```

Fig. 3.6.

We are now ready to define the data format for our Feynman traces. First we define the binary tree nodes and unsimplified traces (this requires fixing the maximum number of gamma-matrices which can occur at the same time in an unsimplified trace). The definitions are given in fig. 3.5.

Next we set up the memory management (fig. 3.6). In addition to a table of nodes, we need a table in which to store the names of the symbols which occur in a calculation. The memory array of symbols is called symbol_table, and the index of the last entry used is called next_sym. When room for storing a symbol is needed, getsym is called: after checking to see if there is a free entry, getsym returns the address of a free slot in symbol_table. The management of the memory array of nodes is identical. The function mknod is used to make a node with the specified operator and branches.

Finally, we have the function, formula3, which implements formula iii) above (fig. 3.7). The undefined macros are specified in the input-reading part of the program. This function is a direct translation of the algorithm given above.

```
node *formula3(expr)
trace expr;
{
node **tr, *p, *q, *r, *pq, *ans, *res;
int ii, ng;

ng = expr. ng;
tr = expr. gp;

if(ng == 0) return mknod(SYMB,0,"4");

p  = tr[ng-1];
q  = tr[ng-2];
pq = mknod(DOT,p,q);

expr. ng = ng-2;
ans = mknod(MULT,pq,formula3(expr));

for (ii = ng-3; ii >= 0; ii- -) {
    r = q;
    q = tr[ii];
    tr[ii] = r;
    pq = mknod(PLUS,q,p);
    res = mknod(MULT,formula3(expr),pq);
    if(ii%2) res = mknod(MULT,mknod(SYMB,0,"-1"),res);
    ans = mknod(PLUS,ans,res);
    }
return ans;
}
```

Fig. 3.7.

In the next section I will describe routines which will read the input and write the output needed to use this program.

4. – Algebraic and interactive programs.

In this section we complete the writing of a program to perform the Dirac algebra which arises in the evaluation of Feynman diagrams. We have designed the internal representation of the algebraic expressions and have written the program which applies the required algebraic rules to an expression. Two capabilities remain to be provided before our program will be useful. Some provision needs to be made for the program to read input. When an expression is input to the program, it will be expressed as a sequence of characters, in a standard mathematical form. The program must translate this representation of an expression into the corresponding internal format. We also need to include some ability to simplify the calculated result. A rudimentary simplifier would employ elementary arithmetic to replace sums of like terms with the corresponding products, replacing $x - x$ by 0 for example. A more sophisticated simplifier would be able to combine and factor sums of polynomial expressions, putting them into the most compact form possible.

We begin by considering the program to read the input expressions. We first need to specify the input format. Symbols will be represented by arbitrary strings of letters and numbers, and mathematical operators will be denoted by special, nonalphanumeric characters. For simplicity, we restrict the set of possible operators to include only those which are generated internally in the course of simplifying the traces. Then the input expression will involve only the operation of addition, multiplication and vector dot product, which will be denoted by the characters "+", "*" and ".", respectively. The operation of subtraction can be replaced with multiplication by -1 and addition; denominators can be represented by multiplication by their reciprocals. The input must also contain traces of products of gamma-matrices, which are represented by $\langle$arg1, arg2, ..., argn$\rangle$. Finally, the input may include parentheses to control the ordering of the operators, in the usual manner.

Note that the specification of the grammar is naturally recursive: an expression can be the sum of terms, each of which is an expression, the product of expressions, and so on. The entire class of expressions can be specified as in fig. 4.1.

In this specification, an entry of the form "name: ..." is a definition of "name". The different possible forms of "name" are separated by the | symbol. The undefined symbols, *i.e.* those not occurring on any left-hand side of a definition, are called tokens. Here the tokens are "+", "*", ",", ".", "(", ")". "$\langle$", "$\rangle$" and "id". The token "id" refers to a character string representing a symbol.

```
input   :   expr ;

expr    :   expr + expr
        |   expr * expr
        |   expr . expr
        |   ( expr )
        |   ⟨ list ⟩
        |   id

list    :   list , expr
        |   expr
```

Fig. 4.1. – Recursive definition of a simple operator-precedence grammar.

This specification is ambiguous: we need to give the precedence and associativity of the operators in order to completely fix it. For instance, the expression $a + b * c$ can mean either $(a + b) * c$ or $a + (b * c)$. The relative precedence of the operators determines the location of the implicit parentheses in an expression. The intuitive content of precedence is simply that one operator has priority over another, and is to be simplified first. Taking the « usual » conventions, we have, from highest to lowest precedence, ( ), then ⟨ ⟩, then . , then *, and finally +. Since * has higher precedence than +, $a + b * c$ means $a + (b * c)$; since . has higher precedence than +, $a . b + c$ means $(a . b) + c$; since ⟨ ⟩ has higher precedence than *, $b * c * \langle x, y, z \rangle$ means $((b * c) * (\text{tr}(x, y, z)))$.

For definiteness, we must fix the meaning of expressions involving a succession of the same operator, even if it is associative. Thus, although $a + b + c$ is not an ambiguous mathematical statement, it must also have implicit parentheses. The conventional choices are to make both + and * left associative, so that $a + b + c$ means $(a + b) + c$.

The process of interpreting the input is usually broken into two phases: a lexical phase, consisting of identifying the tokens, and a syntactical phase, in which the statements are recognized. The division between the two phases is not hard and fast, and the determination of which activities are performed in the first phase by the lexical analyzer and which are performed in the second by the parser can be made for convenience. For our purposes, we can put all of the reading operations into a single function called getoken. When invoked, getoken identifies the next token in the input stream and returns a code indicating which token was found. It also maintains a global variable which points to the character string which was read. If the token was an id, the character string will be stored in the symbol table, and the pointer will indicate the symbol table entry. For the other tokens, the string will just contain the printed character which represents the token.

With getoken for the lexical analyzer, the parser first calls getoken and then takes the appropriate action based on which token is found. This process is repeated until the entire input string has been read. If the new token completes the right-hand side of one of the rules given in fig. 4.1, the parser can use the rule to construct a node in the internal tree. Thus, if an expression of the form expr*expr has been recognized, a node representing the product is constructed. The expr on the left-hand side of the rule then corresponds to the new node. If no rule is applicable, the token must be saved, and the procedure repeated. The tokens and subexpressions which have not yet been placed in the internal tree are stored on a stack.

Consider a simple example: $a + b * c + d;$, where ; is the symbol indicating the end of the input. The actions of the parser are:

```
get a and shift:    push a
get + and shift:    push +
get b and shift:    push b
get * and shift:    push *
get c and shift:    push c
get + and reduce:   make a node n1 = b*c, pop c, *, and b
                    push n1
            shift:  push +.
```

At this point the stack looks like

```
(the beginning of the stack)          a
                                      +
                                      n1
            stack pointer      →      +
```

The parsing continues:

```
get d and shift:    push d
get ; and reduce:   make node n2 = n1+d, pop d, +, and n1
                    push n2
           reduce:  make node n3 = n2+a, pop n2, +, and a
           reduce:  return n3.
```

In order to determine when to shift and when to reduce, we need to ask, for each ordered pair of tokens op1 op2, whether op2 closes the current level, opens a new level, or maintains the current level: if the end of the stack contains the partial expression op1 expr does op2 mark the end of expr or does it initiate a deeper level of nesting? For our system, we have the relations given in fig. 4.2 (for completeness we let represent both the start and finish symbols).

last token	new token	action
(	)	maintains a level
+ * (or ;	(	opens a level
+ *) or id	)	closes a level
+ * (or ;	id	opens a level
+ *) or id	;	closes a level
(	+ * (or id	opens a level
)	+ *) or ;	closes a level
id	+ *) or ;	closes a level
;	+ * (or id	opens a level
using precedence:		
+	*	opens a level
*	+	closes a level
using associativity:		
+	+	closes a level
*	*	closes a level

Fig. 4.2. – Possible sequences of tokens: the end of the stack contains the last token, followed by a node representing an expr.

It is possible to formulate rules which will take one directly from the formal specification of the grammar given above to these relations, and further to build a parser-generator based on these rules. It is not possible to analyze all grammars using this method, but for grammars involving arithmetic operators, our simple-minded procedure will be sufficient. For more general grammars, more sophisticated parsers can be constructed. For further discussion and references, see [9].

It is useful to rewrite this list as a table (fig. 4.3). The choice between shifting and reducing becomes simply a matter of looking in this table to see whether op2 closes op1 or not. If it does not, then the parser shifts. If it does, then the parser reduces op1, exposing a new op1 and the process is repeated. When an op1 is exposed which op2 opens, then op2 is pushed onto the revised stack, and the parser continues. If an op1 op2 pair occurs for

	+	*	(	)	id	;	⟨	⟩	,
+	c	o	o	c	o	c	c	o	c
*	o	o	o	=	o			o	
)	c	c		c		c	c		c
id	c	c		c		c	c		c
;	o	o	o		o			o	
⟨	o	o	o		o				
⟩	o	o	o		o				=
,	c	c		c		c			

Fig. 4.3. – Shift/reduce table for grammar of fig. 4.1.

which the table entry is blank, then a syntax error has been detected. Our parser simply quits, but some sort of error recovery could be attempted.

The process of reducing an operator is fairly simple for our system. If the operator is a trace, then the stack pointer points to the rightmost argument and the last token is ⟨, whose position on the stack the parser has just located. This information is passed to the simplification program, dotrace. Dotrace then copies the arguments off of the stack, making a trace, and calls formula3, which returns the node which is the root of the tree representing the answer, which the parser then puts on the stack. For any other operator, the parser simply builds a node representing the newly found term and puts it on the stack. When the finish token is seen and the stack has been reduced, there should be a single node on the stack, which is the root of the tree representing the full answer.

A program implementing this scheme, using formula3 is presented in the appendix. The output is performed using a simple tree-scanning program which puts parentheses around every subexpression. The design of a more intelligent output program which only prints parentheses when they are needed is left as an exercise for the interested reader.

The final issue regards simplification. The strategy is fairly obvious: given the root of an expression tree, the simplifier walks through the tree and, when it finds a subtree which can be simplified, it creates a new subtree for the simplified expression and inserts it into the main tree. The precise nature of the simplification depends, of course, on the particular operator which appears at the head of each subtree. Consider, for instance, the addition simplifier. When treewalk finds a plus, it calls simplus. Simplus continues climbing the tree until it finds the last contiguous plus, *i.e.* it identifies all of the summands of the current +. The essential function which simplus must perform is that of combining identical terms. For example, the tree

```
root:  PLUS   node1  xxx
node1: PLUS   node2  xxx
node2: PLUS   yyy    zzz
```

where xxx, yyy and zzz are symbols, becomes

```
root:  PLUS   node1  node2
node1: MULT   "2"    xxx
node2: PLUS   yyy    zzz
```

It is desirable, although not strictly necessary, to put the sum into a canonical form. The canonical form involves two aspects. First the commutative property of addition is used to sort the arguments into a standard order. The standard order is arbitrary, it just needs to be unique. The second step is to reorganize the subtree to have the form that none of the left branches have plus for their operator. This step is also for efficiency: the process of identifying the summands is complicated, and, by putting the trees themselves into canonical order, most of the effort can be preserved. The implementation of the simplifier is also left as an exercise for the interested student. Note that the program as it stands is still quite useful: with the addition of a final ;, the output is legal C syntax, and can be used as the input for a numerical integration program.

A number of general-purpose algebra programs are available commercially. The most powerful ones, MACSYMA and SMP, are, roughly speaking, a logical extension of our simple program. In both cases the internal representation is a treelike structure similar to ours, and the simplifier is a program which walks through the nodes of a tree, calling the appropriate simplification routine at each node. These programs understand most sorts of elementary mathematical expressions, and can often be taught simplification rules for algebraic systems which are not already built into them.

In addition to the tree-manipulating functions and the basic arithmetic simplifiers, these programs include implementations of various powerful algorithms for performing complex operations. For instance, the problem of finding the indefinite integral of a function has been solved for those functions whose integral is a rational polynomial function of the integration variable and exponentials and logarithms of such functions. The Risch algorithm procedes from the theorem that the only exponentials which can occur in an indefinite integral must have been present in the integrand, and the logarithms either must have been in the integrand or are logarithms of factors of the denominator, if the integrand was a rational polynomial. The answer is a rational, multivariate polynomial of the integration variable and the exponentials and logarithms. Once the general form of the answer is known, it can be differentiated and set equal to the integrand. The resulting system of simultaneous equations can be solved, yielding the unknown coefficients in the formal solution. Another important problem is the factoring of polynomials. Polynomials with integer coefficients may be factored by first factoring them over a small finite field, where the answer is trivial, and then obtaining a system of equations which gives the extension of the finite field result to the full field of the integers. Arbitrary precision arithmetic can be performed efficiently by reducing the problem to that of finding the convolution of appropriate Fourier transforms. Many interesting algorithms are presented in ref. [8, 10].

As useful, general and extendable as MACSYMA and SMP are, it is important to note that their availability does not eliminate the need for writing one's own to solve a given problem. Indeed, their generality can actually interfere with their ability to solve truly massive problems. An interesting example is the recent calculation of the QCD Born amplitudes for hadron-hadron scattering, which requires the evaluation of hundreds of thousands of

Feynman diagrams [11]. This calculation is far beyond the ability of MACSYMA or SMP to carry out, and required the invention of a new algorithm for performing Dirac algebra and the writing of a special-purpose algebra program to implement the algorithm. I hope that I have convinced you in these lectures that such an undertaking is not too forbidding.

* * *

It is a pleasure to thank N. CABIBBO for organizing this stimulating and interesting summer school, and G. FARRAR for many invaluable comments and suggestions.

APPENDIX

This appendix contains a full listing of the program described in sect. **3** and **4**. The program is contained in five files. The macro definitions (for the token types and for the memory allocation buffers), the new data type definitions and the precedence table for the parser are contained in the header file yy. h. The fine main. c contains the routines for managing the memory and making nodes, for formatting the output as well as the entry routine main. The lexical analyzer and the error recovery routine are contained in the file getok. c. The parser and stack manipulating routines are contained in the file parse. c. The routine which constructs the internal representation of an unsimplified trace and the recursive simplification function are contained in the file trace. c.

The contents of the header file yy. h:

```
#include <stdio. h>
#include <ctype. h>

/* terminals */
#define PLUS      0
#define MULT      1
#define OPRN      2
#define CPRN      3
#define ID        4
#define STOP      5
#define CMA       6
#define OTR       7
#define CTR       8
#define DOT       9

/* pseudo-terminals */
#define SYMB      10
#define NODE      11
```

```
/* data types */
typedef struct{
    int op;
    struct node *left;
    struct node *right;
    } node;

#define MAXTRACE 40           /* largest allowable trace */
typedef struct {
    int ng;
    node *gp[MAXTRACE];
    } trace;

typedef char symbol;

/* table sizes */
#define MAXSTACK 100          /* stack */
#define MAXNODES 10000        /* node table */
#define MAXSYMS      10000    /* symbol table */
#define MAXLEN 10

/* precedence table:           + * ( ) i $ < > , */
#define _b 0
#define _c 1
#define _e 2
#define _o 3
int f0[SYMB] = {_c, _o, _o, _c, _o, _c, _c, _o, _c};
int f1[SYMB] = {_c, _c, _o, _c, _o, _c, _c, _o, _c};
int f2[SYMB] = {_o, _o, _o, _e, _o, _b, _b, _o, _b};
int f3[SYMB] = {_c, _c, _b, _c, _b, _c, _c, _b, _c};
int f4[SYMB] = {_c, _c, _b ,_c, _b, _c, _c, _b, _c};
int f5[SYMB] = {_o, _o, _o, _b, _o, _b, _b, _o, _b};
int f6[SYMB] = {_o, _o, _o, _b, _o, _b, _b, _b, _b};
int f7[SYMB] = {_o, _o, _o, _b, _o, _b, _b, _b, _e};
int f8[SYMB] = {_c, _c, _b, _c, _b, _c, _b, _b, _b};
int *ff[SYMB] = {f0, f1, f2, f3, f3, f4, f5, f6, f7, f8};
```

The contents of the file main.c.:

```
# include yy.h

FILE *infile = stdin;
FILE *outfile = stdout;

extern node *parse( );

main (argc, argv)
int argc;
char *argv[ ];
{
```

```
int y;
y = parse( );
printf("\start: %d", y);
print("\n");
treeprint(y);
}

int next_sym = -1;
char symbol_table[MAXSYMS][MAXLEN];
symbol *getsym( )
{
if( ++next_sym == MAXSYMS ) error("symbol table full");
return( symbol_table+next_sym );
}

int next_node = -1;
node node_table[MAXNODES];
node *getbr( )
{
if( ++next_node == MAXNODES ) error("no free nodes");
return( node_table+next_node );
}

node *mknod(o, l, r)
node *l, *r;
char o;
{
node *node;
node = getbr( );
node->op = o;
node->left = l;
node->right = r;
return node;
}

char tokens[20] = {'+', '*', '(', ')', 0, '$', ',', '<', '>', '.', ':'};

treeprint(nn)
node *nn;
{
if(nn->left == 0) printf(" %s", nn->right);
else {
      printf("(");
      treeprint(nn->left);
      printf(")");

      printf(" %c", tokens[nn->op]);

      printf("(");
      treeprint(nn->right);
      printf(")");
      }
}
```

The contents of the file getok.c:

```
#include yy.h

extern FILE *infile, *outfile;
extern symbol *getsym( );

int yylval;

int getoken( )
{
char c;

while(isspace( (c=fgetc(infile)) ));
if(c==EOF) exit( );
if(isalnum(c)){
    char *loc;
    yylval = loc = getsym( );
    *loc++ = c;
    while(isalnum( (c=fgetc(infile)) )) *loc++ = c;
    *loc = 0;
    ungetc(c, infile);
    return ID;
    }
else {
    yylval = c;
    switch(c){
        case '*': return MULT;
        case '+': return PLUS;
        case '(': return OPRN;
        case ')': return CPRN;
        case ';' return STOP;
        case '<': return OTR;
        case '>': return CTR;
        case ',': return CMA;
        default: printf("\n%c: ", c);
            error("lexical error");
        }
    }
}

error(s)
char *s;
{
fprintf(stderr, " %s", s);
exit( );
}
```

The contents of the file parse. c:

#include yy. h

extern node *getbr( ), *dotrace( );
extern int getoken( ), yylval;

int stack[MAXSTACK], val[MAXSTACK], sp = MAXSTACK;

```
node *parse( )
{
node *node;
int last, this, *stat;

this = STOP;
push(STOP, '$');
for(;;){
    beg:
    last = this;
    this = getoken( );
    switch(ff[last][this]){
        case _b: error("syntax error");
            break;
/* shift */
    case _e:
/* shift */
    case _o: shift:
            push(this, yylval);
            break;
/* reduce */
    case _c: for(stat = stack+sp; stat < stack+MAXSTACK; stat++){
            if(*stat >= SYMB) continue;
            if(ff[*stat][this] == _o) goto shift;
            switch(*stat){
                case ID: node = getbr( );
                    node->left = 0;
                    node->op = stack[sp];
                    node->right = val[sp++];
                    push(NODE, node);
                    break;
                case PLUS:
                case MULT: node = getbr( );
                    node->right = val[sp++];
                    node->op = stack[sp++];
                    node->left = val[sp++];
                    push(NODE, node);
                    break;
                case STOP: if(this!=STOP) goto beg;
                    return val[sp];
```

```
                    case OPRN: sp++;
                        stack[sp] = stack[sp-1];
                        val[sp] = val[sp-1];
                        break;
                    case OTR: if(this!=CTR) goto shift;
                        push(NODE, dotrace(stat));
                        break;
                    case CMA: break;
                    default: error("unknown terminal");
                    }
                }
            error("syntax error");
            }
        }
}
push(term, aux);
int term, aux;
{
if(sp==0) error("push: stack overflow");
stack[--sp] = term;
val[sp] = aux;
}
```

The contents of the file trace. c:

```
#include yy. h

extern node *mknod( );
extern int stack[ ], val[ ], sp;

node *formula3(expr)
trace expr;
{
node **tr, *p, *q, *r, *pq, *ans, *res;
int ii, ng;

ng = expr. ng;
tr = expr. gp;
if(ng == 0) return mknod(SYMB, 0, "4");

p = tr[ng-1];
q = tr[ng-2];
pq = mknod(DOT, p, q);

expr. ng = ng-2;
ans = mknod(MULT, pq, formula3(expr));

for (ii = ng-3; ii >= 0; ii--) {
    r = q;
    q = tr[ii];
    tr[ii] = r;
    pq = mknod(PLUS, q, p);
```

```
        res = mknod(MULT, formula3(expr), pq);
        if(ii"/"2) res = mknod(MULT, mknod(SYMB, 0, "-1"), res);
        ans = mknod(PLUS, ans, res);
    }
return ans;
}
node *dotrace(first)
int *first;
{
trace tr;
node **gammas;
int nn;

nn = (int) (first — stack);
nn = (nn — sp+1)/2;
tr. ng = nn;
gammas = tr. gp;
while(nn){
    switch (stack[sp]){
        case SYMB:
        case NODE: gammas[——nn] = val[sp];
            break;
        case CMA: break;
        default: error("dotrace: unparsed argument");
        }
    sp++;
    }
sp++;
return formula3(tr);
}
```

REFERENCES

[1] R. PEARSON et al.: Santa Barbara preprint NSF-ITP 82-98.
[2] A. OGIELSKY and A. FAST: *Versatile Monte Carlo processor for random systems*, Bell Laboratories, Murray Hill, N. J.
[3] M. LEVINE and S. FRIEND: *Programming the multiple 8087 micro-array-processor*, CMU report CMU-HEP84-9.
[4] J. KOGUT: this volume, p. 315.
[5] E. BROOKS III, G. FOX, M. JOHNSON, S. OTTO, P. STOLORZ, W. ATHAS, E. DE BENEDICTIS, R. FAUCETTE, C. SEITZ and J. STACK: *Phys. Rev. Lett.*, **52**, 2324 (1984).
[6] N. H. CHRIST and A. E. TERRANO: *IEEE Trans. Comp. C*, **33**, 344 (1984).
[7] B. W. KERNIGHAM and D. M. RITCHIE: *The C Programming Language* (Prentice-Hall, Englewood Cliffs, N. J., 1978).
[8] D. E. KNUTH: *The Art of Computer Programming* (Addison-Wesley, Reading, Mass., 1973).
[9] A. H. AHO and J. D. ULLMAN: *Principles of Compiler Design* (Addison-Wesley, Reading, Mass., 1979).
[10] A. H. AHO, J. E. HOPCROFT and J. D. ULLMAN: *The Design and Analysis of Compluter Algorithms* (Addison-Wesley, Reading, Mass., 1974).
[11] G. FARRAR and F. NERI: *Phys. Lett. B*, **130**, 109 (1983).

Selected Topics in Detector Physics.

G. CHARPAK

CERN - Geneva, Switzerland

The evolution of detectors is subjected to two contradictory forces:

The size of the detectors has a tendency to grow. Ten million dollars is the minimum cost of a detector at the intersection of a collider. Five years of construction is reasonable time for building and assembling the components of such a detector. This clearly can kill innovative approaches and force the physicist to freeze the techniques for many years.

On the other hand, questions arising from the theory or from the first results obtained with new accelerators or colliders are creating a demand for new tools. For instance, in the improvement programs for the detectors currently running with the CERN collider, it is surprising and encouraging to see that audacity has still survived in such an environment and that new and relatively risky detectors are envisaged and are being built on a large scale after some small-scale tests, still leaving many questions open. These lectures are not intended to give a detailed technical survey of the detectors at present in operation, but rather to draw attention to those which illustrate how the experimental physicists are adapting themselves to the ever-changing aspects of high-energy physics. I will limit myself to a few subjects—high-accuracy measurements, calorimetry, particle identification—and refer the reader to review articles whenever the subject has already been extensively described.

1. – High-accuracy detectors.

The existence of particles with lifetimes of the order of 10^{-13} s, decaying over lengths inferior to 1 mm in the laboratory, has introduced a strong demand for detectors with very good spatial resolution. In addition, the very high particle multiplicities reached with present machines require also a good multiparticle separation. The ideal detector in this respect is nuclear emulsion. Unfortunately, the nontriggerable feature of this detector makes it unsuitable in most cases, and the first picture of charmed events was obtained from the scanning of 25 l of emulsion, costing about one million Swiss francs, by physicists and scanners of 17 institutes. Whilst it is a valuable approach in the search for new phenomena, it is unsuitable for a systematic study requiring high statistics. This demand for high statistics has triggered off research to improve the spatial accuracy of detectors which visualize complex configurations perfectly: the bubble chamber and the streamer chamber, with some spectacular improvements over the last few years. In parallel, active research effort has been accomplished by the use of solid-state detectors and by the improvement of gaseous detectors.

1'1. *Bubble chambers*. – In order to gain accuracy, the development of the bubbles is limited to much smaller sizes—a few micrometres only. The corresponding loss in the scattered light has led to the use of a readout method, based on holography; this has proved to give enough sensitivity and accuracy for the spatial determination of the bubble position, which can reach 5 μm in diameter [1]. In addition, the depth of focus is much increased over conventional optics and is suitable for the readout of the largest chambers, such as BEBC or the 15-foot Fermilab chamber [2].

However, despite its advantages, the holographic method has some drawbacks in the region of a high-multiplicity vertex because of shadow effects by the multiple tracks. This has led to a development which is valid only for very small chambers of a few litres, and where conventional optics are used to photograph the very tiny bubbles. The lack of light leads to a very small depth of focus, but this is overcome by having a very flat beam. In order to determine vertices of short-lived particles, the only region of interest is the vertex, so that this defect is not too important [3].

Figure 1 shows the accuracy reached at present; the illustrations come from an experiment where several million pictures have already been taken. The 2 l hydrogen chamber can be triggered at a rate of 30 Hz, thus allowing us to take pictures of only those events that are characterized by some additional external trigger. However, whilst this method permits a considerable increase of efficiency over the nuclear-emulsion technique, it is suffering from the fact that the maximum number of particles acceptable in a bubble chamber picture is of the order of 100. This justifies the research on the high-accuracy streamer chambers; these have the advantage of having a short memory (of the order of 1 μs), thus permitting work to be carried out in intense beams with a picture taken only for highly selected events.

1'2. *Streamer chambers*. – In order to gain in accuracy, several steps have been taken:

i) The multiplication is limited to a single avalanche. In most gases the maximum amplification which can be reached by a single avalanche before the streamer propagation is the order of 10^8. By limiting the avalanche to

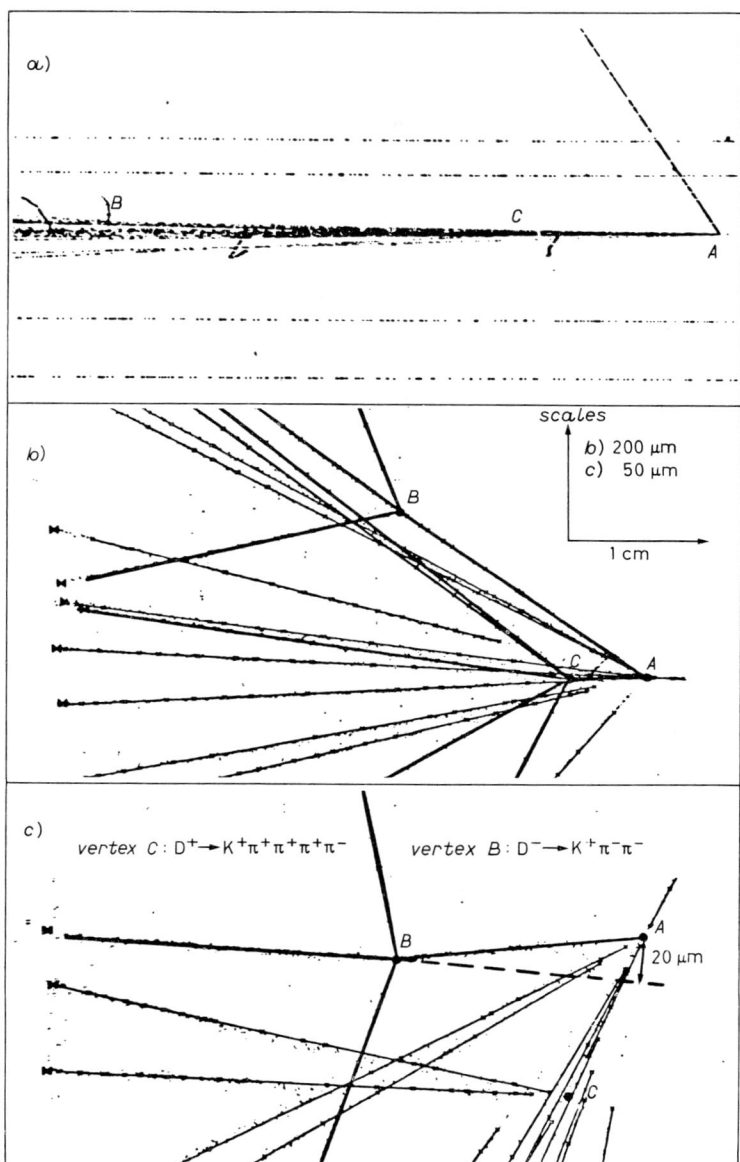

Fig. 1. – *a*) Photograph of event containing two charged decays (vertices *B* and *C*). *b*) HPD digitizations, graphical representation with an enlarged scale. *c*) Charged three-prong decays (vertex *B*) showing an impact parameter of 20 μm for the central track.

the minimum value, the longitudinal spread is limited and also the lateral one. Image intensifiers are used to increase the optical sensitivity, and it is now possible to take photographs of avalanches limited to gains of 10^7.

ii) To prevent the lateral spread of the ionization electrons due to diffusion, the chambers are pressurized up to a few tens of bar. The high pressure is also necessary in order to increase the number of interactions in the gas. This is one disadvantage with respect to bubble chambers, which are still more dense. It is compensated by the possibility of being able to handle much more intense beams.

The price we pay for this high pressure is the necessity of having very intense fields of short duration; voltages of half a million volt per centimetre during a few nanoseconds are necessary, thus limiting the size of such chambers. The improvements have not been spectacular and are still far from those reached in bubble chambers: about 80 μm (FWHM). However, more unconventional methods are being used which may well lead to spectacular improvements. They consist in detecting the avalanches by scattering light. They exploit the change in the optical properties of the gas introduced by the energy deposited by the avalanche. Holographic methods have been tried and, more recently, the simple illumination of the chamber by a laser beam [4]. A considerable gain in the light is observed, thus permitting the use of smaller avalanches and, hence, better accuracies [5]. The accuracy in the position of the avalanches is excellent, probably of the order of a few micrometres. However, the diffusion of the electrons during the memory time still limits the spatial accuracy unless the events are produced during short times predictable from the accelerator structure. In order to improve the situation, some groups are studying the possibility of using an electronegative gas, where the ionization electrons are trapped on a heavy ion and their diffusion is thus considerably slowed down, but they can be stripped from the ion by a flash of UV light. I have heard of some success in this direction by the Sandweiss group at Fermilab. It may be that the streamer chamber could reach the same accuracy as the mini-bubble chamber. However, whilst for some specific experiments requiring a high-accuracy live target this may be the solution, a great effort has been made to use purely electronic detectors which are capable of handling event rates much higher than those handled by the visual detectors. These efforts are concentrated in two directions: solid-state detectors and ultimate improvements in the characteristics of multiwire chambers or drift chambers.

1'3. *MWPC, drift chambers and solid-state detectors*. – Multiwire proportional chambers and drift chambers are routine techniques [6]. Several methods by which the avalanche co-ordinates can be read out have been developed. They can ultimately read the centroid of the avalanches with an accuracy limited only by the intrinsic distribution asymmetries in the charge distribution, and which are due to the succession of physical phenomena which give rise to the avalanches. Along a wire, the computed centroid of an avalanche can be obtained with accuracies of a few micrometres. For instance, with X-rays of 1.5 keV giving rise to photoelectrons of a range of about 30 µm, an accuracy of 35 µm has been obtained for the centroid. The limitation on the accuracy of a relativistic particle comes from δ-rays, which create electrical protuberances along a track, or, in the of inclined tracks, from the asymmetric energy distribution in the two halves of a wire chamber owing to Landau fluctuations. An accuracy of about 100 µm can be reached along a wire; however, the two-particle separation, which is connected to the gap width, is limited to about 5 mm with most common chambers.

For reasons of cost, drift chambers are preferred for high accuracies at moderate rates. The accuracy is limited by several factors: the δ-rays, the fluctuations in the distance between primary ionization clusters and the detecting wire structure; all this—even for a track parallel to the wire plane—leads to different path lengths, depending on the position of the electrical lines of forces with respect to the azimuthal angle of their connection to the anode wire [7]. Attempts to overcome this difficulty are being made with the vertex detector planned for the LEP L3 detector [8]. Time separation between electron ionization clusters along a track is expanded by applying a nonsaturating small drift field, the arrival time of the different clusters is measured, and the azimuthal position of the avalanche produced by each cluster is obtained by comparing the pulses induced on wires placed at each side of an anode wire. The aim is to obtain 30 µm accuracy with 100 µm particle separation. Whilst this is less than what can be obtained with a silicon strip detector, one should keep in mind the much easier and cheaper construction of gaseous detectors for large surfaces. This justifies the continuing effort to improve them. Their lower cost makes it possible to have more detecting planes and a redundancy which may partly compensate the « poor » accuracy. A detailed description of the work done to improve the accuracy of gaseous detectors can be found elsewhere [9]. Let me simply quote its conclusions:

« We have reviewed the efforts being undertaken to improve the accuracy recently achieved in tracking high-energy particles. The great flexibility and relative ease of construction of gaseous detectors has encouraged a sizable effort in this field. It seems, however, that in practical cases it will be hard to achieve accuracies better than $(30 \div 40)$ µm. This is compensated partly by the fact that multisampling along a track is easy ».

Solid-state detectors offer considerable potentialities, and two promising structures emerge:

   i) the multistrip detectors, which have demonstrated a one-dimensional accuracy of 5 µm (r.m.s.) with 120 µm multitrack resolution;

ii) the CCDs, which offer an accuracy of 5 μm, with a multiparticle resolution of 40 μm.

The first device permits high counting rates and has been tested up to 10^6 counts per second on a single strip. The second has considerable limitations connected with its readout system but has been used successfully up to about 10^5 counts per second and square centimetre.

These two devices can still be improved. The other approaches that we mentioned illustrate the considerable potentialities of solid-state electronics and may well produce new practical devices.

For the sake of completeness [10], I would like to mention the research undertaken to exploit the properties of superheated superconducting grains for detection.

When metals such as Al, Cd, Ga, Hg, In and Sn are subdivided into little spheres and placed at a temperature lower than a critical temperature T_c, they remain in a superconductive state, even when the magnetic field $H_c(T)$ has been increased beyond the critical value corresponding to T_c and thus up a critical overheating field $H_{SH}(T)$. It is then sufficient to introduce a very small amount of energy into a grain made of these metals to produce a transition to the normal state.

In the superconducting state the magnetic field is, except for a very small layer $\lambda(T)$, completely expelled from the grain. When $T > T_c$, λ goes to infinity. If the grains are placed inside an induction coil, there is a change of flux which introduces a detectable change of charge. The signal is proportional to the size of the grain. The amount of energy to be introduced has to be sufficient for H_{SH} to be inferior, at least at one point of the surface, to the magnetic field seen at that point. This energy threshold ΔE is a function of the specific heat of the metal in the normal state and of the volume of the grain.

Even a grain as small as 500 Å can be kept in a metastable state. Modern microelectronics permits us to speculate on devices where U-shaped coils would be 1 μm wide and spaced every 2 μm. The coils would be filled with grains of 1000 Å diameter. The signal from such a grain would correspond to about five quanta of flux.

Although the intrinsic difficulties connected with very low temperatures have limited the research efforts in this direction, let me mention an attempt that has been made to exploit the specificity of the interactions of small superconducting grains for different problems such as solar-neutrino detection. It may be that, despite the difficulty to build a localization detector competing with available techniques, this device can exploit its uniqueness for some physical problems [10].

2. – Calorimetry.

2˙1. *The various calorimeters.* – In calorimeters, a particle (or a group of particles) is totally absorbed in the detector and produces a pulse proportional to the total energy.

The reader will find a description of the various types of calorimeters in ref. [11].

The technical problems encountered are widely different for electromagnetic or hadronic particles. The choice of a technique can be imposed by a variety of conditions: available space, required energy resolution, time resolution, budget, etc.

The detector can consist of

    i) a homogeneous medium: solid state for low-energy radiations, scintillators such as NaI, BGO, BaF_2, Čerenkov radiators;

    ii) sandwiches of the interacting medium with various detectors: scintillators, gaseous detectors;

    iii) or liquid ionization chambers.

The best accuracies are obtained with homogeneous media: for instance, NaI or BGO can give resolutions of the order of 1.5 % at 1 GeV. However, this is the most expensive method because of the cost of the crystal, and one has often to choose the least expensive way. In the sampling method the best resolution is obtained with calorimeters where the energy loss in the detecting medium is a sizable fraction of the total deposited energy. In this respect, liquid argon and scintillators are better than gaseous detectors. However, the sampling method may have other virtues than just that of lower cost. The difference in the response of most detectors to both minimum ionizing and heavily ionizing particles introduces an undesirable feature: hadrons and electrons with the same energy deposited in the detector deliver signals which may differ by up to 40 %. WILLIS has shown that, if uranium is chosen as the interaction material, the neutrons liberated by the reactions give rise to fissions which enhance the hadron contribution and partly compensate for the loss introduced by the heavily ionizing recoil particles, or by nuclear excitation, or by neutron leakage. This is of great advantage and explains the interest in calorimeters based on stacks of uranium plates (see, for instance, the UA1 improvement program, CERN SPSC/83-48). This section will only deal with recent attempts to improve the calorimeters: the liquid calorimeter working at room temperature; the BaF_2 calorimeters, which offer the first combination of heavy scintillators with gaseous detectors replacing the photomultipliers; and, finally, the high-density time projection chamber (TPC) which has probably the lowest cost per unit volume for hadron calorimeters.

2'2. *Liquid ionization detectors* (LIDs) *at room temperature.* – Dielectric liquids are normally insulators. Excess electrons can either be trapped at a local site or extended. The mobility is determined by the relative population of these two types of states. The energy of the lowest extended state is a mobility edge. It is defined as V_0 relative to the energy of an electron in vacuum.

The enormous difference in mobility in helium and argon dramatically illustrates the role of V_0 in the mobility. In helium the conduction band is at 1.05 eV as a consequence of dominant repulsive interactions. As a result, the electron creates a void of 30 Å around itself. In argon the conduction band is at 0.20 eV because of the dominance of long-range polarization interactions. The electron moves relatively freely.

TABLE I. – *Values for electron affinity V_0 for various liquids.*

Liquid	T (K)	V_0 (eV)
helium	4	-1.05
neon	25	$+0.67$
argon	84	-0.20
krypton	116	-0.40
xenon	161	-0.67
CH_4	100	-0.25
TMS	295	-0.67
neopentane	295	-0.43
cyclopentane	295	-0.28
isooctane	295	-0.18
N-pentane	295	0.00
N-hexane	295	$+0.04$

The correlation between V_0 and the mobility has been determined for many liquids, including hydrocarbons which are normally insulators, where values of V_0 range from $+0.2$ to 0.4; V_0 is measured simply by comparing the photoelectric threshold of a metal in vacuum and immersed in the liquid. Table I shows some selected values of V_0. LIDs with noble liquefied argon have been relatively popular since they permit a large volume and a good granularity. A review of these calorimeters is given elsewhere [12]. Liquid argon is very cheap. However, the constraints of cryogenics more than compensate the low cost of LAr. In fig. 2 it is interesting to see the enormous difference between the behaviour of the liquefied noble gases. Whilst krypton, xenon and argon enjoy a mobility comparable to that of highly compressed gases, the mobility in helium and neon drops by orders of magnitude. It is related to the de Broglie wavelength of thermalized electrons. The mean free path for elastic collision Λ has to be larger than this to permit a high electron mobility. This mobility is connected to Λ by the relation [13]

$$\mu \geqslant 2e/3\sqrt{2/\pi m k T}\, \Lambda\, ,$$

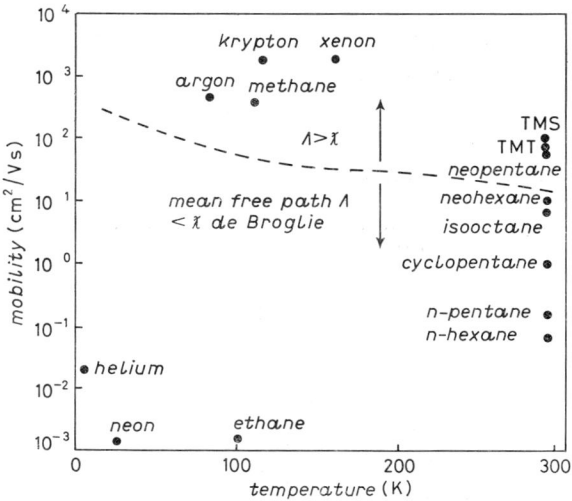

Fig. 2. – Mobility in several liquids. The dashed line separates the region between free and bounded electrons (from ref. [14]).

where kT is the Boltzmann factor, and e and m are the electron charge and mass. In fig. 2 this boundary condition is represented by the dashed line. We see that not only three noble gases—argon, krypton and xenon—are above this line, but also methane, which has been considered by some physicists as a good candidate for gigantic detectors (10^4 T) for proton decay; but, more interesting, some liquids at room temperature are above the line. For practical reasons, such as lack of toxicity, special interest is attached to trimethylsilane (TMS). TMT [$(SnCH_3)_4$] could also be of great interest for electromagnetic calorimeters but it is highly toxic. The drift velocity in TMS grows linearly up to 20 kV/cm, where inelastic collisions set in. A value of 2 kV/mm is very practical for sampling calorimeters where a thin layer of liquid, of the order of 1 mm, has to be used between plates. At this value the drift velocity is $2 \cdot 10^6$ cm/s, which corresponds to a drift time of 50 ns/mm. This is four times faster than in LAr. The first positive results obtained with TMS have been presented by ENGLER and KEIM, and are very impressive and promising [14]. The pulse height in TMS is four to five times smaller than in LAr. For high-energy calorimetry this is not a serious problem. Practical calorimeters can now be envisaged with independent building blocks of $(10 \times 10 \times 10)$ cm³, for instance with uranium foils. The granularity is limited to the size of the block, but the position of the shower, obtained from centroid measurement, is probably much better. The UA1 improvement program envisages a 350 ton ^{238}U calorimeter. The great advantage is that the small radiation length of uranium (3.2 mm) and its very high density (18.6) permit the same calorimeter to be envisaged for electromagnetic or hadronic showers. However, while TMS is quite an attractive solution, some groups are actively investigating the possible use of low-resistivity silicon detectors. These cost much less than the high-purity crystal detectors and have the considerable advantage of not requiring any precaution after construction [15].

2'3. *Liquid photocathodes and the coupling of BaF_2 to wire chambers.* – The photoelectric threshold of substances dissolved in liquids is related to the photoionization potential IP of the solution by factors expressing the energy required to bring into the liquid the liberated electron and the recoiling positive ion:

$$I_{th} = IP + V_0 + P_+ .$$

The electron affinity V_0 has been defined above; P_+ is the polarization energy of positive ion and is a calculable quantity depending on the optical dielectric constant ε_{op} and the radius of the molecule.

ANDERSON, who has done the pioneering work of introducing tetrakis-(dimethylamine)ethylene (TMAE) in photoionizable gaseous detectors, was intrigued by the possibility of using the much lower ionization potential observed for TMAE dissolved in TMS, which was observed to be only 3.56 eV instead of 5.4 eV. He managed to condense a small layer of TMAE and to extract the electrons from this layer and localize them in a wire chamber. For TMAE, $P_+ = -1.26$ eV and V_0 is close to -0.6 eV. The threshold for extracting the photoelectrons from such a thin liquid photocathode is dependent on the physical process involved: either the electrons are liberated in the liquid, thermalized and extracted from the liquid by a strong electric field—in this case the threshold is 3.5 eV; or they are extracted instantly from the liquid by the kinetic energy obtained from the photon, and their range is sufficiently long for them to be ejected. In this case the threshold is higher, since V_0 plays the role of a binding potential which has to be overcome, and it disappears from the value of I_{th}, which is then 4.1 eV. The experiments show that I_{th} is close to 4.3 eV, and the experimental conditions are such that the electric fields are too low for a delayed efficient extraction from the liquid. The study made by ANDERSON in our group at CERN has shown very promising properties of this type of photocathode:

i) in a vacuum the vapour condenses on all surfaces without any need for cooling them and the photoelectric efficiency is of a few percent in the region of 5 eV,

ii) it is easy to collect the electrons from such photocathodes in low-pressure wire chambers where the amplification and localization are in fact easier than at atmospheric pressure,

iii) at least one scintillator, BaF_2, emits radiation which is suitable for efficient detection by this photocathode.

The importance of these findings is illustrated by the observations described elsewhere [16], and which will only be summarized here:

The threshold of absorbed TMAE is sufficiently low, 4.3 eV, to absorb the fast component of BaF_2 photons with an efficiency of a few percent.

The energy resolution is expected to be close to 2%, at 1 GeV, as extrapolated from low-energy measurement. The photoelectron yield of the combination BaF_2-TMAE is, at 45 °C, about 30 keV energy loss per electron. In other words, the energy resolution above 1 GeV is no longer dominated by the photoelectron statistics but by physical effects, as is the case for the better scintillators, NaI, or BGO. Table II shows a comparison of some properties of the three main competing scintillators. The principal advantages of BaF_2 are:

TABLE II. – *Properties of three scintillators.*

	BaF_2	BGO	NaI(Tl)
density (g/cm³)	4.9	7.1	3.7
radiation length (cm)	2.1	1.1	2.6
dE/dx (min) (MeV/cm)	~6	8	4.8
linear attenuation coefficient at 511 keV (cm⁻¹)	0.47	0.92	0.34
peak emission (nm)	225	480	410
	310		
decay constant (ns)	0.6	300	250
	620		
index of refraction	1.56	2.15	1.85
light yield (photons/MeV)	$2 \cdot 10^3$	$2.8 \cdot 10^3$	$4 \cdot 10^4$
	$6.5 \cdot 10^3$		
hygroscopic	no	no	yes

the fast response, which is orders of magnitude better than for the other crystals;

the localization and granularity are determined by the properties of low-pressure wire chambers and not by more expensive and less precise photomultipliers or photodiodes, as far as localization is concerned;

the longitudinal development of showers obtained from the succession of wire planes along the stack of crystals permits a visualization of the shower development which can be decisive for the identification of rare events. This detector is considered as a good candidate for the study of such rare decays as $K^+ \to \pi^+ \nu \nu$, which has a branching ratio of the order of 10^{-10}, the determination of which is of great theoretical importance.

2'4. *The high-density projection chamber.* – The 150 ton calorimeter for DELPHI uses a simpler device [17]. A large gaseous volume is filled with a dense converter which is subdivided in such a way as to have a uniform electric field acting between converter walls, bringing the ionization electrons to a single plane of detecting wires where the pulse height, the time of arrival and localization give a full picture of the shower. For reasons of reliability, each wire is embedded in an independent cell, with cathode pads giving the localization along the tube. The electrons drift over a path of 60 cm. The energy resolution for electromagnetic showers is $T = 15\%$ at 1 GeV in a magnetic field of 1.2 T. The lower value obtained with the simulation comes from a built-in cut-off for electrons below 1 MeV. This is not the only example of ingenious methods invented for the construction of large-size calorimeters using gaseous detectors. Let me mention one other approach to low-cost sampling detectors. The use of extremely cheap wire counters in plastic tubes, the so-called Iarocci tubes [13], which have found their first application in the proton decay experiment under the Mont Blanc. They are based on the use of a new mode of operation of wire counters: the limited Geiger or streamer mode and a clever readout method.

3. – Progress in particle identification.

It is of prime importance, in many experiments, to identify a detected particle. Depending on the energy range, the space and the rejection factors, a great number of methods are available. Here I will refer only to the research done in exploiting the Čerenkov effect. A general description of this problem can be found in ref. [7]. Let me mention that a gigantic ring-imaging Čerenkov (RICH) counter is under construction for the LEP detector DELPHI [19]. It has a surface of about 100 m², and has to match the enormous multiplicity of LEP events. It is based on photosensitive drift chambers, filled with a gas containing TMAE, a vapour which has an ionization potential of only 5.4 eV. At Fermilab an experiment is nearing completion with two RICH counters, which have shown their capacity to separate pions from kaons up to 500 GeV/c. These are based on the properties of multistep avalanche chambers, which are quite adequate for the localization of single photoelectrons [7]. Further research has shown that it is possible to build structures, also based on the multistep chambers, that permit considerable multiplicities to be localized without ambiguity.

4. – Conclusion.

This lecture does not make any pretence at giving a full view of all efforts being undertaken to improve the existing arsenal of detectors. Its aim is merely to show the diversity of the approaches and to point out some of the recent successes. We are far from having exhausted the potential of existing detectors. For instance, in the field of bubble chambers, new ideas are being tackled, from amongst which I would like to quote the following investigations:

High-resolution, continuosly sensitive, heavy-liquid bubble chambers [20]. The idea is to create a small sensitive region by means of a continuously flowing liquid in a converging/diverging nozzle. Because of the spread increase at the neck, the pressure decreases, and the parameters are chosen to make it sensitive at that point. The old tracks are continuously washed away by the flowing liquid, and the analysis shows that rates of up to 10^5/s could be achieved.

Simultaneous operation of a liquid-argon detector as bubble chamber and calorimeter [21]. The idea is to exploit the properties of liquid argon which, if clean enough, is suitable for collection of the ionization electrons, which can deliver fast scintillation flashes and be used as a bubble chamber with its unsurpassed capability of complex event display.

Operation of a liquid-argon chamber under conditions where a 50% duty cycle is possible. Since liquid argon is inexpensive, a 3000 T bubble chamber was investigated as a candidate for proton decay [22].

REFERENCES

[1] A. HERVÉ and K. E. JOHANSSON: preprint CERN-EP/82-28 (1982).
[2] D. R. O. MORRISON: preprint CERN-EP/81-39 (1981); *Proceedings of the Conference on the Application of Holographic Techniques to Bubble Chamber Physics, Rutherford, 1981.*
[3] The LEBC-EHS COLLABORATION: *Neutral D-meson properties in* 360 GeV/c π^-p *interactions*, report CERN-EF/EHS/LEDA 84-4 (1984).
[4] V. ECKARDT, P. LECOQ, S. WENIG and E. WIATROWSKI: *A holographic high-pressure streamer chamber*, CERN/EF/83-18 (1983).
[5] V. ECKARDT and S. WENIG: private communication.
[6] G. CHARPAK and F. SAULI: *Nucl. Instrum. Methods*, **162**, 405 (1979).

[7] F. SAULI: *New developments in gaseous detectors*, in *Techniques and Concepts of High-Energy Physics* II, edited by THOMAS and FARBEL (Plenum Press, New York, N. Y., 1983), p. 303.
[8] A. M. WALENTA: *IEEE Trans. Nucl. Sci.*, NS-**26**, 73 (1979).
[9] G. CHARPAK and F. SAULI: preprint CERN-EP/84-35 (1984), submitted to *Annu. Rev. Nucl. Sci.*
[10] G. WAYSAND: *Détection radioélectrique des neutrinos solaires par chambres de grains supraconducteurs métastables, Rencontre de Cargèse: Astrophysique et interactions fondamentales* (1983).
[11] C. W. FABJAN and E. G. FISCHER: *Rep. Prog Phys.*, **43**, 1003 (1980).
[12] C. BRASSARD: *Nucl. Instrum. Methods*, **162**, 29 (1979).
[13] M. M. COHEN and J LEKNER: *Phys. Rev.*, **158**, 305 (1967).
[14] J. ENGLER and H. KEIM: *A liquid ionization chamber using tetramethylsilane*, Kernforschungszentrum Karlsruhe report, KFK 36-38 (1984).
[15] P.-G. RANCOITA and A. FELDMAN: *Silicon detectors in calorimetry*, INFN/AE-84/1.
[16] D. F. ANDERSON, R. BOUCLIER, G. CHARPAK, S. MAJEWSKI and G. KNELLER: *Nucl. Instrum. Methods*, **217**, 217 (1983).
[17] M. BERGGREN, A. CATTAI, H. G. FISCHER, E. LILLETHUN, F. NAVARRIA, M. PANTER, S. RAGAZZI, L. ROSSI, O. ULLALAND, J. R. URSIN and N. YAMDAGNI: *The DELPHI HPC calorimeter*, in *Proceedings of the III International Conference on Instrumentation for Colliding Beam Physics, Novosibirsk, 1984*.
[18] G. BATTISTONI, E. IAROCCI, M. M. MARSAI, J. NICOLETTI and L. TRASATTI: *Nucl. Instrum. Methods*, **164**, 57 (1979).
[19] The RICH GROUP: *IEEE Trans. Nucl. Sci.*, NS-**31**, No. 2 (1984).
[20] A. HERVÉ: *Ideas for constructing COBC, a high-resolution, continuously sensitive, heavy-liquid bubble chamber accepting $2 \cdot 10^5$ beam tracks per second*, CERN/EF//EHS/TE 80-3 (1980).
[21] J.-C. BESSET, M. BURNS, G. HARIGEL, J. LINDSAY, G. LINSER and F. SCHENK: preprint CERN-EF/82-7 (1982), submitted to *Nucl. Instrum. Methods*.
[22] G. HARIGEL, A. HERVÉ and K. WINTER: *Nucl. Instrum. Methods*, in press.

Introduction to Accelerator Physics.

K. Johnsen

CERN - Geneva, Switzerland

1. – Introduction.

Particle accelerators became an important tool for experimental nuclear physics from around 1930. Since then there has been tremendous progress in the construction of such accelerators with an increase of about one and a half orders of magnitude in beam energy per decade, as illustrated by the updated Livingston plot shown in fig. 1.

Just looking at this chart one notices a very important feature of the progress. The progress of each type of accelerator has saturated fairly quickly, whereas new ideas have been advanced regularly and have been the main contributors to the rapid progress. Two startling examples can be noticed: The invention of the alternating-gradient focusing in the early fifties and the application of colliding beams in the sixties and onwards. Although technological advances in accelerator components have been important in the applications of the various ideas, the main steps forward have in the past come through inventions that lie within the field of accelerator physics.

These lectures will be mainly devoted to basic, and fairly elementary, aspects of accelerator physics. For the more sophisticated aspects the reader can be referred, for instance, to various courses in the process of being organized by the CERN Accelerator School. Technological problems may occasionally be touched upon as part of examples of the utilization of the accelerator physics knowledge gained.

In order to have some further limits in the presentation, we shall mainly concentrate on aspects related to circular machines, and in particular synchrotrons and storage rings. However, it is believed that the knowledge thus gained will nevertheless be useful as a good starting point for those who also want to understand the behaviour of other types of accelerators.

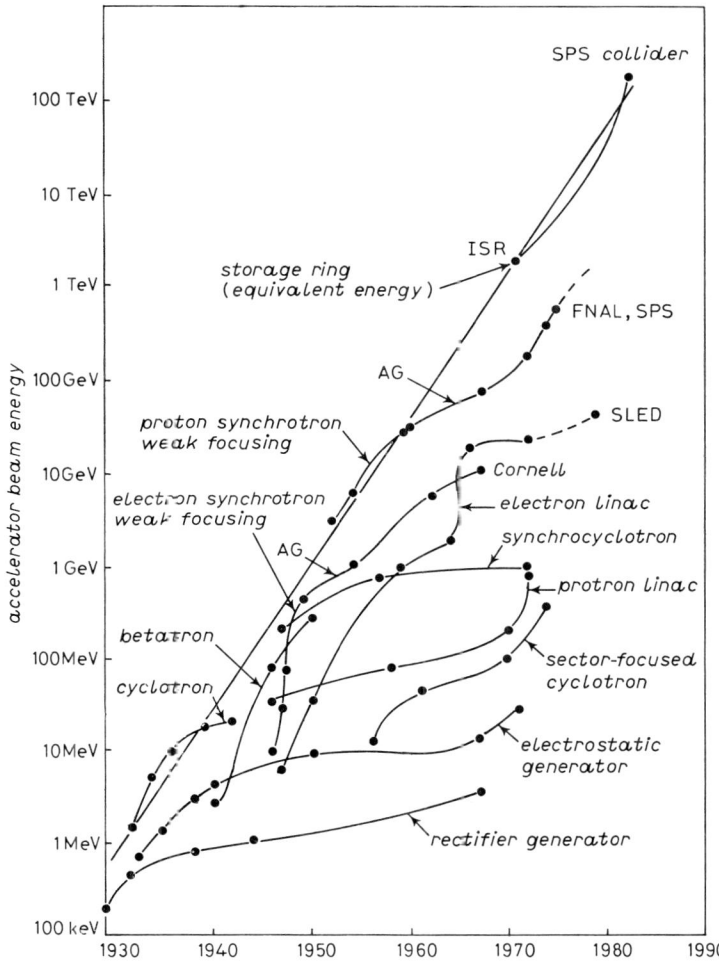

Fig. 1. – Livingston plot showing the maximum energy as a function of year for various accelerator technologies.

2. – Short description of a few circular accelerators.

2'1. *The cyclotron.* – In 1924 ISING proposed for the first time a particle accelerator that would give particles more energy than that provided by the maximum voltage in the system, and in 1928 WIDERÖE built the first linear accelerator by using radiofrequency acceleration over two gaps.

In 1930 LAWRENCE proposed the application of the same resonance principle, but now inside a homogeneous magnetic field such that the particles would be bent back to the same r.f. gap twice for each period of the radiofrequency (see fig. 2).

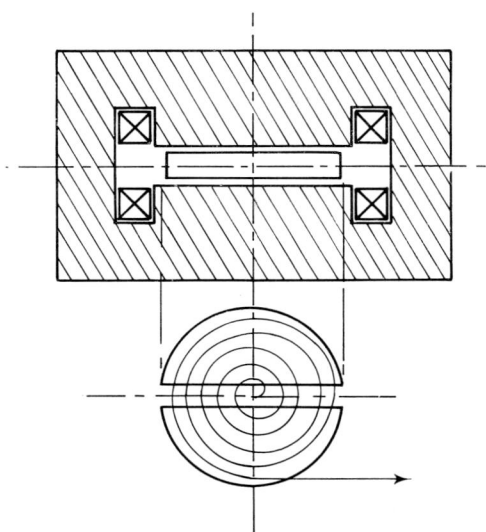

Fig. 2. – The cyclotron.

In a homogeneous magnetic field B, a particle with charge e moves in a circle of radius

$$\varrho = p/eB \,, \tag{1}$$

where p is the momentum of the particle. The angular frequency of the rotation is

$$\omega_c = eB/m \,, \tag{2}$$

where

$$m = m_0(1-\beta^2)^{-\frac{1}{2}} \,, \tag{3}$$

m_0 being the rest mass of the particle. This frequency is called the cyclotron frequency.

The resonance condition in the cyclotron is obtained by choosing the radio-frequency equal to the cyclotron frequency so that particles that pass the gap near the peak of the r.f. voltage will continue to pass near the peak at every half-turn, moving on ever-increasing half-circles. With a fixed frequency, the particles should continue to spiral until they reach the edge of the magnetic field or until they become relativistic and slip back with respect to the gap voltage (see (2) and (3)). In 1931 LIVINGSTON demonstrated the principle by accelerating H_2^+ ions to 80 keV. However modest this result may seem now, it was a very important milestone. Only a year afterwards, a 1.2 MeV cyclotron was working and LAWRENCE followed up the success by further steps and, by 1939, had completed the so-called « Croker » cyclotron that gave 20 MeV deuterons and 40 MeV doubly charged helium ions. LAWRENCE was awarded the Nobel prize for his cyclotron work. Similar cyclotrons were built in many other laboratories. All of these made great contributions to nuclear physics.

However, by the late thirties, it had become clear that the normal cyclotron had been pushed close to its limit in energy, *i.e.* the limit due to the change in cyclotron frequency because of relativistic effects (cf. (2)).

The remedy was to modulate the applied r.f. to keep in step with the cyclotron frequency. In 1945, this synchronous acceleration was proposed independently by MCMILLAN and VEKSLER, who also proved that the particles would remain azimuthally stable, an essential condition to carry them to much higher energies. These stability conditions will be discussed in a later section. The application of the method to cyclotrons made it possible for them to reach much higher energies and many synchro-cyclotrons, as this kind of cyclotron was named, were built for energies in the (100÷1000) MeV range.

2˙2. *The betatron.* – The cyclotron could accelerate protons and heavier ions, but for electrons it was useless. In 1940, KERST invented and constructed the first circular electron accelerator, the betatron, which produced 2.3 MeV electrons. The acceleration in this device is achieved by the electric field induced by the change in magnetic flux going through the circular electron orbit. By that time, the theory of transverse stability of a particle beam was sufficiently well developed that the guide field was carefully shaped and given a radial gradient in order to provide vertical and horizontal stability (more about this later). The largest betatron ever made was also built by KERST and reached 300 MeV in 1950.

2˙3. *The synchrotron.* – However, by this time the synchrotron had been invented and this almost put the betatron out of business for nuclear-physics research (although not for medical purposes). In connection with the cyclotron I have already mentioned synchronous acceleration invented by MCMILLAN and VEKSLER. In their original papers they both proposed the electron synchrotron as an application of their principle, which is illustrated in fig. 3. The guide field is similar to that of the betatron, with a bending field to keep the particles on a circular orbit and an appropriate radial field gradient to achieve vertical and horizontal stability. Acceleration is by r.f. voltage operated on a harmonic of the revolution frequency. As the particle energy increases, the field is also increased at a rate that makes the particle orbit (more precisely, the so-called closed orbit) the same at all energies, and the radiofrequency is changed in synchronism with the revolution frequency. For synchrotrons this means an upwards modulation of the frequency, in contrast to the downwards modulation in the synchrocyclotron.

Several laboratories started at once to build electron synchrotrons. GOWARD

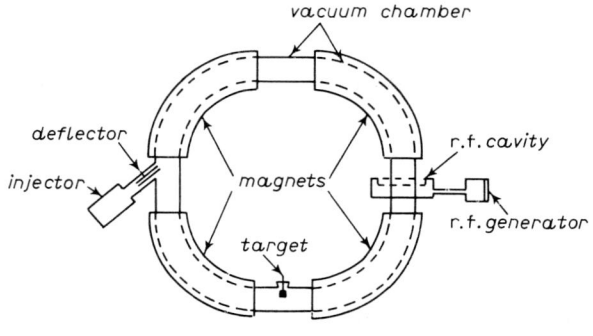

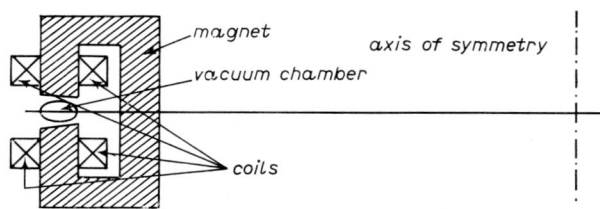

Fig. 3. – The synchrotron.

and BARNES in England were the first ones to make a synchrotron work. They had a betatron magnet which they converted to a small synchrotron in early 1946.

The synchrotorn has been tremendously important for nuclear physics and elementary-particle physics and, ever since its invention, larger and larger machines have been built. The largest electron synchrotron in operation at present is the 12 GeV machine at Cornell, but there are several others in the world in the range 5 to 10 GeV.

The first analysis of a proton synchrotron was published by OLIPHANT, GOODEN and HYDE in 1947, and the construction of such a machine of 1 GeV was started soon afterwards, which became operational in 1953. However, in the meantime, this group had been overtaken by one at the Brookhaven National Laboratory who completed the 3 GeV Cosmotron in 1952. Many larger machines followed all over the world, the largest now being the 800 GeV synchrotron at the Fermi National Laboratory and the 500 GeV SPS at CERN.

2˙4. *Collider beams.* – The next large step in energy was made by colliding-beam devices, as seen from fig. 1.

In physics with fixed-target accelerators, only the centre-of-mass energy is «useful» in an interaction. When relativistic particles of energy E bombard a stationary target this is

(4a) $$E_{c.m.} \simeq (2m_0 c^2 E)^{\frac{1}{2}},$$

which is only a fraction of the total particle energy. If, on the contrary, two particles with energy E are made to collide head on, we get

(4b) $$E_{c.m.} = 2E,$$

i.e. all the available energy can be made use of in the interaction.

The purpose of colliders is just to make such head-on collisions possible with a useful rate of interactions. In the case of electron storage rings, there is an additional benefit from the fact that this is the only way of creating collisions between electrons and positrons, with the resultant very clean interaction conditions.

The history of colliding-beam devices begins in 1956 when the group at the Midwestern Universities Research Association (MURA), USA, put forward the idea of particle stacking in circular accelerators. Of course, people who worked with particle accelerators had speculated earlier about the possibilities of obtaining high centre-of-mass energies with colliding beams. But such ideas were unrealistic with the particle densities available in normal accelerator beams. The invention of particle stacking changed the picture significantly, and opened up the possibility of making two intense beams collide with sufficiently high interaction rates that experimentation would be feasible in an energy range otherwise unattainable by known techniques except at enormous cost.

In electron storage rings, the characteristics of the magnets can be determined in a way such that the synchrotron radiation can make the beams shrink in all dimensions, and by this means intense electron beams can be accumulated.

In proton rings the synchrotron radiation is so small that it is neither of disturbance nor of help, and the MURA Group proposed in their original papers momentum stacking as a way of collecting the required intense beams.

For collecting intense antiproton beams stochastic cooling, invented by S. VAN DER MEER at CERN, has been used successfully in recent years.

The first storage rings to operate were the 2×250 MeV electron ring ADA at Frascati and the 2×500 MeV Princeton-Stanford ring. Both started operating in 1961. Novosibirsk followed shortly afterwards with VEPP-1 in 1963 and later with several other ones. Orsay made ACO, and Frascati constructed ADONE, which was the first storage ring with an energy above 1 GeV in each beam. The most successful electron storage rings have been SPEAR (1972) and DORIS (1974), both at around 2×4 GeV. These two machines, as you all know, have had a tremendous impact on high-energy physics, first through the discovery of the J/ψ particle at SPEAR (but at the same time by TING at Brookhaven) and then through the study of the properties of this new particle at both SPEAR and DORIS. This work has been a determining factor in the authorization of new storage-ring projects at both places (2×15 GeV PEP at SLAC and 2×19 GeV PETRA at DESY).

The first colliding-beam facility for protons was the 2×31 GeV Intersecting Storage Rings which became operational at CERN in 1971. This machine worked as theory predicted, and was the highest-energy machine in the world until the SPS at CERN started operating as a proton-antiproton collider in 1981 at 2×270 GeV. For this facility stochastic cooling was for the first time applied to accumulate sufficiently intense antiproton beams.

With this introductory background we shall now go through some simple accelerator physics aspects related to the kind of circular accelerators described above. For further reading on the subject the reader can be referred to:

Proceedings of the First Course of the International School of Particle Accelerators of the « Ettore Majorana » Centre for Scientific Culture, Erice, 10-22 Novembre 1976, printed as CERN Report 77-13.

Lectures on the Physics of Electron Storage Rings given by M. SANDS to the Scuola Internazionale di Fisica « E. Fermi » (Varenna) June 1969 and also printed as SLAC-121 UC-28 (ACC).

3. – Transverse motion in weak-focusing accelerators.

We take a very simple example of a circular accelerator with a magnet gap in which the field does not change azimuthally, *i.e.*

$$\partial/\partial\theta = 0 ,$$

and we also assume symmetry about the median plane.

Let us choose a circle of radius R_0 and with the constant field B_0 in the median plane. Then a particle of momentum (see (1))

$$p_0 = eB_0 R_0$$

follows this circular orbit if injected tangentially onto it. This orbit we call the equilibrium orbit for p_0 (often also called « closed orbit »). It is now important to study the motion of particles deviating from this orbit to see if the system is stable (focusing).

3'1. *The radial motion.* – The equation for the radial motion is

$$(5) \qquad \frac{dp_\varrho}{dt} = - ev_0 B_z + p_0 v_0/\varrho .$$

We introduce s as the co-ordinate along the equilibrium orbit, i.e.

(6)
$$d/dt = v_0 \, d/ds ,$$

and get

$$\frac{dp_\varrho}{ds} - p_0/\varrho + eB_z = 0 .$$

We introduce the deviations

$$\Delta B_z = B_z - B_0 , \qquad x = \varrho - R_0$$

and assume $x \ll R_0$, which always gives a good approximation.

This together with the expression for the equilibrium orbit gives

$$\frac{dp_\varrho}{ds} + p_0 \frac{x}{R_0^2} + \frac{p_0 \Delta B_z}{R_0 B_0} = 0 .$$

With good approximation

$$p_\varrho = p_0 \frac{dx}{ds} .$$

Consequently

(7)
$$\frac{1}{p_0} \frac{d}{ds} \left(p_0 \frac{dx}{ds} \right) + \frac{x}{R_0^2} + \frac{1}{R_0} \frac{\Delta B_z}{B_0} = 0 .$$

We can expand the magnetic field in a series

(8)
$$B_z = B_0 + x \frac{\partial B_z}{\partial x} + \dots .$$

We shall for our purpose use only two terms, i.e. we linearize the equation, and we further introduce the convention

(9)
$$n = -\frac{R_0}{B_0} \frac{\partial B_z}{\partial x}$$

and get the following equation for small radial deviations in a linear field:

(10)
$$\frac{1}{p_0} \frac{d}{ds} \left(p_0 \frac{dx}{ds} \right) + \frac{1-n}{R_0^2} x = 0 .$$

3˙2. *The vertical motion.* – In an analogous way we find for the vertical motion

$$\frac{1}{p_0}\frac{\mathrm{d}}{\mathrm{d}s}\left(p_0\frac{\mathrm{d}z}{\mathrm{d}s}\right) - \frac{1}{R_0}\frac{B_\varrho}{B_0} = 0 \,. \tag{11}$$

Linearized

$$\frac{1}{p_0}\frac{\mathrm{d}}{\mathrm{d}s}\left(p_0\frac{\mathrm{d}z}{\mathrm{d}s}\right) + \frac{n}{R_0^2}z = 0 \,. \tag{12}$$

3˙3. *Solutions.* – The coefficients in (10) and (12) change slowly with time and the adiabatic (or WKB) solution can be applied. This solution can be written as follows, valid in both planes:

$$x, z = K_{x,z}\sqrt{\frac{\lambdabar_{\beta x,z}}{p_0}}\cos\left(\int\frac{\mathrm{d}s}{\lambdabar_{\beta x,z}} + \varphi_{x,z}\right), \tag{13}$$

where the indices refer to one or the other plane depending on which plane one considers, K and φ are integration constants and

$$\lambdabar_{\beta_x} = \frac{R_0}{\sqrt{1-n}}, \qquad \lambdabar_{\beta_z} = \frac{R_0}{\sqrt{n}}, \qquad \text{where } \lambdabar_\beta = \lambda_\beta/2\pi \,. \tag{14}$$

This equation is valid for weak-focusing cyclotrons, betatrons and synchrotrons. For betatrons and synchrotrons the solution can be simplified because in such machines R_0 and n are kept constant during the acceleration. This gives

$$x, z = \frac{K_{x,z}}{\sqrt{p_0}}\cos\left(\frac{s}{\lambdabar_{\beta_{x,z}}} + \varphi_{x,z}\right). \tag{15}$$

3˙4. *Stability.* – Stability can be secured only if the solutions are oscillatory, i.e. λ_β must be real for both planes.

Horizontal λ_β is real if

$$1 - n > 0, \qquad\qquad n < 1 \,. \tag{16a}$$

Vertical λ_β is real if

$$n > 0 \,. \tag{16b}$$

This means oscillatory solutions for both planes if

$$0 < n < 1 \tag{16c}$$

or

$$0 < -\frac{1}{B_0}\frac{\mathrm{d}B}{\mathrm{d}x} < R_0^{-1} \,. \tag{16d}$$

It should be noted that overall stability is achieved by a small overlapping at the extremes of the stability regions for the two planes, and this small overlapping is all due to the effect of the centrifugal forces in the horizontal plane. Since it means working near the limit of stability in both planes, the focusing is weak, and this is why such focusing is in fact called « weak focusing ».

In the region of oscillatory motion it can be seen from the solutions that the wavelength of the oscillation can be written

$$\lambda_\beta = 2\pi \lambdabar_\beta = \begin{cases} 2\pi R_0/\sqrt{1-n} & \text{horizontally}, \\ 2\pi R_0/\sqrt{n} & \text{vertically}. \end{cases}$$

(17a)
(17b)

This oscillation is for historical reasons called the *betatron oscillation* and the wavelength is called the *betatron wavelength*. The number of oscillations per revolution is called the *betatron tune* given as

$$(18) \qquad Q = R_0/\lambdabar_\beta .$$

It can be noted from (14) that in a weak-focusing machine $Q < 1$.

Let us also look at the amplitudes of the oscillations

$$(19a) \qquad \hat{x}, \hat{z} \propto (\lambdabar_{\beta x,z}/p_0)^{\frac{1}{2}},$$

which in the case of constant λ_β (*i.e.* betatrons and synchrotrons) becomes

$$(19b) \qquad \hat{x}, \hat{z} \propto p_0^{-\frac{1}{2}} .$$

This shows that the amplitudes decrease with increasing momentum. This important property is called *adiabatic damping*.

3`5. *Acceptance/emittance*. – We rewrite (13) in a simplified manner by introducing

$$\xi = x, z , \qquad \lambdabar_\beta = \lambdabar_{\beta x,z}$$

and get

$$(20) \qquad \xi = K(\lambdabar_\beta/p_0)^{\frac{1}{2}} \cos\left(\int \mathrm{d}s/\lambdabar_\beta + \varphi\right) .$$

For the slope of the particle orbit we get correspondingly

$$(21) \qquad \frac{\mathrm{d}\xi}{\mathrm{d}s} = \xi' \simeq K(\lambdabar_\beta p_0)^{-\frac{1}{2}} \sin\left(\int \mathrm{d}s/\lambdabar_\beta + \varphi\right)$$

and combining the two gives

(22) $$\xi^2 + \lambda_\beta^2 \xi'^2 = K^2 \lambda_\beta / p_0 \,.$$

This means that in a (ξ, ξ')-plane the particles trace elliptical orbits (fig. 4).

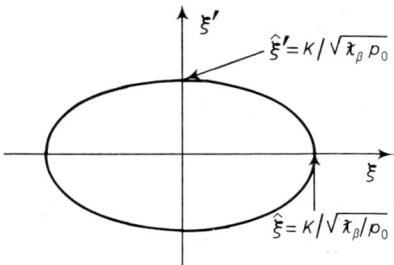

Fig. 4. – Phase-space trajectory.

The area of this ellipse is $\pi K^2 p_0^{-1}$ and the « shape » (ratio of diameters) is λ_β. In a given machine there is a maximum possible amplitude a (e.g. given by the vacuum chamber or other restrictions). The *area* of the corresponding ellipse is called the *acceptance*

(23) $$A = \pi a^2 / \lambda_\beta \,.$$

It should be noted that acceptance is a property of the accelerator.

Let us then look at some beam properties. The particles in a beam oscillate with different amplitudes and phases. The particle with the largest amplitude $\hat{\xi}$ we assume follows the ellipse drawn in fig. 4. All other particles follow orbits of the same shape but located inside. In such a case the area inside the drawn « maximum » ellipse is a measure of the size of the beam, and we call this area the emittance of the beam

(24) $$E = \pi \hat{\xi}^2 / \lambda_\beta \,,$$

which is an important beam property. If we now take into account (19a), we notice

(25) $$E \propto p_0^{-1} \,,$$

which means that the emittance diminishes with increasing momentum, i.e. the beam shrinks away from the vacuum chamber.

Instead of ξ' we could have chosen as variable the corresponding transverse momentum

(26) $$p_t = p_0 \, d\xi / ds \,.$$

In this case we would have obtained an invariant emittance very often written in normalized form as

(27) $$E_n = \beta\gamma E = \text{invariant}.$$

Correspondingly for the acceptance

(28) $$A_n = \beta\gamma A \propto p_\bullet.$$

A practical consequence of the expressions for acceptance and emittance is that in machine designs it is from this point of view desirable to choose a high energy for the injection into the accelerator.

The simple-minded picture of a beam size used above for the emittance definition does not quite fit a practical beam which has more like a Gaussian distribution of amplitudes. In practice one, therefore, uses standard deviations of amplitude distributions to describe a beam emittance. Unfortunately, the experts cannot agree on what value to use. Some use *one* standard deviation, some use *two*, which results in a factor four difference in the numerical values.

3˙6. *Momentum compaction*. – So far we have discussed deviations in position and slope of orbits for particles all having the same momentum p_0. We shall now look at the deviation of the equilibrium orbit for a particle of momentum deviation Δp, assuming the same simple magnetic field satisfying $\partial/\partial\theta = 0$. From (1) we get (see fig. 5)

$$\frac{\Delta B}{B_0} = \frac{\Delta p}{p_0} - \frac{\Delta\varrho}{R_0}.$$

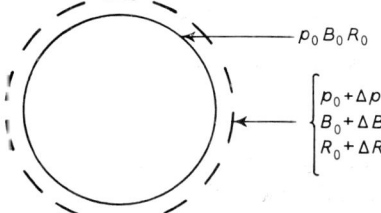

Fig. 5. – Orbit separation from a momentum deviation.

We define the *momentum compaction* as

(29) $$\alpha = \frac{d\varrho}{dp}\frac{p_0}{R_0}$$

and get

$$\frac{\Delta B}{B_0} = \frac{\Delta \varrho}{R_0}(\alpha^{-1}-1).$$

If we now make use of (9), we get

(30) $$\alpha = (1-n)^{-1}$$

and from (14) and (18) follows

(31) $$\alpha = Q_x^{-2}.$$

We shall later come to more complicated systems where α is difficult to calculate and in which (31) does not hold exactly, but can still be used as a useful approximation.

3'7. *Summary of the important concepts introduced so far.* – This is as far as we shall go in the analysis of the transverse particle motion in weak-focusing circular accelerators. The system studied has been a rather simplified one, but with that restriction the analysis could be fairly rigorous, and has given us the opportunity of introducing some very important concepts of accelerator physics, a summary of which is given below.

Concept of transverse stability/instability.

The transverse oscillations are historically called betatron oscillations (irrespective of the type of accelerator), their wavelength, the betatron wavelength, and the number of such oscillations per revolution, the betatron tune.

Betatron oscillation amplitudes are adiabatically damped as the particles are being accelerated.

Concept of acceptance of an accelerator.

Concept of emittance of a beam.

Concept of momentum compaction.

All these concepts will reappear in one form or another in the continuation of these lectures, but in most cases with less stringent derivation of formulae.

4. – Introduction to alternating-gradient focusing.

4'1. *A segment of a magnet as a focusing element.* – Instead of considering a whole circumference of a machine, we now consider only a segment (normally relatively short), but over this segment we assume, as before, that the gradient

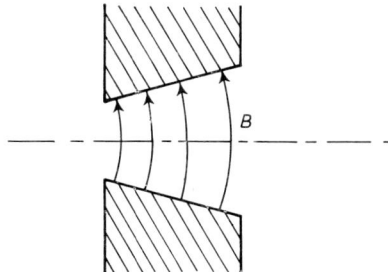

Fig. 6. – Magnetic field in a segment.

is constant (fig. 6). To simplify the analysis we further disregard the acceleration (*i.e.* assume p_0 constant) and get a differential equation with constant coefficients. From (10), (12) and (14) we get

$$\frac{d^2\xi}{ds^2} + \lambda_\beta^{-2}\xi = 0 \,. \tag{32}$$

In general we assume no restriction on the magnitude of the field gradient. If λ_β is real, the element is focusing, if λ_β is imaginary, the element is defocusing. Since we consider only a segment, the solution is most conveniently written in matrix form

$$\begin{pmatrix}\xi\\\xi'\end{pmatrix} = \begin{pmatrix} \cos s/\lambda_\beta & \lambda_\beta \sin s/\lambda_\beta \\ -\lambda_\beta^{-1}\sin s/\lambda_\beta & \cos s/\lambda_\beta \end{pmatrix}\begin{pmatrix}\xi_0\\\xi_0'\end{pmatrix}, \tag{33a}$$

$$= T\begin{pmatrix}\xi_0\\\xi_0'\end{pmatrix}. \tag{33b}$$

The matrix T is called the transfer matrix over the segment. It should be noted that its determinant is equal to unity.

If one has a series of different elements, one gets the solution for the whole system by multiplying the transfer matrices for the elements in beam order.

The most typical and representative elements are

bending magnet with gradient,

quadrupole lens,

point lens,

drift length (field free).

For the simple approach in these lectures the first two are mathematically equivalent. The third element is an approximation of the two first ones, but often a very good approximation, which we shall make much use of from now on. The fourth element occurs in all practical systems between the magnetic elements.

We shall more specifically derive the transfer matrices for the last two elements.

4˙1.1. Point lens. A point lens is an element in which the length l satisfies

$$l/\lambda_\beta \ll 1 \quad \text{but finite while} \quad \lambda_\beta \to 0 .$$

From the expression of the transfer matrix we then get

$$(34) \qquad T \simeq \begin{pmatrix} 1 & 0 \\ -l/\lambda_\beta^2 & 1 \end{pmatrix}.$$

From the optical analogy we recognize

$$(35a) \qquad \text{focal length} \quad f = \lambda_\beta^2/l ,$$

$$(35b) \qquad \text{focal strength} \quad \delta = l/\lambda_\beta^2 ,$$

i.e.

$$(36) \qquad T \simeq \begin{pmatrix} 1 & 0 \\ -\delta & 1 \end{pmatrix}.$$

4˙1.2. Drift length. From the general expression of T we get the T for a drift length by putting $s = L$ and $\lambda_\beta = \infty$

$$(37) \qquad T = \begin{pmatrix} 1 & L \\ 0 & 1 \end{pmatrix}.$$

4˙1.3. The relation between vertical and horizontal focusing of an element with high field gradient. We shall now use the point-lens approximation. Remembering (9), we see that a high field gradient means $|n| \gg 1$. If we look at an element with positive n we get (remembering (14) and (35))

δ_V positive, i.e. vertical focusing,

δ_H negative, i.e. horizontal defocusing,

and *vice versa* if we would look at an element with negative n.

This result is fundamental (in general easily derived from Maxwell's equations): *A transverse magnetic field that is focusing in one plane is defocusing in the other, and vice versa.* This is clearly seen in fig. 7, where a quadrupole field has been sketched.

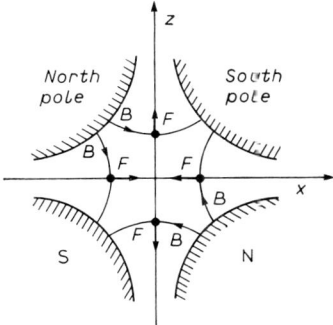

Fig. 7. – Forces in a quadrupole field.

In subsect. **3'4** we saw what looked like an exception to this in a weak-focusing machine. In that case the small effect of the centrifugal forces made a small overlap of otherwise separate stability regions.

4'2. *Doublet.* – Figure 8 shows a combination of a focusing lens and a defocusing lens of equal strengths. I leave it as an exercise for the students to prove that the parameters L and δ can be chosen such that the combination has an overall focusing property, and this for both of the cases shown.

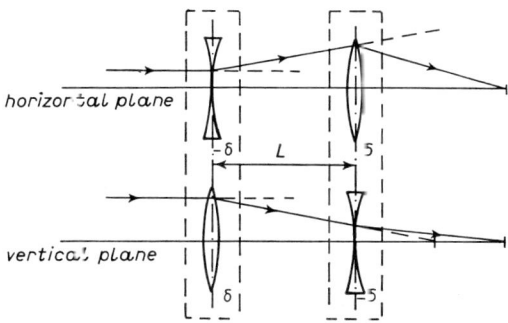

Fig. 8. – Focusing doublet.

If these are magnetic lenses, we have already seen that, if the upper case is chosen for the horizontal plane, we get automatically the lower case for the vertical plane. A focusing arrangement like this is called a doublet.

From this starting point it is natural to investigate combinations of more positive and negative lenses. The next step would be a triplet, which is a combination much used in beam transport systems.

We shall, however, go straight to a system which essentially is an infinite number of plus and minus lenses, as this is close to the practical case of an alternating-gradient (AG) accelerator.

4'3. *Simple description of an* AG *accelerator.* – An alternating-gradient accelerator consists of magnetic elements to bend the particles 360° with superimposed focusing elements that alternate between strong positive and negative gradients.

To a first approximation these elements are arranged into a periodic structure around the machine (see fig. 9).

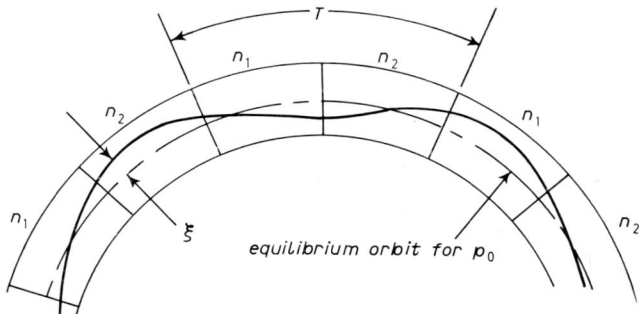

Fig. 9. – Simple example of AG accelerator.

Earlier we have developed the transfer matrix for the kind of elements that appear in a period.

The transfer matrix for a period is obtained by multiplying the individual elements. For the time being we are only interested in the transfer matrix for the period, which we call T.

4'4. *Stability criterion.* – Particles going through an AG accelerator see what effectively is an infinite series of transfer matrices T (see fig. 10):

$$(38) \qquad \begin{pmatrix} \xi_1 \\ \xi_1' \end{pmatrix} = T \begin{pmatrix} \xi_0 \\ \xi_0' \end{pmatrix}, \qquad \begin{pmatrix} \xi_\eta \\ \xi_\eta' \end{pmatrix} = T^n \begin{pmatrix} \xi_0 \\ \xi_0' \end{pmatrix}.$$

The use of a similarity transformation gives

$$(39) \qquad S^{-1} \begin{pmatrix} \xi_1 \\ \xi_1' \end{pmatrix} = S^{-1} T S S^{-1} \begin{pmatrix} \xi_0 \\ \xi_0' \end{pmatrix}$$

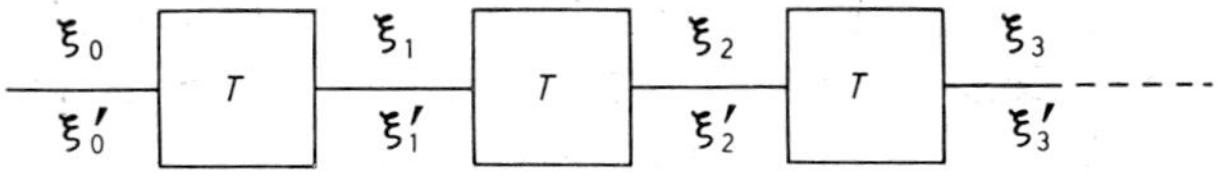

Fig. 10. – Matrix representation of periodic structure.

and we require S to diagonalize T

(40) $$S^{-1} T S = \begin{pmatrix} d_1 & 0 \\ 0 & d_2 \end{pmatrix}.$$

A similarity transformation does not change the trace nor the determinant

(41a) $$d_1 + d_2 = \text{tr } T,$$

(41b) $$d_1 d_2 = |T| = 1.$$

Consequently

(42) $$d^2 - d \text{ tr } T + 1 = 0,$$

(43) $$d = \frac{\text{tr } T}{2} \pm \sqrt{\left(\frac{\text{tr } T}{2}\right)^2 - 1}.$$

After the n-th passage

(44) $$T^n = S D S^{-1} \ldots = S D^n S^{-1},$$

$$\begin{pmatrix} \xi_\eta \\ \xi'_\eta \end{pmatrix} = S D^n S^{-1} \begin{pmatrix} \xi_0 \\ \xi'_0 \end{pmatrix}.$$

For the deviation ξ to remain bounded, however large we choose n, we require

$$|d_1| \not> 1, \quad |d_2| \not> 1.$$

Since $d_1 d_2 = 1$ this means

(45) $$|d_1| = |d_2| = 1,$$

which is satisfied only if

(46) $$\left|\frac{\text{tr } T}{2}\right| \leqslant 1.$$

This criterion must be satisfied both for the horizontal and the vertical planes to assure full stability of the transverse particle motion.

4˙5. *Simple example of an* AG *system.* – Through a simple example we are now in a position to demonstrate the importance of the findings so far. We choose a system of focusing and defocusing point lenses with drift length L in between (fig. 11).

This is a *good* approximation with separate elements for focusing and bending

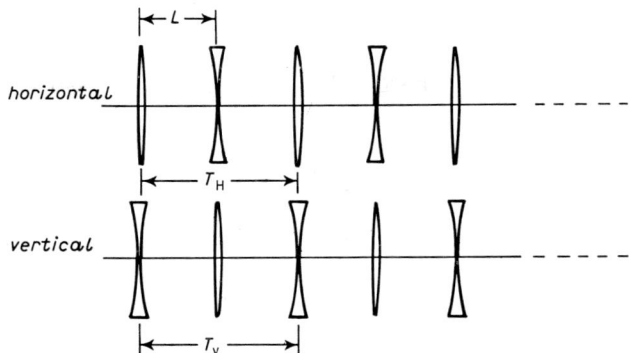

Fig. 11. – Thin-lens model of a periodic structure.

and a *fair* approximation when focusing and bending are combined:

$$(47) \qquad T_H = \begin{pmatrix} 1 & L \\ 0 & 1 \end{pmatrix} \begin{pmatrix} 1 & 0 \\ -\delta_1 & 1 \end{pmatrix} \begin{pmatrix} 1 & L \\ 0 & 1 \end{pmatrix} \begin{pmatrix} 1 & 0 \\ \delta_2 & 1 \end{pmatrix}.$$

Simple calculations give the trace and the stability requirement:

$$(48) \qquad \left| \frac{\operatorname{tr} T_H}{2} \right| = \left| 1 - (\delta_1 - \delta_2)L - \tfrac{1}{2}\delta_1\delta_2 L^2 \right| < 1.$$

By putting the expression inside the bars equal to $+1$ we get one limiting curve:

$$(49) \qquad \delta_2 L = \frac{\delta_1 L}{1 - \tfrac{1}{2}\delta_1 L}.$$

By putting the same expression equal to -1 we get the other limiting curve:

$$(50) \qquad \delta_1 L = 2.$$

These two curves are drawn and appropriately marked in fig. 12. Horizontal stability is assured between the curves and instability outside.

For the vertical motion we get similarly

$$(51) \qquad T_V = \begin{pmatrix} 1 & L \\ 0 & 1 \end{pmatrix} \begin{pmatrix} 1 & 0 \\ \delta_1 & 1 \end{pmatrix} \begin{pmatrix} 1 & L \\ 0 & 1 \end{pmatrix} \begin{pmatrix} 1 & 0 \\ -\delta_2 & 1 \end{pmatrix},$$

$$(52) \qquad \left| \frac{\operatorname{tr} T_V}{2} \right| = \left| 1 - (\delta_2 - \delta_1)L - \tfrac{1}{2}\delta_1\delta_2 L^2 \right| < 1$$

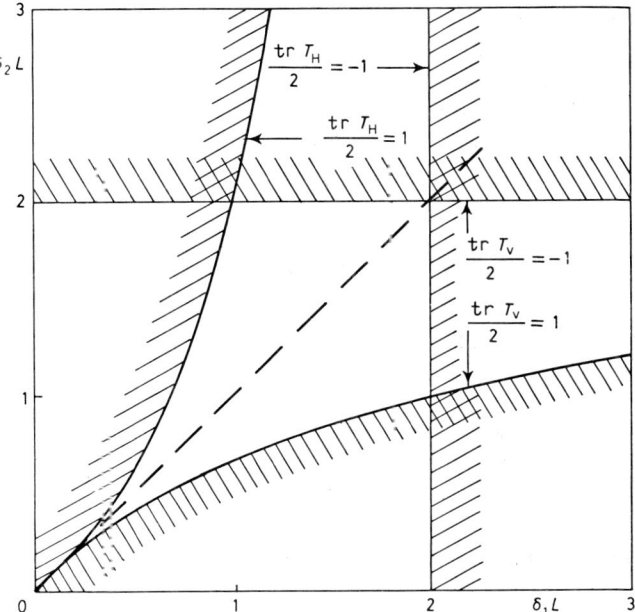

Fig. 12. – The stability diagram («necktie»).

and the corresponding curves for the values ± 1 are

$$\delta_1 L = \frac{\delta_2 L}{1 - \tfrac{1}{2}\delta_2 L} \qquad \text{for } -1, \tag{53}$$

$$\delta_2 L = 2 \qquad \text{for } -1. \tag{54}$$

The corresponding curves are also drawn and marked in fig. 12 with vertical stability between them and instability outside.

Similar diagrams can be calculated and drawn for any kind of alternating-gradient structure, only normally they require more work than this simple example. The important conclusion can be drawn that the accelerator parameters must be chosen such that the «working point» is well inside the region that is stable for both planes and well away from any of the stability limits.

However, there are further complications. Most accelerators do not have a simple periodicity but have the magnets arranged in further superperiods. The effects of these can be calculated by similar (only more complicated) methods to the one above and result in broad instability bands through the «necktie». The «working point» must be kept well away from such instabilities.

In addition, unavoidable imperfections lead to the real periodicity being the whole machine circumference. Again this can be calculated by fairly simple means and leads to a network of new (relatively narrow) instability bands through the «necktie». The widths of these are determined by the magnitude of the imperfections. Tight tolerances must be imposed to give sufficiently wide stability regions between the instability bands.

Finally the effect of nonlinearities should be mentioned, although their treatment is outside the scope of these lectures. They create further stopbands to be avoided, and tight tolerances on the shape of the focusing fields are required.

4'6. *Comments on the elements of the transfer matrix over a period of an AG structure.* – Since the determinant of the transfer matrix is always unity, it contains only three independent quantities. It is customary to write the transfer matrix as

$$(55) \qquad T = \begin{pmatrix} \cos\mu + \alpha\sin\mu & \beta\sin\mu \\ -\dfrac{1+\alpha^2}{\beta}\sin\mu & \cos\mu - \alpha\sin\mu \end{pmatrix}.$$

With this choice of parameters we notice

$$(56) \qquad \cos\mu = \frac{\operatorname{tr} T}{2}.$$

It can further be shown (not quite trivial) that μ represents the phase advance of the (nonsinusoidal) betatron oscillation over a period represented by T. This means

$$(57) \qquad \mu = 2\pi Q/N_p = L_p/\lambda_\beta,$$

where N_p is the number of periods around the machine and L_p is the length of a period.

Since N_p can be chosen much larger than unity, we can also arrive at

$$Q \gg 1,$$

which illustrates that AG focusing can be made much stronger than weak focusing where we always get $Q < 1$. This is the reason why AG focusing is often called strong focusing. It is also important to notice that the larger R_0 the larger one can choose N_p. This means that we can keep the focusing strong even in very large accelerator rings.

For a discussion of β we make the simplifying assumption that the period has two symmetry points and we consider T between such a point in one period to the same point in the next. (It should be noted that most accelerators have such symmetry points.) This results in

$$\alpha = 0$$

and the transfer matrix

$$T = \begin{pmatrix} \cos \mu & \beta \sin \mu \\ -\beta^{-1} \sin \mu & \cos \mu \end{pmatrix}, \tag{58}$$

which resembles the transfer matrix found earlier for sinusoidal oscillations in weak focusing. In fact, β can be interpreted as the « local » λ_β and is a direct measure of how strong the focusing is. If one wants to squeeze the beam locally, one makes β small at that point, which is carried to the extreme in special low-β insertions in many storage rings.

In a given structure $\hat{\beta}$ has its maximum in a focusing element. If we call this $\hat{\beta}$, we find the acceptance of a given aperture as (see also (23) in subsect. 3'5)

$$A = 2\pi a^2 / \hat{\beta}. \tag{59}$$

This illustrates the importance of keeping $\hat{\beta}$ small, *i.e.* keep the focusing strong in order to save aperture. The limit on how far one can go in this direction is in the end given by achievable tolerances.

5. – Longitudinal particle motion.

In the past few sections the discussions of the transverse motion were generally valid for several types of accelerators. In the present section we shall largely use synchrotronlike rings as the basis for the analysis.

In a synchrotron (fig. 3) one or more accelerating gaps are fed by a r.f. voltage with frequency and phase such that the particles get a positive energy increment each time they pass the gap. The magnetic field follows to keep the equilibrium orbit unchanged.

In analogy with the transverse motion there is a stability problem for particles deviating from the motion of the equilibrium, or phase stable, particle for which the frequency is adjusted. We shall in the present section analyse this longitudinal deviation.

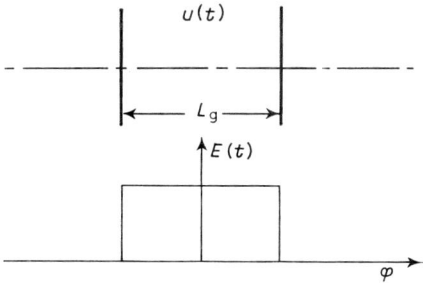

Fig. 13. – Electric field in accelerating gap.

5˙1. *Travelling-wave representation of the accelerating field*. – We assume one accelerating gap with an accelerating voltage given by (see also fig. 13)

$$u(t) = \hat{u} \cos \int \omega \, dt \, . \tag{60}$$

The longitudinal electric field as a function of φ is

$$E(t) = \begin{cases} u(t)/L_g & \text{for } |\varphi| < \dfrac{L_g}{2R}, \\ 0 & \text{for } \dfrac{L_g}{2R} < |\varphi| < \pi. \end{cases}$$

(61a)

(61b)

This field, with a space periodicity of 2π in φ, can be Fourier analysed with the result (detailed derivation omitted)

$$E(t) = \frac{\hat{u}}{2\pi R} \left(\cos \int \omega \, dt + \sum_{n=1}^{\infty} \left[\cos\left(\int \omega \, dt + n\varphi\right) + \cos\left(\int \omega \, dt - n\varphi\right) \right] \right), \tag{62}$$

i.e. two sets of rotating waves. All wave components act as a.c. fields on the particles (with zero average effect) except one satisfying the condition

$$\int \omega \, dt - M\varphi = \text{const}$$

or

$$M\dot{\varphi} = \omega, \qquad M\Omega = \omega, \tag{63}$$

where Ω is the rotational frequency of the equilibrium particle given by

$$\Omega = v/R \, . \tag{64}$$

This means that the only field component of interest for analysing the longitudinal particle motion is

$$E = \frac{\hat{u}}{2\pi R} \cos \left(M\varphi - \int \omega \, dt \right). \tag{65}$$

M represents the number of r.f. cycles per particle revolution and is called the harmonic number.

It should be noted that, although it was convenient to assume a short single gap for the acceleration, it is immaterial for the analysis how the field component (65) is set up, whether by one or many gaps or even travelling-wave structures.

5˙2. *The equations for the longitudinal motion.* – We introduce the abbreviation

(66) $$M\varphi - \int \omega \, dt = \theta .$$

The motion is then given by

$$\frac{dp}{dt} = \frac{e\hat{u}}{2\pi R} \cos \theta .$$

The amplitude is chosen such that the phase-stable particle stays still with respect to the travelling wave at a point $\theta = \theta_0$ (see fig. 14), and, therefore, experiences the force

(67) $$\frac{dp_0}{dt} = \frac{e\hat{u}}{2\pi R} \cos \theta_0 .$$

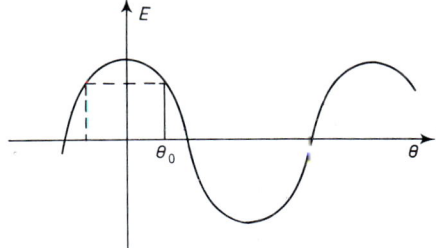

Fig. 14. – Travelling-wave representation of accelerating field.

Deviations from the phase-stable particle can then be written

(68) $$\frac{d\Delta p}{dt} = \frac{dp}{dt} - \frac{dp_0}{dt} = \frac{e\hat{u}}{2\pi R} \left(\cos(\theta_0 + \Delta\theta) - \cos \theta_0 \right) .$$

Some manipulations give

(69) $$\frac{\dot{\Delta\theta}}{\omega} = \frac{\Delta v}{v_0} - \frac{\Delta R}{R_0} .$$

Simple relativity calculations give

(70) $$\frac{\Delta v}{v_0} = \gamma^{-2} \frac{\Delta p}{p_0} ,$$

where γ, as usual, is given by

(71) $$\gamma = m/m_0 .$$

The momentum compaction factor was defined by (29) in subsect. **3·6**. We use the same definition

$$\frac{\Delta R}{R_0} = \alpha \frac{\Delta p}{p_0} \, . \tag{72}$$

We combine (69), (70) and (72) and get

$$\Delta p = \frac{p_0}{\omega} (\gamma^{-2} - \alpha)^{-1} \dot{\Delta\theta} \, . \tag{73}$$

Since

$$p_0 = m R_0 \omega M^{-1} = \frac{E_0}{c^2} R_0 \omega M^{-1} \, , \tag{74}$$

where E_0 is the total energy of the phase-stable particle, we get

$$\Delta p = \frac{E_0}{c^2} M^{-1} R_0 (\gamma^{-2} - \alpha)^{-1} \dot{\Delta\theta} \, , \tag{75}$$

which introduced into (68) gives

$$\frac{\mathrm{d}}{\mathrm{d}t} \left\{ \frac{E_0}{c^2} R_0 M^{-1} (\gamma^{-2} - \alpha)^{-1} \dot{\Delta\theta} \right\} = \frac{e\hat{u}}{2\pi R_0} [\cos(\theta_0 + \Delta\theta) - \cos\theta_0] \, . \tag{76}$$

This is the basic phase equation valid for all amplitudes, and the equation to be used in particular to study the important problems related with the boundaries (separatrices) between stable and unstable regions.

We shall consider only the special case of small deviations in $\Delta\theta$, *i.e.* we use the linearized approximation of (76)

$$\frac{\mathrm{d}}{\mathrm{d}t} \left\{ \frac{E_0}{c^2} R_0 M^{-1} (\gamma^{-2} - \alpha)^{-1} \dot{\Delta\theta} \right\} + \frac{e\hat{u}}{2\pi R_0} \sin\theta_0 \, \Delta\theta = 0 \, . \tag{77}$$

5·3. *Stability. Transition energy.* – We first use (77) for some important stability considerations. Equation (77) has oscillatory solutions if the two coefficients have the same sign. If, on the contrary, the coefficients have opposite signs, the solutions are exponentiallike and are definitely unstable. This means that stability can be obtained only if

$$0 < \theta_0 < \pi/2 \qquad \text{for } \gamma < \alpha^{-\frac{1}{2}}, \tag{78a}$$

or

$$0 > \theta_0 > -\pi/2 \qquad \text{for } \gamma > \alpha^{-\frac{1}{2}}. \tag{78b}$$

(Strictly speaking one should put π instead of $\pi/2$ in the above expressions, but since we want acceleration and not deceleration $\pi/2$ is more appropriate as the limit.)

The above conditions show that, at a given energy, there is a transition between the conditions (78a) and (78b), given by

$$(79) \qquad E_{\rm tr} = \alpha^{-\frac{1}{2}} m_0 c^2$$

The momentum compaction α is a property of the focusing structure. For weak focusing we showed in subsect. 3'6 that

$$(31) \qquad \alpha = Q_x^{-2} \, .$$

Since in weak-focusing rings $Q_x < 1$, we see that in such machines the transition energy is below the rest energy and is, therefore, nonphysical, *i.e.* in weak-focusing machines it is always (78b) that applies as the stability condition.

This is different for AG focusing rings. Although (79) is not strictly correct in such machines, it can be shown to represent a fair approximation for most regular AG structures. Since in such structures Q is normally well above unity, the transition energy falls in the physical range. In electron machines, however, injection energy is always above transition and again it is only (78b) that applies. For proton accelerators one often has to choose the machine parameters such that $E_{\rm tr}$ falls somewhere in between injection and top energy. To maintain stability on either side of transition, the stable phase has to be jumped from θ_0 to $-\theta_0$ just as the energy passes through $E_{\rm tr}$, as shown in fig. 15.

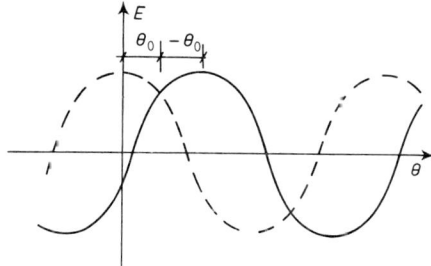

Fig. 15. – Phase jump at transition.

If this is done at the right moment, the linear phase equation can be written in a form valid both below and above transition

$$(80) \qquad \frac{\rm d}{{\rm d}t}\left\{\frac{E_0}{c^2} R_0 M^{-1} |\gamma^{-2} - \alpha|^{-1} \dot{\Delta\theta}\right\} + \frac{e\hat{u}}{2\pi R_0} \sin|\theta_0| \Delta\theta = 0 \, .$$

5˙4. *The adiabatic solution.* – For that part of the energy range where the coefficients in the differential equation change little over a radian of the oscillation the adiabatic solution of (80) applies

$$(81) \qquad \Delta\theta \simeq A \left[\frac{2\pi M}{m_0 e \hat{u} \sin|\theta_0|} \frac{|\gamma^{-2} - \alpha|}{\gamma} \right]^{\frac{1}{4}} \sin \int \omega_s \, dt ,$$

where ω_s is called the synchrotron oscillation frequency and given by

$$(82) \qquad \omega_s = \left(\frac{e\hat{u} \sin|\theta_0|}{2\pi R_0} \frac{M}{m_0 R} \frac{|\gamma^{-2} - \alpha|}{\gamma} \right)^{\frac{1}{2}} .$$

In a similar way and with the same approximation one also gets the expression for the momentum deviation

$$(83) \qquad \Delta p \simeq A \frac{\sin \theta_0}{\sin |\theta_0|} [\ \ ]^{-\frac{1}{4}} \cos \int \omega_s \, dt ,$$

where the contents of the square bracket are given in full in eq. (81).

It is important to notice that

$$\omega_s \to 0 \quad \text{as} \quad E \to E_{\text{tr}}$$

and that this property of the solution comes from the fact that what can be considered as an « equivalent mass » in (80) tends towards infinity at E_{tr}. This means that near transition the system is not only slow but also « stiff ». This facilitates the « gymnastics » needed for the phase jump at transition.

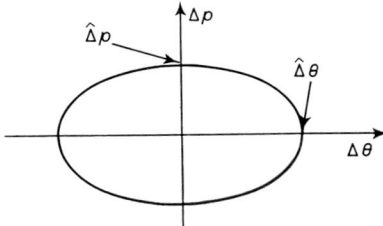

Fig. 16. – Longitudinal phase-space trajectory.

Let us now look at the particle trajectories in the $\Delta\theta$, Δp phase plane as given by (81) and (82). It is readily seen that these trajectories are ellipses with slowly varying shapes (fig. 16). The area inside the ellipse

$$\pi \Delta\hat{p} \Delta\hat{\theta} = \pi A^2 = \text{invariant} ,$$

i.e. it does not change with energy, whereas the shape does. In analogy with the transverse emittance, the longitudinal emittance is defined as the area of the ellipse in fig. 16 corresponding to the particle with the largest excursion within the beam.

The conservation of area holds also when the amplitudes are so large that the nonlinear equation must be used. We shall not treat this case any further in these lectures.

The adiabatic solutions (81) and (83) are clearly not valid near transition. Solutions can be found fairly easily, but also this is considered outside the scope of these introductory lectures.

As already mentioned, most large proton synchrotrons have to go through transition. However, the transition is normally rather low in their energy range and it is, therefore, of some interest to look at the amplitude changes with energy for the case $E \gg E_{tr}$.

For this case (81) gives

$$(84) \qquad \Delta\hat{\theta} \propto \gamma^{-\frac{1}{4}},$$

i.e. the phase amplitude is damped.

From (83) we get in the same way

$$(85) \qquad \Delta\hat{p} \propto \gamma^{\frac{1}{4}},$$

which means that this quantity increases with energy. However,

$$\frac{\Delta\hat{p}}{p} \propto \gamma^{-\frac{3}{4}}$$

or

$$(86) \qquad \Delta R \propto \gamma^{-\frac{3}{4}},$$

i.e. the corresponding radial excursions are damped.

From this it is seen that with increasing energy the particles tend to concentrate in phase around θ_0 and to decrease the radial excursion due to phase oscillations. This general decrease in amplitude is an important additional element in the stability of the beam during its acceleration.

6. – Representative examples of accelerator systems for hadrons.

The most typical examples of large accelerator systems based on the principles described in the previous examples are found at CERN and at Fermilab (USA). Fermilab has reached the highest fixed-target energy of 800 GeV, *i.e.* $E_{c.m.} \approx 40$ GeV. CERN has the highest centre-of-mass energy of 540 GeV.

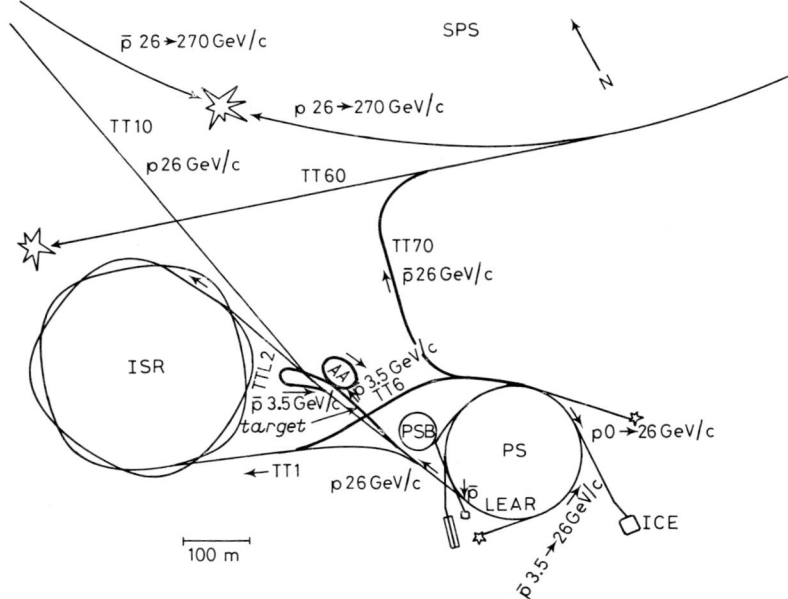

Fig. 17. – Major accelerator facilities at CERN.

Figure 17 illustrates the CERN facilities consisting of

PS at 25 GeV, *i.e.* $E_{c.m.} \simeq 7$ GeV;

SPS at 500 GeV, *i.e.* $E_{c.m.} \simeq 30$ GeV;

ISR with $E_{c.m.} \simeq 63$ GeV (closed December 1983);

SPS Collider with $E_{c.m.} \simeq 540$ GeV.

In the following lecture the ISR and the SPS Colliders served as representative examples in a description of typical beam behaviour in systems with extreme requirements on beam qualities. The ISR description was based on another talk given last January and published as CERN Report 84-13 and the SPS Collider description was based on an introductory talk to the CAS p$\bar{\text{p}}$ Course in October 1983 and published in the Proceedings of that Course, CERN 84-15.

Since these reports are generally available, their content is not reproduced in the present proceedings.

7. – Some special features of electron machines.

A charge that is being accelerated radiates photons. In any kind of accelerator the radiation due to the longitudinal acceleration is negligible. However, this is not so with the radiation due to the transverse acceleration provided by the guide field to keep the particles on a circular path.

The radiated power from a charged particle moving in a circle of radius ϱ is

$$(87) \qquad P_r = \tfrac{2}{3} c r_0 m_0 c^2 \gamma^4 / \varrho^2 \, ,$$

where r_0 is the classical radius of the particle. From this we see an important difference between protons and electrons. If we take the same energy and the same bending radius for the two particles, we get

$$(88) \qquad \frac{P_r(\text{electron})}{P_r(\text{proton})} = \left[\frac{m_0(\text{electron})}{m_0(\text{proton})} \right]^{-4} \simeq 10^{13} \, .$$

In proton accelerators this radiation has been negligible so far (although it cannot quite be disregarded for some accelerators planned for the future), whereas the radiation plays an important role both in the beam dynamics and in the practical design of electron machines, and is in fact the main limitation in achievable energy of circular electron machines.

7˙1. *Radiated power and associated losses*. – The radiated energy per turn per particle

$$(89) \qquad \Delta E = P_r 2\pi \varrho / c = \frac{4\pi}{3} r_0 m_0 c^2 \gamma^4 / \varrho \, ,$$

$$(90) \qquad \qquad\qquad = C \gamma^4 / \varrho \, .$$

(It is worth noting that C is independent of type of particle.)

This radiated energy has to be supplied by the accelerating structure of the machine and requires for large accelerators much more r.f. voltage than needed for the acceleration.

Some numerical examples for *electrons* may be interesting

$$(91) \qquad \Delta E(\text{keV}) = 88.5 \, E^4(\text{GeV})/\varrho(\text{m}) \, .$$

For LEP at 55 GeV and $\varrho = 3100$ m we get $\Delta E = 260$ MeV/turn. At 85 GeV the formula gives 1.3 GeV/turn.

It is also important to consider the total radiated power from a beam of N particles. This can be expressed as

$$(92) \qquad N f \Delta E = N f C \gamma^4 / \varrho = (I/e) C \gamma^4 / \varrho \, ,$$

where f is the revolution frequency. For LEP at 55 GeV beam energy and a beam current of 3 mA the total radiated power (two beams) becomes 1.6 MW.

At 85 GeV it approaches 10 MW. This power has to be supplied by the r.f. system, but more importantly has to be disposed of in a way that is not damaging to machine components or experiments.

In addition to the radiated power, the associated ohmic loss in the r.f. structure has to be supplied. This is given by the expression

$$P_c = \frac{(k\Delta E/e)^2}{Z_s L_c} = \left(\frac{k}{e}\right)^2 C^2 \frac{\gamma^8}{\varrho^2 Z_s L_c}, \tag{93}$$

where k is a factor representing the overvoltage required for phase stability and bunch shape, Z_s is the shunt impedance per unit length of the accelerating structure and L_c the length of this structure. This loss has a very strong energy dependence even if one lets L_c increase with maximum energy and dominates in large machines. As an example we take LEP again, which has a total estimated ohmic loss in the r.f. structure of about 12 MW at 55 GeV. This starts becoming prohibitive if one would like to approach an energy of about 100 GeV.

This kind of loss gives the main incentive for the development of superconducting r.f. structures. If this development proves successful, the radiation limit in LEP may be lifted to (120÷130) GeV per beam.

To construct circular electron accelerators for much higher energies seems doubtful. This is the main reason why much interest in the development of linear colliders has risen recently, with the first such device under construction at SLAC.

7'2. *Radiation damping.* – The radiation has a significant effect on the beam dynamics of the electrons in circular machines. We shall in the present subsection consider some of the consequences of this without going into details.

7'2.1. The vertical betatron oscillations. The physical consequences of the radiation are most easily seen for the vertical betatron oscillation. For this we refer to fig. 18, which represents a typical situation of a momentum vector at a r.f. gap. The line OA represents the momentum of a particle as it enters the r.f. gap. AB represents the momentum the particle has lost due to radiation after it left the gap the previous time. In order to make up for the lost momentum, the particle must gain the momentum AC as it now crosses

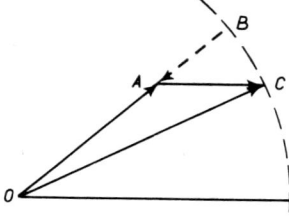

Fig. 18. – Damping of betatron oscillations.

the gap, and it, therefore, leaves the gap with the momentum OC. This has as a consequence that the slope of the particle orbit has been reduced by the angle between OA and OC. This happens to a larger or smaller extent to all particles (with the exception of when a particle crosses the r.f. gap with only longitudinal momentum). This illustrates that on the average all particles get their vertical betatron oscillation amplitudes reduced by the combined effect of radiation and compensation of the radiation loss. The vertical betatron oscillations are being damped.

The damping can be represented by

$$\dot{z} \propto \exp[-\alpha_z t], \tag{94}$$

where α_z is called the vertical damping factor and can be written as

$$\alpha_z = \frac{1}{2} \frac{P_r}{E} J_z. \tag{95}$$

The coefficient J_z can be shown to be unity in any machine with a perfect median plane.

7˙2.2. The horizontal betatron oscillation. In the horizontal plane the radiation has a similar effect, but the situation is somewhat complicated by the small jump in closed orbit that the particle experiences as it passes the r.f. gap. This gives a superimposed antidamping effect. We write the damping factor in the same way as for the vertical plane

$$\alpha_x = \frac{1}{2} \frac{P_r}{E} J_x, \tag{96}$$

but in this case J_x is a more complicated function of the focusing and bending structure of the machine, and can easily become negative (antidamping). Some electron synchrotrons have negative J_x, which is acceptable because of the short time needed for the acceleration. For electron storage rings, in which long beam lifetimes are required, the structure must be chosen such that J_x becomes positive. Separate function lattices give this situation automatically.

7˙2.3. Longitudinal oscillations. Again we cannot go through the calculations. One arrives at a longitudinal damping factor that can be written in a similar form to the two transverse ones

$$\alpha_s = \frac{1}{2} \frac{P_r}{E} J_s, \tag{97}$$

where J_s again depends on the focusing and bending structure.

7'2.4. The relation between the three partition numbers J. For machines with perfect median planes (no vertical bending of the closed orbit) the following relations can be derived:

$$(98a) \qquad J_z = 1 ,$$

$$(98b) \qquad J_x + J_s = 3 .$$

More generally, a relation valid for all machines can be derived:

$$(99) \qquad J_x + J_z + J_s = 4 ,$$

which is the reason why $J_{x,z,s}$ are called partition numbers. For a perfect separated-function machine one would find

$$(100) \qquad J_z = 1 , \quad J_x \simeq 1 , \quad J_s \simeq 2 .$$

The radiation properties can be much influenced by specially designed wigglers or undulators. This is used in storage rings to get a desired damping in special energy ranges, for instance at injection, where the radiation damping is used for the accumulation of intense electron beams.

Of course, the beam size does not shrink to zero, but it is in the end determined by a balance against other effects. The most important ones are quantum fluctuations of the radiation, beam-beam effects, intrabeam scattering, etc. Some of these, and other ones, are similar to effects playing important roles also in hadron colliders.

7'2.5. Summary of the differences between electron and proton machines. The fundamental differences between electron machines and proton machines can be summed up as follows.

Final beam properties in an electron machine are independent of injection conditions. The radiation causes the beam to have only a short-term memory.

Final beam properties in a proton machine depend critically on what you do to the beam from the ion source to the final beam condition. The fact that Hamiltonian dynamics apply (in particular Liouville's theorem) gives the beam a very long memory.

8. – Prospects for the future.

The last lecture of this series was devoted to the prospects for high-energy accelerators in the future. A similar lecture, but somewhat more complete, was given at a course on General Accelerator Physics organized at Orsay during the period September 3-14, 1984, and I refer to the Proceedings from this course (to be published as a CERN Yellow Report) for the text of this lecture.

PROCEEDINGS OF THE INTERNATIONAL SCHOOL OF PHYSICS
« ENRICO FERMI »

Course I
Questioni relative alla rivelazione delle particelle elementari, con particolare riguardo alla radiazione cosmica
edited by G. PUPPI

Course II
Questioni relative alla rivelazione delle particelle elementari, e alle loro interazioni con particolare riguardo alle particelle artificialmente prodotte ed accelerate
edited by G. PUPPI

Course III
Questioni di struttura nucleare e dei processi nucleari alle basse energie
edited by C. SALVETTI

Course IV
Proprietà magnetiche della materia
edited by L. GIULOTTO

Course V
Fisica dello stato solido
edited by F. FUMI

Course VI
Fisica del plasma e applicazioni astrofisiche
edited by G. RIGHINI

Course VII
Teoria della informazione
edited by E. R. CAIANIELLO

Course VIII
Problemi matematici della teoria quantistica delle particelle e dei campi
edited by A. BORSELLINO

Course IX
Fisica dei pioni
edited by B. TOUSCHEK

Course X
Thermodynamics of Irreversible Processes
edited by S. R. DE GROOT

Course XI
Weak Interactions
edited by L. A. RADICATI

Course XII
Solar Radioastronomy
edited by G. RIGHINI

Course XIII
Physics of Plasma: Experiments and Techniques
edited by H. ALFVÉN

Course XIV
Ergodic Theories
edited by P. CALDIROLA

Course XV
Nuclear Spectroscopy
edited by G. RACAH

Course XVI
Physicomathematical Aspects of Biology
edited by N. RASHEVSKY

Course XVII
Topics of Radiofrequency Spectroscopy
edited by A. GOZZINI

Course XVIII
Physics of Solids (Radiation Damage in Solids)
edited by D. S. BILLINGTON

Course XIX
Cosmic Rays, Solar Particles and Space Research
edited by B. PETERS

Course XX
Evidence for Gravitational Theories
edited by C. MØLLER

Course XXI
Liquid Helium
edited by G. CARERI

Course XXII
Semiconductors
edited by R. A. SMITH

Course XXIII
Nuclear Physics
edited by V. F. WEISSKOPF

Course XXIV
Space Exploration and the Solar System
edited by B. Rossi

Course XXV
Advanced Plasma Theory
edited by M. N. Rosenbluth

Course XXVI
Selected Topics on Elementary Particle Physics
edited by M. Conversi

Course XXVII
Dispersion and Absorption of Sound by Molecular Processes
edited by D. Sette

Course XXVIII
Star Evolution
edited by L. Gratton

Course XXIX
Dispersion Relations and Their Connection with Causality
edited by E. P. Wigner

Course XXX
Radiation Dosimetry
edited by F. W. Spiers and G. W. Reed

Course XXXI
Quantum Electronics and Coherent Light
edited by C. H. Townes and P. A. Miles

Course XXXII
Weak Interactions and High-Energy Neutrino Physics
edited by T. D. Lee

Course XXXIII
Strong Interactions
edited by L. W. Alvarez

Course XXXIV
The Optical Properties of Solids
edited by J. Tauc

Course XXXV
High-Energy Astrophysics
edited by L. Gratton

Course XXXVI
Many-Body Description of Nuclear Structure and Reactions
edited by C. Bloch

Course XXXVII
Theory of Magnetism in Transition Metals
edited by W. Marshall

Course XXXVIII
Interaction of High-Energy Particles with Nuclei
edited by T. E. O. Ericson

Course XXXIX
Plasma Astrophysics
edited by P. A. Sturrock

Course XL
Nuclear Structure and Nuclear Reactions
edited by M. Jean and R. A. Ricci

Course XLI
Selected Topics in Particle Physics
edited by J. Steinberger

Course XLII
Quantum Optics
edited by R. J. Glauber

Course XLIII
Processing of Optical Data by Organisms and by Machines
edited by W. Reichardt

Course XLIV
Molecular Beams and Reaction Kinetics
edited by Ch. Schlier

Course XLV
Local Quantum Theory
edited by R. Jost

Course XLVI
Physics with Storage Rings
edited by B. Touschek

Course XLVII
General Relativity and Cosmology
edited by R. K. Sachs

Course XLVIII
Physics of High Energy Density
edited by P. Caldirola and H. Knoepfel

Course IL
Foundations of Quantum Mechanics
edited by B. d'Espagnat

Course L
Mantle and Core in Planetary Physics
edited by J. Coulomb and M. Caputo

Course LI
Critical Phenomena
edited by M. S. Green

Course LII
Atomic Structure and Properties of Solids
edited by E. Burstein

Course LIII
Developments and Borderlines of Nuclear Physics
edited by H. Morinaga

Course LIV
  Developments in High-Energy Physics
  edited by R. R. Gatto

Course LV
  Lattice Dynamics and Intermolecular Forces
  edited by S. Califano

Course LVI
  Experimental Gravitation
  edited by B. Bertotti

Course LVII
  History of 20th Century Physics
  edited by C. Weiner

Course LVIII
  Dynamic Aspects of Surface Physics
  edited by F. O. Goodman

Course LIX
  Local Properties at Phase Transitions
  edited by K. A. Müller and A. Rigamonti

Course LX
  C*-Algebras and their Applications to Statistical Mechanics and Quantum Field Theory
  edited by D. Kastler

Course LXI
  Atomic Structure and Mechanical Properties of Metals
  edited by G. Caglioti

Course LXII
  Nuclear Spectroscopy and Nuclear Reactions with Heavy Ions
  edited by H. Faraggi and R. A. Ricci

Course LXIII
  New Directions in Physical Acoustics
  edited by D. Sette

Course LXIV
  Nonlinear Spectroscopy
  edited by N. Bloembergen

Course LXV
  Physics and Astrophysics of Neutron Stars and Black Holes
  edited by R. Giacconi and R. Ruffini

Course LXVI
  Health and Medical Physics
  edited by J. Baarli

Course LXVII
  Isolated Gravitating Systems in General Relativity
  edited by J. Ehlers

Course LXVIII
  Metrology and Fundamental Constants
  edited by A. Ferro Milone, P. Giacomo and S. Leschiutia

Course LXIX
  Elementary Modes of Excitation in Nuclei
  edited by A. Bohr and R. A. Broglia

Course LXX
  Physics of Magnetic Garnets
  edited by A. Paoletti

Course LXXI
  Weak Interactions
  edited by M. Baldo Ceolin

Course LXXII
  Problems in the Foundations of Physics
  edited by G. Toraldo di Francia

Course LXXIII
  Early Solar System Processes and the Present Solar System
  edited by D. Lal

Course LXXIV
  Development of High-Power Lasers and their Applications
  edited by C. Pellegrini

Course LXXV
  Intermolecular Spectroscopy and Dynamical Properties of Dense Systems
  edited by J. Van Kranendonk

Course LXXVI
  Medical Physics
  edited by J. R. Greening

Course LXXVII
  Nuclear Structure and Heavy-Ion Collisions
  edited by R. A. Broglia, R. A. Ricci and C. H. Dasso

Course LXXVIII
  Physics of the Earth's Interior
  edited by A. M. Dziewonski and E. Boschi

Course LXXIX
  From Nuclei to Particles
  edited by A. Molinari

Course LXXX
  Topics in Ocean Physics
  edited by A. R. Osborne and P. Malanotte Rizzoli

Course LXXXI
  Theory of Fundamental Interactions
  edited by G. Costa and R. R. Gatto

Course LXXXII
Mechanical and Thermal Behaviour of Metallic Materials
edited by G. CAGLIOTI and A. FERRO MILONE

Course LXXXIII
Positron Solid-State Physics
edited by W. BRANDT and A. DUPASQUIER

Course LXXXIV
Data Acquisition in High-Energy Physics
edited by G. BOLOGNA and M. VINCELLI

Course LXXXV
Earhquakes: Observation, Theory and Interpretation
edited by H. KANAMORI and E. BOSCHI

Course LXXXVI
Gamow Cosmology
edited by F. MELCHIORRI and R. RUFFINI

Course LXXXVII
Nuclear Structure and Heavy-Ion Dynamics
edited by L. MORETTO and R. A. RICCI

Course LXXXVIII
Turbulence and Predictability in Geophysical Fluid Dynamics and Climate Dynamics
edited by M. GHIL, R. BENZI and G. PARISI

Course LXXXIX
Highlights of Condensed-Matter Theory
edited by F. BASSANI, F. FUMI and M. P. TOSI

Course XC
Physics of Amphiphiles: Micelles, Vesicles and Microemulsions
edited by V. DEGIORGIO and M. CORTI

Course XCI
From Nuclei to Stars
edited by A. MOLINARI and R. A. RICCI

TIPOGRAFIA COMPOSITORI - BOLOGNA